D1634067

S/O 'ML
NO INF.

**Books are to be returned on
the last date below.**

ANDERSONIAN LIBRARY
☆
WITHDRAWN
FROM
LIBRARY
STOCK
☆
UNIVERSITY OF STRATHCLYDE

EXS 77

bib547270

Stress-Inducible Cellular Responses

Edited by U. Feige
R. I. Morimoto
I. Yahara
B. S. Polla

Birkhäuser Verlag
Basel · Boston · Berlin

Editors

Dr. U. Feige
Department of Pharmacology
AMGEN Center
Mail Stop 15-2-A-224
1840 De Havilland Drive
Thousand Oaks, CA 91320-1789
USA

Dr. R. I. Morimoto
Department of Biochemistry,
Molecular and Cell Biology
Northwestern University
2153 Sheridan Road
Evanston, IL 60208-3500
USA

Dr. I. Yahara
Department of Cell Biology
The Tokyo Metropolitan Institute
of Medical Science
Honkomagome 3-18-22, Bunkyo-Ku
Tokyo 113
Japan

Dr. B. S. Polla
Laboratoire de Physiologie Respiratoire
Université Paris 5
UFR Cochin Port-Royal
24, rue du Faubourg Saint-Jacques
F-75014 Paris
France

Library of Congress Cataloging-in-Publication Data
A CIP catalogue record for this book is available from the library of Congress,
Washington D.C., USA

Deutsche Bibliothek Cataloging-in-Publication Data
EXS – Basel; Boston; Berlin: Birkäuser.
    Früher Schriftenreihe
    Fortlaufende Beil. zu: Experientia
77. Stress inducible cellular responses. – 1996
Stress inducible cellular responses / ed. by U. Feige. – Basel;
Boston; Berlin: Birkhäuser, 1996
    (EXS; 77)
    ISBN 3-7643-5205-1 (Basel . . .)
    ISBN 0-8176-5205-1 (Boston)
NE: Feige, Ulrich [Hrsg.]

The publisher and editor can give no guarantee for the information on drug dosage and administration contained in this publication. The respective user must check its accuracy by consulting other sources of reference in each individual case.
The use of registered names, trademarks etc. in this publication, even if not identified as such, does not imply that they are exempt from the relevant protective laws and regulations or free for general use.
This work is subject to copyright. All rights are reserved, whether the whole or part of the material is concerned, specifically the rights of translation, reprinting, re-use of illustrations, recitation, broadcasting, reproduction on microfilms or in other ways, and storage in data banks. For any kind of use permission of the copyright owner must be obtained.

© 1996 Birkhäuser Verlag, PO Box 133, CH-4010 Basel, Switzerland
Printed on acid-free paper produced from chlorine-free pulp. TCF ∞
Printed in Germany
ISBN 3-7643-5205-1
ISBN 0-8176-5205-1
9 8 7 6 5 4 3 2 1

Contents

Foreword

"Stress-Inducible Cellular Responses" are essential for survival of cells of all species under adverse conditions. At the molecular level this is accomplished by a number of essential proteins all of which are involved in various aspects of cellular homeostasis through protective or adaptive functions. Interestingly, molecules such as heat shock proteins have properties as molecular chaperones and are involved in multiple stages of a protein biogenesis beginning with synthesis and involvement in the subsequent events of folding, translocation, and degradation. Heat shock proteins have a critical role to stabilize folding intermediates and to prevent protein aggregation. In addition, stress-proteins serve as targets for immune responses in immune homeostasis and during infections.

The term Stress Response reflects the rapid growth and breadth of this field which includes the molecular and cellular response to drugs, UV irradiation, oxidative stress, and environmental toxins. Radical scavengers such as superoxide dismutases and inducible regulatory proteins of metallic ion status such as ferritin and protein disulfide isomerases are also considered within the frame of stress proteins and represent a new and growing class of stress responses.

As our understanding on the regulation of heat shock genes and the function of heat shock proteins has grown, we now realize potential practical applications of heat shock proteins in toxicology, for the diagnosis and prognosis of disease states in medicine, and eventually therapy of clinical conditions associated with an increased oxidative burden. The roles of heat shock proteins in modulation of immune responses and during viral infection are also reviewed. The detailed understanding of the heat shock and stress response has very recently opened up new areas of importance such as cellular responses to complex changes in the environment.

This book, Stress-Inducible Cellular Responses, deals with different aspects of heat shock proteins and more generally with stress-related inducible gene expression as a pleiotropic adaptive response to stress. It aims to offer a textbook-like overview of the field that will be useful not only to experts in the immediate field, but additionally to a broad range of biomedical researchers including physiologists, pharmacologists, cell biologists, biochemists, as well as physicians. We hope that the topics covered in this monograph act to stimulate new experimental approaches to fields bordering the classical heat shock field.

We would like to thank all contributing authors for their effort, flexibility and rapid responses to all requests, which allowed us to proceed from the idea to completion of the book in a very short time.

Ulrich Feige,   *Ichiro Yahara,*
Richard I. Morimoto,   *Barbara S. Polla*   February, 1996

Stress-Inducible Cellular Responses
ed. by U. Feige, R.I. Morimoto, I. Yahara and B. Polla
© 1996 Birkhäuser Verlag Basel/Switzerland

Introduction

I. Yahara

The Tokyo Metropolitan Institute of Medical Science, Department of Cell Biology, Honkomagome 3-18-22, Bunkyo-Ku, Tokyo 113, Japan

Natural selection has no analogy with any aspect of human behavior. If one wanted to use a comparison, however, one would have to say that this process resembles not engineering but tinkering, bricolage we say in French.
   As pointed out by Claude Levi-Strauss, none of the materials at the tinkerer's disposal has a precise and definite function. Each can be used in different ways.

François Jacob, The Possible and the Actual

Since living organisms appeared on the earth, they have been exposed to environments which altered gradually and sometimes rapidly. They adapted to environments by creating mutants and natural selection among them. This process generally took a long time, however. Stress responses are another strategy that organisms have acquired for the purpose of protecting themselves from environmental alterations. This response can be induced quickly and repeatedly and is, therefore, particularly useful for organisms living under rapidly altering environments. Such stress responses are commonly seen in all organisms from bacteria to higher vertebrates, suggesting that the response originated probably shortly after primitive cells appeared on the earth. Those that had defects in the responses must have been eliminated during evolution. This book deals with stress-inducible cellular responses in various respects that were originally evolved as cellular defense mechanism and were later utilized in various cellular processes.

   Cellular components targeted by harmful environments and poisonous substances are mainly proteins and DNA. The production of heat shock proteins is the main event of heat shock response and other stress responses. Heat shock proteins protect functional proteins from irreversible denaturation caused by heat and other stresses and assists them in renaturation. Elaborate mechanisms such as SOS response in procaryotes and DNA repair systems in eucaryotes are induced to cope with DNA damages caused by various insults such as replication errors or irradiation. Interestingly, living organisms came to utilize these cell defense mechanisms even under non-stressful conditions. Heat shock proteins are expressed constitutively at certain levels in cells growing under normal conditions and elicit their functions as molecular chaperones to mediate assembly processes of cellular structures and transport of proteins to membraneous organella. It would be reasonable to suppose that heat shock proteins such as IIsp70 played important roles in evolution of organella from symbiotic procaryotes and became essential for eucaryotes, although these proteins except GroE are mostly nonessential for procaryotes. The DNA repair mechanisms are used also for homologous recombination in meiosis that enhances genetic diversity.

Oxygen is highly toxic to living organisms because it oxidizes and inactivates functional biomolecules. Under oxidative stress, lipids as well as proteins and nucleic acids are damaged by reactive oxygen species. Living organisms came to use oxygen, however, for producing energy by respiration. Defense mechanisms against oxidative stress have thus necessarily evolved. On the other hand, oxidative stress was utilized for killing parasites by phagocytic cells.

Various biological responses including growth response and cell differentiation might have evolved from stress responses. In the primitive prokaryotes that did not perform cell-to-cell interactions, environmental factors including nutrients must have been solely the sources of biological responses. It would be possible that mechanisms operating in stress responses might be modified and utilized in other biological responses which are elaborated particularly in higher organisms. In fact, various stresses including nutrient starvation are known to induce cell differentiation in lower cucaryotes. In addition, there is some evidence which suggests that certain common signal transduction mechanisms, for example, the cascade reactions involving the MAP kinase family, mediate both stress and growth responses.

In multicellular organisms, cell-to-cell interactions always take place. Some of them are mediated by direct cell contacts and others are indirectly mediated by humoral factors such as cytokines and hormones. All of these interactions are also the sources of stress responses because neighboring cells are evidently part of cell environments. Stress responses must have also affected to a large extent evolution of multicellular organisms.

It is now evident that stress responses and stress proteins as products of the responses influenced evolution in various aspects and played important roles in determining features of living organisms. For this reason, this book is useful not only for molecular biologists, biochemists, cell biologists and physiologists working in related fields, but also for those working in clinical fields and applied sciences.

Part I
Functions of stress proteins in unstressed cells

Introduction

I. Yahara

*The Tokyo Metropolitan Institute of Medical Science, Department of Cell Biology,
Honkomagome 3-18-22, Bunkyo-Ku, Tokyo 113, Japan*

In cells growing under normal conditions, proteins are continuously produced, transported to cell compartments where they express their functions, and assembled into complexes if necessary, after which they function for indicated periods and finally are consumed. Tertiary structure of protein appears to be thermodynamically determined, but the process of protein folding is still a matter of dispute. It has become clear, however, that immature secretory proteins are retained in endoplasmic reticulum by forming complexes with stress proteins functioning as molecular chaperones. Recently, a body of evidence has accumulated that folding of nascent polypeptides is transiently prevented also by forming complexes with constitutively expressed isoforms of stress proteins. The complexes are not dissociated until their translations are completed, assembled and, if indicated, transported into membraneous organella. Molecular chaperones assist their substrates in the complexes in folding, assembly and appropriate processing. In addition, dissociation of supramolecular complexes such as clathrin coats occurs only in the presence of molecular chaperones. Finally, degradation of proteins also has been reported to occur in a stress protein-dependent manner. Thus, it is evident that cellular proteins are taken care of by stress proteins from the cradle to the grave. This part deals with all of these problems.

Stress-Inducible Cellular Responses
ed. by U. Feige, R.I. Morimoto, I. Yahara and B. Polla
© 1996 Birkhäuser Verlag Basel/Switzerland

Normal protein folding machinery

D. Hartman and M. J. Gething

Department of Biochemistry and Molecular Biology, University of Melbourne, Parkville, Victoria 3052, Australia

Summary. A highly conserved protein folding machine has been maintained in the cytosol of both prokaryotic and eukaryotic organisms and in eukaryotic mitochondria. Homologous components of this machinery have also been identified in other organelles such as the endoplasmic reticulum in which HSP70 and DnaJ-like homologs reside. The high degree of conservation presumably reflects the proficiency with which these molecules have evolved to mediate the folding of proteins to their native functional states.

Introduction

The optimal functioning and maintenance of any cell is explicitly dependent upon the presence of correctly folded proteins. Although Anfinsen's postulate still holds true that the information necessary to specify the correct three-dimensional conformation of a protein resides entirely within its primary structure (Anfinsen, 1973), it is now evident that auxiliary proteinacious machines mediate protein folding *in vivo*. Both authentic catalysts of protein folding, such as protein disulphide isomerase and peptidyl proyl *cis-trans* isomerase, and "passive" folding proteins, collectively called molecular chaperones, whose catalytic properties remain to be established, have been described (reviewed in Gething and Sambrook, 1992). Molecular chaperones encompass a large ubiquitous group of proteins whose common function is to stabilise aggregation-prone folding intermediates and facilitate the acquisition of the functional native state. The synthesis of many molecular chaperones is enhanced in cells experiencing or recovering from protein perturbing stresses, such as heat shock, and hence molecular chaperones are also referred to as stress proteins, or more commonly, heat shock proteins (HSP).

The molecular chaperone machinery of the *E. coli* cytosol is the most thoroughly dissected protein folding apparatus and can be used as a paradigm for folding in other prokaryotes, and in the cytosol, mitochondria and chloroplast of eukaryotic organisms which possess functionally homologous machineries. A hierarchical model of chaperone mediated protein folding can be proposed in which progressively more highly structured folding intermediates interact first with DnaK (hsp70) and/or DnaJ (hsp40) and are then transferred, in a process aided by GrpE, to GroEL (hsp60) and GroES (hsp10) which mediate the final stages of folding. This chapter will describe first the *E. coli* cytosolic folding apparatus and then the functionally homologous folding apparatuses of the eukaryotic mitochondria and cytosol.

The *E. coli* cytosolic protein folding machinery

The DnaK, DnaJ and GrpE Machinery

The *dnaK* and *dnaJ* genes form an operon and encode polypeptides of approximately 69 kDa and 41 kDa, respectively (reviewed by Georgopolous et al. 1990, 1994). DnaK behaves predominantly as a monomeric protein in solution while DnaJ behaves as a homodimer (Georgopolous et al., 1990). The *grpE* gene is monocistronic and encodes a polypeptide of 22 kDa which, in solution, behaves as a dimer (Schonfeld et al., 1995). DnaK and DnaJ are integral components of the protein folding machinery in prokaryotic cells (Liberek et al., 1991 a, 1991 b; Langer et al., 1992 a; Hendrick et al., 1993; Szabo et al., 1994). They appear to stabilise early folding intermediates and suppress aggregation, a process which competes kinetically with productive folding. GrpE, on the other hand, acts as a polypeptide release factor, enabling the folding intermediate to be passed onto the GroEL/GroES system before final folding (Langer et al., 1992 a). Essential to the chaperone function of DnaK is its ability to bind non-native folding intermediates and to discharge them upon ATP binding (Liberek et al., 1991 a, Palleros et al., 1993). Both DnaJ and GrpE modulate the ATPase activity of DnaK and in this way contribute intimately to the chaperone function of DnaK (Liberek et al., 1991b). The weak ATPase activity of DnaK is stimulated 2 fold by DnaJ alone and 50 fold in the joint presence of both DnaJ and GrpE. Specifically, DnaJ accelerates the rate of hydrolysis of DnaK-bound ATP whereas GrpE acts as a nucleotide exchange factor by increasing the release of DnaK-bound nucleotides. Studies with DnaJ mutants including one that contains only the N-terminal 108 residues of the protein indicate that this highly-conserved N-terminal J region is both necessary and sufficient for stimulation of the ATPase activity of DnaK (Wall et al., 1994).

The substrate specificities of DnaK and DnaJ differ
Although both DnaK and DnaJ can bind non-native polypeptides and stabilise them against aggregation, the protein conformations recognised by DnaK and DnaJ appear to be markedly distinct (Langer et al., 1992 a). DnaK binds efficiently to reduced and carboxy-methylated-α-lactalbumin (RCMLA), a protein which maintains an extended conformation devoid of stable secondary structural elements, whereas DnaJ has very low affinity for this particular molecule. In contrast, DnaJ interacts with the so-called molten-globule form of α-lactalbumin; a compact state that contains

considerable secondary structure and a fluctuating hydrophobic core. Similarly, DnaJ, but not DnaK, is able to interact with casein, a protein which inherently shares attributes of a compact folding intermediate. Peptide binding experiments indicate that while DnaK is fully competent to bind short synthetic peptides (Gragerov et al., 1994, see below), DnaJ is not (B. Bukau, personal communication).

   In vitro experiments demonstrate that DnaJ can mediate the interaction of DnaK with some non-native polypeptides. Although a high molar excess of DnaK is required to suppress the aggregation of rhodanese diluted from denaturant, the addition of DnaJ to levels which are itself not enough to inhibit aggregation, markedly potentiates the ability of DnaK to suppress aggregation (Langer et al., 1992a). DnaK and DnaJ have also been shown to act synergistically to avert the *in vitro* aggregation of luciferase (Schroder et al., 1993). In the case of both rhodanese and luciferase, DnaJ alone suppresses the aggregation process more efficiently than DnaK alone. *In vivo* evidence also demonstrates that the synergistic actions of DnaK and DnaJ are required to prevent protein aggregation. In a *rhoH* background in which heat shock proteins are not induced at elevated temperatures, over-expression of DnaK and DnaJ together, but not either chaperone alone, is required to maintain solubility of newly-synthesised proteins (Gragerov et al., 1992). Recently, Wall et al. (1995) reported that the conserved G/F motif of DnaJ is necessary for the activation of the substrate binding properties of DnaK. Finally, reactivation of heat-inactivated polypeptides can also be mediated by the DnaK, DnaJ and GrpE system (Skowyra et al., 1990; Ziemienowicz et al., 1993). *In vitro* reactivation of heat-inactivated luciferase requires the coordinated action of DnaK, DnaJ and GrpE as omission of any component abolishes reactivation (Schroder et al., 1993). Likewise, all three components are required for the efficient *in vivo* reactivation of heat-inactivated luciferase.

The peptide-binding specificities of DnaK and other HSP70 chaperones
The general properties of peptides that bind to individual HSP70 proteins have been defined using sequence analysis of a bulk population of randomly-synthesized heptapeptides that bind to the endoplasmic reticulum chaperone, BiP (Flynn et al., 1991) or by employing affinity panning of libraries of bacteriophages that display peptides to identify individual sequences that bind with high affinity to BiP (Blond-Elguindi et al., 1993), DnaK (Gragerov et al., 1994) or bovine hsc70 (Hightower et al., 1994). These studies identified peptides with a high content of aromatic and hydrophobic amino acids and in the case of BiP, exposed a motif best described as Hy (W/X) HyXHyXHy, where Hy is a large hydrophobic or aromatic amino acid (most frequently Trp, Leu or Phe), W is Trp, and X is any amino acid. Basic amino acids were usually well tolerated, particularly by DnaK and hsc70, while the presence of acidic residues was generally unfavourable for binding. When the peptide binding specificities of the

three different HSP70 family members were directly compared (Fourie et al., 1994), it was found that peptides that bound with high affinity to all three chaperones contained the binding motif already identified for BiP, i.e., stretch of at least seven amino acids that was enriched in large hydrophobic and aromatic residues. The motif has a high degree of redundancy, consistent with the capacity of HSP70 proteins to recognize a wide variety of nascent polypeptides that share no obvious sequence similarity, and it contains a preponderance of residues whose side chains would normally be buried in the interior of a folded protein, consistent with the ability of these chaperones to discriminate accurately between folded and unfolded structures. Peptides that showed negligible affinity for any of the three chaperones either lacked large hydrophobic and aromatic residues or were enriched in acidic residues. Despite the common preference for a heptameric motif rich in hydrophobic residues, specificity differences between the HSP70 proteins were frequently observed, the most extreme example being a peptide rich in basic residues that bound with high affinity to DnaK, intermediate affinity to hsc70, and negligible affinity to BiP (Fourie et al., 1994). Thus, HSP70 proteins can exhibit common or exclusive binding specificities, depending on the peptide sequence.

DnaK and DnaJ may interact with nascent polypeptides on the ribosome
Experimental evidence indicates that the chaperone machinery interacts with nascent polypeptides on the ribosome. DnaJ can be crosslinked to nascent ribosome-bound polypeptides as short as 55 residues (Hendrick et al., 1993). Indeed, cotranslational binding of DnaJ during a translation reaction resulted in an arrest of folding that could be subsequently removed by the addition of DnaK and GrpE (Hendrick et al., 1993). Interestingly, under the ATP-rich conditions of the translation reaction, a stable interaction between the nascent polypeptide and DnaK was not observed. Conceivably, it may be DnaJ that binds nascent chains at an early stage of translation in order to maintain a productive folding conformation. However, since it is well established that DnaK is able to bind extended peptides in the absence of DnaJ, it is possible that whether DnaK or DnaJ first contacts a particular nascent chain depends on the rate and extent of secondary structure formation by the growing polypeptide. Thus DnaK and DnaJ would function cooperatively to maintain folding competence, playing varying roles depending on the nature (i.e., the sequence and folding characteristics) of each polypeptide substrate.

What are the conformations of DnaK- and DnaJ-bound substrates?
Fundamental to a complete understanding of the role the chaperone system performs in protein folding is the elucidation of the conformations of substrates bound to the chaperones. Nuclear magnetic resonance studies indicate that a peptide bound to DnaK maintains an extended conformation

(Landry et al., 1992). Tryptophan fluorescence emission experiments are consistent with DnaJ-bound rhodanese adopting a conformation in which its tryptophan residues are still in a non-native environment, but in a more hydrophobic environment than the fully unfolded state (Langer et al., 1992 a). DnaJ-bound rhodanese shows significant 1-anilino-naphtalene-8-sulphonate (ANS) fluorescence which is indicative of collapsed folding intermediates containing loosely packed secondary structure. Even higher ANS fluorescence is observed with DnaK/DnaJ-bound rhodanese. In contrast to native and aggregated rhodanese, DnaJ and DnaK/DnaJ-bound rhodanese are very susceptible to protease digestion. Therefore the concerted actions of DnaK and DnaJ most likely result in the stabilisation of early folding intermediates in states that contain an as yet undefined degree of secondary structure and no ordered tertiary structure. In probably the majority of cases the DnaK/DnaJ-bound folding intermediate is passed onto the GroEL/GroES system, in a transfer mediated by GrpE, for final folding (Langer et al., 1992 a). Nevertheless, for some proteins, the folding intermediate released from DnaK/DnaJ is competent to fold into the native conformation without further intervention by chaperones (Szabo et al., 1994).

The GroEL/GroES chaperonin complex

The *groEL* and *groES* genes form an operon and encode polypeptides of approximately 57 kDa and 10 kDa respectively (reviewed by Georgopolous et al., 1990). GroEL forms a homotetradecamer in solution which is arranged into two stacks of seven-membered rings that possess a sevenfold axis of symmetry. The two rings are arranged in a back-to-back orientation which, under the electron microscope, resembles two donut-shaped objects stacked on top of each other with a large central channel. Recently, the three-dimensional structure of a mutant form of GroEL, whose functional properties are indistinguishable from that of wild-type GroEL, has been determined (Braig et al., 1994). Each subunit is comprised of three distinct domains: a large equatorial domain which constitutes the foundations of the molecule by providing the majority of side-to-side contacts between subunits in the ring along with all of the contacts between both rings; an apical domain which forms the entrance of the central channel; and an intermediate domain which provides a covalent connection between the equatorial and the apical domain. The central channel has a diameter of 47 Å, suggesting that its volume is sufficient to accommodate a molten globule-like folding intermediate of approximately 50–60 kDa. A mutational analysis of GroEL that relates the functional properties of the protein to its crystal structure identified a potential polypeptide-binding site on the inside surface of the apical domain, facing the central channel, consisting of hydrophobic residues (Fenton et al., 1994). These same residues are essential for the binding of the co-chaperonin GroES, which itself exists as a homoheptamer that is arranged into a single seven-membered ring-shaped structure (reviewed by Georgopolous et al., 1990). Native GroES has a highly mobile and accessible polypeptide loop whose mobility and accessibility are lost upon formation of the GroES/GroEL complex (Landry et al., 1993). Amino acid substitutions present in eight independently isolated mutant groES alleles map in the mobile loop.

The affinity of GroEL for unfolded polypeptides is modulated by ATP and GroES

GroEL mediates the efficient folding of a polypeptide via a series of ATP-dependent binding and release reactions (reviewed by Frydman and Hartl, 1994). GroEL has an ATPase activity which is inhibited by GroES but stimulated by unfolded substrates (Martin et al., 1991). Cooperativity of ATP binding and hydrolysis at 14 sites on the GroEL oligomer is mediated by the action of GroES (Bochkareva et al., 1992, Jackson et al., 1993). The substrate affinity of GroEL is also regulated by nucleotides. Nucleotide-free and ADP-bound forms of GroEL have high affinity for unfolded polypeptide substrates whereas the ATP-bound form has a markedly lower substrate affinity (Bochkareva et al., 1992; Jackson et al., 1993; Martin et al., 1993). Martin et al. (1993) proposed a reaction cycle in which GroES and unfolded substrate polypeptide counteract each others' effect on GroEL. They suggest that GroES stabilises GroEL in the ADP-bound state while substrate binding triggers release of ADP and GroES. ADP/ATP exchange then permits the reassociation of GroES with concomitant hydro-lysis of ATP which in turn discharges the GroEL-bound substrate. If the released polypeptide is unable to proceed to the native state it is rebound by the chaperonin machinery for further rounds of binding and release. How-ever, the assertion that substrate induces dissociation of the GroEL/GroES complex has been challenged (Todd et al., 1994).

Requirement for GroES

Many, but not all, polypeptide substrates require GroES for efficient release from GroEL. Tightly-bound substrates may need GroES together with ATP for dissociation whereas for weakly-bound substrates, ATP alone may suffice (Schmidt et al., 1994a; Schmidt and Buchner, 1992). An inter-mediate situation may be observed for other GroEL-bound substrates; their dissociation may not be totally dependent upon GroES, although their release is enhanced in its presence. Proteins whose release does not require GroES are characterised by their inherent ability to fold to the native state spontaneously *in vitro*. Recently Schmidt et al. (1994b) reported that the requirement for GroES *in vitro* may be dependent upon the folding environment in which the experiment is performed. Proteins which were reconstituted under conditions in which unassisted spontaneous refolding did not occur were found to be dependent upon GroES, however, if the same proteins were reconstituted under conditions in which spontaneous refolding could occur then the requirement for GroES was removed. The requirement for GroES is therefore not strictly an intrinsic property of the substrate polypeptide but also reflects the folding environment. Evidence also exists that, in some cases, GroES may contribute to the folding process beyond just assisting the release of GroEL-bound substrates. GroEL-bound folding intermediates of rhodanese (Martin et al., 1991), RUBISCO (Baneyx and Gatenby, 1992) and tubulin (Frydman et al., 1992) can be partially discharged by the addition of ATP alone, however, a significant percentage of the released proteins aggregate. If GroES is present, aggregate formation is attenuated indicating the potentiality of a further role for GroES in folding extending beyond dissociation alone. The ability to cross-link GroES and pre-β-lactamase to each other when simultaneously bound by GroEL is consistent with this possibility (Bochkareva and Girshovich, 1992).

Does folding occur in the central cavity of GroEL?
Whether polypeptide folding occurs while the substrate is bound within the central cavity of GroEL or proceeds in the bulk solution following dissociation is currently a topic generating significant discussion and dissent. The existence of a large central cavity in GroEL prompted Martin et al. (1993) to speculate that folding could proceed within this orifice at effectively infinite dilution thereby eliminating unproductive folding events. Electron microscopy studies indicate that substrate polypeptides indeed reside, at least in part, within the central cavity of GroEL (Langer et al., 1992 b). Mutational studies are likewise consistent with the substrate binding site residing inside the central channel (Fenton et al., 1994). However, it does not necessarily follow that folding is completed within the cavity or that the process is committed to occur within a single GroEL oligomer.

Weissman et al. (1994) used GroEL mutants that can bind non-native folding intermediates but not release them, to tackle this issue. In a series of elegant experiments, it was established that upon dissociation of a GroEL-substrate complex by the addition of GroES and ATP, the released substrate could be trapped by the discharge-defective GroEL mutant. Since GroEL does not bind to the native state of the polypeptide, the discharged molecule was therefore still in a non-native state indicating that the substrate is released into the bulk solution to attempt folding. Consistent with this observation, the rate of release of GroEL-bound substrate was significantly faster than the rate of refolding. Crosslinking experiments indicated that non-native folding intermediates could be rebound by a GroEL particle distinct from the one from which it had just been released, providing evidence that folding does not necessarily proceed within a single GroEL particle. Interestingly, tryptophan fluorescence and proteolytic susceptibility experiments indicate that the rebound substrate adopts a similar conformation as the original GroEL-substrate complex. Single nucleotide turnover experiments are likewise consistent with the folding intermediate being discharged in a non-native form (Todd et al., 1994). How then would GroEL mediate the folding process under this scenario? Conceivably, GroEL could mediate folding by binding aggregation-prone folding intermediates and unfolding them, thereby providing the polypeptide with a further chance to reach the native state. The released polypeptide would undergo kinetic partitioning; a portion of the released substrate would proceed to the native state while another portion of the substrate would progress down an unproductive folding pathway leading to kinetically trapped or aggregation prone intermediates. These non-native intermediates would then again be recognised and bound by GroEL to perpetuate the cycle. Indeed, nuclear magnetic resonance experiments indicate that the entire secondary structure of cyclophilin is disrupted when bound to GroEL (Zahn et al., 1994).

Contradictory evidence has been provided by Hartl and colleagues who report the demonstration of successive rounds of protein folding by GroEL without intermittent release of unfolded polypeptide into the bulk solution (F.-U. Hartl, personal communication).

What are the conformations of GroEL-bound substrates?
Although a considerable degree of progress has been achieved in determining the mechanistic details of chaperonin action, much still remains to be ascertained regarding the conformation of GroEL-bound folding intermediates. Since GroEL lacks tryptophan residues, intrinsic fluorescence measurements have been employed to investigate the conformations of GroEL-bound substrates which themselves contain tryptophan (Martin et al., 1991). The fluorescence properties of GroEL-bound dihydrofolate reductase (DHFR) are consistent with a folding intermediate containing little significant ordered tertiary structure and tryptophan residues inhabiting an environment that is intermediate in hydrophobicity between the native and fully denatured state. Consistent with a flexible conformation, GroEL-bound DHFR is more susceptible to proteinase K digestion than native DHFR. ANS fluorescence experiments also favour a folding intermediate that resembles the molten globule state. Numerous other GroEL-bound folding intermediates are likewise consistent with the molten globule state (Hayer-Hartl et al., 1994; Robinson et al., 1994).

Greater than 50% of the proteins denatured by the addition of 5 M guanidine-HII to an *E. coli* cell lysate, are able to form isolatable binary complexes with GroEL while the native form of these same proteins are not (Viitanen et al., 1992a). Evidently, GroEL has high substrate plasticity in regard to non-native folding intermediates while, at the same time, having very little affinity for native proteins. GroEL does not bind short peptides, or polypeptides such as RCMLA that maintain an extended conformation

(Langer et al., 1992a). At present, however, it is unclear what structural features of non-native substrates do confer GroEL recognition. Clearly, the promiscuity exhibited by GroEL precludes distinct primary structural motifs. Based on nuclear magnetic resonance experiments, Landry et al. (1992) speculated that side-chain hydrophobicity and an innate capacity to form an α-helix may be important determinants for GroEL binding. By subjecting unfolded rhodanese complexed to GroEL to limited proteolysis followed by sequence analysis of the protected fragments, Hlodan et al. (1995) identified protected species of 11000 and 7000 daltons originating from the two homologous domains of the protein. The shorter fragment contained one hydrophobic and one amphipathic helix that map to the domain interface in folded rhodanese, while the larger fragment contained the homologous regions from the other domain, as well as an additional hydrophobic helix. However, proteins that contain only β-sheet secondary structural elements are also recognised by GroEL (Schmidt and Buchner, 1992).

The DnaK/DnaJ/GrpE and GroEL/GroES systems cooperate in polypeptide folding

Both *in vivo* and *in vitro* experiments demonstrate that the DnaK/DnaJ/ GrpE and GroEL/GroES systems function in a cooperative manner to mediate protein folding. Using purified components, Langer et al. (1992a) devised an *in vitro* protein folding machinery that could mediate the reconstitution of chemically denatured rhodanese. In the presence of MgATP, DnaK and DnaJ acted synergistically to effectively suppress the aggregation of denatured rhodanese. Transfer of the non-native folding intermediate to the GroEL/GroES system for final folding was mediated by the addition of GrpE. Inactive ribosome-bound rhodanese can be activated and released by the addition of the complete chaperone machinery consisting of DnaK, DnaJ, GrpE, GroEL, GroES and ATP (Kudlicki et al., 1994a). Efficient reactivation was dependent on the presence of each of the chaperone proteins. DnaK, DnaJ and GrpE appear to function before GroEL and GroES in this reaction (Kudlicki et al., 1994b). *In vivo* experiments utilising a *rhoH* strain containing plasmids expressing GroEL, GroES, DnaK, DnaJ and combinations thereof, also indicate that these chaperones function synergistically to ensure proper polypeptide folding (Gragerov et al., 1992). From these experiments a hierarchical model of chaperone action can be postulated in which progressively more complex folding intermediates interact first with DnaK and/or DnaJ and are transferred, via GrpE, to GroEL and GroES which mediate the final stages of folding (Fig. 1).

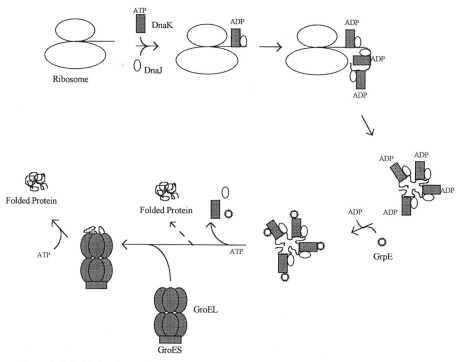

Figure 1. Model for the pathway of chaperone-mediated folding in the cytosol of bacteria. (Modified from Langer et al., 1992 a). For details see the text.

The mitochondrial protein folding apparatus

The protein folding apparatus of the prokaryotic cytosol has been conserved in the mitochondria of eukaryotic organisms. Mitochondrial homologues of GroEL (hsp60), GroES (hsp10, DnaK (hsp70, Ssc1), DnaJ (Mdj1) and GrpE (Mge1) have all been identified (for references, see below). Although they have as yet not been as intensively characterised *in vitro* as their prokaryotic counterparts, there is substantial evidence demonstrating their importance in mitochondrial protein biogenesis.

Mitochondrial hsp60

Direct evidence that mitochondrial hsp60 performs indispensable functions in protein folding was obtained in experiments employing yeast strains harbouring temperature-sensitive lethal mutations in the gene encoding hsp60 (Cheng et al., 1989; Hallberg et al., 1993). At the non-permissive temperature, assembly of newly-synthesised mitochondrially-localised proteins is impaired leading to the formation of insoluble aggregates. The interaction of hsp60 with newly-translocated mitochondrial preproteins was directly investigated using *in vitro* mitochondrial import and folding assays (Ostermann et al., 1989). The denatured artificial precursor protein PreSu9-DHFR could be rapidly translocated into isolated *Neurospora crassa* mitochondria upon which the signal sequence was removed and the mature size protein adopted a protease-resistant conformation indicative of the folded state. However, under conditions of ATP depletion, the translocated protein remained in a protease-sensitive conformation which could be recovered as a high molecular mass complex that co-fractionated with hsp60 during gel filtration chromatography. Furthermore, Su9-DHFR could be co-immunoprecipitated with anti-hsp60 antibodies. A temperature-sensitive yeast mutant (*mif4*), which loses hsp60 function at the non-permissive temperature, was utilised to demonstrate a role for mitochondrial hsp60 in the prevention of heat denaturation (Martin et al., 1992). When expressed at the non-permissive temperature of 30 °C, the model protein mOTC-DHFR was extracted from *mif4* mitochondria in an insoluble fraction devoid of enzymatic activity. However, when expressed at the permissive temperature in the *mif4* mutant or at 30 °C in a wild-type strain, mOTC-DHFR was extracted in a soluble enzymatically active form. Since the aggregated mOTC-DHFR could have arisen from denaturation of existing mOTC-DHFR or from an inability of newly imported mOTC-DHFR to fold, further experiments were performed to demonstrate that hsp60 could indeed protect pre-existing mOTC-DHFR from heat denaturation. Thus, hsp60 plays a role not only in the initial acquisition of the functional state, but also in the maintenance of this state once it is acquired.

Mitochondrial hsp10

Mitochondria contain a polypeptide (cpn10 or hsp10) that is functionally equivalent to GroES (Lubben et al., 1990; Hartman et al., 1992). This protein is able to functionally substitute for bacterial GroES in the chaperonin-dependent *in vitro* reconstitution of chemically-denatured RUBISCO (Lubben et al., 1990) or ornithine transcarbamoylase (Hartman et al., 1992). Hsp10 closely resembles GroES in its heat stress-inducibility and subunit and oligomeric molecular mass (Hartman et al., 1992). Like bacterial GroEL and GroES, mitochondrial hsp60 and hsp10 are able to form an isolatable binary complex in the presence of ATP (Viitanen et al., 1992 b). Mitochondrial hsp10 is also able to complex with bacterial GroEL and in doing so inhibit its ATPase activity in a fashion similar to bacterial GroES, further indicating functional homology between the bacterial and mitochondrial systems. However, bacterial GroES is unable to mediate refolding with mitochondrial hsp60 because of their inability to form a complex. The gene encoding mitochondrial hsp10 from *S. cerevisiae* has been cloned and disrupted revealing hsp10 to be an essential gene (Rospert et al., 1993). Studies with mitochondria isolated from a yeast strain harbouring a temperature-sensitive *hsp10* mutation demonstrate that Hsp10 is required for the folding and assembly of proteins imported into the matrix compartment (Hohfeld and Hartl, 1994). The temperature-sensitive mutations mapped to a domain of the Hsp10 polypeptide that corresponds to the mobile loop region of GroES, and result in a reduced binding affinity of hsp10 for hsp60 at the non-permissive temperature.

Mitochondrial hsp70

Mitochondrial hsp70 (Ssc1) plays a direct role in folding and assembly of proteins that are encoded by the mitochondrial genome, with the availability of Ssc1 function influencing the pattern of proteins synthesized in yeast both *in vivo* and *in vitro* (Hermann et al., 1994). Ssc1 acts to maintain the var1 protein, the only mitochondrially-encoded subunit of mitochondrial ribosomes, in an assembly competent state, and also helps to facilitate assembly of mitochondrially-encoded subunits of the ATP synthase complex.

Many studies have analysed the role of Ssc1 during translocation of nuclear-encoded mitochondrial preproteins into the matrix compartment. Ssc1 associates with a precursor protein that is partially translocated into yeast mitochondria (Scherer et al., 1990), binding to both the presequence and mature portions of the protein (Voos et al., 1993). Kang et al. (1990) employed experiments similar to those used to study mitochondrial hsp60 to show that a protease resistant state of preSu9-DHFR was not attained upon translocation into mitochondria isolated from a *ssc1*[ts] strain incubated at the non-permissive temperature. However, the protease resistant state was acquired upon import into wild-type mitochondria. Finally, using anti-hsp70 antibodies for immunoprecipitation experiments, a direct physical interaction between imported Su9-DHFR and heat inactivated Ssc1 could be observed. Recently, several groups have reported that during an early step in the translocation process, complexes can be isolated that include the precursor protein, Ssc1 and MIM44 (Isp45), a component of the mitochondrial import machinery that is localized on the matrix side of the mitochondrial inner membrane (Kronidou et al., 1994; Rassow et al., 1994; Schneider et al., 1994). MIM44 is thought to recruit Ssc1 to the import apparatus.

Manning-Krieg et al. (1991) also studied the *in vitro* import, folding and assembly of Mas2p to determine if there is any co-operation between mitochondrial hsp70 and hsp60 in protein folding, analogous to the situation with DnaK and GroEL in the prokaryotic cytosol. Employing anti-hsp70 and anti-hsp60 antibodies in immunoprecipitation experiments, it was concluded that newly-imported Mas2p first interacted with hsp70 then with hsp60. Following interaction with hsp60, Mas2p reached its protease resistant conformation and assembled with its partner subunit Mas1p.

Mitochondrial DnaJ

Since the sequential action of hsp70 and hsp60 is apparently conserved be-
tween the cytosol of prokaryotes and the mitochondria of eukaryotes, it was
anticipated that the search for mitochondrial homologues of DnaJ and
GrpE would be fruitful. A gene encoding a mitochondrial DnaJ homologue
(*MDJ1*) was identified during DNA sequence analysis of clones from a
S. cerevisae genomic library (Rowley et al., 1994). Disruption of *MDJ1* leads
to a petite phenotype, loss of mitochondrial DNA and inviability at 37 °C.
Furthermore, the folding of newly imported Su9-DHFR into mitochondria
isolated from a Δ*mdj1* strain, was shown to be markedly impaired. Likewise, the amount of enzymatically active luciferase produced in Δ*mdj1*
mitochondria following *in vitro* import was substantially reduced compared to control mitochondria highlighting the integral role performed by
Mdj1 during protein folding. Finally, heat treatment of mitochondria
derived from a Δ*mdj1* strain, resulted in a reduction of the amount of
protease-resistant imported test protein Su9-DHFR when compared to
levels from wild-type mitochondria. Mdj1 is therefore, like hsp60, important both for mitochondrial protein folding and for protection against heat
denaturation.

Mitochondrial GrpE

The gene encoding mitochondrial GrpE (also called Mge1 and Yge1) from
Saccharomyces cerevisiae was found in the Genbank database on the basis
of its similarity to *E. coli* GrpE (Bolliger et al., 1994; Laloraya et al., 1994).
Mge1 is a 24 kDa soluble mitochondrial matrix protein that associates with
mitochondrial hsp70 (Ssc1p) (Laloraya et al., 1994). Analogous to the
situation with bacterial DnaK and GrpE, the complex between Mge1 and
mitochondrial hsp70 is stable in the presence of salt but completely disrupted by the addition of ATP (Bolliger et al., 1994; Nakai et al., 1994;
Voos et al., 1994). Mge1 can functionally substitute for GrpE in complementing the temperature-sensitive phenotype of a *grpE* mutant (Ikeda
et al., 1994). It is therefore likely that Mge1 performs similar protein folding functions as its bacterial counterpart. Cells with reduced levels of Mge1
or mutant alleles accumulate precursors of some mitochondrial preproteins
(Nakai et al., 1994; Laloraya et al., 1994), while studies with temperature-
sensitive *Mge1* strains indicate that Mge1 modulates the function of mt-
hsp70 in facilitating import from the cytosol of nuclear encoded mitochondrial precursors and may also be required for maturation and folding
of some preproteins following import (Laloraya et al., 1995; Westerman
et al., 1995).

The concerted action of mitochondrial molecular chaperones
and the biogenesis of mitochondrial proteins

Since the chaperone components of the bacterial cytosol have been conserved in eukaryotic mitochondria, it is reasonable to suggest a model for protein folding in mitochondria based on the bacterial chaperone machinery. Mitochondrial preproteins exit the mitochondrial import machinery in an extended conformation similar to that of polypeptides exiting the ribosome. The preproteins interact with mitochondrial hsp70 even during their translocation and this association may continue as the polypeptides are extruded into the matrix compartment, maintaining them in a somewhat unfolded conformation. Mdj1 does not appear to play any role in assisting Ssc1 during the process of translocation, but is required for folding of newly-imported precursor proteins. Finally, in a process mediated by Mge1, polypeptides may be passed onto hsp60 and hsp10 for final folding.

The protein folding machinery of the eukaryotic cytosol

Cytosolic chaperonin complex

Recent studies have uncovered a eukaryotic cytosolic folding complex distantly related to eubacterial GroEL, but more closely related to the TF55 protein/thermosome complex of archaebacteria (reviewed by Horwich and Willison, 1993; Kim et al., 1994). Cytosolic chaperonin complexes (variously named TCP-1 complex, chromobindin A or TRiC, hereafter collectively referred to as CCT) were originally isolated from bovine and rabbit reticulocytes as large molecular assemblies of 800 to 970 kDa constructed of several polypeptides whose molecular masses measured from 52 to 65 kDa (Gao et al., 1992; Frydman et al., 1992). Electron microscopic images display these oligomeric complexes as large toroidal shaped structures containing a central cavity reminiscent of the GroEL molecule (Gao et al., 1992; Frydman et al., 1992). The CCT complex is now known to be a 960 kDa hetero-oligomeric double-torus-like structure composed of 16 subunits each of molecular mass ~60 kDa (Kubota et al., 1995a). The complex contains at least eight different, but related, subunit species encoded by different genes and is abundant in the cytosol of eukaryotic cells.

CCT mediates the folding of both actin and tubulin *in vitro* (Frydman et al., 1992; Gao et al., 1992; Yaffe et al., 1992; Melki et al., 1993) and also interacts with newly-synthesized actin and tubulin *in vivo* (Sternlicht et al., 1993). Studies of temperature-sensitive mutants of yeast CCT subunits indicate that each of four different subunit species are involved in the folding and assembly of actin and tubulin in the eukaryotic cytosol (Ursic and Culbertson, 1991; Chen et al., 1994; Miklos et al., 1994; Ursic et al., 1994;

Vinh et al., 1994). Several other proteins are folded or bound by CCT *in vitro*, including firefly luciferase (Frydman et al., 1992, 1994) and a neuro-filament peptide (Roobol and Carden, 1993). In addition, CCT is involved in the assembly of the hepatitis B virus capsid (Lingappa et al., 1994). However, some denatured proteins, such as cap-binding protein, cyclin B, H-ras and c-myc, show low affinities for CCT *in vitro* (Melki and Cowan, 1994).

Formation of a binary complex between CCT and its substrate *in vitro* is independent of the presence of ATP (Gao et al., 1992; Frydman et al., 1992). Discharge of the binary complex leading to the attainment of the native conformation is, however, dependent upon ATP. Melki and Cowan (1994) postulate that CCT binds target proteins in its ADP-bound state with release being controlled by the relatively low substrate affinity of CCT in the ATP-bound state. ATP hydrolysis therefore acts as a molecular switch converting CCT between two conformational states differing in substrate affinity. At present, very little has been ascertained regarding the conformations of CCT-bound substrates except that they appear to be in a more protease-sensitive state than the native forms.

Do cytosolic homologues of GroES exist?

As yet, no cytosolic GroES homologues have been identified, however, the CCT-mediated folding of assembly competent α- and β-tubulin has been reported to be absolutely dependent upon two other protein co-factors termed co-factor A and co-factor B (along with GTP) (Gao et al., 1993). Release of α- and β-tubulin from CCT can be mediated by Mg-ATP alone, however the discharged substrates do not reach their native conformations. Co-factor A and co-factor B therefore participate directly in the folding reaction and are not required merely for discharge of the binary complex. Co-factor A has been purified and cloned but shows no significant sequence homology to any known protein. The monomeric mass of co-factor A is approximately 13 kDa while that of the native protein is approximately 28 kDa implicating a dimer as the functional unit. The indispensable function performed by co-factor A is, at present, not clearly defined, however, co-factor A has been shown to stimulate the ATPase activity of CCT.

Cytosolic HSP70 and HSP40 proteins

Eukaryotic genomes encode multiple cytosolic HSP70 species whose synthesis can either be strictly inducible, constitutive or constitutive and inducible (reviewed by Becker and Craig, 1994; Boorstein et al., 1994). Cytosolic isoforms have been reported to be associated with nascent chains emerging from the ribosome (Beckmann et al., 1990; Nelson et al., 1992), to be required for the folding of newly-synthesized polypeptides (see below, Frydman et al., 1994; Levy et al., 1995) and to be involved in stabilizing nascent polypeptides prior to their translocation into the endoplasmic reticulum and mitochondria (Deshaies et al., 1988).

Multiple cytosolic DnaJ (hsp40) homologues that interact with and modulate the ATPase activities and polypeptide binding properties of cytosolic HSP70s have been identified (reviewed by Cyr et al., 1994). Members of the DnaJ-like protein family are structurally diverse, containing different combinations of the three conserved domains identified in *E. coli* DnaJ as well as frequently containing additional sequences (Silver and Way, 1993; Cyr et al., 1994). Different DnaJ-like proteins have developed specialized functions in a variety of intracellular events, including protein translocation across membranes, protein folding, translation initiation, gene expression and growth control (Cyr et al., 1994). For each of these functions the DnaJ-like proteins apparently recruit partner HSP70 molecules via their interaction with the conserved DnaJ domain.

The dynamics of the cytosolic protein folding apparatus

Experiments examining the folding of luciferase as it emerges from the ribosome in a rabbit reticulocyte lysate have illuminated the dynamics of the eukaryotic cytosolic protein folding apparatus (Frydman et al., 1994). These experiments implicate hsp70, hsp40 and CCT as components of a high molecular mass assembly that is associated with the nascent chain as it exits the ribosome. Immunodepletion experiments indicated that all three chaperones were required for efficient folding of luciferase. Further experiments were consistent with a model in which hsp70 and hsp40 molecules first interact with the elongating complex before the addition of CCT. Moreover, hsp40 may mediate the loading of hsp70 onto the polypeptide as it emerges from the ribosome. In all of these aspects, but with CCT replacing GroEL/hsp60, the process of post-translational folding in the eukaryotic cytosol closely resembles that occurring in *E. coli* or in the mitochondrial matrix. However, no homologue of GrpE has yet been identified in the cytosol of eukaryotes.

The endoplasmic reticulum contains DnaK and DnaJ homologues

Although the endoplasmic reticulum does not appear to contain homologues of GrpE, GroEL or GroES, it does possess homologues of DnaK termed BiP in mammalian cells (Munro and Pelham, 1986; Kosutsumi et al., 1989) or Kar2p in yeast (Rose et al., 1989; Normington et al., 1989) as well as two proteins with homology to DnaJ called Sec63p (Sadler et al., 1989) and Scj1 (Schlenstedt et al., 1995).

In mammalian cells BiP has been shown to interact transiently with newly-synthesised secretory proteins and more permanently with mal-folded or unassembled polypeptides (reviewed by Gething and Sambrook, 1992). Like other HSP70 proteins, BiP has a weak ATPase activity that can be stimulated by unfolded polypeptides and synthetic peptides (Flynn et al., 1989). As discussed earlier, the substrate specificity of murine BiP has been thoroughly investigated (Blond-Elguindi et al., 1993).

Experiments in yeast have delineated separate roles for BiP in protein translocation across the ER membrane (Brodsky et al., 1995) and in folding in the ER lumen (Simons et al., 1995) that parallel the roles played by Ssc1 in mitochondria. BiP's role in translocation is aided by the DnaJ homologue Sec63 (Brodsky et al., 1995), an ER membrane protein that spans the bilayer three times (Sadler et al., 1989) with the J domain oriented into the ER lumen (Feldheim et al., 1992) where it probably interacts with BiP in a manner analogous to that of DnaJ with DnaK (Scidmore et al., 1993). The second DnaJ homologue, Scj1, is a soluble protein of the ER lumen whose structure more closely resembles that of DnaJ (Silver and Way, 1993). Scj1 probably cooperates with BiP in assisting the folding in the ER of newly-synthesized polypeptides (Schlenstedt et al., 1995).

Further details of the endoplasmic reticulum chaperone system will be presented in the chapter by Wei and Hendershot. (this volume)

References

Anfinsen, C.B. (1973) Principles that govern the folding of protein chains. *Science* 182: 223–230.

Baneyx, F. and Gatenby, A.A. (1992) A mutation in GroEL interferes with protein folding by reducing the rate of discharge of sequestered polypeptides. *J. Biol. Chem.* 267:11637–11644.

Beckman, R.P., Mizzen, L.A. and Welch, W.J. (1990) Interaction of hsp70 with newly synthe-sized proteins: Implications for protein folding and assembly. *Science* 248:850–854.

Blond-Elguindi, S., Cwirla, S.E. Dower, W.J. Lipshutz, R.J. Sprang, S.R. Sambrook, J.F. and Gething M.J. (1993) Affinity panning of a library of peptides displayed on bacteriophages reveals the binding specificity of BiP. *Cell* 75:717–728.

Bochkareva, E.S. and Girshovich, A.S. (1992) A newly synthesized protein interacts with GroES on the surface of chaperonin GroEL. *J. Biol. Chem.* 267:25672–25675.

Bochkareva, E.S., Lissen, N.M., Flynn, G.C., Rothman, J.E. and Girshovich, A.S. (1992) Posi-tive cooperativity in the functioning of molecular chaperone GroEL. *J. Biol. Chem.* 267: 6796–6800.

Bolliger, L., Deloche, O., Glick, B.S., Georgopolous, C., Jeno, P., Kronidou, N., Horst, M., Morishima, N. and Schatz, G. (1994) A mitochondrial homolog of bacterial GrpE interacts with mitochondrial hsp70 and is essential for viability. *EMBO J.* 13:1998–2006.

Braig, K., Otwinowski, Z., Hegde, R., Boisvert, D.C., Joachimiak, A., Horwich, A.L. and Sigler, P.B. (1994) The crystal structure of the bacterial chaperonin GroEL at 2.8 A. *Nature* 371: 578–586.

Brodsky, J.L., Goeckeler, J. and Schekman, R. (1995) Sec63 and BiP are required for both co- and post-translational protein translocation into yeast microsomes. *Proc. Natl. Acad. Sci. USA* 92:9642–9646.

Chen, X., Sullivan, D.S. and Huffaker, T.C. (1994) Two yeast genes with similarity to TCP-1 are required for microtubule and actin function *in vivo. Proc. Natl. Acad. Sci. USA* 91: 9111–9115.

Cheng, M.Y., Hartl, F.-U., Martin, J., Pollock, R.A., Kalousek, F., Neupert, W., Hallberg, E.M., Hallberg, R.L. and Horwich, A.L. (1989) Mitochondrial heat-shock protein hsp60 is essential for assembly of proteins imported into yeast mitochondria. *Nature* 337: 620–625.

Feldheim, D., Rothblatt, J. and Schekman, R. (1992) Topology and functional domains of Sec63p, an endoplasmic reticulum membrane protein required for secretory protein translocation. *Mol. Cell Biol.* 12: 3288–3296.

Fenton, W.A., Kashi, Y., Furtak, K. Horwich, A.L. (1994) Residues in chaperonin GroEL required for polypeptide binding and release. *Nature* 371: 614–619.

Flynn, G.C., Chappell, T.G. and Rothman, J.E. (1989) Peptide binding and release by proteins implicated as catalysts of protein assembly. *Science* 245: 385–390.

Flynn, G.C., Pohl J., Flocco, M.T. and Rothmann. J.E. (1991) Peptide-binding specificity of the molecular chaperone BiP. *Nature* 353: 726–730.

Fourie, A.M., Sambrook, J.F. and Gething, M.-J.H. (1994) Common and divergent peptide binding specificities of hsp70 molecular chaperones. *J. Biol. Chem.* 269: 30470–30478.

Frydman, J., Nimmesgern, E., Erdjument-Bromage, H., Wall, J.S., Tempst, P. and Hartl, F.-U. (1992) Function in protein folding of TRiC, a cytosolic ring complex containing TCP-1 and structurally related subunits. *EMBO J.* 11: 4767–4778.

Frydman, J., Nimmesgern, E., Ohtsuka, K. and Hartl, F.-U. (1994) Folding of nascent polypeptide chains in a high molecular mass assembly with molecular chaperones. *Nature* 370: 111–117.

Gao, Y., Thomas, J.O., Chow, R.L., Lee, G.-H. and Cowan, N.J. (1992) A cytoplasmic chaperonin that catalyzes β-action folding. *Cell* 69: 1043–1050.

Gao, Y., Vainberg, I.E., Chow, R.I. and Cowan, N.J. (1993) Two cofactors and cytoplasmic chaperonin are required for the folding of α- and β-tubulin. *Mol. Cell Biol.* 13: 2478–2485.

Gao, Y., Melki, R., Walden, P.D., Lewis, S.A., Ampe, C., Rommelaere, H., Vandekerckhove, J. and Cowan, N.J. (1994) A novel cochaperonin that modulates the ATPase activity of cytoplasmic chaperonin. *J. Cell Biol.* 125: 989–996.

Georgopolous, C.P., Ang, D., Liberek, K. and Zylic, M. (1990) Properties of the *Escherichia coli* heat shock proteins and their role in bacteriophage lambda growth. *In:* R.I. Morimoto, A. Tissieres and C. Georgopolous (eds): *Stress proteins in biology and medicine.* Cold Spring Harbor Laboratory Press, Cold Spring Harbor, NY, pp 191–221.

Georgopoulos, C., Liberek, K. Zylicz, M. and Ang. D. (1994) Properties of the heat shock proteins of *Escherichia coli* and the autoregulation of the heat shock response. *In:* R.I. Morimoto, A. Tissieres and C. Georgopoulos (eds): *The Biology of Heat Shock Proteins and Molecular Chaperones.* Cold Spring Harbor Laboratory Press, NY, pp 209–249.

Gething, M.J. and Sambrook. J.F. (1992) Protein folding in the cell. *Nature* 355: 33–45.

Gragerov, A., Nudler, E., Komissarova, N., Gaitanaris, G.A., Gottesman, M.E. and Nikiforov, V. (1992) Cooperation of GroEL/GroES and DnaK/DnaJ heat shock proteins in preventing protein misfolding in *Escherichia coli. Proc. Natl. Acad. Sci. USA* 89: 10341–10344.

Gragerov, A., Zeng, L. Zhao, X. Burkholder, W. and Gottesman. M.E. (1994) Specificity of DnaK-peptide binding. *J. Mol. Biol.* 235: 848–854.

Hallberg, E.M., Shu, Y. and Hallberg, R.L. (1993) Loss of mitochondrial hsp60 function: Nonequivalent effects on matrix-targeted and intermembrane-targeted proteins. *Mol. Cell. Biol.* 13: 3050–3057.

Hartman, D.J., Hoogenraad, N.J., Condron, R. and Hoj, P.B. (1992) Identification of a mammalian 10-kDa heat shock protein, a mitochondrial chaperonin 10 homolog essential for the assisted folding of trimeric ornithine transcarbamoylase *in vitro. Proc. Natl. Acad. Sci. USA* 89: 3394–3398.

Hayer-Hartl, M.K., Ewbank, J.J., Creighton, T.E. and Hartl, F.U. (1994) Conformational specificity of the chaperonin GroEL for the compact folding intermediates of alpha-lactalbumin. *EMBO. J.* 13: 3192–3202.

Hendrick, J.P., Langer, T., Davis, T.A., Hartl, F.-U. and Wiedmann, M. (1993) Control of folding and membrane translocation by binding of the chaperone DnaJ to nascent polypeptides. *Proc. Natl. Acad. Sci. USA* 90: 10216–10220.

Herman, J., Stuart, R., Craig, E. and Neupert, W. (1994) Mitochondrial heat shock protein 70, a molecular chaperone for proteins encoded by mitochondrial DNA. *J. Cell Biol.* 127: 893–902.

Hightower, L.E., Sadis, S.E. and Takenaka, I.M. (1994) Interactions of vertebrate hsc70 and hsp70 with unfolded proteins and peptides. *In:* A. Morimoto, A. Tissieres and C. Georgopoulos, (eds): *The Biology of Heat Shock Proteins and Molecular Chaperones.* Cold Spring Harbor Laboratory Press, Cold Spring Harbor NY, pp 179–207.

Hlodan, R., Tempst, P. and Hartl, F.U. (1995) Binding of defined regions of a polypeptide to GroEL and its implications for chaperonin-mediated protein folding. *Nature Struct. Biol.* 2:587–595.

Hohfeld, J. and Hartl, F.U. (1994) Role of the chaperonin cofactor hsp10 in protein folding and sorting in yeast mitochondria. *J. Cell Biol.* 126:305–315.

Horwich, A.L. and Willison, K. (1993) Protein folding in the cell: Functions of two families of molecular chaperones, hsp60 and TF55-TCP1. *Philos. Trans. R. Soc. Lond. B. Biol. Sci.* 339:313–325.

Ikeda, E., Yoshida, S., Mitsuzawa, H., Uno, I. and Toh-e, A. (1994) *YGE1* is a yeast homologue of *Escherischia coli grpE* and is required for the maintenance of mitochondrial function. *FEBS LETT.* 339:265–268.

Jackson, G.S., Staniforth, R.A., Halsall, D.J., Atkinson, T., Holbrook, J.J., Clarke, A.R. and Burston, S.G. (1993) Binding and hydrolysis of nucleotides in the chaperonin catalytic cycle: implications for the mechanism of assisted protein folding. *Biochemistry* 32:2554–2563.

Kang, P.-J., Ostermann, J., Shilling, J., Neupert, W., Craig, E.A. and Pfanner, N. (1990) Requirement for hsp70 in the mitochondrial matrix for translocation and folding of precursor proteins. *Nature* 348:137–143.

Kim, S., Willison, K.R. and Horwich, A.L. (1994) Cytosolic chapeonin subunits have a conserved ATPase domain but diverged polypeptide-binding domains. *Trends Biochem. Sci.* 19:543–548.

Kozutsumi, Y., Normington, K., Press, E., Slaughter, C., Sambrook, J. and Gething, M.J. (1989) Identification of immunoglobulin heavy chain binding protein as glucose-regulated protein 78 on the basis of amino acid sequence, immunological cross-reactivity, and functional activity. *J. Cell Sci. Suppl.* 11:115–137.

Kronidou, N.G., Oppliger, W., Bolliger, L., Hannavy, K., Glick, B., Schatz, G. and Horst, M. (1994) Dynamic interaction between Isp45 and mitochondrial hsp70 in the protein import system of yeast mitochondrial inner membrane. *Proc. Natl. Acad. Sci. USA* 91:12818–12822.

Kubota, H., Hynes, G. and Willison, K. (1995) The chaperonin containing t-complex polypeptide 1 (TCP-1): Multisubunit machinery assisting in protein folding and assembly in the eukaryotic cytosol. *Eur. J. Biochem.* 230:13–16.

Kudlicki, W., Odom, O., Kramer, G. and Hardesty, B. (1994a) Activation and release of enzymatically inactive, full-length rhodanese that is bound to ribosomes as peptidyl-tRNA. *J. Biol. Chem.* 269:16549–16553.

Kudlicki, W., Odom, O.W., Kramer, G. and Hardesty, B. (1994b) Chaperone-dependent folding and activation of ribosome-bound nascent rhodanese. *J. Mol. Biol.* 244:319–331.

Laloraya, S., Gambil, B.D. and Craig, E.A. (1994) A role for a eukaryotic GrpE-related protein, Mge1 in protein translocation. *Proc. Natl. Acad. Sci. USA* 91:6481–6485.

Laloraya, S., Dekker, P., Voos, W., Craig, E.A. and Pfanner, N. (1995) Mitochondrial GrpE modulates the function of matrix Hsp70 in translocation and maturation of proteins. *Mol. Cell Biol.* 15:7098–7105.

Landry, S.J., Jordan, R., McMacken, R. and Gierasch, L.M. (1992) Different conformations for the same polypeptide bound to chaperones DnaK and GroEL. *Nature* 355:455–457.

Landry, S.J., Zeilstra-Ryalls, J., Fayet, O., Georgopoulos, C. and Gierasch, L.M. (1993) Characterization of a functionally important mobile domain of GroES. *Nature* 364:255–258.

Langer, T., Lu, C., Echols, H., Flanagen, J., Hayer, M. and Hartl, F.-U. (1992a) Successive action of DnaK, DnaJ and GroEL along the pathway of chaperone-mediated protein folding. *Nature* 356:2683–689.

Langer, T., Pfeifer, G., Martin, J., Baumeister, W. and Hartl, F.-U. (1992b) Chaperonin-mediated protein folding: GroES binds to one end of the GroEL cylinder, which accommodates the protein substrate within its central cavity. *EMBO J.* 11:4757–4765.

Levy, E., McCarty, J., Bukau, B. and Chirico, W. (1995) Conserved ATPase and luciferase refolding activities between bacteria and yeast Hsp70 chaperones and modulators. *FEBS Lett.* 368:435–440.

Liberek, K., Skowyra, D, and Zylicz, M. (1991a) The *Escherichia coli* DnaK chaperone, the 70-kDa heat shock protein eukaryotic equivalent, changes conformation upon ATP hydrolysis, thus triggering its dissociation from a bound target protein. *J. Biol. Chem.* 266: 14491–14496.

Liberek, K., Marszalek, J., Ang, D. and Georgopolous, C. (1991b) *Escherichia coli* DnaJ and GrpE heat shock proteins jointly stimulate ATPase activity of DnaK. *Proc. Natl. Acad. Sci. USA* 88: 2874–2878.

Lingappa, J.R., Martin, R.L., Wong, M.L., Gamen, D., Welch, W.J. and Lingappa, V.R. (1994) A eukaryotic cytosolic chaperonin is associated with a high molecular weight intermediate in the assembly of hepatitis B virus capsid, a multimeric particle. *J. Cell Biol.* 125: 99–111.

Lubben, T.H., Gatenby, A.A., Donaldson, G.K., Lorimer, G.H. and Viitanen, P.V. (1990) Identification of a groES-like chaperonin in mitochondria that facilitates protein folding. *Proc. Natl. Acad. Sci. USA* 87: 7683 –7687.

Manning-Krieg, U.C., Scherer, P.E. and Schatz, G. (1991) Sequential action of mitochondrial chaperones in protein import into the matrix. *EMBO J.* 10: 3273–3280.

Martin, J., Langer, T., Boteva, R., Schramel, A., Horwich, A.L. and Hartl, F.-U. (1991) Chaperonin-mediated protein folding at the surface of GroEL through a "molten globule"-like intermediate. *Nature* 352: 36–42.

Martin, J., Horwich, A.L. and Hartl, F.-U. (1992) Prevention of protein denaturation under heat stress by the chaperonin hsp60. *Science* 258: 995–998.

Martin, J., Mayhew, M., Langer, T. and Hartl, F.U. (1993) The reaction cycle of GroEL and GroES in chaperonin-assisted protein folding. *Nature* 366: 228–233.

Melki, R. and Cowan, N.J. (1994) Facilitated folding of actins and tubulins occurs via a nucleotide-dependent interaction between cytoplasmic chaperonin and distinctive folding intermediates. *Mol. Cell Biol.* 14: 2895–2904.

Miklos, D., Caplan, S., Martens, D., Hynes, G., Pitluk, Z., Brown, C., Barrell, B., Horwich, A.L. and Willison, K. (1994) Primary structure and function of a second essential member of heterooligomeric TCP1 chaperonin complex of yeast. *Proc. Natl. Acad. Sci. USA* 91: 2743–2747.

Munro, S. and Pelham. H.R.B. (1986) An hsp70-like protein in the ER: Identity with the 78 kd glucose-regulated protein and immunoglobulin heavy chain binding protein. *Cell* 46: 291–300.

Nakai, M., Kato, Y., Ikeda, E., Toh-e, A. and Endo, T. (1994) YGE1p, a eukaryotic GrpE homologue, is localized in the mitochondrial matrix and interacts with mitochondrial hsp70. *Biochem. Biophys. Res. Comm.* 200: 435–441.

Nelson, R.J., Ziegelhoffer, T., Nicolet, C., Werner-Washburne, M. and Craig, E.A. (1992) The translation machinery and 70 kDa heat shock protein cooperate in protein synthesis. *Cell* 71: 97–105.

Normington, K., Kohno, K., Kozutsumi, Y., Gething, M.J. and Sambrook, J. (1989) *S. cerevisiae* encodes an essential protein homologous in sequence and function to mammalian BiP. *Cell* 57: 1223–1236.

Ostermann, J., Horwich, A.L., Neupert, W. and Hartl, F.-U. (1989) Protein folding in mitochondria requires complex formation with hsp60 and ATP hydrolysis. *Nature* 341: 125–130.

Palleros, D.R., Reid, K., Shi, L., Welch, W.J. and Fink, A.L. (1993) ATP-induced protein-Hsp70 complex dissociation requires K+ but not ATP hydrolysis. *Nature* 365: 664–666.

Pfanner, N. and Meijer, M. (1995) Protein pulling: Pulling in the proteins. *Curr. Biol.* 5: 132–135.

Rassow, J., Maarse, A., Krainer, E., Kubrich, M., Muller, H., Meijer, M., Craig, E. and Pfanner, N. (1994) Mitochondrial protein import: Biochemical and genetic evidence for interaction of matrix hsp70 and the inner membrane protein MIM44. *J. Cell Biol.* 127: 1547–1556.

Robinson, C.V., Grob, M., Eyles, S.J., Ewbank, J.J., Mayhew, M., Hartl, F.-U., Dobson, C.M. and Radford, S.E. (1994) Conformation of GroEL-bound α-lactalbumin probed by mass spectrometry. *Nature* 372: 646–651.

Roobol, A. and Carden, M.J. (1993) Identification of chaperonin particles in mammalian brain cytosol and t-complex polypeptide 1 as one of their components. *J. Neurochem.* 60: 2327–2330.

Rose, M.D., Misra, L.M. and Vogel, J.P. (1989) *KAR2*, a karyogamy gene, is the yeast homologue of the mammalian BiP/GRP78 gene. *Cell* 57: 1211–1221.

Rospert, S., Junne, T., Glick, B.S. and Schatz, G. (1993) Cloning and disruption of the gene encoding yeast mitochondrial chaperonin 10, the homolog of *E. coli* groES. *FEBS Lett.* 335:358–360.

Rowley, N., Prip-Buus, C., Westermann, B., Brown, C., Schwarz, E., Barrell, B. and Neupert, W. (1994) Mdj1p, a novel chaperone of the DnaJ family, is involved in mitochondrial biogenesis and protein folding. *Cell* 77:249–259.

Sadler, I., Chiang, A., Kurihara, T., Rothblatt, J., Way, J. and Silver, P. (1989) A yeast gene important for protein assembly into the endoplasmic reticulum and the nucleus has homology to DnaJ, an *Escherichia coli* heat shock protein. *J. Cell Biol.* 109:2665–1675.

Schlenstedt, G., Harris, S., Risse, B., Lill, R. and Silver, P. (1995) A yeast homolog, Scj1p, can function in the endoplasmic reticulum with BiP/Kar2p via a conserved domain that specifies interactions with hsp70s. *J. Cell Biol.* 129:979–988.

Scherer, P., Kreig, U., Hwang, S., Vestweber, D. and Schatz, G. (1990) A precursor protein partially translocated into yeast mitochondria is bound to a 70kd mitochondrial stress protein. *EMBO J.* 9:4315–4322.

Schmidt, M. and Buchner, J. (1992) Interaction of GroE with an all-β-protein. *J. Biol. Chem.* 267:16829–16833.

Schmidt, M., Bucheler, U., Kaluza, B. and Buchner, J. (1994a) Correlation between the stability of the GroEL-protein ligand complex and the release mechanism. *J. Biol. Chem.* 269:27964–27972.

Schmidt, M., Buchner, J., Todd, M.J., Lorimer, G.H. and Viitanen, P.V. (1994b) On the role of groES in the chaperonin-assisted folding reaction. *J. Biol. Chem.* 269:10304–10311.

Schneider, H.-C., Berthold, J., Bauer, M.F., Dietmeier, K., Guiard, B., Brunner, M. and Neupert, W. (1994) Mitochondrial Hsp70/MIM44 complex facilitates protein import. *Nature* 371:768–774.

Schroder, H., Langer, T., Hartl, F.-U. and Bukau, B. (1993) DnaK, DnaJ and GrpE form a cellular chaperone machinery capable of repairing heat-induced protein damage. *EMBO J.* 12:4137–4144.

Schonfeld, H.-J., Schmidt, D., Schroder, H. and Bukau, B. (1995) The DnaK chaperone system of *Escherichia coli:* Quaternary structures and interactions of the DnaK and DnaJ components. *J. Biol. Chem.* 270:2183–2189.

Scidmore, M., Okamura, H.H. and Rose, M.D. (1993) Genetic interactions between *Kar2* and *Sec63*, encoding eukaryotic homologues of DnaK and DnaJ in the endoplasmic reticulum. *Mol. Biol. Cell* 4:1145–1159.

Silver, P. and Way, J.C. (1993) Eukaryotic DnaJ homologues and the specificity of Hsp70 activity. *Cell* 74:5–6.

Simons, J.F., Ferro-Novick, S., Rose, M.D. and Helenius, A. (1995) Bip/Kar2p serves as a molecular chaperone during carboxypeptidase Y folding in yeast. *J. Cell Biol.* 130:41–49.

Skowyra, D., Georgopoulos, C. and Zylicz, M. (1990) The *E. coli* DnaK gene product, the hsp70 homolog, can reactivate heat-inactivated RNA polymerase in an ATP-hydrolysis-dependent manner. *Cell* 62:939–944.

Sternlicht, H., Farr, G.W., Sternlicht, M.L., Driscoll, J.K., Willison, K. and Yaffe, M.B. (1993) The t-complex polypeptide 1 complex is a chaperonin for tubulin and actin *in vivo. Proc. Natl. Acad. Sci. USA* 90:9422–9426.

Szabo, A., Langer, T., Schroder, H., Flanagan, J., Bukau, B. and Hartl, F.U. (1994) The ATP hydrolysis-dependent reaction cycle of the *Escherichia coli* Hsp70 system DnaK, DnaJ and GrpE. *Proc. Natl. Acad. Sci. USA* 91:10345–10349.

Todd, M.J., Viitanen, P.V. and Lorimer, G.H. (1994) Dynamics of the chaperonin ATPase cycle: Implications for facilitated protein folding. *Science* 265:659–666.

Ursic, D. and Culbertson, M.R. (1991) The yeast homolog to mouse *Tcp-1* affects microtubule-related processes. *Mol. Cell Biol.* 11:2629–2640.

Ursic, D., Sedbrook, J.C., Himmel, K.L. and Culbertson, M.R. (1994) The essential yeast Tcp1 protein affects actin and microtubules. *Mol. Biol. Cell* 5:1065–1080.

Viitanen, P.V., Gatenby, A.A. and Lorimer, G.H. (1992a) Purified chaperonin 60 (groEL) interacts with the nonnative states of a multitude of *Escherichia coli* proteins. *Protein Sci.* 1:363–369.

Viitanen, P.V., Lorimer, G.H., Seetharam, R., Gupta, R.S., Oppenheim, J., Thomas, J.O. and Cowan, N.J. (1992b) Mammalian mitochondrial chaperonin 60 functions as a single toroidal ring. *J. Biol. Chem.* 267:695–698.

Vinh, D.B.-N. and Drubin, D.G. (1994) A yeast TCP-1 like protein is required for actin function *in vivo. Proc. Natl. Acad. Sci USA* 91:9116–9120.

Voos, W., Gambill, B.D., Guiard, B., Pfanner, N. and Craig, E.A. (1993) Presequence and mature parts of preproteins strongly influence the dependence of mitochondrial protein import on heat shock protein 70 in the matrix. *J. Cell Biol.* 129:119–123.

Voos, W., Gambill, B.D., Laloraya, S., Ang, D., Craig, E. and Pfanner, N. (1994) Mitochondrial GrpE is present in a complex with hsp70 and preproteins in transit across membranes. *Mol. Cell Biol.* 14:6627–6634.

Wall, D., Zylicz, M. and Georgopoulos, C. (1994) The NH_2-terminal 108 amino acids of the *Escherichia coli* DnaJ protein stimulate the ATPase activity of DnaK and are sufficient for lambda replication. *J. Biol. Chem.* 269:5446–5451.

Wall, D., Zylicz, M. and Georgopoulos, C. (1995) The conserved G/F motif of the DnaJ chaperone is necessary for the activation of the substrate binding properties of the DnaK chaperone. *J. Biol. Chem.* 270:2139–2144.

Weissman, J., Kashi, Y., Fenton, W.A. and Horwich, A.L. (1994) GroEL-mediated protein folding proceeds by multiple rounds of binding and release of nonnative forms. *Cell* 78:693–702.

Westerman, B., Prip-Buus, C., Neupert, W. and Schwarz, E. (1995) The role of the GrpE homologue, Mge1p, in mediating protein import and protein folding in mitochondria. *EMBO J.* 13:1998–2006.

Yaffe, M.B., Farr, G.W., Miklos, D., Horwich, A.L., Sternlicht, A.L. and Sternlicht, H. (1992) TCP1 complex is a molecular chaperone in tubulin biogenesis. *Nature* 358:245–248.

Zahn, R., Spitzfaden, C., Ottiger, M., Wuthrich, K. and Pluckthun, A. (1994) Destabilisation of the complete protein secondary structure on binding to the chaperone GroEL. *Nature* 368:261–265.

Ziemienowicz, A., Skowyra, D., Zeilstra-Ryalls, J., Fayet, O. and Georgopoluos, C. (1993) Both the *Escherichia coli* chaperone systems, GroEL/GroES and DnaK/DnaJ/GrpE, can reactivate heat-treated RNA polymerase. Different mechanisms for the same activity. *J. Biol. Chem.* 268:25425–25431.

Stress-Inducible Cellular Responses
ed. by U. Feige, R.I. Morimoto, I. Yahara and B. Polla
© 1996 Birkhäuser Verlag Basel/Switzerland

Roles for hsp70 in protein translocation across membranes of organelles

D. M. Cyr[1] and W. Neupert[2]

[1] *Department of Cell Biology, School of Medicine, University of Alabama at Birmingham, Birmingham, AL 35294-0005, USA*
[2] *Institut für Physiologische Chemie der Universität München, Goethestrasse 33, D-80336 München, Germany*

Summary. The family of hsp70 molecular chaperones plays an essential and diverse role in cellular physiology. Hsp70 proteins appear to elicit their effects by interaction with polypeptides that present domains which exhibit non-native conformations at distinct stages during their life in the cell. Work pertaining to the functions of hsp70 proteins in driving protein translocation across membranes is reviewed herein. Hsp70 proteins function to deliver polypeptides to protein translocation channels, unfold polypeptides during transit across membranes and drive the translocation process. All these reactions are facilitated in an ATP-dependent reaction cycle with the assistance of different partner proteins that modulate the function of hsp70.

Introduction

The life cycle of a eukaryotic protein includes its synthesis on ribosomes, sorting to a specific subcellular compartment, folding to an active conformation and finally its turnover. Different members of the hsp70 family of molecular chaperones have been observed to help facilitate each of these stages in protein metabolism (Gething and Sambrook, 1992). The general mechanism for hsp70 protein function involves its ATP-dependent binding to and release from polypeptides that assume non-native conformations (Hartl et al., 1994): Transient complex formation between hsp70 and polypeptides appears to stabilize the conformation of substrate proteins and prevent their misfolding or aggregation.

Hsp70 proteins interact with nascent polypeptides on ribosomes to initiate their folding (Beckmann et al., 1990; Frydman et al., 1994). Upon release from hsp70 some polypeptides are directed to pathways for translocation across the membranes of organelles (Stuart et al., 1994a; Höhfeld and Hartl, 1994b). Hsp70 proteins have been implicated in facilitating aspects of protein translocation into chloroplasts, the endoplasmic reticulum (ER), lysosomes, mitochondria, the nucleus and peroxisomes (reviewed in Brodsky, 1996; Dice et al., 1994; Lithgow et al., 1995; Moore and Blobel, 1994; Soll and Alefsen, 1993; Subramani, 1993).

Studies on protein translocation into mitochondria have provided many of the seminal observations concerning the functions of hsp70 proteins in many different aspects of protein metabolism. In this review we outline the basic steps of the mitochondrial protein import pathway and provide key examples where different hsp70 proteins act to catalyze different discrete steps common to all protein translocation reactions. Particular attention will be paid the partner proteins or co-chaperones that function with hsp70 proteins in mitochondrial biogenesis. In this manner the reader is provided with a concise view of how interactions of preproteins with hsp70 can facilitate their unidirectional transfer across cellular membranes.

The mitochondrial protein import pathway

The majority of mitochondrial proteins are encoded by the nucleus and synthesized on cytoplasmic ribosomes as precursor proteins. Studies with *Saccharomyces cerevisiae* indicate that the assembly of nuclear encoded proteins into mitochondria requires the action of several different classes of cytosolic and mitochondrial molecular chaperones (Tab. 1). Mitochondrial

Table 1. Hsp70 homologs and co-chaperones that assist in protein translocation across mitochondrial membranes in *S. cerevisiae*

Chaperone	Cellular location	Essential for viability	Reference
Ssb1/Ssb2 [a]	Cytosol	No [b]	Nelson et al., 1992
Ssa1/Ssa2	Cytosol	Yes	Deshaies et al., 1988
Ydj1	Cytosol	No [c]	Caplan and Douglas, 1991 Atencio and Yaffe, 1992
Ssc1 (mt-hsp70)	Mitochondrial matrix	Yes	Craig et al., 1989 Kang et al., 1991
Mdj1	Mitochondrial matrix	No [d]	Rowley et al., 1994
Mge1	Mitochondrial matrix	Yes	Ikeda, E. et al., 1994
hsp60	Mitochondrial matrix	Yes	Cheng et al., 1989
hsp10	Mitochondrial matrix	Yes	Rospert et al., 1993 Höhfeld and Hartl, 1994a
Mim44	Mitochondrial matrix	Yes	Maarse et al., 1992

Ssa, Ssb and Ssc denote proteins from different hsp70 subfamilies. Ydj1 and Mdj1 are homologs of the *E. coli* protein DnaJ. Mge1 is a homolog of the *E. coli* protein GrpE. Hsp60 and hsp10 are heat shock proteins of 60 and 10 KD in size, respectively. Mim44 is a 44 KD inner membrane protein that is part of the mitochondrial import apparatus.

[a] Ssb1/Ssb2p are not specifically required for protein translocation into mitochondria, but are included here because they are proposed to interact with polypeptides during their synthesis on ribosomes.
[b] Deletion strains are cold-sensitive for growth.
[c] Deletion strains are viable at 23 °C, but not at 37 °C.
[d] Deletion strains are temperature-sensitive for growth on fermentable carbon sources and non-viable at all temperatures on non-fermentable carbon sources.

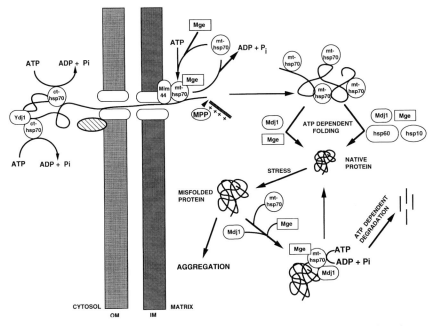

Figure 1. Model for the functions of hsp70 and its co-chaperones in the biogenesis and turn-over of mitochondrial proteins. Ct-hsp70 represents the Ssa1 and Ssa2 proteins. MPP denotes the matrix processing peptidase responsible for cleavage of the mitochondrial presequence. The presequence is represented by the shaded and positively charged region of the precursor protein. Table 1 has definitions for the other terms used in this figure.

proteins are typically synthesized in a precursor form containing a transient amino-terminal presequence which serves as a targeting signal. Protein import into mitochondria normally occurs via a post-translational mecha-nism along a multi-step pathway (Fig. 1) which begins with the interaction of cytosolic hsp70 proteins (ct-hsp70) with nascent precursor proteins during their synthesis on polyribosomes (Beckmann et al., 1990; Frydman et al., 1994; Nelson et al., 1992). Interaction of hsp70 proteins with newly synthesized precursor proteins are assisted by the cytosolic DnaJ homolog Ydj1 (Atencio and Yaffe, 1992; Caplan et al., 1992; Cyr et al., 1994). By working together, hsp70 and DnaJ family members are thought to maintain precursor proteins in loosely folded conformations that are competent for passage across mitochondrial membranes (Caplan et al., 1992; Chirico, 1992; Deshaies et al., 1988). Targeting of proteins to mitochondria invol-ves the specific recognition of the presequence by receptor proteins located on the surface of the mitochondrial outer membrane (Mayer et al., 1995; Kiebler et al., 1993; Lithgow et al., 1995). Translocation across the outer and inner mitochondrial membranes is mediated by independent translo-cation machineries in the respective membranes (Berthold et al., 1995). Passage of the presequence into mitochondria is dependent on both a membrane potential, $\Delta\Psi$, across the inner membrane, and ATP hydrolysis in the matrix. Upon entry into the matrix, the presequence is cleaved from the precursor protein by the matrix processing peptidase (MPP). Complete translocation of precursors into the matrix is dependent on ATP and the mitochondrial hsp70 cognate, Ssc1p (mt-hsp70: Gambill et al., 1993; Glick et al., 1993; Stuart et al., 1994a). Upon completion of translocation into the matrix, mt-hsp70 and the mitochondrial DnaJ homolog Mdj1 act on the precursor protein to facilitate its folding to an active conformation (Rowley et al., 1994). These proteins may act alone or in combination with hsp60 and hsp10 to facilitate this reaction (Hendrick and Hartl., 1995; Fig. 1). Finally, after the productive life of a mitochondrial protein is over, mt-hsp70 and Mdj1 also facilitate the delivery of misfolded or denatured proteins to the protein degradation machinery of mitochondria (Wagner et al., 1994). Thus, hsp70 proteins located in the cytosol and mitochondrial matrix chaperone many different stages in the life cycle of a mitochondrial protein. Details concerning the functions of different hsp70 family members, in particular mt-hsp70, in protein translocation reactions will be provided below.

Interactions of cytosolic hsp70 family members with nascent precursor proteins

The cytosol of *S. cerevisiae* contains two different subfamilies of hsp70 proteins, the Ssa and Ssb proteins (Tab. 1). These different subfamilies of hsp70 proteins exhibit about 60% identity to each other, but play different roles in protein metabolism (Craig et al., 1994). The Ssb hsp70 proteins interact with nascent, presumably unfolded, polypeptides early in their synthesis, as they emerge from the large subunit of the ribosome, and help facilitate the elongation of polypeptide chains (Nelson et al., 1992). The Ssa hsp70 proteins also interact with nascent polypeptides, but apparently at a later stage in protein synthesis, possibly after they have assumed some aspect of secondary structure (Beckmann et al., 1990; Chirico, 1992). Ssa hsp70 proteins help facilitate post-translational translocation of proteins into mitochondria (Deshaies et al., 1988). The Ssb hsp70 proteins on the other hand, do not appear to be required for protein translocation (Craig et al., 1994). These observations suggest that the different subfamilies of cytosolic hsp70 proteins interact with the same substrate proteins in sequential order with the Ssb proteins binding first, followed by the Ssa proteins. Interaction with the Ssb proteins would be important for the rapid synthesis of mitochondrial precursors or other proteins, while interaction with the Ssa proteins would ensure that proteins are delivered to mitochondria in a transport competent conformation.

    What determines the order by which these different cytosolic hsp70 proteins act? Proximity to the nascent chain is one possibility. Through interaction with a component of the ribosome the Ssb proteins might be preferentially targeted to the exit channel in the large subunit of the ribosome. This would be analogous to the targeting of mt-hsp70 to the mitochondrial inner membrane by Mim44 (see below). The Ssa hsp70 proteins are probably not targeted to the ribosome and may only come in contact with partially folded polypetides that appear at late stages of protein synthesis. The binding and release of some partially folded substrates to hsp70 proteins is assisted by the action of the *E. coli* DnaJ protein (Ang et al., 1991; Liberek et al., 1991) and its eukaryotic homologs (Cyr et al., 1994). The cytosol of *S. cerevisiae* contains at least two different DnaJ homologs, one of which is Ydj1 (Atencio and Yaffe, 1992; Caplan and Douglas, 1992). Ydj1 has been shown to interact with Ssa1 and Ssa2, but appararently not the Ssb proteins, to regulate their chaperone function (Cyr et al., 1992; Cyr and Douglas, 1994; Cyr, 1995). Ydj1 appears to function with the Ssa hsp70 proteins in protein translocation events. Mutations in Ydj1 cause defects in mitochondrial protein import and protein translocation into the ER (Atencio and Yaffe, 1992; Caplan et al., 1992). Ydj1 may serve to facilitate the binding of partially folded precursor proteins to Ssa hsp70 proteins and thereby improve their ability to be translocated across membranes. Ydj1 may also assist in determining the order in which the Ssa and Ssb hsp70 proteins interact with nascent precursor proteins.

In spite of the above discussion, it should be noted that the requirement for cytosolic hsp70 proteins in protein translocation across mitochondrial membranes is not universal. A recent study from the Schatz laboratory has investigated the cytosolic ATP/hsp70 requirements for protein translocation into mitochondria (Wachter et al., 1994). They observe that only a subset of precursor proteins that are prone to aggregation require cytosolic ATP to maintain their translocation competence. In contrast, the import of all proteins into the matrix appears to require the action of mt-hsp70 (Stuart et al., 1994a). The details of mt-hsp70 function in protein translocation across the inner membrane of mitochondria will be discussed in the sections below.

Mt-hsp70 facilitates several aspects of protein translocation across the inner membrane of mitochondria

Recent results from two independent types of experiments have shed light on the importance of mt-hsp70 for protein import into mitochondria. These studies revealed mt-hsp70 to perform essential functions at multiple steps of the import pathway. The activity of mt-hsp70 in an *in vitro* import assay was modulated via two different experimental approaches: (i) modulation of matrix ATP concentrations to levels that adversely affect the ATP dependent action of mt-hsp70 and (ii) generation of conditional mutations in the SSC1 gene, encoding mt-hsp70 in *Saccharomyces cerevisiae*, that lead to temperature sensitive (ts) defects in mt-hsp70 function. In isolated mitochondria matrix ATP levels can be easily depleted using a combination of energy poisons (Cyr et al., 1993; Glick et al., 1993; Stuart et al., 1994b). This procedure allows matrix ATP to be stringently reduced to experimentally definable levels, without affecting the $\Delta\Psi$ (membrane potential). Characterization of translocation intermediates that accumulate in "ATP depleted mitochondria" allows the determination of specific steps in the import pathway that are dependent upon mt-hsp70 (Cyr et al., 1993; Glick et al., 1993; Hawang and Schatz, 1989; Stuart et al., 1994b).

Genetic manipulation of mt-hsp70 action was also achieved following the production of ts yeast mutants carrying mutations in the SSC1 gene (Gambill et al., 1993; Kang et al., 1990; Voos et al., 1993): Two such mutants have been characterized, *ssc1-2* and *ssc1-3*, both of which displayed an inhibition of protein import into mitochondria *in vivo* when the cells were shifted to the non-permissive temperature. Such mutants have proved invaluable for the study of mt-hsp70 action as their import ts phenotype can be induced *in vitro*, when isolated mitochondria from the mutant strains are exposed to non-permissive temperature prior to import studies.

The results stemming from both the study of the energetics of protein import and the *in vitro* analysis of the ts *ssc1* mutants have indicated an involvement of mt-hsp70 at at least three distinct stages of mitochondrial import (Glick et al., 1993; Neupert et al., 1990; Stuart et al., 1994a): (i) Mt-hsp70 interacts with mitochondrial targeting sequences upon their exposure to the matrix and thereby stabilizes them on the *trans*-side of the inner membrane. This important mt-hsp70 action serves to make the initial import step irreversible and represents the first step of commitment for the precursor in the import process. (ii) Mt-hsp70, by binding to matrix exposed parts of preproteins, serves to secure the unfolding of tightly-folded segments of preproteins on the cis-side of the outer membrane. (iii) Through a series of binding and release cycles to additional domains of the preprotein, mt-hsp70 action is required for completion of translocation across the inner membrane. Details of the different aspects of mt-hsp70 function in these sub-reactions of the mitochondrial protein import pathway are provided below.

Unidirectional transfer of the presequence across the inner membrane requires both $\Delta\Psi$ and mt-hsp70

The mitochondrial presequence is transmitted across the inner membrane in a $\Delta\Psi$ dependent step. For some time it was thought that this step was solely dependent upon the $\Delta\Psi$ (Schleyer and Neupert, 1985). Recent reports that analyze protein translocation intermediates reveal that mt-hsp70 is also required in this process (Cyr et al, 1993; Hwang and Schatz, 1989; Glick et al., 1993; Ungermann et al., 1994). The presequence can oscillate in the import channels and mt-hsp70 is required to bind and trap it in the matrix.

The efficiency by which mt-hsp70 binds the incoming chain is enhanced by its targeting to the import channel. Mt-hsp70 was recently demonstrated to form a 1:1 complex with Mim44 (Schneider et al., 1994; Rassow et al., 1994; Kronidou et al., 1994), an inner membrane component of the import apparatus (Maarse et al., 1992). If formation of the Mim44/mt-hsp70 complex was interfered with, then import intermediates that were arrested with only their presequence in the matrix were observed to diffuse backwards out of mitochondria (Ungermann et al., 1994). Thus, the binding of mt-hsp70 to the presequence of incoming chains serves to lock incoming chains in the import channels and confer unidirectionality on the import process.

Mt-hsp70 activity is required for unfolding of precursors outside the mitochondria

Precursor proteins cannot cross membranes in a folded conformation (Eilers and Schatz 1986). During import it appears that segments of 50 amino acid residues of incoming precursor proteins span the import channels in an extended conformation (Rassow et al., 1990). These constraints require that precursor proteins destined for the matrix do not fold tightly upon translation in the cytosol. Folding of precursor proteins is prevented by a number of factors in the cytosol. These include hindrance of folding by the targeting signal or presequence, the lack of bound prosthetic groups which sometimes form an integral part of the mature functional enzyme and, as discussed previously, interaction of cytosolic chaperones with precursors. On the other hand, it is clear that subdomains of some preproteins do indeed fold tightly while in the cytoplasm and yet these preproteins are imported efficiently into mitochondria (Glick et al., 1993). How does the import machinery cope with such folded domains? Does a mechanism exist to "unfold" domains on precursor proteins after they engage the import machinery? The answer to this question is yes, and mt-hsp70 plays an important role in the process of preprotein unfolding.

During the course of studying the energy requirements of the import of the precursor of cytochrome b_2 (pb_2), it was demonstrated that mt-hsp70 activity is required for the unfolding of tightly folded segments in the mature part of this precursor outside of mitochondria. Cytochrome b_2 (L-lactate dehydrogenase) is located in the intermembrane space and contains both heme and flavin as prosthetic groups. The initial 100 amino acid residues of the mature cytochrome b_2 polypeptide chain constitute a tightly-folded structure, termed the cytochrome b_5 or heme binding domain (Xia and Matthews, 1990). The heme binding domain folds in the cytosol prior to its translocation across the outer mitochondrial membrane (Glick et al., 1993).

The import of cytochrome b_2 displayed a very strong requirement for matrix ATP (Glick et al., 1993; Stuart et al., 1994b; Gruhler et al., 1995). In the absence of matrix ATP, the precursor accumulated as an unprocessed species on the outer surface of mitochondria. Similar results were obtained if import was performed using mitochondria prepared from mutant strains of yeast with defects in mt-hsp70 (Voos et al., 1993). The necessity for matrix ATP for cytochrome b_2 import reflected a dependence on mt-hsp70. Furthermore, if the precursor of cytochrome b_2 was unfolded in 8 M urea prior to import, it could be imported very efficiently into both ATP-depleted mitochondria and into mt-hsp70 mutant mitochondria (Glick et al., 1993; Stuart et al., 1994b; Voos et al., 1993).

If subdomains on precursor proteins fold into stable structures prior to entering the import channel, further import of the polypeptide is prevented when the folded domain comes in contact with the outer membrane. ATP-dependent binding of mt-hsp70 to matrix exposed parts can facilitate the unfolding of portions of import intermediates that are outside of mitochondria. However, in some cases the binding of mt-hsp70 to incoming chains is insufficient to drive the unfolding of polypeptides inserted into the import channel (Eilers and Schatz, 1986). This occurs when folded domains are stabilized by the binding of ligands (Eilers and Schatz, 1986; Voos et al., 1993) and when folding is tight (Stein et al., 1994). The thermo-dynamic barriers that must be overcome to disrupt tightly folded domains on import intermediates appear lower than the energy released through the hydrolysis of ATP by hsp70 proteins. Hence, mt-hsp70 probably does not convert the energy generated from ATP hydrolysis into a force that pulls apart folded domains on translocation intermediates. Instead, mt-hsp70 appears to unfold import intermediates by binding them and shifting the equilibrium of folding reactions away from the native state.

Mt-hsp70 and Mim44 cooperate to drive the completion of preprotein translocation into mitochondria

Mt-hsp70 and Mim44 cooperate in the import process to drive polypep-tides that are stably inserted across the inner membrane into the matrix. This process occurs in a reaction cycle that involves the interaction of the incoming chain with both mt-hsp70 and Mim44 (Kronidou et al., 1994; Rassow et al., 1994; Schneider et al., 1994; Ungermann et al., 1994). The molecular details of this reaction cycle are yet to be elucidated (Glick, 1995; Pfanner and Meijer, 1995), but there is evidence to support the pro-posal that mt-hsp70 and Mim44 function as a molecular ratchet to drive the import process (Berthold et al., 1995; Schneider et al., 1994; Simon et al., 1992; Ungermann et al., 1994; Ungermann et al., 1996). As stated above, this reaction cycle is initiated with the $\Delta\Psi$ driven insertion of the pre-sequence across the inner membrane (Fig. 2). Through interaction with

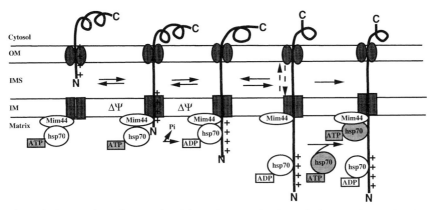

Figure 2. Proposed reactions catalyzed by mt-hsp70 and Mim44 during protein translocation across the mitochondrial inner membrane. The shaded ovals and squares denote the inner (IM) and outer (OM) membrane protein translocation machineries. The plus signs (+) mark regions of the mitochondrial presequence. The dashed arrows denote oscillation of the incoming chain in the import channel. Processing of the presequence is not denoted, but appears to occur after release of mt-hsp70 from the incoming chain at the last stage of this model.

Mim44 regions of the incoming chain are delivered to mt-hsp70 that is localized to the matrix face of the inner membrane. Mt-hsp70 then hydrolyzes ATP to ADP which stabilizes the complex between it and the incoming chain (Hartl et al., 1994). The conformational change in mt-hsp70 that is coupled to ATP hydrolysis appears to destabilize its interaction with Mim44 (Ungermann et al, 1996). This causes the mt-hsp70/precursor protein complex to dissociate from Mim44 which unlocks the incoming chain and allows it to diffuse reversibly in the import channel. Diffusion out of the membranes is limited by the association of mt-hsp70 with the incoming chain. When movement of a segment of the incoming chain has proceeded deep enough into the matrix a second Mim44 bound mt-hsp70 binds. This again locks the chain in the import apparatus and confers net movement on the translocation reaction cycle. Release of ADP from mt-hsp70 in a Mge dependent step then allows for the recycling of mt-hsp70 and the continuation of the import process (Fig. 1; Bolliger et al., 1994; Laloraya et al., 1994; Pfanner and Meijer, 1995; Westermann et al., 1995). Repeated cycles of Mim44 mediated binding and release of mt-hsp70 from import intermediates drive the completion of the import process.

Hsp70 acts with partner proteins to facilitate mitochondrial protein import

The above discussion illustrates examples where hsp70 proteins function in discrete reactions to drive the movement of proteins across the lipid bilayers of mitochondria. The targeting of hsp70 proteins to function in this diverse array of import steps is conferred by its interaction with co-chaperone proteins. Ydj1 facilitates the interaction of ct-hsp70 with nascent chains and Mim44 targets mt-hsp70 to incoming chains as they emerge in the matrix space. Mdj1 assists mt-hsp70 in at least two aspects of mitochondrial protein metabolism. Mdj1 assists in the folding of newly imported proteins (Rowley et al., 1993) and also helps deliver misfolded proteins to proteolytic enzymes of the matrix (Suzuki et al., 1994; Wagner et al., 1994). Mdj1 appears to elicit its effects by assisting in both the binding and release of proteins from mt-hsp70 (Wagner et al., 1994). The decision between the delivery of polypeptides to the folding machinery versus the proteolytic system appears to involve a kinetic partitioning between the two different systems.

Although Mdj1 can interact with newly imported chains, it does not appear to play a role in protein translocation reactions (Rowley et al., 1993): Apparently, Mim44 substitutes for Mdj1 in the import process. Mim44 has weak homology to DnaJ (Rassow et al., 1994) and is a new protein that can be classified as a co-chaperone. Mim44 forms weak complexes with incoming chains (Schneider et al., 1994) and this interaction probably maintains them in an extended conformation. In this manner Mim44 presents mt-hsp70 with a substrate polypeptide in a conformation that it recognizes with high affinity (Gething and Sambrook, 1992). Mt-hsp70 binds the presequence with lower affinity than mature segments of incoming chains (Ungermann et al., 1996). This makes both the targeting and chaperone functions of Mim44 critical in assisting mt-hsp70 in trapping the presequence in the matrix at early stages of import (Ungermann et al., 1996). Thus, via different mechanisms, co-chaperone proteins play an essential role in specifying the types of reactions catalyzed by cytosolic and matrix forms of hsp70 in the mitochondrial import pathway.

Multiple roles for hsp70 in post-translocational protein import into the ER

A membrane bound co-chaperone protein also plays an important role in facilitating post-translational protein translocation into the lumen of the ER in *S. cerevisiae* (Brodsky, 1996). The DnaJ homolog Sec63, an integral membrane protein, contains a domain with high homology to the amino-terminus of *E. coli* DnaJ (Sadler et al, 1989; Rothblatt et al., 1989). Sec63 is essential for cell viability and contains a DnaJ-like region which is localized to an internal loop of the protein that extends into the lumen of the ER (Feldheim et al., 1992). Similar to mitochondrial import, post-translational protein translocation into the ER requires the action of an hsp70 family member, BiP, on the trans-side membrane. BiP and Sec63 have been demonstrated to interact via biochemical and genetic methods (Brodsky and Schekman, 1993; Scidmore et al., 1993). Mutations in the DnaJ-domain inactivate its activity in protein translocation suggesting that interactions between BiP and Sec63 are mediated by this region. Similar to Mim44, is appears that Sec63 helps target the activity of an hsp70 protein, BiP, to a specific site in the cell. Lumenial components of the ER are required to confer unidirectionality on this translocation reaction (Nicchitta and Blobel, 1993). The BiP/Sec63 complex appears to play a major role in trapping proteins in the ER lumen and driving the overall translocation reaction (Brodsky, 1996). In addition, interactions between Sec63 and BiP may also serve to facilitate the assembly of the proteins that comprise the translocation apparatus (Sanders et al., 1992).

It is clear that some aspects of protein translocation into the ER and mitochondria which are catalyzed by hsp70 proteins are similar. However, in the case of the ER, the Sec63/BiP complex plays the additional role of regulating the assembly of the translocation channel (Sanders et al., 1992). The protein import channels of mitochondria are dynamic structures (Berthold et al., 1995). Mim44 is in a complex with possibly four additional polypeptides that comprise the import channel of the inner membrane (Berthold et al., 1995). Whether interactions of mt-hsp70 with Mim44 serve to regulate the import channels of mitochondria is under investigation.

Conclusions

From the study of protein translocation into mitochondria paradigms for the general functions of hsp70 proteins have been elucidated. There are numerous parallels between the mechanisms for mitochondrial protein import and reaction sequences for protein translocation into the ER, chloroplasts and for protein secretion from E. coli (Brodsky, 1996; Höhfeld and Hartl, 1994b; Soll and Alefsen, 1993; Wickner, 1994). However, hsp70 is not the only molecule that can facilitate protein translocation reactions (Economou and Wickner, 1994; Schmitt et al., 1995; Wickner, 1994). SecA protein plays an active role in driving segments of polypeptides across bacterial membranes during protein secretion in E. coli. (Economou and Wickner, 1994; Kim et al., 1994). Hsp78, a mitochondrial matrix protein, which was recently characterized as a molecular chaperone protein, can substitute for mt-hsp70 in driving specific stages of protein translocation into the mitochondrial matrix (Schmitt et al., 1995).

More than one mechanism exists by which hsp70 proteins facilitate the transfer of proteins across intracellular membranes (Dice et al., 1994; Hutchison et al., 1994; Shi and Thomas, 1992; Subramani, 1993). Hsp70 proteins recognize specific sequences in damaged proteins destined for degradation in lysosomes and help target them to the organelle (Dice et al., 1994). In addition, hsp70 proteins are localized to the lumen of lysosomes and may play a role in the internalization of cellular proteins. How hsp70 proteins enter the lumen of lysosomes remains a mystery (Dice et al., 1994).

Nuclear import occurs through a pore assembly in the nuclear envelope that allows access by a wide range of macromolecular particles to the matrix (Moore and Blobel, 1994). Cytosolic hsp70 and DnaJ chaperone proteins play a role in steroid receptor signaling through the nuclear membrane by facilitation the assembly of aporeceptor complexes in the cytosol (Caplan et al., 1995; Kimura et al., 1995). A role for hsp70 in driving the transport of proteins through the nuclear pore has not been suggested.

Protein translocation into peroxisomes also requires ct-hsp70, but in contrast to the mitochondrial and ER systems, polypeptides appear to traverse the membrane of this organelle in a folded conformation (Walton et al., 1994). So far there is no evidence for a protein conducting pore in peroxisomal membranes or an hsp70 homolog in the lumen of these organelles. The role of hsp70 in protein translocation into peroxisomes is a mystery. How proteins are assembled into macromolecular complexes within peroxisomes is also unknown. Why an apparently different mechanism for protein translocation into peroxisomes as compared to mitochondria and the ER has evolved is a topic of current interest. The identification of components that comprise protein import machinery of peroxisomes should help address some of these issues (Subramani, 1993; Heyman et al., 1994; Marzioch et al., 1994). Future studies on organelle biogenesis will elucidate new and novel mechanisms through which hsp70 and its co-chaperones facilitate the transfer of proteins across cellular membranes.

Acknowledgements
Work quoted from the laboratory of Walter Neupert was supported by Grants from the Sonderforschungsbereich 184, the Fonds der Chemischen, and the Human Frontier Science Program Organization (HFSPO). Douglas M. Cyr received support through a long-term fellowship from the HFSPO.

References

Ang, D., Liberek, K., Skowyra, D., Zylicz, M. and Georgopoulos, C. (1991) Biological role and regulation of the universally conserved heat shock proteins. *J. Biol. Chem.* 266:24233–24236.

Atencio, D.P. and Yaffe, M.P. (1992) Mas5, a yeast homolog of DnaJ involved in mitochondrial protein import. *Mol. Cell. Biol.* 12:283–291.

Beckmann, R.P., Mizzen, L. and Welch, W. (1990) Interaction of hsp70 with newly synthesized proteins: Implications for protein folding and assembly. *Science* 248:850–856.

Berthold, J., Bauer, M.F., Schneider, H.-C., Klaus, C., Dietmeier, K., Neupert, W., and Brunner, M. (1995) The MIM complex mediates preprotein translocation across the mitochondrial inner membrane and couples it to the mt-hsp70/ATP driving system. *Cell* 81:1085–1093.

Bolliger, L., Deloche, O., Glick, B.S., Georgopoulos, C., Jeno, P., and Kronidou, N. and Schatz, G. (1994) A mitochondrial homolog of bacterial GrpE interacts with mitochondrial hsp70 and is essential for viability. *EMBO J.* 13:1998–2006.

Brodsky, J.L. and Schekman, R. (1993) A Sec63p-BiP complex from yeast is required for protein translocation in a reconstituted proteoliposome. *J. Cell Biol.* 123:1355–1363.

Brodsky, J.L. (1996) Post-translocational protein translocation: not all hsc70s are created equal. *Trend. Biochem. Sci.* 21:122–126.

Caplan, A.J. and Douglas, M.G. (1991) Characterization of YDJ1: A yeast homolog of the *E. coli* dnaJ gene. *J. Cell Biol.* 114:609–622.

Caplan, A.J., Cyr, D.M. and Douglas, M.G. (1992) YDJ1 facilitates polypeptide translocation across different intercellular membranes by a conserved mechanism. *Cell* 71:1143–1155.

Caplan, A.J., Langley, E., and Wilson, E.M. and Vidal, J. (1995) Hormone-dependent transactivation by the human androgen receptor is regulated by a dnaJ protein. *J. Biol. Chem.* 270:5251–5257.

Cheng, M.Y., Hartl, F.-U., Martin, J., Pollock, R.A., Kalusek, F., Neupert, W., Hallberg, E.M., Hallberg, R.L. and Horwich, A.L. (1989) Mitochondrial heat-shock protein hsp60 is essential for assembly of proteins imported into yeast mitochondria. *Nature* 337:620–625.

Chirico, W.J. (1992) Dissociation of complexes between 70 Kda stress proteins and presecretory proteins is facilitated by a cytosolic factor. *Biochem. Biophys. Res. Commun.* 189:1150–1156.

Craig, E.A., Kramer, S., Shilling, J., Werner-Washburne, M., Holmes. S., Kosic-Smither, J. and Nicolet, C.M. (1989) SSC1, an essential member of the *S. cerevisiae* HSP70 multigene family, encodes a mitochondrial protein. *Mol. Cell. Biol.* 9:3000–3008.

Craig, E.A., Baxter, B.K., Becker, J., Halladay, J. and Zeigelhoffer, T. (1994) Cytosolic hsp70s of *Saccharomyces cerevisiae*: Roles in protein synthesis, protein translocation, proteolysis and regulation. *In*: R.I. Morimoto, A. Tissieres and C. Georgopoulos (eds): *The Biology of Heat Shock Proteins and Molecular Chaperones,* Cold Spring Harbor Laboratory Press, NY, pp 31–52.

Cyr, D.M., Lu, X. and Douglas, M.G. (1992) Regulation of eukaryotic hsp70 function by a DnaJ homolog. *J. Biol. Chem.* 267:20927–20931.

Cyr, D.M., Stuart, R.A. and Neupert, W. (1993) A matrix ATP requirement for presequence translocation across the inner membrane of mitochondria. *J. Biol. Chem.* 268:23751–23754.

Cyr, D.M. and Douglas, M.G. (1994) Differential regulation of hsp70 subfamilies by the eukaryotic DnaJ homolog YDJ1. *J. Biol. Chem.* 269:9798–9804.

Cyr, D.M., Langer, T. and Douglas, D.M. (1994) DnaJ-like proteins: molecular chaperones and specific regulators of hsp70. *Trends Biochem. Sci.* 19:176–181.

Cyr, D.M. (1995) Cooperation of the molecular chaperone Ydj1 with specific hsp70 homologs to suppress protein aggregation. *FEBS Letters* 359:129–132.

Deshaies, R.B., Koch, B., Werner-Washburne, M., Craig, E.A. and Schekman, R. (1988) A subfamily of stress proteins facilitates translocation of secretory and mitochondrial precursor polypeptides. *Nature* 332:800–805.

Dice, J.F., Agarraberes, F., Kirven-Brooks, M., Terlecky, L.J. and Terlecky, S.R. (1994) Heat shock 70-Kd proteins and lysosomal proteolysis. *In:* R.I. Morimoto, A. Tissieres and C. Georgopoulos (eds): *The Biology of Heat Shock Proteins and Molecular Chaperones,* Cold Spring Harbor Laboratory Press, NY, pp 137–152.

Economou, A. and Wickner, W. (1994) SecA promotes preprotein translocation by undergoing ATP-driven cycles of membrane insertion and deinsertion. *Cell* 78:835–842.

Eilers, M. and Schatz, G. (1986) Binding of a specific ligand inhibits import of a purified precursor protein into mitochondria. *Nature* 322:228–232.

Feldheim, D., Rothblatt, J. and Schekman, R. (1992) Topology and functional domains of Sec63p, an endoplasmic reticulum membrane protein required for secretory protein translocation. *Mol. Cell. Biol.* 12:3288–3296.

Frydman, J., Nimmesgern, E., Ohtsuka, K. and Hartl, F.-U. (1994) Folding of nascent polypeptide chains in a high molecular mass assembly with molecular chaperones. *Nature* 370:111–117.

Gambill, B.D., Voos, W., Kang, P.J., Miao, B., Langer, T., Craig, E.A. and Pfanner, N. (1993) A dual role for mitochondrial heat shock protein 70 in membrane translocation of preproteins. *J. Cell. Biol.* 123:119–126.

Gething, M.-J. and Sambrook, J. (1992) Protein folding in the cell. *Nature* 355:33–45.

Glick, B.S., Wachter, C., Reid, G.A. and Schatz, G. (1993) Import of cytochrome b_2 to the mitochondrial intermembrane space – the tightly folded heme-binding domain makes import dependent upon matrix ATP. *Protein Sci.* 2:1901–1917.

Gruhler, A., Ono, H., Guiard, B., Neupert, W. and Stuart, R.A. (1995) A novel intermediate on the import pathway of cytochrome b_2 into mitochondria: evidence for conservative sorting. *EMBO J.* 14:1349–1359.

Hartl, F.-U., Holdan, R. and Langer, T. (1994) Molecular chaperones in protein folding: the art of avoiding sticky situations. *Trends Biochem. Sci.* 19:20–25.

Hendrick, J.P. and Hartl, F.-U. (1993) Molecular chaperone functions of heat shock proteins. *Annu. Rev. Biochem.* 62:349–384.

Heyman J.A., Monosov, E. and Subramani, S. (1994) Role of the PAS1 gene of *Pichia pastoris* in peroxisome biogenesis. *J. Cell Biol.* 127:1259–1273.

Höhfeld, J. and Hartl, F.-U. (1994a) Requirement of the chaperonin cofactor Hsp10 for protein sorting in yeast mitochondria. *J. Cell Biol.* 126:305–315.

Höhfeld, J. and Hartl, F.-U. (1994b) Post-translational protein import and folding. *Curr. Opin. Cell Biol.* 6:499–509.

Hutchison, K.A., Dittmar, K.D., Czar, M.J. and Pratt, W.B. (1994) Proof that hsp 70 is required for assembly of the glucocorticoid receptor into a heterocomplex with hsp90. *J. Biol. Chem.* 269:50432–5049.

Hwang, S.T. and Schatz, G. (1989) Translocation of proteins across the mitochondrial inner membrane, but not into the outer membrane, requires nucleoside triphosphates in the matrix. *Proc. Natl. Acad. Sci. USA* 86:8432–8436.

Ikeda, E., Yoshida, S., Mitsuzawa, H., Uno, I. and Toh-e, A. (1994) YGE1 is a yeast homologue of *Escherichia coli* grpE and is required for maintenance of mitochondrial functions. *FEBS Lett.* 39:265–268.

Kang, P.-J., Ostermann, J., Shilling, J., Neupert, W., Craig, E.A. and Pfanner, N. (1990) Requirement for hsp70 in the mitochondrial matrix for translocation and folding of precursor proteins. *Nature* 348:137–143.

Kiebler, M., Becker, K., Pfanner, N. and Neupert, W. (1993) Mitochondrial protein import: specific recognition and membrane translocation of preproteins. *J. Membrane Biol.* 135:191–207.

Kim, Y.J., Rajapandi, T. and Oliver, D. (1994) SecA protein is exposed to the periplasmic surface of the *E. coli* inner membrane in its active state. *Cell* 78:845–852.

Kimura, Y., Yahara, I. and Lindquist, S. (1995) Role of the protein chaperone YDJ1 in establishing Hsp90-mediated signal transduction pathways. *Science* 268:1362–1365.

Kronidou, N.G., Oppliger, W., Bolliger, L. Hannavy, K., Glick, B.S. and Schatz, G. (1994) Dynamic interaction between Isp45 and mitochondrial hsp70 in the protein import system of the yeast mitochondrial inner membrane. *Proc. Nat. Acad. Sci. USA* 91:12818–12822.

Laloraya, S., Gambill, B.D. and Craig, E.A. (1994) A role for a eukaryotic GrpE-related protein, Mge1p, in protein translocation. *Proc. Nat. Acad. Sci. USA* 91:6481–6485.

Liberek, K., Marszalek, J., Ang, D., Georgopoulos, C. and Zylicz, M. (1991) *Escherichia coli* DnaJ and GrpE heat shock proteins jointly stimulate ATPase activity of DnaK. *Proc. Natl. Acad. Sci. USA* 88:2874–2878.

Lithgow, T., Glick. B.S. and Schatz, G. (1995) The protein import receptor of mitochondria. *Trends. Biochem. Sci.* 20:98–101.

Mayer, A., Neupert, W. and Lill, R. (1995) Mitochondrial import: reversible binding of the presequence at the *trans*-side of the outer membrane drives partial translocation and unfolding. *Cell* 80:127–137.

Maarse, A.C., Blom, J., Grivell, L.A. and Meijer, M. (1992) MPI1, an essential gene encoding a mitochondrial membrane protein, is possibly involved in protein import into yeast mitochondria. *EMBO J* 11:3619–3628.

Marzioch, M., Erdmann, R., Veenhuis, M. and Kunau, W.H. (1994) PAS7 encodes a novel yeast member of the WD-4a0 protein family essential for import of 3-oxoacyl-CoA thiolase, a PTS2-containing protein, into peroxisomes. *EMBO J.* 13:4908–4918.

Moore, M.S. and Blobel, G. (1994) A G protein involved in nucleocytoplasmic transport: the role of Ran. *Trends. Biochem. Sci.* 19:211–216.

Nelson, R.J., Zeigelhoffer, T., Nicolet, C., Werner-Washburne, M. and Craig E.A. (1992) The translation machinery and seventy kilodalton heat shock protein cooperate in protein synthesis. *Cell* 71:97–105.

Neupert, W., Hartl, F.-U., Craig, E.A. and Pfanner, N. (1990) How do polypeptides cross mitochondrial membranes? *Cell* 63:447–450.

Nicchitta, C.V. and Blobel, G. (1993) Lumenal proteins of the mammalian endoplasmic reticulum are required to complete protein translocation. *Cell* 73:989–998.

Pfanner, N. and Meijer, M. (1995) Protein sorting: Pulling in the proteins. *Curr. Biol.* 5:132–135.

Rassow, J., Hartl, F.-U., Guiard, B., Pfanner, N. and Neupert, W. (1990) Polypetides traverse the mitochondrial envelope in an extended state. *FEBS Lett.* 275:190–194.

Rassow, J., Maarse, A.C., Krainer, F., Kubrich, M., Muller, M., Meijer, M., Craig, E.A. and Pfanner, N. (1994) Mitochondrial protein import: biochemical and genetic evidence for the interaction of matrix hsp70 and the inner membrane protein Mim44. *J. Cell Biol.* 127:1547–1556.

Rospert, S., Junne, T., Glick, B.S. and Schatz, G. (1993) Cloning and disruption of the gene encoding yeast mitochondrial chaperonin 10, the homolog of *E. coli* groES. *FEBS Lett.* 335:358–360.

Rothblatt, J.A., Deshaies, R.J., Sanders S.L., Daum, G. and Schekman, R. (1989) Multiple genes are required for proper insertion of secretory proteins into the endoplasmic reticulum in yeast. *J. Cell Biol.* 109:2641–2652.

Rowley, N., Prip-Buus, C., Westermann, B., Brown, C., Schwarz, E., Barrel, B. and Neupert, W. (1994) Mdj1, a novel chaperone of the DnaJ family, is involved in mitochondrial biogenesis and protein folding. *Cell* 77:249–259.

Sadler, I., Chiang, A., Kurihara, T., Rothblatt, J., Way, J. and Silver, P. (1989) A yeast gene important for protein assembly into the endoplasmic reticulum and the nucleus has homology to DnaJ, an *Escherichia coli* heat shock. *J. Cell Biol.* 109:2665–2675.

Sanders, S.L., Whitfield, K.M., Vogel, J.P., Rose, M.D. and Schekman, R.W. (1992) Sec61p and BiP directly facilitate polypeptide translocation into the ER. *Cell.* 69:353–365.

Schleyer, M. and Neupert, W. (1985) Transport of proteins into mitochondria: translocational intermediates spanning contact sites between outer and inner membranes. *Cell* 43:339–350.

Schmitt, M., Neupert, W. and Langer, T. (1995) Hsp78, a Clp-homologue within mitochondria, can substitute for chaperone functions of mt-hsp70. *EMBO J:* 14:3434–3444.

Schneider, H.C., Berthold, J., Bauer, M.F., Dietmeier, K., Guiard, B., Brunner, M. and Neupert, W. (1994) Mitochondrial Hsp70/MIM44 complex facilitates protein import. *Nature* 371:768–774.

Scidmore, M.A., Okamura, H.H. and Rose, M.D. (1993) Genetic interactions between Kar2 and Sec63, encoding eukaryotic homologues of DnaK and DnaJ in the endoplasmic reticulum. *Mol. Biol. Cell.* 4:1145–1159.

Shi, Y. and Thomas, J.O. (1992) The transport of proteins into the nucleus requires the 70-kilodalton heatshock protein or its cytosolic cognate. *Mol. Cell. Biol.* 12:2186–2192.

Simon, S.M., Peskin, C.S. and Oster, G.F. (1992) What drives the translocation of proteins? *Proc. Nat. Acad. Sci. USA* 98:3770–3774.

Soll, J. and Alefsen, H. (1993) The protein import apparatus of chloroplasts. *Physiol. Plant* 87:433–440.

Stein, I., Peleg, Y.Y., Even-Ram, S. and Pines, O. (1994) The single translation product of the FUM1 gene (fumarase) is processed in mitochondria before being distributed between the cytosol and mitochondria in *Saccharamyces cerevisia. Mol. Cell Biol.* 14:4770–4778.

Stuart, R.A., Cyr, D.M., Craig, E.A. and Neupert, W. (1994a) Mitochondrial molecular chaperones: their role in protein translocation. *Trends Biochem. Sci.* 19:87–92.

Stuart, R.A., Gruhler, A., van der Klei, I.J., Guiard, B., Koll, H. and Neupert, W. (1994b) The requirement of matrix ATP for the import of precursor proteins into the mitochondrial matrix and intermembrane space. *Eur. J. Biochem.* 220:9–18.

Subramani S. (1993) Protein import into peroxisomes and biogenesis of the organelle. *Ann. Rev. Cell Biol.* 9:445–478.

Suzuki, C.K., Suda, K., Wang, N. and Schatz, G. (1994) Requirement of the yeast gene LON in mitochondrial proteolysis and maintenance of respiration. *Science* 364:1250–1253.

Ungermann, C., Neupert, W. and Cyr, D.M. (1994) The role of hsp70 in conferring unidirectionality on protein translocation into mitochondria. *Science* 266:1250–1253.

Ungermann, C., Guiard, B., Neupert, W. and Cyr, D.M. (1996) The $\Delta\Psi$ and hsp70/Mim44 dependent reaction cycle driving early steps of protein translocation into mitochondria. *EMBO J.* 15:735–744.

Voos, W., Gambill, D.B., Guiard, B., Pfanner, N. and Craig, E.A. (1993) Presequence and mature portion of preproteins strongly influence the dependence of mitochondrial protein import on the heat shock 70 protein in the matrix. *J. Cell Biol.* 123:109–118.

Wagner, I., Arlt, H., van Dyck, L., Langer, T. and Neupert, W. (1994) Molecular chaperones cooperate with PIM1 protease in the degradation of misfolded proteins in mitochondria. *EMBO J.* 13:5135–5145.

Walton, P.A., Wendland, M., Subramani, S., Rachubinski, R.A. and Welch, W.J. (1994) Involvement of 70-kD heat-shock proteins in peroxisomal import. *J. Cell Biol.* 125: 1037–1046.

Wachter, C., Schatz, G. and Glick, B.S. (1994) Protein import into mitochondria: the requirement for external ATP is precursor-specific whereas intramitochondrial ATP is universally needed for translocation into the matrix. *Mol. Cell. Biol.* 5: 465–474.

Westermann, B., Prip-Buus, C., Neupert, W. and Schwarz, E. (1995) The role of the GrpE homologue, Mge1p, in mediating protein import and protein folding in mitochondria. *EMBO J.* 14: 3452–3460.

Wickner, W. (1994) How ATP drives proteins across membranes. *Science* 266: 1197–1198.

Xia, Z. and Matthews, F.S. (1990) Molecular structure of flavocytochrome B2 at 2.4 Å resolution. *J. Mol. Biol.* 212: 837–863.

Stress-Inducible Cellular Responses
ed. by U. Feige, R.I. Morimoto, I. Yahara and B. Polla
© 1996 Birkhäuser Verlag Basel/Switzerland

Protein folding and assembly in the endoplasmic reticulum

J. Wei and L.M. Hendershot

Department of Tumor Cell Biology, St. Jude Children's Research Hospital, 332 N. Lauderdale, Memphis, TN 38105, USA

Summary. The newly synthesized protein emerging through the ER membrane enters a unique environment for folding and assembly. Unlike the cytosol, the ER provides an oxidizing environment, has high levels of calcium, and contains enzymes for N-linked glycosylation. The growing nascent polypeptide chain is in many cases modified co-translationally with N-linked sugars and begins to fold while still attached to the ribosome. Disulfide bond formation stabilizes the tertiary structure of the protein. The *in vivo* folding and assembly of nascent proteins requires a delicate balance between allowing folding to occur and preventing incorrect interactions that would ultimately lead to improper folding and/or aggregation. In the past several years, two groups of proteins that interact transiently with incompletely folded and assembled proteins in the ER have been identified and characterized. The first group consists of enzymes that promote or stabilize protein folding. The second is composed of proteins termed "molecular chaperones" that bind transiently to nascent polypeptides and apparently prevent misfolding by masking those regions that could lead to incorrect interactions between protein domains or aggregation.

The ER represents a unique environment for protein folding

Some proteins refold spontaneously *in vitro* after denaturation, demonstrating that primary sequence information can be sufficient to guide correct folding. These studies suggested that protein folding is initiated by collapse of hydrophobic regions into the interior of the molecule, followed by formation of secondary structures such as α-helices, and is stabilized by covalent and non-covalent side chain interactions. Although there is evidence for some proteins that the *in vitro* folding pathways mimic the normal *in vivo* pathways, the optimal conditions for *in vitro* folding are quite different from those encountered *in vivo*. While *in vitro* refolding experiments use full-length proteins, nascent polypeptides are modified and begin to fold co-translationally, allowing interactions between regions that might be less favored in the context of the whole protein. *In vitro* refolding experiments are performed with purified, dilute proteins and are often inefficient; *in vivo* folding occurs efficiently at both very high and very low protein concentrations with the majority of newly synthesized proteins reaching a mature functional configuration.

In addition to the problems encountered by all proteins attempting to fold in a cell, the environment of the ER provides additional constraints. The nascent chain translocates into an environment that contains millimolar

Table 1. Properties of ER chaperones and folding enzymes

	M.W.	Other names	Calcium binding	Recognition structure on substrate	Release	Yeast gene	Activity	Co-regulation with BiP
BiP	78 (s)[a]	GRP78	?[b]	hydrophobic amino acids	ATP	S. cerevisiae (E)[c] S. pombe (E)	Folding/retent	
GRP94	94 (s)	Endoplasmin, ERp99, CaBP4	+				?	Yes
Calnexin	88 (m)	p88, IP90	+	monoglucosylated N-linked sugars	Glu II	S. cerevisiae (N-E) S. pombe (E)	Folding/retent./ quality control	
Calreticulin	60 (s)	CaBP3, calregulin	+					Yes
ERp72	72 (s)	CaBP2	+	cysteine residues (?)			Degrad/ disulf. isomeras	Yes
hsp47	47 (s)			procollagen	pH↓		?	
PDI	54 (s)	ERp59	?	cysteine residues		S. cerevisiae (E) S. pombe (E)	Disulf. Isomerase	Yes

[a] s = soluble protein, m = integral membrane protein.
[b] Conflicting data in literature.
[c] E = essential, N-E = non-essential.

concentrations of calcium, is oxidizing, and represents a concentrated soup of many different proteins in various stages of folding and assembly. As it is translocated, the vast majority of newly synthesized ER proteins are modified by a complex set of sugars via N-linkage to asparagine residues. The oxidizing environment causes disulfide bonds to form between cysteine residues which act to stabilize folding and to covalently attach protein sub-units. In many proteins, the correct disulfide linkages do not occur between sequential cysteines but instead occur between disparate cysteines that are brought into close proximity after folding occurs. Thus, the potential for incorrect disulfide bond formation in a protein that is being extruded one amino acid at a time into the ER seems staggering. It is clear that the ER has evolved mechanisms to allow protein folding to occur in this environment, and in fact, agents that alter these conditions have deleterious effects on protein folding and secretion.

Thus, the *in vivo* folding and assembly of nascent proteins requires a delicate balance between allowing folding to occur and preventing incorrect interactions that would ultimately lead to improper folding and/or aggregation. In the past several years, proteins that interact transiently with incompletely folded and assembled proteins in the ER have been identified and characterized (Tab. 1). Some of these associated proteins are enzymes which catalyze reactions that promote or stabilize protein folding (e.g., PDI and PPIase), Others, termed "molecular chaperones" (e.g., BiP, GRP94, ERp72, calnexin, hsp47), bind transiently to nascent polypeptides and appear to stabilize protein folding intermediates by either preventing incorrect interactions between protein domains or promoting correct ones. Two of these chaperones (BiP, GRP94) represent the ER homologues of heat shock proteins (HSP70 and HSP90), and thus information obtained for cytosolic homologues provides insights into their functions. Others have no homology to heat shock proteins but are coordinately regulated by the same stress stimuli that transcriptionally activate BiP and GRP94 and can be considered to be stress-inducible proteins.

The role of BiP/GRP78 in protein folding in the ER

Immunoglobulin (Ig) heavy chain binding protein (BiP) was first identified to be stably and non-covalently associated with the unassembled, non-transported Ig heavy chains in Abelson virus transformed pre-B cell lines (Haas and Wabl, 1983), and was subsequently shown to bind transiently to assembling Ig intermediates but not to completely assembled Ig molecules (Bole et al., 1986). Since this time, BiP has been found stably associated with many other ER proteins (Gething and Sambrook, 1992) that are unable to acquire their transport competent conformation such as unassembled protein subunits, mutant, misfolded proteins, and glycosylation-defective proteins. BiP binding to wild-type proteins is restricted to incompletely folded intermediates or unassembled subunits, with BiP release occurring just prior to or upon completion of folding and assembly. In the case of Ig heavy chains, BiP binds to the C_H1 domain, which is the site for covalent attachment of Ig light chains. Mutant heavy chains that lack the C_H1 domain can be secreted as incompletely assembled molecules, suggesting that BiP plays a role in catalyzing the assembly or blocking the secretion of nascent heavy chains in the ER. cDNA cloning revealed that BiP is closely related to the HSP70 family of proteins and identical to GRP78, a protein that is induced by glucose starvation and other stresses that affect protein folding in the ER. BiP and other soluble resident ER proteins contain a four amino acid (KDEL) motif at their carboxyterminus which is responsible for their retention in the ER.

All HSP70 proteins can be divided into two functional domains: (1) the N-terminal ATP binding domain and (2) the C-terminal protein binding domain. The binding of ADP in the N-terminal pocket stabilizes protein binding to the C-terminal portion, whereas binding of ATP changes the conformation of the protein binding domain causing the release of bound proteins (Palleros et al., 1993). The fact that BiP interacts with so many different proteins that do not contain any obvious conserved sequence motifs suggests that BiP does not recognize a unique linear amino acid sequence. Studies using randomly generated peptides revealed that peptides containing hydrophobic amino acids are the most effective stimulators of BiP's ATPase activity (Flynn et al., 1991). Affinity screening of a peptide display library, revealed that BiP preferentially bound to peptides with the heptameric motif of HyXHyXHyXHy, where Hy is a bulky aromatic or hydrophobic residue and X is any amino acid (Blond Elguindi et al., 1993). The alternating patter of Hy residues in the binding motif is compatible with BiP binding to proteins when they are in an extended conformation, causing the bulky aromatic/hydrophobic side chains to lie on one side of the protein and to presumably point into the protein binding site on BiP. The results of these two studies are consistent with the idea that BiP binds to extended hydrophobic stretches on nascent or unassembled polypeptides, which would become inaccessible after folding and assembly were complete. Although BiP binds to both nascent proteins and protein subunits, it is not clear whether BiP binds to unfolded regions of the protein in both cases, or whether it is able to associate with both linear hydrophobic regions and hydrophobic faces on a folded protein subunit.

Like all HSP70 proteins, BiP binds ATP tightly and has a weak ATPase activity (Kassenbrock and Kelly, 1989) that promotes its release from bound proteins (Gething and Sambrook, 1992). Depletion of cellular ATP inhibits protein folding and results in prolonged association of some proteins with BiP. These results, together with the recent demonstration of ATP transport into the ER (Clairmont et al., 19932), support a role for ATP in BiP's *in vivo* function. However, conditions that are optimal for *in vitro* ATP hydrolysis do not exist in the ER. This has led investigators to propose that either co-stimulatory molecules exist, similar to those found in bacteria, or that nucleotide binding, not hydrolysis, is sufficient for HSP70 release from proteins (Palleros et al., 1993).

The three-dimensional structure of an ATP hydrolyzing fragment of hsc70, has been solved (Flaherty et al., 1990). BiP presumably forms a similar ATP-binding structure, because mutation of BiP residues corresponding to those in hsc70 that are implicated in nucleotide interactions severely inhibit BiP's ATPase activity and its ability to release bound proteins *in vitro* with ATP. The *in vivo* expression of mammalian BiP ATPase mutants (Hendershot et al., 1996) and yeast BiP (Kar2) mutants that map to the ATP binding domain (Simons et al., 1995) block the *in vivo* folding of substrate proteins, providing the first direct evidence for BiP's involvement in protein folding. The stable binding of BiP mutants to wild-type proteins supports the hypothesis that BiP undergoes multiple cycles of ATP-mediated binding and release to unfolded regions on proteins and that folding occurs while the protein is not bound to BiP. If the protein is unable to fold or has not completed folding, BiP would rebind. This would explain both the transient binding of BiP to wild-type proteins and its stable (or "continuous") binding to mutant proteins.

The yeast BiP homologue was identified and found to be an essential gene. Either inactivation or depletion of yeast BiP results in the accumulation of untranslocated proteins in the cytosol (Gething and Sambrook, 1992). *In vitro* reconstitution of these deficient yeast microsomes with purified BiP restores translocation (Brodsky and Schekman, 1993). Genetic studies revealed that yeast BiP interacts with sec63, and ER membrane protein with a DnaJ-like domain, to translocate nascent proteins across the ER membrane (Scidmore et al., 1993). Several attempts have been made to demonstrate a similar role for mammalian BiP in translocation (Nicchitta and Blobel, 1993; Gorlich and Rapaport, 1993), but thus far the data have been contradictory and inconclusive.

N-linked glycosylation, calnexin, and protein folding

The majority of proteins that enter the ER are co-translationally modified by addition of a complex carbohydrate structure ($Glu_3Man_9GlcNAc_2$) to asparagine residues (N-linked glycosylation). Almost immediately the glucose molecules are removed one at a time by glucosidase I and II which reside in the ER. If a protein remains in the ER, a single glucose can be readded by the enzyme UDP-glucosyltransferase (UDP-GT) (Sousa et al., 1992) only to be trimmed again by glucosidase II. Alternatively, while the protein is still in the ER, the nonglucosylated sugar complex can be acted on by mannosidase I which removes four of the nine mannoses. Once more than two of the mannose sugars are removed, the sugar complex is no longer a substrate for UDP-GT. The reasons for this intricate addition and removal of glucose moieties that are not a component of the mature glyco-protein has remained one of the puzzles of ER protein biosynthesis until very recently.

In 1991, calnexin was identified as a transmembrane 90 kD calcium bind-ing, ER phosphoprotein (Wada et al., 1991). It transiently associates with many nascent proteins and more stably to incompletely assembled proteins and blocks their transport to the Golgi (Bergeron et al., 1994). Replacement of the transmembrane domain of class I heavy chains with a GPI-linkage abrogated binding of the truncated protein to calnexin and addition of the class I transmembrane domain to a non-calnexin binding protein resulted in a hybrid protein with calnexin binding properties (Bergeron et al., 1994), suggesting that the transmembrane domain of proteins contained the calnexin binding site. However, conflicting data has been obtained in other similar experiments.

Simultaneously, the binding of proteins to calnexin in a hepatoma cell line was found to require both the presence of N-linked carbohydrate moieties and incomplete folding of the target proteins (Ou et al., 1993). Inhibitors of glucosidase I and II inhibited calnexin binding to influenza HA and VSV G viral proteins, and a temperature-sensitive mutant of VSV G protein, which accumulates as a monoglucosylated protein at non-per-missive temperatures, remained associated with calnexin (Hammond et al., 1994), suggesting that monoglucosylated residues on nascent glycopro-teins provide the recognition site for calnexin. These data, coupled with the observation that UDP-GT prefers unfolded proteins as a substrate (Sousa et al., 1992) has led to the following hypothesis (Hebert et al., 1995). Nas-cent glycoproteins enter the ER and are acted upon by glucosidase I and II before translocation and folding are complete. After glucosidase II exposes the monoglucosylated sugar complex, calnexin is able to bind the nascent protein, and removal of the remaining glucose by glucosidase II results in the release of calnexin from the glycoprotein. If the protein is incom-pletely folded, UDP-GT would bind and add back a single glucose residue, which would restore calnexin binding. The process would continue until folding was complete. This model is extremely compelling and offers an explanation for the perplexing addition, removal, and readdition of glucose residues that are not ultimately a component of the mature glycoprotein. However, data showing that removal of the entire N-linked sugar complex by endo H does not release a protein from calnexin implies that either calnexin binds through both sugar and protein interactions, or that the individual protein calnexin binds to determines which interaction is more stable.

The calnexin gene of *S. pombe* was recently identified and shown to be essential (Jannatipour and Rokeach, 1995). Interestingly, UDP-GT activity was also demonstrated in this organism (Fernandez et al., 1994) suggesting a similar mechanism for recognition might occur in *S. pombe* and under-scoring the importance of this process. It is not presently known if calnexin actually assists in protein folding or whether it primarily monitors the folding process and acts to prevent proteins form escaping from the ER prematurely. A completely folded protein that still bears a $Glu_1Man_{7-9}GlcNAc_2$ modification should be a substrate for calnexin binding until the glucose is removed. The fact that HA, a calnexin substrate, folds normally and is secreted in the presence of glucosidase inhibitors (Hammond et al., 1994) would be more compatible with a role for calnexin in quality control instead of protein folding.

Other resident ER proteins that may act as chaperones

Calreticulin, a major ER calcium binding protein, possesses a C-terminal KDEL tail which restricts its movement from the ER and is thought to play a role in calcium storage (Smith and Koch, 1989). Calreticulin binds to multiple ER proteins in a protein overlay assay (Burns and Michalak, 1993), and it associates transiently with a precursor of myeloperoxidase (MPO) but not to the mature protein (Nauseef et al., 1995). Interestingly, this binding was dependent on the glycosylation of MPO, suggesting that calreticulin could associate with target proteins via a mechanism that is similar to that described for calnexin.

GRP94 was originally identified as a protein that was coordinately regulated by agents that induced BiP synthesis (Lee, 1992). Evidence for its involvement in ER protein folding has been difficult to obtain and limited to only a few proteins. Like BiP, GRP94 appears to preferentially associate with unassembled proteins (Melnick et al., 1992) and binds ATP (Clairmont et al., 1992), but unlike BiP, GRP94 cannot be released from proteins *in vitro* with ATP (Melnick et al., 1994). The mechanism for the release of GRP94 from proteins is not currently known.

Unlike the molecular chaperones discussed thus far, hsp47 is only found in the ER of collagen secreting cells and apparently binds, transiently to newly synthesized procollagen molecules and more stably to conformationally abnormal ones (Nakai et al., 1992). The depletion of hsp47 by antisense oligonucleotides directed to the translation initiation site causes a reduction in the amount of full-length procollagen chains and an increase in shorter procollagen peptides that are associated with peptidyl t-RNA (Sauk et al., 1994). This suggests that hsp47 is involved in procollagen translocation and is reminiscent of the role of yeast BiP in this process. Procollagen:hsp47 complexes can be dissociated *in vitro* by decreasing the pH (Nakai et al., 1992), which has led investigators to speculate that hsp47 might dissociate as the complex moves through the increasingly acidic secretory pathway. A similar mechanism has been hypothesized for the dissociation of BiP, and other resident ER proteins, from the KDEL receptor, which leads to the retrograde transport of KDEL-bearing proteins back to the ER.

ER folding enzymes

Unlike other cellular organelles, the ER is oxidizing which promotes the formation of disulfide bonds between cysteine residues. PDI, a multifunctional resident ER protein that is structurally related to thioredoxin (Freedman, 1989), catalyzes the formation and isomerization of disulfide bonds both *in vitro* (Puig and Gilbertz, 1994; Lilie et al., 1994) and in isolated microsomes. Mutational analyses of PDI revealed that the cysteines in the two N-terminal thioredoxin-like regions are required for the catalytic activity of PDI (Lyles and Gilbert, 1994). The protein binding site resides in a highly acidic region in the C-terminus of PDI (Noiva et al., 1993) and preferentially binds to peptides that contain cysteine residues (Morjana and Gilbert, 1991). A yeast homologue of PDI has been identified that catalyzes the formation and isomerization of disulfide bonds and been found to be an essential gene (LaMantia et al., 1991).

Two other resident ER proteins, ERp72 and ERp61, have been identified that have homology to PDI and possess disulfide isomerization activity (Rupp et al., 1994; Kozaki et al., 1994). ERp72 can catalyze the reappearance of biologically active Fab fragments and RNase after they have been denatured and reduced (Rupp et al., 1994). Both mammalian PDI and ERp72 complemented yeast PDI mutants, but ERp61 did not (Gunther et al., 1993). Recently both ERp61 and ERp72 were found to have cysteine protease activity (Urade et al., 1993). It has been well documented that proteins failing to successfully fold and assemble are often degraded in the ER (Klausner and Sitia, 1990), but the components of the ER degradation system have not been identified. ERp72 and ERp61 are potential candidates for the ER proteolytic machinery for the following reasons: (1) the ER degradation signal for some proteins requires the presence of an unpaired cysteine (Stafford and Bonifacino, 1991), (2) both of these proteins may recognize cysteine residues because they are inhibited by cysteine protease inhibitors (Urade et al., 1993), and (3) both proteins are up-regulated in cells with high secretory capacity (Kozaki et al., 1994) and by conditions that lead to the accumulation of unfolded proteins in the ER (see below). However, at this time there are no data to indicate that either of these proteins plays a role in protein degradation *in vivo*.

Two families of cytosolic immunophilins have been identified that possess peptidyl prolyl *cis-trans* isomerase (PPIase) activity. Recently, ER homologues of both the cyclophilin family (Hasel et al., 1991) and the FK506 binding proteins (Jin et al., 1991) have been isolated and shown to possess PPIase activity. The ER cyclophilin was shown to refold RNaseT1 in an *in vitro* assay (Bose et al., 1994), and the FKBP-13 mRNA was regulated coordinately with other ER foldases and chaperones (Bush et al., 1994).

Interactions between chaperones and folding enzymes

Although a host of ER chaperones have been identified and characterized, it is presently unclear whether the various ER chaperones are functionally redundant, with each one recognizing a different structure on unfolded proteins to ensure they are all recognized, or whether these chaperones might act coordinately or sequentially to fold ER proteins. Only a few recent studies have begun to approach this question. A study on the binding of GRP94 and BiP to Ig LCs revealed that BiP bound to a less oxidized form of LC, and GRP94 bound to the more oxidized form (Melnick et al., 1994), providing evidence for sequential binding of these chaperones. Several proteins have been identified that bind to both BiP and calnexin, and studies to determine the order of their binding have been conflicting. Pulse-chase experiments revealed that BiP bound earlier to a less mature form of VSV G protein, and that calnexin bound later to a more oxidized form (Hammond and Helenius, 1994). However, examination of the interactions of calnexin and BiP with DTT-reduced thyroglobulin, found the opposite sequence of binding (Kim and Arvan, 1995). Although the structure that UDP-GT binds on unfolded proteins has not been determined, because both BiP and UDP-GT recognize unfolded proteins and probably undergo multiple cycles of binding and release, it is possible that they compete for the same region. Thus, at present it is not at all clear whether these chaperones are binding independently or in an ordered fashion, nor is it known if each protein will have a different cohort of chaperones that bind and whether these chaperones will bind in a certain order based on the folding pathway of the individual protein.

There are data to suggest cooperativity or ordered interactions between an ER chaperone (BiP) and a folding enzyme (PDI). Studies that examined the oxidation state of proteins that bind to BiP found these proteins were often only partially oxidized when bound (Gething and Sambrook, 1992; Knittler et al., 1995; Melnick et al., 1994), and manipulation of the ER redox potential prevents disulfide bond formation and results in the stable association of proteins with BiP. Both yeast BiP (Simons et al., 1995) and mammalian BiP ATPase (Hendershot et al., 1996) mutants block the *in vivo* oxidation of their substrate proteins, implying that BiP release must occur for PDI to catalyze disulfide bond formation.

Regulation of chaperone synthesis

Most of the ER foldases and chaperones are coordinately induced by agents that affect protein folding (Tab. 1), suggesting that a common mechanism exists to regulate their synthesis. Understanding the regulation of these proteins has for the most part progressed from the two ends: (1) delineation of lumenal ER sensing mechanisms and (2) identification of *cis* acting elements that regulate the individual genes.

The regulation of BiP by unfolded proteins in the ER was directly shown by expressing viral mutant proteins and their wild-type counterparts and testing the effects on BiP synthesis. Wild-type, secreted proteins did not induce BiP synthesis, whereas mutant, misfolded proteins which accumulated in the ER did (Kozutsumi et al., 1988). The ability of a misfolded protein to stimulate BiP synthesis is dependent on its binding to BiP in the ER. Although most of the ER chaperones are up-regulated in response to misfolded or unassembled proteins, there is some data to suggest that the occupancy of BiP, not other chaperones, is what the cell monitors. Expression of BiP from a heterologous promoter down-regulates both endogenous BiP and the other ER chaperones, and prevents their induction by stress conditions (Dorner et al., 1992). Conversely, over-expression of either ERp72 or PDI from a heterologous promoter does not down-regulate the endogenous Erp72 or PDI, nor does it block the induction of other ER chaperones (Dorner et al., 1992). We do not know whether the cell simply monitors the amount of free BiP and synthesizes more when it falls below a certain level, or if the mechanism to monitor BiP occupancy is more complex. Unlike BiP bound to proteins, free BiP dimers are phosphorylated and ADP-ribosylated (Freiden et al., 1992), which could be used to monitor the levels of free BiP. Alternatively, BiP dimers could bind to a transmembrane receptor in the ER, cause it to dimerize, and transduce a negative signal for the transcription of ER chaperones. It is not known how or if unfolded proteins that do not associate with BiP cause the induction of the particular ER chaperone they require.

The *cis* acting elements that control the transcription of BiP have been delineated through a series of 5′ deletions, internal deletions, and linker scanning mutagenesis (Lee, 1992) and found to contain functionally redundant elements. The activation of BiP and GRP94 by malfolded proteins, glycosylation inhibitors, and calcium ionophores map to the same regions, suggesting that they share a common signal for induction. Within this region is a 36 nucleotide sequence termed the grp core that is highly conserved in both BiP and GRP94 promoters in species ranging from humans to yeast. The distal portion contains a CCAAT motif which contributes to basal expression and a proximal region that contains regulatory elements for stress induction. Integration of tandem copies of the rat BiP promoter into the genome of a Chinese hamster ovary cell line inhibited the induction of BiP, GRP94, and ERp72, suggesting that the same factors regulate all three promoters. A near consensus cAMP response element (CRE) was identified by DNA footprinting analyses in the rat BiP promoter that bound protein complexes in gel mobility shift assays. These proteins were immunologically distinct from the 42 kDa CRE binding protein but were able to bind to a conventional CRE, suggesting that they might belong to the CREB/P family.

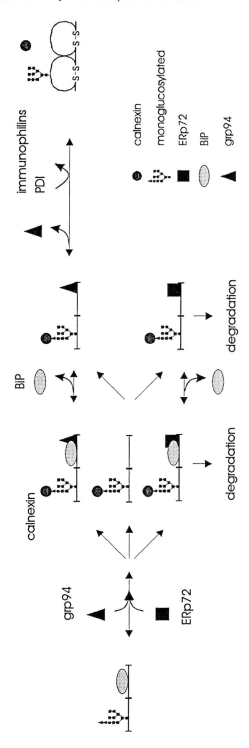

Figure 1. Model for interaction of multiple chaperones with a folding protein.

The inability to link transcriptional regulation of ER chaperones and the sensing mechanisms for unfolded proteins in the ER was remedied with the delineation of the yeast BiP promoter (Mori et al., 1992). Two groups independently fused this element to a reporter gene, LacZ, and screened for yeast cells bearing this construct that did not turn blue after tunicamycin treatment. This genetic screen identified an ER transmembrane serine/threonine kinase that appears to be a proximal component of the BiP signaling pathway (Cox et al., 1993; Mori et al., 1993) and that is essential for the survival of yeast cells under conditions that promote the accumulation of unfolded proteins in the ER. Interestingly, induction of mammalian ER chaperones is inhibited by serine/threonine kinase inhibitors and enhanced by serine/threonine phosphatase inhibitors, demonstrating the involvement of a serine/threonine kinase in their regulation. However, to date no mammalian homologue has been identified.

Conclusions

A whole consortium of ER chaperones and folding enzymes have been identified that apparently bind to nascent proteins transiently and allow them to fold efficiently in a very complex environment. The isolation of yeast homologues that are often required for viability attests to their essential role in this process. Perhaps the most interesting question at this time is how these multiple, coordinately regulated proteins recognize a nascent protein, interaction with one another, allow the protein to fold, and then release it. The cytoplasmic chaperones act as multiprotein, dynamic complexes, and it is conceivable that some of their ER homologues will act in a similar way (Fig. 1). The other pressing question is how the cell monitors protein folding in the ER and responds by either up-regulating or down-regulating the ER chaperones. Although progress has been made, there are still large gaps in our understanding of this signal transduction pathway. The power of yeast genetics, the isolation of enzymatically active components, and the establishment of cell-free systems for folding are likely to provide exciting answers to these questions in the near future.

References

Bergeron, J.J., Brenner, M.B., Thomas. D.Y. and Williams, D.B. (1994) Calnexin: a membrane-bound chaperone of the endoplasmic reticulum. *Trends. Biochem Sci.* 19:124–128.

Blond Elguindi, S., Cwirla, S.E., Dower, W.J., Lipshutz, R.J., Sprang, S.R., Sambrook, J.F. and Gething, M.J. (1993) Affinity panning of a library of peptides displayed on bacteriophages reveals the binding specificity of BiP. *Cell* 75:717–728.

Bole, D.G., Hendershot, L.M. and Kearney, J.F. (1986) Posttranslational association of immunoglobulin heavy chain binding protein with nascent heavy chains in nonsecreting and secreting hybridomas. *J. Cell. Biol.* 102:1558–1566.

Bose, S., Mucke, M. and Freedman, R.B. (1994) The characterization of a cyclophilin-type peptidyl prolyl cis-trans-isomerase from the endoplasmic-reticulum lumen. *Biochem J.* 300:871–875.

Brodsky, J.L. and Schekman, R. (1993) A Sec63p-BiP complex from yeast is required for protein translocation in a reconstituted proteoliposome. *J. Cell Biol.* 123:1355–1363.

Burns, K. and Michalak, M. (1993) Interactions of calreticulin with proteins of the endoplasmic and sarcoplasmic reticulum membranes. *FEBS Lett.* 318:181–185.

Bush, K.T., Hendrickson, B.A. and Nigam, S.K. (1994) Induction of the FK506-binding protein, FKBP13, under conditions which misfold proteins in the endoplasmic reticulum. *Biochem J.* 303:705–708.

Clairmont, C.A., De Maio, A. and Hirschberg, C.B. (1992) Translocation of ATP into the lumen of rough endoplasmic reticulum-derived vesicles and its binding to luminal proteins including BiP (GRP78) and GRP94. *J. Biol. Chem.* 267:3983–3990.

Cox, J.S., Shamu, C.E. and Walter, P. (1993) Transcriptional induction of genes encoding endoplasmic reticulum resident proteins requires a transmembrane protein kinase. *Cell* 73:1197–1206.

Dorner, A.J., Walsey, L.C. and Kaufman, R.J. (1992) Overexpression of GRP78 mitigates stress induction of glucose regulated proteins and blocks secretion of selective proteins in Chinese hamster ovary cells. *EMBO J.* 11:1563–1571.

Fernandez, F.S., Trombetta, S.E., Hellman, U. and Parodi, A.J. (1994) Purification to homogeneity of UDP-glucose: glycoprotein glucosyltransferase from *Schizosaccharomyces pombe* and apparent absence of the enzyme from *Saccharomyces cerevisiae. J. Biol. Chem.* 269:30701–10706.

Flaherty, K.M., DeLuca Flaherty, C. and McKay, D.B. (1990) Three-dimensional structure of the ATPase fragment of a 70 K heat-shock cognate protein. *Nature* 346:623–628.

Flynn, G.C., Pohl, J., Flocco, M.T. and Rothman, J.E. (1991) Peptide-binding specificity of the molecular chaperone BiP. *Nature* 353:726–730.

Freedman, R.B. (1989) Protein disulfide isomerase: multiple roles in the modification of nascent proteins. *Cell* 57:1069–1072.

Freiden, P.J., Gaut, J.R. and Henershot, L.M. (1992) Interconversion of three differentially modified and assembled forms of BiP. *EMBO J.* 11:63–70.

Gething, M.J. and Sambrook, J. (1992) Protein folding in the cell. *Nature* 355:33–45.

Gorlich, D. and Rapaport, T.A. (1993) Protein translocation into proteoliposomes reconstituted from purified components of the endoplasmic reticulum membrane. *Cell* 75:615–630.

Gunther, R., Srinivasan, M., Haugejorden, S., Green, M., Ehbrecht, I.M. and Kuntzel, H. (1993) Functional replacement of the *Saccharomyces cerevisiae* Trg1/Pdi1 protein by members of the mammalian protein disulfide isomerase family. *J. Biol. Chem.* 268:7728–7732.

Haas, I.G. and Wabl, M. (1983) Immunoglobulin heavy chain binding protein. *Nature* 306:387–389.

Hammond, C. and Helenius, A. (1994) Folding of VSV G protein: sequential interaction with BiP and calnexin. *Science* 266:456–458.

Hammond, C., Braakman, I. and Helenius, A. (1994) Role of N-linked oligosaccharide recognition, glucose trimming, and calnexin in glycoprotein folding and quality control. *Proc. Natl. Acad. Sci. USA* 91:913–917.

Hasel, K.W., Glass, J.R., Godbout, M. and Sutcliffe, J.G. (1991) An endoplasmic reticulum-specific cyclophilin. *Mol. Cell Biol.* 11:3484–3491.

Hebert, D.N., Foellmer, B. and Helenius, A. (1995) Glucose trimming and reglucosylation determine glycoprotein association with calnexin in the endoplasmic reticulum. *Cell* 81:425–433.

Hendershot, L.M., Wei, J.-Y., Gaut, J.R., Melnick, J., Aviel, S. and Argon, Y. (1996) Inhibition of immunoglobulin folding and secretion by dominant negative BiP ATPase mutants; *PNAS* 93:5269–5274.

Jannatipour, M. and Rokeach, L.A. (1995) The *Schizosaccharomyces pombe* homologue of the chaperone calnexin is essential for viability. *J. Biol. Chem.* 270:4845–4853.

Jin, Y.J., Albers, M.W., Lane, W.S., Bierer, B.E., Schreiber, S.L. and Burakoff, S.J. (1991) Molecular cloning of a membrane-associated human FK506- and rapamycin-binding protein, FKBP-13. *Proc. Natl. Acad. Sci. USA* 88:6677–6681.

Kassenbrock, C.K. and Kelly, R.B. (1989) Interaction of heavy chain binding protein (BiP/GRP78) with adenine nucleotides. *EMBO J.* 8:1461–1467.

Kim, P.S. and Arvan, P. (1995) Calnexin and BiP act a sequential molecular chaperones during thyrogobulin folding in the endoplasmic reticulum. *J. Cell Biol.* 128:29–38.

Klausner, R.D. and Sitia, R. (1990) Protein degradation in the endoplasmic reticulum. *Cell* 62:611–614.

Knittler, M.R., Dirks, S. and Haas, I.G. (1995) Molecular chaperones involved in protein degradation in the endoplasmic reticulum: Quantitative interaction of the heat shock cognate protein BiP with partially folded immunoglobulin light chains that are degraded in the endoplasmic reticulum. *Proc. Natl. Acad. Sci. USA* 92:1764–1768.

Kozaki, K., Miyaishi, O., Asai, N., Iida, K., Sakata, K., Hayashi, M., Nishida, T., Matsuyama, M., Shimizu, S., Kaneda, T. and Saga, S. (1994) Tissue distribution of ERp61 and association of its increased expression with IgG production in hybridoma cells. *Exp. Cell Res.* 213:348–358.

Kozutsumi, Y., Segal, M., Normington, K., Gething, M.J. and Sambrook, J. (1988) The presence of malfolded proteins in the endoplasmic reticulum signals the induction of glucose-regulated proteins. *Nature* 332:462–464.

LaMantia, M., Miura, T., Tachikawa, H., Kaplan, H.A., Lennarz, W.J. and Mizunaga, T. (1991) Glycosylation site binding protein and protein disulfide isomerase are identical and essential for cell viability in yeast. *Proc. Natl. Acad. Sci. USA* 88:4453–4457.

Lee, A.S. (1992) Mammalian stress response: induction of the glucose-regulated protein family. *Curr. Opin. Cell Biol.* 4:267–273.

Lilie, H., McLaughlin, S., Freedman, R. and Buchner, J. (1994) Influence of protein disulfide isomerase (PDI) on antibody folding *in vitro*. *J. Biol. Chem.* 269:14290–14296.

Lyles, M.M. and Gilbert, H.F. (1994) Mutations in the thioredoxin sites of protein disulfide isomerase reveal functional nonequivalence of the N- and C-terminal domains. *J. Biol. Chem.* 269:30946–30952.

Melnick, J., Aviel, S. and Argon, Y. (1992) The endoplasmic reticulum stress protein GRP94, in addition to BiP, associates with unassembled immunoglobulin chains. *J. Biol. Chem.* 267:21303–21306.

Melnick, J., Dul, J.L. and Argon, Y. (1994) Sequential interaction of the chaperones BiP and Grp94 with immunoglobulin chains in the endoplasmic reticulum. *Nature* 370:373–375.

Mori, K., Sant, A., Kohno, K., Normington, K., Gething, M.J. and Sambrook, J.F. (1992) A 22 bp cis-acting element is necessary and sufficient for the induction of the yeast Kar2 (BiP) gene by unfolded proteins. *EMBO J.* 11:2583–2593.

Mori, K., Wenzhen, M., Gething, M.J. and Sambrook, J. (1993) A transmembrane protein with a cdc2+/CDC28-related kinase activity is required for signaling form the ER to the nucleus. *Cell* 74:743–756.

Morjana, N.A. and Gilbert, H.F. (1991) Effect of protein and peptide inhibitors on the activity of protein disulfide isomerase. *Biochemistry* 30:4985–4990.

Nakai, A., Satoh, M., Hirayoshi, K. and Nagata, K. (1992) Involvement of the stress protein HSP47 in procollagen processing in the endoplasmic reticulum. *J. Cell. Biol.* 117:903–914.

Nauseef, W.M., McCormick, S.J. and Clark, R.A. (1995) Calreticulin functions as a molecular chaperone in the biosynthesis of myeloperoxidase. *J. Biol. Chem.* 270:4741–4747.

Nicchitta, C.V. and Blobel, G. (1993) Lumenal proteins of the mammalian endoplasmic reticulum are required to complete protein translocation. *Cell* 73:989–998.

Noiva, R., Freedman, R.B. and Lennarz, W.J. (1993) Peptide binding to protein disulfide isomerase occurs at a site distinct form the active sites. *J. Biol. Chem.* 268:19210–19217.

Ou, W.-J., Cameron, P.H., Thomas, D.Y. and Bergeron, J.J.M. (1993) Association of folding intermediates of glycoproteins with calnexin during protein maturation. *Nature* 364:771–776.

Palleros, D.R., Reid, K.L., Shi, L., Welch, W.J. and Fink, A.L. (1993) ATP-induced protein-Hsp70 complex dissociation requires K⁺ but not ATP hydrolysis. *Nature* 365:664–666.

Puig, A. and Gilbert, H.F. (1994) Anti-chaperone behavior of BiP during the protein disulfide isomerase-catalyzed refolding of reduced denatured lysozyme. *J. Biol. Chem.* 269:25889–25896.

Rupp, K., Birnbach, U., Lundstrom, J., Van, P.N. and Soling, H.D. (1994) Effects of CaBP2, the rat analog of ERp72, and of CaBP1 on the refolding of denatured reduced proteins. Comparison with protein disulfide isomerase. *J. Biol. Chem.* 269:2501–2507.

Sauk, J.J., Smith, T., Norris, K. and Ferreira, L. (1994) Hsp47 and the translation-translocation machinery cooperate in the production of alpha 1(I) chains of type I procollagen. *J. Biol. Chem.* 269:3941–3946.

Scidmore, M.A., Okamura, H.H. and Rose, M.D. (1993) Genetic interactions between Kar2 and Sec63, encoding eukaryotic homologues of DnaK and DnaJ in the endoplasmic reticulum. *Mol. Biol. Cell* 4:1145–1159.

Simons, J.F., Ferro-Novick, S., Rose, M.D. and Helenius, A. (1995) BiP/Kar2p serves as a molecular chaperone during carboxypeptidase Y folding in yeast. *J. Cell. Biol.* 130:41–49.

Smith, M.J. and Koch, G.L. (1989) Multiple zones in the sequence of calreticulin (CRP55, calregulin, HACBP), a major calcium binding ER/SR protein. *EMBO J.* 8:3581–3586.

Sousa, M.C., Ferrero Garcia, M.A. and Parodi, A.J. (1992) Recognition of the oligosaccharide and protein moieties of glycoproteins by the UDP-Glc:glycoprotein glucosyltransferase. *Biochemistry* 31:97–105.

Stafford, F.J. and Bonifacino, J.S. (1991) A permeabilized cell system identifies the endoplasmic reticulum as a site of protein degradation, *J. Cell. Biol.* 115:1225–1236.

Urade, R., Takenaka, Y., and Kito, M. (1993) Protein degradation by ERp72 from rat and mouse liver endoplasmic reticulum. *J. Biol. Chem.* 268:22004–22009.

Wada, I., Rindress, D., Cameron, P.H., Ou, W.J., Doherty, J.J., Louvard, D., Bell, A.W. Dignard, D., Thomas, D.Y. and Bergeron, J.J. (1991) SSR alpha and associated calnexin are major calcium binding proteins of the endoplasmic reticulum membrane. *J. Biol. Chem.* 266:19599–19610.

Wherever possible review articles have been cited instead of the original papers. We sincerely apologize to the people who actually did the studies.

Stress-Inducible Cellular Responses
ed. by U. Feige, R.I. Morimoto, I. Yahara and B. Polla
© 1996 Birkhäuser Verlag Basel/Switzerland

Involvement of molecular chaperones in intracellular protein breakdown

M.Y.S. Sherman[1] and A.L. Goldberg[2]

[1] *Boston Biomedical Research Institute, 20 Staniford St., Boston, MA 02114, USA*
[2] *Department of Cell Biology, Harvard Medical School Boston, 25 Shattuck St., Boston, MA 02115, USA*

Summary. In all cells and organelles, there exist multiple molecular chaperones, which not only can facilitate the proper folding, transport and assembly of multimeric structures, but also appear to function in intracellular protein degradation. Recent findings in *E. coli* indicate that the major chaperones of the Hsp70 (DnaK) and Hsp60 (GroEL) families and their cofactors (DnaJ, GrpE or GroEL and Trigger Factor) associate with certain short-lived proteins (e.g. mutant polypeptides or regulatory proteins) and promote their degradation by the ATP-dependent proteases, La (lon or ClpP). Moreover, ATPases of ClpA/B family not only function in ATP-dependent proteolysis in association with the Clp protease, but by themselves can facilitate or act as chaperones in protein assembly. In eukaryotes, Hsp70 and their cofactors, the DnaJ homologs, are essential for the ubiquitination of certain abnormal and regulatory proteins and in the breakdown of certain polyubiquitinated polypeptides by 26S proteasome. It is likley that the chaperones function in proteolysis either as elements that faciliate the recognition of unfolded proteins or that the chaperones partially unfold substrates to make them more susceptible to proteases or ubiquitinating enzymes.

Introduction

In response to heat shock or to other harsh conditions (e.g., oxidative stress or exposure to heavy metals), both prokaryotic and eukaryotic cells induce the synthesis of a limited number of proteins (termed heat-shock proteins or hsps) that enhance survival under such stressful conditions. Earlier studies have indicated that in bacteria (Parsell et al., 1989; Goff et al., 1985) and eukaryotic cells (Ananthan et al., 1986) the common intracellular signal for the expression of the various heat shock genes under stressful conditions is the accumulation of damaged proteins in the cell. For example, even at normal temperatures, the production of highly abnormal polypeptides induced by incorporation of amino acid analogs or puromycin or by expression of recombinant proteins that fail to fold in the cytosol causes the induction of the heat-shock response (Goff et al., 1985). Similarly, microinjection of denatured polypeptides, but not the native proteins, into frog oocytes will cause induction of hsps (Ananthan et al., 1986). In fact, the main function of these stress-induced proteins appears to be to prevent the intracellular accumulation of such unfolded, potentially toxic polypeptides. Among the major group of proteins that are specifically induced under these conditions in bacterial cells are intracellular proteases

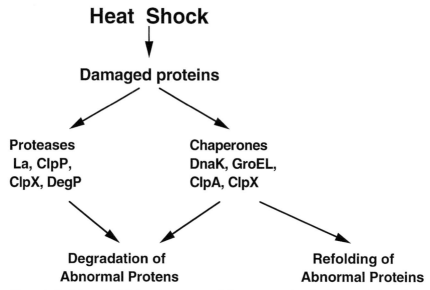

Figure 1. Both heat shock induced proteases and chaperones are involved in protein degradation.

(Fig. 1) that catalyze the selective degradation of such abnormal proteins (Goldberg, 1992), proteases La (lon) (Goff et al., 1984; Philips et al., 1984), Clp (Ti) (Kroh and Simon, 1990), and HtrA (DegP) (Lipinska et al., 1990). Among the hsps in eukaryotic cells are ubiquitin (Bond and Schlesinger, 1985) and certain ubiquitin-conjugating enzymes, E2s, that are necessary for degradation of abnormal proteins (Hershko and Ciechanover, 1992; Seufert and Jentsch, 1990). The other major group of heat-shock proteins are molecular chaperones (e.g., members of Hsp70 or Hsp60 families) (Fig. 1), which can prevent the aggregation of denatured proteins (Todd et al., 1994; Becker and Craig, 1994; Hartl et al., 1994) and promote the refolding of heat-damaged polypeptides. Molecular chaperones are known to specifically recognize unfolded structures in polypeptides (either hydrophobic sequences or molten globule state), to prevent aggregation, and to promote protein refolding (Gething and Sambrook, 1992; Weissman et al., 1994). The increased level of hsps thus enhances the cell's capacity to refold the damaged proteins and to rapidly eliminate irreparable denatured polypeptides.

Recent findings have shown that in addition to their role in protein refolding, molecular chaperones serve an essential role in the degradation of certain abnormal polypeptides in both prokaryotic and eukaryotic cells. Thus, if members of Hsp70, Hsp60, and Clp families fail to refold a denatured protein, they can facilitate its degradation by the cell's proteolytic machinery (Fig. 2); for example, by acting as enzymatic cofactors that solubilize the substrate, partially unfold it, or prevent its aggregation.

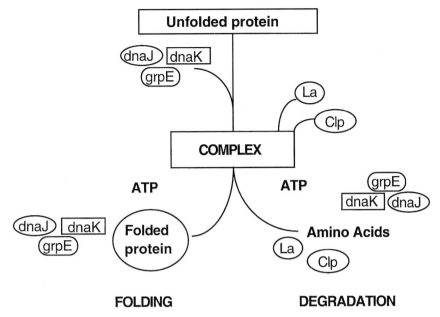

Figure 2. DnaK and its cofactors bind to unfolded protein and either fold it, or if failed, promote the degradation of this protein by the ATP-dependent proteases.

By serving as cofactors for the cell's degradative machinery, the chaperones help assure that unfolded polypeptides do not accumulate in the cell. Such abnormal proteins can themselves be toxic, and if they accumulate, bound to chaperones, they might also cause a continual consumption of ATP in futile folding cycles. Since the same chaperones are essential for the initial folding of newly synthesized polypeptides, their functioning in the degradation of misfolded proteins also provides an important link between the cell's biosynthetic and degradative systems, and would thus comprise an elegant "quality control" mechanism for cellular proteins.

Protein degradation in prokaryotes and mitochondria

The degradation of most abnormal proteins in *E. coli* is catalyzed by the very large ATP-hydrolyzing proteases. Most, if not all, of them are Hsps. Particularly important are protease La (Ion) and protease Clp (Ti), which cleave proteins and ATP in a coupled reaction. Protease La (Ion) is a homopolymer of probably six subunits, and each subunit contains peptidase and ATP-hydrolyzing domains (Goldberg, 1992; Goldberg et al., 1994). One important feature of this enzyme that aids in the selective degradation of abnormal proteins is an allosteric region specific for unfolded polypeptides. Binding of substrates to this regulatory domain activates its ATPase and

proteolytic activities (Waxman and Goldberg, 1982, 1986) leading to substrate hydrolysis. Unlike protease La, the Clp protease consists of two components, one contains six large subunits with ATPase activity (ClpA) and
the other contains seven small subunits with peptidase activity (ClpP)
(Kessel et al., 1995). This ATPase component is also activated by protein
substrates, which trigger proteolysis (Hwang et al., 1988; Kessel et al.,
1995; Murizi et al. 1990; Thompson and Maurizi, 1994). The association
between the ATPase and peptidase components is very weak, and the
complex easily dissociates during isolation. The ClpA ATPase represents a
family of homologous proteins (Gottesman et al., 1990), several of which
(e.g., ClpX) can function in protein degradation together with the ClpP
peptidase (Gottesman et al., 1994).

Protease La (Ion) appears responsible for the rapid degradation of most
abnormal proteins and also of a number of short-lived regulatory proteins
(Goldberg, 1992; Goldberg et al., 1994), while protease Clp is involved in
the rapid breakdown of normally stable polypeptides upon nitrogen
starvation (Damerau and St. John, 1993) and certain fusion polypeptides
(Gottesman et al., 1990). Recently, a membrane-associated ATP-dependent
protease, HflB (FtsH), has been shown to function in degradation of certain
proteins *in vivo* and in cell-free systems (Herman et al., 1993; Tomoyasu et
al., 1995). This protease is homologous to certain subunits of the major
ATP-dependent protease in the eukaryotic cytosol, the 26S proteasome
complex, and to membrane-bound proteases in mitochondria (Tauer et al.,
1994). HflB is involved in the rapid degradation of the phage CII protein
(Herman et al., 1993; Banuett et al., 1986) and of the heat-shock regulated
transcription factor, σ^{32} (Tomayasu et al. 1995; Herman et al., 1995). This
latter process is critical in determining the rate of expression of hs genes
(see below). Another potentially important ATP-dependent protease was
identified genetically in *E. coli*, HflA (Cheng et al., 1988; Noble et al.,
1993), a protease distantly related to ClpP. In addition, two more protease-
related proteins were discovered in the *E. coli* genome sequencing project,
HslV (Chuang et al., 1993), which bears strong homology to the 20S proteasome in eukaryotic cells and HslU, ATPase resembling the Clp-like
ATPase. Although their function *in vivo* has not been uncovered, these two
hs genes are adjacent and appear to form an ATP-dependent proteolytic
complex (Rohrwild et al., 1996).

A role for DnaK, DnaJ and GrpE in proteolysis

The first evidence that the degradation of an individual protein requires the
function of chaperones came from the observation that inactivation of DnaK
and its cofactors caused a dramatic stabilization of the heat shock transcription factor, σ^{32} (Straus et al., 1990; Tilly et al., 1989). This protein is one of
the most short-lived ones in the cell ($t_{1/2} = 1$ min.), and its stabilization

at high temperatures or under other harsh conditions leads to an induction of the whole group of heat-shock genes (Yura et al., 1990). Moreover, DnaK and its cofactors, DnaJ and GrpE, were found associated with σ^{32} in the cell (Gamer et al., 1992). These observations suggested that DnaK, when bound to σ^{32}, facilitates the rapid degradation of this protein at normal temperature, apparently by the HflB protease (Tomayasu et al., 1995). This binding of DnaK appears to be critical for the induction of hsps by denatured proteins. When such damaged proteins accumulate upon heat shock, they bind to free DnaK. Consequently, less DnaK is available for binding to σ^{32}, and the free transcription factor, a σ^{32}, is no longer rapidly hydrolyzed by HflB. The resulting accumulation of σ^{32} causes transcription of the heat-shock genes (Craig and Gross, 1991). The subsequent expression of more proteases and chaperones in the stressed cell leads to the elimination of the damaged proteins, the restoration of rapid degradation of σ^{32}, and the termination of the heat-shock response.

In addition to this specialized regulatory role, DnaK, DnaJ, and GrpE were also found to be essential for the rapid degradation of the mutant polypeptide, PhoA61, a variant of alkaline phosphatase which cannot be secreted due to a mutation in its signal sequence (Michaelis et al., 1986). In the reducing environment of the cytosol PhoA61 cannot form disulfide bridges and is not active. At 30°C, PhoaA61 is relatively stable and found in the cell not associated with chaperones. However, at 37° or 41°C, the degradation of this protein is rapid, and up to 30% of the cell's PhoA61 becomes associated with DnaK (Sherman and Goldberg, 1992). Proteases La, Clp and DnaK cofactor, GrpE, are also present in these complexes. Both La and Clp proteases contribute to the breakdown of PhoA61, since it was partially stabilized in both the *lon* and *clp*P mutants (G. Goldmacher, personal communication). A *dna*K deletion completely stabilized PhoA61, which indicates that this chaperone is absolutely required for degradation by either proteases La or Clp. In contrast to this deletion, the missense mutation, *dna*K756, enhanced PhoA61 degradation. This DnaK variant can associate with PhoA61, but this complex (unlike that containing wild-type DnaK) does not dissociate in the presence of ATP. This dissociation of DnaK requires the nucleotide exchange factor, GrpE, and in a GrpE ts-mutant, the amount of PhoA61 associated with DnaK increased, as did degradation.

Thus, complex formation with DnaK appears to be essential for the ATP-dependent degradation of PhoA61, and the continued association of the mutant protein with DnaK, and perhaps DnaJ seems to facilitate attack by protease La or ClpP. DnaK, DnaJ and GrpE were shown to bind also to normal alkaline phosphatase and to promote its translocation through the *E. coli* membrane (Wild et al., 1992). Presumably, the failure of the chaperones to promote this transport process (or to promote the polypeptide folding), as occurs with PhoA61 or with deletion of the PhoA61 signal sequence (H.C. Huang et al., unpublished observations), leads to a prolonged association with the chaperones, which somehow appears to favor proteolytic attack (Fig. 2). This model (e.g., a form of kinetic partitioning) accounts for the stimulation of proteolysis by the DnaK756 strain discussed above.

Mitochondria contain proteins homologous to DnaK, DnaJ and GrpE which also function in the translocation of proteins and their subsequent assembly into active form (Pfanner et al., 1990; Stuart et al., 1994). Recent findings indicate that in mitochondria, as in *E. coli*, the degradation of certain abnormal proteins requires these chaperones (Wagner et al., 1994) and the mitochondrial homolog of protease La, Pim (Suzuki et al., 1994). In this work, W. Neupert and his colleagues constructed plasmids expressing bovine lactalbumin and cytochrome b_2-DHFR fusion protein with a mitochondrial targeting sequence in yeast (Wagner et al., 1994). These polypeptides were transported into mitochondria, and were rapidly degraded there by the Pim protease. Mutations in either mitochondrial Hsp70 (DnaK), or Mdj1 (DnaJ) (Rowley et al., 1994) prevented this degradation (Wagner et al., 1994). Moreover, both lactalbumin and b_2-DHFR were found to associate with the chaperones, and formation of such complexes appeared to be essential for their breakdown.

Interestingly, mutations in the mitochondrial chaperones that cause the formation of a nondissociating complex with the abnormal proteins reduced their breakdown, in contrast to our data with PhoA61 in *E. coli*. Therefore, the breakdown of these mitochondrial proteins requires not only the formation of a complex with the chaperones, but also the dissociation of the complexes. Most probably, the chaperones release these unfolded proteins in a soluble form which is particularly susceptible for digestion by the mitochondria's ATP-dependent protease (see below). It is presently unclear why with PhoA61 in *E. coli*, the complexes with nondissociating DnaK favored degradation (Sherman and Goldberg, 1992), while with b_2-DHFR in mitochondria dissociation from the chaperones is necessary for rapid breakdown (Wagner et al., 1994). Apparently, the function of the chaperones in degradation is substrate specific, and different polypeptides may be preferentially attacked by proteases in either the chaperone-bound conformation or upon dissociation. (Alternatively, this difference may be due to some additional unknown feature of the two nondissociating DnaK mutants used.)

A role of GroEL, GroES and trigger factor in proteolysis

Like DnaK, GroEL, the homolog of mitochondrial Hsp60, is a very abundant protein in *E. coli*, and upon heat shock, its level can reach 7–10% of cell proteins. GroEL also appears to be essential for the rapid degradation of certain abnormal proteins in *E. coli* (Kandror et al., 1994). However, its structure, binding properties and enzymatic activities are very different form those of DnaK. Presumably, these polypeptides are ones whose degradation does not involve DnaK, DnaJ, or grpE (although only one such polypeptide has been studied thus far). These findings emerged from studies of the breakdown of the recombinant fusion protein, CRAG, so named because it contains a truncated protein A region flanked by short unfolded domains (12 residues from the cro repressor at its amino terminus and 14 from β-galactosidase at its carboxyl terminus) (Murby et al., 1991). This fusion protein becomes associated with the major hsps, GroEL, DnaK and GrpE, *in vivo* and *in vitro* (Sherman and Goldberg, 1991). When expressed in *E. coli*, CRAG is degraded with a half-life of 20 min by an ATP-dependent process that requires the ClpP subunit of protease Ti (Kandror et al., 1994). CRAG (31 kDa) was found in cells in multiple forms: some intact CRAG molecules are associated with DnaK in complexes that are not related to degradation, and large fragments of CRAG are found in cells associated with GroEL. These polypeptides are of 16–26 kDa, and contain the normal N-terminal region of CRAG. Pulse-chase experiments showed that these fragments are short-lived intermediates in CRAG breakdown; i.e., they were generated rapidly by proteolysis and then disappeared during the course of CRAG degradation. Thus, CRAG's degradation by ClpP does not appear to be by a processive mechanism. The production of these fragments form CRAG and their subsequent degradation required the binding of CRAG to GroEL, its GroES-dependent dissociation, and cleavage by ClpP protease. These conclusions are based on a variety of experimental observations, including the finding that the rate of CRAG breakdown correlated with the amount of CRAG associated with GroEL, that overproduction of wild-type GroEL accelerated degradation, and that both the generation of CRAG fragments and their disappearance were inhibited in the *GroES* and *ClpP* mutants. Therefore, it is likely that upon GroES-catalyzed release from GroEL, CRAG'S carboxyl terminal domain is attacked by the ClpP protease, and then the fragments reassociate with GroEL, where presumably they undergo unfolding and then are released, allowing further proteolysis (Fig. 3). This model, involving repeated cycles of GroEL binding, release and limited proteolysis, resembles recent findings concerning the mechanism of GroEL/ES-induced folding of polypeptide (Weissman et al., 1994), which proceeds through repeated cycles of protein binding and release so as to provide additional chances for successful folding.

One exciting, unexpected development was the discovery of another polypeptide present in these "degradative complexes" containing CRAG fragments and GroEL. This protein was identified as "Trigger Factor" (TF), a protein originally isolated by W. Wickner and colleagues, who showed that it binds to certain unfolded polypeptides and promotes their translocation into isolated membrane vesicles (Wickner, 1989; Guthric and Wickner, 1990). However, subsequent *in vivo* studies indicated that this protein does not play an essential role in translocation of these same polypeptides in the cell (Wickner, 1989). We found that one TF molecule is associated with each GroEL 14-mer containing either a CRAG or a CRAG fragment, but GroEL-TF complexes were also found in cells not expressing CRAG. The

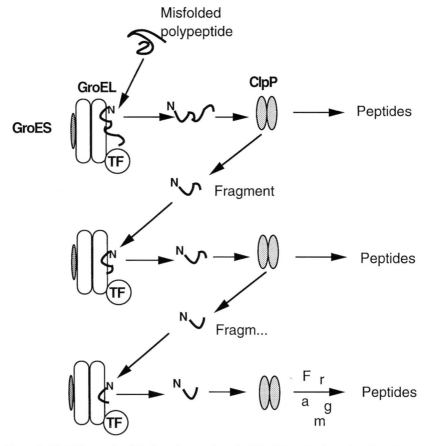

Figure 3. Multiple rounds of binding of a protein to GroEL-TF, GroES-dependent dissociation, and cleavage by the ClpP lead to protein degradation.

presence of TF in the degradative complexes containing GroEL and CRAG fragments suggested that TF may function with GroEL in proteolysis. To test this possibility, we analyzed strains with altered levels of expression of the *tig* (TF) gene (Guthric and Wickner, 1990), and showed that a reduction in the cellular content of TF led to a substantial decrease in the rate of CRAG breakdown, while increased TF content enhanced markedly the degradation of this protein (Kandror et al., 1995). Thus, TF, like GroEL, appears to be a rate-limiting factor in the proteolytic pathway.

What could be the role of the TF that is associated with GroEL in protein degradation? In principle, TF may facilitate the formation or dissociation of the CRAG-GroEL complexes, or it may change the properties of GroEL, to favor its function in degradation instead of folding and transport. In our recent studies, TF was shown to promote *in vivo* GroEL's ability to bind both to CRAG and to other unfolded proteins (Kandror et al., 1996). Upon

TF induction, much more of the cell's GroEL could bind to affinity columns containing CRAG or a variety of other unfolded proteins. In contrast, in cells containing low levels of TF, only a trace amount of GroEL could bind to these columns (Kandror et al., 1996). This surprising ability of TF to promote GroEL binding to abnormal proteins may account for its capacity to stimulate GroEL-dependent degradation of CRAG. Since TF can promote GroEL binding to proteins, TF may act as a "cochaperone" more generally; for example, in GroEL-mediated folding or transport. Recently, TF has been reported to possess an isoprolyl isomerase activity that is insensitive to cyclosporin or FK506. It is possible that TF increases the binding of abnormal proteins to GroEL by catalyzing the isomerization of critical proline residues on these substrates.

Clp ATPases function as molecular chaperones

Another group of heat-shock proteins that is important in protein degradation is the Clp family of ATPases, which can function both as molecular chaperones and also as subunits of proteolytic complexes. Members of this protein family contain one or two ATPase domains and form multisubunit complexes (Gottesman et al., 1990). These polypeptides have been highly conserved in evolution, and related proteins are present in both prokaryotes and eukaryotes (Gottesman et al., 1990). In eukaryotic cells, ClpA and B homologs are found in the cytosol and mitochondria (Gottesman et al., 1990). A ubiquitous member of the Clp ATPase family, Hsp104, the homolog of *E. coli*, ClpB, is strongly heat inducible, and was shown to be primarily responsible for protection of eukaryotic cells against heat damage (Parsell et al., 1991). Upon heat shock in the *hsp104* mutant in yeast, proteins aggregate, and cells die. Moreover, pure Hsp104 was able to protect proteins against aggregation, and even to promote the refolding of certain unfolded polypeptides (Parsell et al., 1994). Therefore, Hsp104 appears to function as a chaperone, although its biochemical mechanisms have been studied only to a limited extent. ClpA, the first known member of the Clp ATPase family was originally described as a component of the Clp (Ti) protease (Hwang et al., 1988; Gottesman et al., 1990), and its protein-activated ATPase activity was shown to control the peptidase activity of ClpP and to be essential for hydrolysis of protein substrates. However recently, ClpA was reported to be able to function as a chaperone in the disassembly of P1phage replication initiator protein, RepA (Wickner et al., 1994). For activation of DNA binding, the RepA dimer must be converted to monomers (Wickner et al., 1992), and this reaction is catalyzed *in vivo* by DnaK and DnaJ (Wickner et al., 1991). Recently, it was discovered that the ClpA protein, in the absence of ClpP peptidase, can also catalyze this dissociation of RepA dimers, allowing RepA activation. However, in the presence of both ClpA and ClpP, RepA is rapidly destroyed. Therefore, ClpA can either activate RepA, or promote its degradation depending on whether the protease subunit is available. This ability of ClpA to promote RepA dimer dissociation is an interesting *in vitro* phenomenon that demonstrates the ability of ClpA, to function as a chaperone, but it probably does not occur in the cell, since *dnaK* and *dnaJ* mutations prevent this dissociation even in the presence of functional ClpA (Wickner et al., 1991, 1992).

The degradation of certain proteins by ClpP is independent of ClpA, but instead requires the function of a homologous ATPase, ClpX (Gottesman et al., 1994; Woitkowiak et al., 1994). The pair of ClpP and ClpX proteins was shown both genetically and biochemically to degrade the bacteriophage lambda O replication protein, and thus to prevent phage propagation (Gottesman et al., 1994; Woitkowiak et al., 1994). Interestingly, in the absence of ClpP, ClpX protects lambda O from heat-induced aggregation, disaggregates performed lambda O aggregates, and even promotes the efficient binding of lambda O to its DNA recognition sequence (Wawrzynow et al., 1995). As with other chaperones (e.g., DnaK, GroEL), the ClpX catalyzed disaggregation reaction required ATP hydrolysis. Therefore, the function of the Clp proteases requires the association of the peptidase component, ClpP, and interchangeable ATPase components, ClpA, ClpX and their homologs. By themselves, these ATPases can either promote the refolding of the denatured proteins, or in complexes with ClpP promote hydrolysis of the abnormal proteins. At present, it is unclear what determines the choice between these two fates, whose biological consequences are as different as life and death. At the molecular level, however, these roles may be very similar, since the structural features that favor digestion of a polypeptide (e.g., being monomeric, soluble, nonaggregated, and unfolded) are also critical for its proper folding.

Interestingly, degradation of CRAG polypeptide by ClpP did not require any of the known *E. coli* ATPases of the Clp family, i.e., ClpA, ClpB, or ClpX (Kandror et al., 1994). Thus, the degradation of CRAG may involve an unknown ATPase subunit. Alternatively, the chaperones GroEL/ES may function in CRAG degradation in place of the ATPase cofactors of ClpP to unfold the substrate. When ClpA and ClpP function as a proteolytic complex, the polypeptide appears to be injected directly into the protease and is digested in a processive fashion (i.e., without the accumulation of intermediates). By contrast, in the case of CRAG breakdown by GroEL, ClpP was not detected in a complex with GroEL, and proteolysis was not processive (i.e., polypeptide intermediates in the degradative pathway were detected in the cell).

The ATP-dependent proteases in prokaryotes and mitochondria can thus be viewed as a complex of an endoprotease component and an ATPase component that functions as a chaperone in maintaining substrate in a conformation susceptible for the proteolytic attack. However, the ATPase components are also responsible for substrate recognition. Because the association between these components sometimes is very loose, a single peptidase (ClpP) can cooperate with different substrate-specific ATPase in the degradation of different polypeptides. It is even possible that certain proteins have to associate sequentially with different chaperones before complete digestion is possible, just as the proteins transported into mitochondrial matrix first bind to Hsp70, and then are passed to Hsp60 before acquiring the active conformation (Pfanner et al., 1990; Stuart et al., 1994). Interestingly, some ATP-dependent proteases (e.g., protease La from bacteria (Goff et al., 1984; Wawrzynow et al., 1995) and its homolog from mitochondria (Suzuki et al., 1994), the membrane-bound bacterial protease HflB (Tomoyasu et al., 1993), and the mitochondrial protease Yta10 (Tauer et al., 1994), seem to be fusion proteins containing a peptidase domain and ATPase domain in a single subunit. However, these ATPases are not only functioning as "chaperones." In the case of protease La and its homologs, ATP-binding to the enzyme is necessary also for the formation of the protease's active site, and ATP hydrolysis leads to an inactivation of the enzyme; this mechanism helps prevent proteolysis in the cell until an appropriate substitute binds to the enzyme which stimulates ATP-binding and activation of the protease (Simons et al., 1995). For protease ClpAP, the different subunits also control each other's activity. The P subunit by itself reduces ATPase activity and the ATPase component also regulates the proteolytic site. The binding of ATP allows larger substrates to be digested by the peptidase, which in the absence of ClpA can only degrade small peptides (Thompson and Maurizi, 1994; Gottesman et al., 1990). Thus, ATP hydrolysis has critical function both in regulating the proteolytic activity and as a chaperone affecting substrate's susceptibility to proteolysis.

Role of chaperones in protein breakdown in eukaryotes

Like *E. coli*, eukaryotic cells rapidly degrade various abnormal proteins (e.g., mutant, analog-containing or damaged polypeptides) by an ATP-requiring process (Goldberg, 1992; Hershko and Ciechanover, 1992; Jentsch, 1992; Varshavsky, 1992). However, this ATP-dependent proteolytic pathway, which is present in both the eukaryotic cytosol and nucleus, is much more complex than that in *E. coli*. The first steps in this process involve the covalent attachment of multiple ubiquitin molecules to the protein substrate, which marks it for rapid degradation by the 26S (1500 kDa) proteasome complex (Goldberg, 1992; Hershko and Ciechanover, 1992; Ciechanover and Schwartz, 1994). Ubiquitin-conjugation involves the ATP-dependent ubiquitin-activating enzyme (E1), a family of Ub-carrier proteins (E2), and a family of ubiquitin ligases (E3) that link the activated ubiquitins to the protein substrates (Finley and Chau, 1991). However, it is unclear how the latter enzymes (E2 and E3) recognize different unfolded proteins or short-lived normal proteins for conjugation.

Recently, we obtained evidence that molecular chaperones (e.g., the eukaryotic homologs of DnaJ (Caplan et al., 1993) and several different Hsp70s, which are close homologs of *E. coli* DnaK, also function in the ubiquitination of certain proteins with abnormal conformation. The cytosolic DnaJ homologs and/or Hsp70s are known to bind to nascent (Beckman et al., 1990; Frydman et al., 1994) or heat-damaged polypeptides and to promote their folding. However, in the case of unsuccessful folding, as may occur frequently with mutant or heat-damaged polypeptides or with certain short-lived normal polypeptides, the hsps can promote the modification of these proteins by the ubiquitin-conjugating enzymes, E2 and E3.

In order to test if the molecular chaperones might play a role in the ubiquitin-dependent pathway (Jentsch, 1992), we chose to undertake genetic studies in yeast. Such studies of the chaperones are somewhat more complicated in yeast than in *E. coli*, because in the *S. cerevisiae* cytosol, there are at least six different Hsp70s: the SSA family includes four closely related species of Hsp70, and the SSB group two (Craig et al., 1993). Ssa proteins include the major heat-inducible species, and were shown to promote the translocation of precursor polypeptides into the endoplasmic reticulum and mitochondria (Deshaies et al., 1988). Ssb proteins, which are associated with the ribosome (Nelson et al., 1992), seem to serve an important role in the initial folding of nascent polypeptides. These Hsp70s probably work in concert with the two DnaJ homologs in the cytoplasm, Ydj and Sis proteins (Luke et al., 1991; Caplan and Douglas, 1991, 1992). Normally, in protein transport and folding, Ssa seems to function together with Ydj and Ssb with Sis (Caplan and Douglas, 1992; Caplan et al., 1993); however, there seems to be some overlap in their functions and cofactor requirements (Caplan and Douglas, 1992; Caplan et al., 1993). Because of this redundancy in Hsp functions, which can complicate analysis of the effects of mutants, most of our experiments have focused on the roles in protein degradation of the two DnaJ homologs.

Hsp70 and DnaJ mutations reduce degradation of short-lived proteins

In the yeast mutant, in which two major Hsp70 genes (SSA1 and 2) were inactivated by insertions, the degradation of both short-lived normal proteins and analog-containing polypeptides was much slower than in wild-type cells. A similar reduction in the breakdown of normal short-lived and analog containing proteins was observed in an *ssb1ssb2* mutant, where both SSB genes were inactivated by insertion mutations (Nelson et al., 1992), and in ts-mutants in both DnaJ homologs, *ydj1* and *sis1*. In contrast, degradation of the long-lived cell proteins, which occurs largely in the yeast vacuole (McCracken and Cruse, 193), was not reduced in any of these mutants at either temperature (Lee et al., 1995). Thus, these effects appear to be very selective. Since the rapid breakdown of the short-lived normal and analog-containing proteins in yeast requires ubiquitin and the proteasome (Seufert and Jentsch, 1990; Jentsch, 1992), the chaperones appear to be involved in this pathway.

An important step for understanding the mechanism of the chaperone's action, was to find individual short-lived proteins whose breakdown required these chaperones. After a screen of a number of short-lived protein fusions, which are rapidly hydrolyzed by the ubiquitin-proteasome pathway, we found that a short-lived fusion protein Ubiquitin-Pro-β-galactosidase (Ub-P-β-gal) ($t_{1/2}$ = 5 min) was markedly stabilized in the *ydj* mutant at the nonpermissive temperature. This fusion protein appears to be recognized as an unfolded polypeptide, and ubiquitin conjugation to it requires the E2, Ubc4, which is a heat-shock protein (Seufert and Jentsch, 1990; Jentsch, 1992). However, in the *ydj* mutant, the rapid degradation of two other short-lived recombinant proteins, R-β-galactosidase and L-β-galactosidase, both of which are substrates of the "N-end rule" pathway, was not reduced. Thus, the inhibition of degradation in the chaperone mutants was specific to certain substrates of the ubiquitin-proteasome pathway, which can function normally in the degradation of other polypeptides.

The DnaJ and Hsp70 family of chaperones are known to help prevent protein aggregation and to promote their solubilization and unfolding. However, in the degradation of Ub-P-β-gal, Ydj cannot be simply preventing the aggregation of this fusion protein or maintaining it in a soluble form, because in the *ydj1* mutant at the nonpermissive temperature, Ub-P-β-gal was soluble and fully active. Nevertheless, in the wild-type cells, the Ydj1 directly associates with the Ub-P-β-gal fusion protein. Isolation of Ub-P-β-gal by affinity chromatography, together with associated proteins, showed that the Ydj1 protein (and some of the major species of Ssa) were bound to Ub-P-β-gal, but not with the stable wild-type β-galactosidase. Thus, these chaperones appear to be able to play a role in recognition of substrates for ubiquitin-dependent degradation.

Involvement of Ydj1 and Hsp70 in protein ubiquitination

To test if the *ydj1* mutation affects ubiquitination of substrates, we compared the total level of ubiquitinated proteins in the wild-type cells and in the *ydj1* mutant. In this mutant, the total level of ubiquitinated proteins was reduced by at least 50% below the level in wild type cells. More importantly, the level of ubiquitinated derivatives of Ub-P-β-gal was markedly reduced in the *ydj1* mutant at the nonpermissive temperature when proteolysis was blocked. Therefore, the Ydj1 protein appears to play an essential role in the ubiquitination of certain short-lived polypeptides. Most likely Ydj1, probably in cooperation with Hsp70s, specifically binds to certain abnormal proteins and maintains them in a partially unfolded form which favors the attachment of ubiquitin moieties by the critical E2S and E3S.

Recent experiments with mammalian cell extracts also indicate that Hsp70s are essential for ubiquitin-conjugation to certain model proteins. In these studies, Ciechanover and colleagues (1995) found that depletion of Hsp70s from reticulocyte extracts with the anti-Hsp70 antibody strongly reduced ubiquitination of ^{125}I-labeled actin and glyceraldehyde-3-phosphate dehydrogenase (GAPDH), but not of other substrates. It is quite possible that in these experiments both actin and GAPDH were denatured or oxidatively damaged upon iodination (a common consequence using standard methods for iodination), and thus could be recognized by the ubiquitination enzymes as "abnormal" polypeptides. It is noteworthy that in reticulocyte cell extracts, as in our genetic experiments in yeast, the chaperones were not necessary for ubiquitination of the N-end rule substrates (e.g., lysozyme). These authors also reported that Hsp70-dependent ubiquitination of actin required E1, a specific E2, and E3. These data suggest that Hsp70 functions in the ubiquitination of the protein substrate in cooperation with ubiquitin ligase, E3. However further studies with purified preparations will be necessary to uncover the precise biochemical role for Hsp70 and its cofactor, the DnaJ homolog, in the ubiquitination of abnormal proteins.

The DnaJ homolog Sis is necessary for degradation of certain ubiquitinated proteins

As mentioned above, a ts-mutation in the other DnaJ homolog present in yeast cytosol, Sis1, also dramatically reduced the breakdown of abnormal proteins. However, unlike Ydj1 which is necessary for protein ubiquitination, Sis1 appears to function in a subsequent step in the pathway (Lee, D.H., manuscript in preparation). A search for the individual substrates that require Sis1 for their ubiquitin-dependent degradation, showed that as in *ydj1* mutant, the breakdown of Ub-P-β-galactosidase was strongly reduced in the *sis1* mutant. However, in contrast to the *ydj1* mutation which blocked ubiquitination, *sis1* caused an accumulation of both Ub-P-β-galactosidase and its ubiquitin-conjugated derivatives. Therefore, Sis1 is not required for the ubiquitination of Ub-P-β-galactosidase, but rather for the degradation of the multiubiquitinated derivatives of this protein by the 26S proteasome complex. Accordingly, in the *sis1* mutant, we observed a tremendous increase in the total cell content of ubiquitin conjugates at the nonpermissive temperature. The effect of the *sis1* mutation was specific to Ub-P-β-galactosidase and to certain other substrates, and no inhibition of the rapid degradation of R-β-galactosidase (the N-end rule substrate) or Cln3-β-galactosidase (see below) was observed in these strains. These data indicate that the *sis1* mutation does not cause a general defect in the functioning of the 26S proteasome. Perhaps, this DnaJ homolog, together with Hsp70s binds to certain polyubiquitinated proteins and facilitates their association with the 26S proteasome by promoting the partial unfolding of substrates.

Accordingly, we did find the Sis1 protein associated with the affinity-purified fraction of protein-ubiquitin conjugates. In the future, it will be important to clarify whether Sis1 binds to the substrate protein or to the polyubiquitin chain, and what are the features of the substrate that make its degradation dependent on Sis1.

Role of Ydj1 in phosphorylation of G1 cyclin Cln3

In addition to abnormal polypeptides, a variety of normal proteins with important regulatory functions (Bonifacino and Lippincott-Schwartz, 1991; Ciechanover and Schwartz, 1994) are destroyed rapidly by the ubiquitin-dependent pathway. Many of these proteins (e.g., cyclins, IkB, rate-limiting enzymes) contain special sequences (degradation signals) that target them for rapid degradation under specific cellular conditions. These degradation signals are believed to be the sites of interaction of these normal short-lived proteins with the ubiquitin conjugating enzymes (E2) or ubiquitin ligases (E3). If molecular chaperones are involved in the recognition of abnormal proteins for ubiquitination, it seemed also possible that the chaperones bind to the degradation signals on normal short-lived proteins, and facilitate their ubiquitination. Therefore, we recently tested whether Ydj1 may function in the ubiquitin-dependent breakdown of a normal short-lived protein, Cln3 cyclin. Cln3 is the most important regulator of the G1 to S-phase transition, since it regulates transcription of other G1 cyclins (Tyers et al., 1993), and of S-phase cyclins (Schwob and Nasmyth, 1993; Epstein and Cross, 1992; Amon et al., 1994). Recently, Yaglom et al. have found that the rapid breakdown of Cln3 (half-life of about 5 min (Yaglom et al., 1995) proceeds through the ubiquitin pathway. The critical event that triggers ubiquitination of this important regulator of the cell cycle is the phosphorylation of Cln3 (Yaglom et al., 1995), which in turn is dependent on the major cell cycle regulatory kinase, p34[Cdc28] (Yaglom et al., 1995).

To test whether chaperones are involved in Cln3 degradation, we compared the half-life of this cyclin in the wild-type and in the *ydj1* mutant (Yaglom et al., 1996). At the nonpermissive temperature, the degradation of Cln3 was markedly reduced in the mutant. Unexpectedly, in the *ydj1* mutant, stabilization of Cln3 did not lead to an accumulation of the phosphorylated form of the protein, as occurs in the mutants defective in ubiquitination (Yaglom et al., 1995). This finding suggested that Ydj1 is necessary for the phosphorylation of Cln3, which triggers its ubiquitination. Indeed, using a special deletion mutant of Cln3 that is stable, but can be phosphorylated normally, we have found that even in the absence of degradation, the phosphorylation of this polypeptide requires a functional Ydj1 (Yaglom et al., 1996). Furthermore, Ydj1 and the major Hsp70s of the Ssa family were found to bind *in vivo* to the degradation signal segment of Cln3. Together, these observations suggested that the binding of these chaperones to this region is required for Cln3 phosphorylation, which then leads to its ubiquitination. Moreover, the *ydj1* mutation did not affect the phosphorylation of other substrates of p34[Cdc28] kinase. The specific defect in Cln3 phosphorylation in the *ydj1* mutant, and the finding that the Ydj1 protein associates with Cln3 within close proximity of the phosphorylation site, argues strongly that this chaperone affects Cln3 rather than the kinase. Presumably, Ydj1 together with Hsp70s, enhances Cln3 phosphorylation by maintaining this substrate in a conformation that promotes its recognition by p34[cdc28] kinase (e.g. by exposing a previously hidden phosphorylation site). Therefore, Ydj1 chaperone seems to function in proteolysis in at least two steps: in phosphorylation that targets for ubiquitination, and in ubiquitination itself.

Thus, the two DnaJ homologs, Ydj and Sis, can function at multiple distinct steps in the proteolytic pathway: Ydj is necessary for phosphorylation of certain regulatory proteins that signal their ubiquitination; it is also essential for ubiquitination of certain abnormal polypeptides; Sis on the other hand plays a critical role in the later step of degradation of the ubiquitinated polypeptides (Fig. 4).

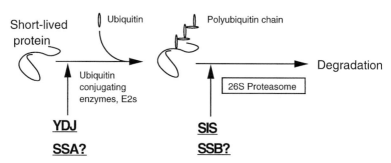

Figure 4. Ydj1 is necessary for ubiquitination of certain polypeptides, while Sis is involved in the degradation of polyubiquitinated polypeptides by the 26S proteasome.

Role of BiP in protein degradation in the Endoplasmic Reticulum

Many secretory or membrane proteins that are transported into the ER do not reach the cell surface, but undergo rapid degradation in the ER lumen (Bonifacino and Lippincott-Schwartz, 1991). These unstable proteins include misfolded (e.g., nonglycosylated) polypeptides (Hammond and Helenius, 1994) and polypeptides that could not properly assemble into multicomponent complexes (Bonifacino and Lippincott-Schwartz, 1991). Although most of the proteins entering the secretory pathway are transported through the Golgi apparatus to the cell surface, a large fraction of many newly synthesized polypeptides, including apolipoprotein B (Boren et al., 1991) or CFTR (Harris and Argent, 1993; Welsh et al., 1993), undergo rapid breakdown in the ER. Although proteolytic systems in the ER have not been characterized biochemically, there is some evidence that the Hsp70 homolog in ER, BiP, plays a role in this degradative process (Knittler and Haas, 1992; Knittler et al., 1995; Cotner and Pious, 1995). Originally, BiP was shown to facilitate import of polypeptides into ER lumen (Sanders et al., 1992; Scidmore et al., 1993), to promote their proper folding and to prevent aggregation (Simons et al., 1995; Hammond and Helenius, 1994). In these processes, BiP functions in cooperation with several other ER chaperones, including calnexin (Hammond and Helenius, 1994), Grp94 (Kuznetsow et al., 1994; Melnik et al., 1994), and probably Erp72 (Kuznetsov et al., 1994).

Recently, Knittler and colleagues obtained data indicating that BiP is critical for the breakdown of light chain of IgG (Knittler et al., 1992). Normally, IgG light chain is assembled in the ER with the heavy chain, which is then transported out of the cell. However, there are cell lines that synthesize light chains in the absence of heavy chains (Knittler and Haas, 1992) in which the light chain is degraded within the ER with a half-life of about 20 min. BiP was found associated with this protein prior to its degradation. Moreover, the half-life of the light chain corresponds to the half-life of its association with BiP. Therefore, it is likely that BiP is involved in targeting the light chain for degradation in the ER (Knittler et al., 1992). As we found for protein breakdown in the cytosol, this requirement for the Hsp70s for degradation was specific for certain substrates; for example, the breakdown of hemagglutinin in the ER lumen does not require its association with BiP (Doyle et al. 1986; Gething and Sambrook, 1992). Thus far, the roles of BiP and other ER chaperones have not been systematically studied by genetic approaches. These new insights into the role of Hsp70 and its cofactors in *E. coli*, mitochondria and the eukaryotic cytosol should lead to analogous genetic and biochemical studies of the protein turnover in the ER.

Conclusion In all cells and organelles, there exist several types of molecular chaperones, which have evolved efficient mechanisms to promote the proper folding, transport, and assembly of multimeric structures. However, if successful folding is impossible, as must occur often following heat damage or with mutant proteins, these molecular chaperones have also evolved the capacity to promote the rapid elimination of the unfolded, potentially damaging polypeptides. The rapid degradation of such proteins would appear to have clear selective advantage for the cell, since it would also prevent the prolonged wasteful consumption of ATP in futile efforts of the chaperones to fold such proteins. In any case, this linkage of the chaperones and proteolytic machinery constitutes an elegant "quality control mechanism for cell proteins."

Note. Originally, the term "molecular chaperone" was coined by analogy to the traditional role of "a chaperone", whose role was to accompany and protect a young lady from harm (specifically, form events that may lead to "social degradation" or "a fate worse than death"). This term seemed very appropriate with the demonstration that Hsps of the Hsp70, Hsp60, and ClpA families can prevent protein aggregation and promote proper maturation, transport, or assembly. However, since these same molecules can also actively contribute to the complete degradation of a protein, the term "chaperone" seems misleading or deceptive for this role. Perhaps a better term for this function of the Hsps would be a "molecular undertaker".

References

Amon, A., Irniger, S. and Nasmyth, K. (1994) Closing the cell cycle circle: G2 cyclin proteolysis initiated at mitosis persists until the activation of G1 cyclins in the next cycle. *Cell* 77:1037–1050.

Ananthan, J., Goldberg, A. and Voellmy, R. (1986) Abnormal proteins serve as eukaryotic stress signals and trigger the activation of heat shock genes. *Science* 232:500–503 (1986).

Banuett, F., Hoyt, M., McFarlane, L., Echols, H. and Herskowitz, I. (1986) hflB, a new *Escherichia coli* locus regulating lysogeny and the level of bacteriophage lambda cII protein. *J. Mol. Biol.* 187:213–224.

Becker, J. and Craig, E (1994) Heat-shock proteins as molecular chaperones. *Eur. J. Biochem.* 219:11–23.

Beckman, R., Mizzen, L. and Welch, W. (1990) Interaction of hsp70 with newly synthesized proteins: implication for protein folding and assembly. *Science* 248:850–854.

Bond, U. and Schlesinger, M. (1985) Ubiquitin is a heat shock protein in chicken embryo fibroblasts. *Mol. Cell Biol.* 5:949–956.

Bonifacino, J. and Lippincott-Schwartz, J. (1991) Degradation of proteins within the endoplasmic reticulum. *Curr. Opin. Cell Biol.* 3:592–600.

Boren, J., White, A., Wettesten, M., Scott, J., Graham, L. and Olofsson, S. (1991) The molecular mechanism for the assembly and secretion of ApoB-100-containing lipoproteins. *Progress in Lipid Research* 30:205–218.

Caplan, A. and Douglas, M. (1991) Characterization of YDJ1: a yeast homologue of bacterial dnaJ protein. *J. Cell Biol.* 114:609–621.

Caplan, A., Cyr, D. and Douglas, M. (1992) YDJ1p facilitates polypeptide translocation across different intracellular membrane by a conserved mechanism. *Cell* 71:1143–1155.

Caplan, A., Cyr, D. and Douglas, M. (1993) Eukaryotic homologues of *Escherichia coli* dnaJ: A diverse protein family that functions with hsp70 stress proteins. *Mol. Biol. Cell* 4:555–563.

Cheng, H., Muhlrad, P., Hoyt, M. and Echols, H. (1988) Cleavage of the cII protein of phage lambda by purified HflA protease: control of the switch between lysis and lysogeny. *Proc. Natl. Acad. Sci. USA* 85:7882–7886.

Chuang, S., Burland, V., Plunkett, G.E., Daniels, D. and Blattner, F. (1993) Sequence analysis of four new heat-shock genes constituting the hslTS/ibpAB and hslVU operons in *Escherichia coli*. *Gene* 134:1–6.

Ciechanover, A. (1995) Cold Spring Harbor Symposium. Abstract.

Ciechanover, A. and Schwartz, A.L. (1994) The ubiquitin-mediated proteolytic pathway-mechanisms of recognition of the proteolytic substrates and involvement in the degradation of native cellular proteins. *FASEB J.* 8:182–191.

Cotner, T. and Pious, D. (1995) HLA-DR beta chains enter into an aggregated complex containing GRP-78/BiP prior to their degradation by the pre-Golgi degradative pathway. *J. Biol. Chem.* 270:2379–2386.

Craig, E., Gambill, B. and Nelson, R. (1993) Heat shock proteins: Molecular chaperones of protein biogenesis. *Microbiol. Rev.* 51:402–414.

Damerau, K. and St. John, A. (1993) Role of Clp protease subunits in degradation of carbon starvation proteins in *Escherichia coli*. *J. Bacteriol.* 175:53–63.

Deshaies, R., Koch, B., Werner-Washburne, M., Craig, E. and Scheckman, R. (1988) A subfamily of stress proteins facilitates translocation of secretory and mitochondrial precursor polypeptides. *Nature* 332:800–808.

Doyle, C., Sambrook, J. and Gething, M. (1986) Analysis of progressive deletions of the transmembrane and cytoplasmic domains of influenza hemagglutinin. *J. Cell Biol.* 103:1193–1204.

Craig, E. and Gross, C. (1991) Is hsp70 the cellular thermometer? *TIBS* 16:135–140.

Epstein, C. and Cross, F. (1992) CLB5: a novel B cyclin from budding yeast with a role in S-phase. *Genes Dev.* 6:1695–1707.

Finley, D. and Chau, V. (1991) Ubiquitination. *Ann. Rev. Cell Biol.* 7:25–69.

Frydman, J., Nimmesgern, E., Ohtsuka, K. and Hartl, F.-U. (1994) Folding of nascent polypeptide chains in high molecular weight assembly weight molecular chaperones. *Nature* 370:111–116.

Gamer, J., Bujard, H. and Bukau, B. (1992) Physical interaction between heat shock proteins DnaK, DnaJ, and GrpE and the bacterial heat shock transcription factor σ^{32}. *Cell* 69:833–842.

Gething, M. and Sambrook, J. (1992) Protein folding in the cell. *Nature* 355:33–45.

Goff, S., Casson, L. and Goldberg, A. (1984) Heat shock regulatory gene htpR influences rates of degradation and expression of the Ion gene in *E. coli. Proc. Natl. Acad. Sci. USA* 81:6647–6651.

Goff, S. and Goldberg, A. (1985) Production of abnormal proteins in *E. coli* stimulates transcription of Ion and other heat shock genes. *Cell* 41:587–595.

Goldberg, A. (1992) The mechanism and function of ATP-dependent protease in bacterial and animal cells. *Eur. J. Biochem.* 203:9–23.

Goldberg, A.L., Moerschell, R.P., Chung C.H. and Maurizi, M.R. (1994) ATP-dependent protease La (lon) from *Escherichia coli. Methods in Enzymology* 244:350–375.

Gottesman, S., Clark, W. and Maurizi, M. (1990) The ATP-dependent Clp protease of *Escherichia coli*. Sequence of clpA and identification of a Clp-specific substrate. *J. Biol. Chem.* 265:7886–7893.

Gottesman, S., Squires, C., Pichersky, E., Carrington, M., Hobbs, M., Mattick, J., Dalrymple, B., Kuramitsu, H., Shiroza, T., Foster, T. (1990) Conservation of the regulatory subunit for the Clp ATP-dependent protease in prokaryotes and eukaryotes. *Proc. Natl. Acad. Sci. USA* 87:3513–3517.

Gottesman, S., Clark, W., de Crecy-Lagard, V. and Maurizi, M. (1994) ClpX, an alternative subunit for the ATP-dependent Clp protease of *Escherichia coli*. Sequence and *in vivo* activities. *J. Biol. Chem.* 268:22618–22626.

Guthrie, B. and Wickner, W. (1990) Trigger factor depletion or overproduction causes defective cell division but does not block protein export. *J. Bacteriol.* 172:5555–5562.

Hammond, C. and Helenius, A. (1994) Folding of VSV G protein: sequential interaction with BiP and calnexin. *Science* 266:456–458.

Harris, A. and Argent, B. (1993) The cystic fibrosis gene and its product CFTR. *Seminars in Cell Biology.* 4:37–44.

Hartl, F., Hlodan, R. and Langer, T. (1994) Molecular chaperones in protein folding: the art of avoiding sticky situations. *TIBS* 19:20–25.

Herman, C., Ogura, T., Tomoyasu, T., Hiraga, S., Akiyama, Y., Ito, K., Thomas, R., D'Ari, R. and Bouloc, P. (1993) Cell growth and lambda phage development controlled by the same essential *Escherichia coli* gene, ftsH/hflB. *Proc. Natl. Acad. Sci. USA* 90:10861–10865.

Herman, C., Thevenet, D., D'Ari, R. and Bouloc, P. (1995) Degradation of sigma 32, the heat shock regulator in *Escherichia coli*, is governed by HflB. *Proc. Natl. Acad. Sci. USA* 92:3516–3520.

Hershko, A. and Ciechanover, A. (1992) The ubiquitin system for protein degradation. *Ann. Rev. Biochem.* 61:761–807.

Hwang, B., Woo, K., Goldberg, A. and Chung, C. (1988) Protease Ti, a new ATP-dependent protease in *E. coli*, contains protein-activated ATPase and proteolytic functions in distinct subunits. *J. Biol. Chem.* 263:8727–8734.

Jentsch, S. (1992) The ubiquitin-conjugation system. *Ann. Rev. Genet.* 26:179–207.

Kandror, O., Busconi, L., Sherman, M. and Goldberg, A. (1994) Rapid degradation of an abnormal protein in *E. coli* involves the chaperones groEL and groES. *J. Biol. Chem.* 269:23575–23582.

Kandror, O., Sherman, M. and Goldberg, A. (1995) Trigger factor is involved together with groEL in the rapid degradation of abnormal protein in *E. coli. EMBO J.* 14:6021–6028.

Kandror, O., Sherman, M.Y., Moerschel, R. and Goldberg, A.L. (1996) GroEL associates with trigger factor in *Escherichia coli* and promotes GroEL binding to certain unfolded proteins. Submitted.

Kessel, M., Maurizi, M., Kim, B., Kocsis, E., Trus, B., Singh, S. and Steven, A. (1995) Homology in structural organization between *E. coli* ClpAP protease and the eukaryotic 26S proteasome. *J. Mol. Biol.* 250:587–594.

Knittler, M. and Haas, I. (1992) Interaction of BiP with newly synthesized immunoglobulin light chain molecules: cycles of sequential binding and release. *EMBO J.* 11:1573–1581.

Knittler, M., Dirks, S. and Haas, I. (1995) Molecular chaperones involved in protein degradation in the endoplasmic reticulum: quantitative interaction of the heat shock cognate protein BiP with partially folded immunoglobulin light chains that are degraded in the endoplasmic reticulum. *Proc. Natl. Acad. Sci. USA* 92:1764–1768.

Kroh, H. and Simon, L. (1990) The ClpP component of Clp protease is the sigma32-dependent heat shock protein F21.5. *J. Bacteriol.* 172:6026–6034.

Kuznetsov, G., Chen, L. and Nigam, S. (1994) Several endoplasmic reticulum stress proteins, including ERp72, interact with thyroglobulin during its maturation. *J. Biol. Chem.* 269: 22990–22995.

Lee, D., Sherman, M. and Goldberg, A. (1996) Involvement of the dnaJ-homolog, Ydj1p and hsp70s of SSA family in the ubiquitin-dependent degradation of abnormal proteins in yeast; *Mol. Cell. Biol., in press.*

Lipinska, B., Zylicz, M. and Georgopoulos, C. (1990) The htrA (degP) protein, essential for *E. coli* survival at high temperatures, is an endopeptidase. *J. Bacteriol.* 172:1791–1797.

Luke, M., Sutton, A. and Arndt, K. (1991) Characterization of Sis1, a *Saccharomyces cerevisiae* homologue of bacterial dnaJ protein. *J. Cell. Biol.* 114:623–638.

Maurizi, M., Clark, W., Katayama, Y., Rudikoff, S., Pumphrey, J., Bowers, B. and Gottesman, S. (1990) Sequence and structure of ClpP, the proteolytic component of the ATP-dependent Clp protease of *E. coli. J. Biol. Chem.* 265:12536–12545.

McCracken, A. and Kruse, K. (1993) Selective protein degradation in the yeast exocytic pathway. *Mol. Biol. Cell* 4:729–736.

Melnick, J., Dul, J. and Argon, Y. (1994) Sequential interaction of the chaperones BiP and GRP94 with immunoglobulin chains in the endoplasmic reticulum. *Nature* 370:373–375.

Michaelis, S., Hunt, J. and Beckwith, J. (1986) Effects of signal sequence mutations on the kinetics of alkaline phosphatase export to the periplasm of *E. coli. J. Bacteriol.* 167: 160–167.

Murby, M., Cedergren, L., Nilsson, J., Nygren, P., Hammarberg, B., Nilsson, B., Enfors, S. and Uhlen, M. (1991) Stabilization of recombinant proteins from proteolitic degradation in *E. coli* using dual affinity fusion strategy. *Biotechnol. & Applied Biochem.* 14:336–346.

Nelson, R., Ziegelhoffer, T., Nicolet, C., Werner-Washburne, M. and Craig, E. (1992) The translation machinery and 70kd heat shock protein cooperate in protein synthesis. *Cell* 71:97–105.

Noble, J., Innis, M., Koonin, E., Rudd, K., Banuett, F. and Herskowitz, I. (1993) The *Escherichia coli* hflA locus encodes a putative GTP-binding protein and two membrane proteins, one of which contains a protease-like domain. *Proc. Natl. Acad. Sci. USA* 90:10866–10870.

Parsell, D. and Sauer, R. (1989) Induction of heat shock-like response by unfolded protein in *E. coli*: dependence on protein level not protein degradation. *Genes Dev.* 3:1226–1232.

Parsell, D., Sanchez, Y., Stitzel, J. and Lindquist, S. (1991) Hsp104 is a highly conserved protein with two essential nucleotide-binding sites. *Nature* 353:270–273.

Parsell, D., Kowal, A., Singer, M. and Lindquist, S. (1994) Protein disaggregation mediated by heat-shock protein Hsp104. *Nature* 372:475–478.

Pfanner, N., Ostermann, J., Rassow, J., Hartl, F. and Neupert, W. (1990) Stress proteins and mitochondrial protein import. *Anton. Leeuwenhoek Int. J. Gen. M.* 58:191–193.

Philips, T., Van Bogelen, R. and Neidhardt, F. (1984) Lon gene product of *E. coli* is a heat shock protein. *J. Bacteriol.* 159:283–287.

Rohrwild, M., Coux O., Huang, H.C., Moerschell, R.P., Yoo, S.J., Seol, S.H., Chung, C.H. and Goldberg, A.L. (1996) HslV-HslU: a novel ATP-dependent protease complex in *Escherichia coli* related to the encaryotie proteasome. *Proc. Natl. Acad. Sci. USA* 93: 5808–5813.

Rowley, N., Prip-Buus, C., Westermann, B., Brown, C., Schwarz, E., Barrell, B. and Neupert, W. (1994) Mdj1p, a novel chaperone of the DnaJ family, is involved in mitochondrial biogenesis and protein folding. *Cell.* 77:249–259.

Sanders, S., Whitfield, K., Vogel, J., Rose, M. and Schekman, R. (1992) Sec61p and BiP directly facilitate polypeptide translocation into the ER. *Cell* 69:353–365.

Schwob, E. and Nasmyth, K. (1993) CLB5 and CLB6, an new pair of B cyclins involved in S-phase and mitotic spindle formation in *S. cerevisiae. Genes Dev.* 7:1160–1175.

Scidmore, M., Okamura, H. and Rose, M. (1993) Genetic interactions between Kar2 and Sec63, encoding eukaryotic homologues of DnaK and DnaJ in the endoplasmic reticulum. *Mol. Biol. Cell* 4:1145–1159.

Seufert, W. and Jentsch, S. (1990) Ubiquitin-conjugating enzymes UBC4 and UBC5 mediate selective degradation of short-lived and abnormal proteins. *EMBO J.* 9:543–550.

Sherman, M. and Goldberg, A. (1991) Formation *in vitro* of complexes between an abnormal fusion protein and the heat shock proteins from *E. coli* and yeast mitochondria. *J. Bacteriol.* 173:7249–7256.

Sherman, M. and Goldberg, A. (1992) Involvement of the chaperonin dnaK in the rapid degradation of a mutant protein in *E. coli. EMBO J.* 11:71–77.

Simons, J., Ferro-Novick, S., Rose, M. and Helenius, A. (1995) BiP/Kar2p serves as a molecular chaperone during carboxypeptidase Y folding in yeast. *J. Cell Biol.* 130:41–49.

Straus, D., Walter, W. and Gross, C. (1990) DnaK, dnaJ, and grpE heat shock proteins negatively regulate heat shock gene expression by controlling the synthesis and stability of sigma 32. *Genes Dev.* 4:2202–2209.

Stuart, R., Cyr, D., Craig, E. and Neupert, W. (1994) Mitochondrial molecular chaperones: their role in protein translocation. *Trends Biochem. Sci.* 19:87–89.

Suzuki, C., Suda, K., Wang, N. and Schatz, G. (1994) Requirement for the yeast gene LON in intramitochondrial proteolysis and maintenance of respiration. *Science* 264:891–891.

Tauer, R., Mannhaupt, G., Schnall, R., Pajic, A., Langer, T. and Feldmann, H. (1994) YTA10p, a member of a novel ATPase family in yeast, is essential for mitochondria function. *FEBS Lett.* 353:197–200.

Thompson, M. and Maurizi, M. (1994) Activity and specificity of *Escherichia coli* ClpAP protease in cleaving model peptide substrates. *J. Biol. Chem.* 269:18201–18208.

Tilly, K., Spence, J. and Georgopoulos, C. (1989) Modulation of stability of the *E. coli* heat shock regulatory factor sigma. *J. Bacteriol.* 171:1585–1589.

Todd, M., Viitanen, P. and Lorimer, G. (1994) Dynamics of the chaperonin ATPase cycle: implications for facilitated protein folding. *Science* 265:659–666.

Tomoyasu, T., Yuki, T., Morimura, S., Mori, H., Yamanaka, K., Niki, H., Hiraga, S. and Ogura, T. (1993) The *Escherichia coli* FtsH protein is a prokaryotic member of a protein family of putative ATPases involved in membrane functions, cell cycle control, and gene expression. *J. Bacteriol.* 175:1344–1351.

Tomoyasu, T., Gamer, J., Bukau, B., Kanemori, M., Mori, H., Rutman, A., Oppenheim, A., Yura, T., Yamanaka, K., Niki, H., Hiraga, S. and Ogura, T. (1995) *Escherichia coli* FtsH is a membrane bound, ATP-dependent protease which degrades the heat-shock transcription factor sigma(32). *EMBO J.* 14:2551–2560.

Tyers, M., Tokiwa, G. and Futcher, B. (1993) Comparison of the *Saccharomyces cerevisiae* G1 cyclins: Cln3 may be an upstream activator of Cln1, Cln2 and other cyclins. *EMBO J.* 12:1955–1968.

Varshavsky, A. (1992) The N-end rule. *Cell* 69:725–735.

Wagner, I., Arlt, H., van Dyck, L., Langer, T. and Neupert, W. (1994) Molecular chaperones cooperate with PIM1 protease in the degradation of misfolded proteins in mitochondria. *EMBO J.* 13:5135–5145.

Wawrzynow, A., Wojtkowiak, D., Marszalek, J., Banecki, B., Jonsen, M., Graves, B., Georgopoulos, C. and Zylicz, M. (1995) The ClpX heat-shock protein of *Escherichia coli*, the ATP-dependent substrate specificity component of the ClpP-ClpX protease, is a novel molecular chaperone. *EMBO J.* 14:1867–1877.

Waxman, L. and Goldberg, A. (1982) Protease La from *E. coli*. hydrolyzes ATP and proteins in a linked fashion. *Proc. Natl. Acad. Sci. USA* 79:4883–4887.

Waxman, L. and Goldberg, A. (1986) Selectivity of intracellular proteolysis: protein substrates activate the ATP-dependent protease (La). *Science* 232:500–503.

Weissman, J., Kashi, Y., Fenton, W. and Horwich, A. (1994) GroEL-mediated protein folding proceeds by multiple rounds of binding and release of nonnative forms. *Cell* 78:693–702.

Welsh, M., Denning, G., Ostedgaard, L. and Anderson, M. (1993) Dysfunction of CFTR bearing the delta F508 mutation. *J. Cell Sci.* 17:235–239.

Wickner, W. (1989) Secretion and membrane assembly. *Trends Biol. Sci.* 14:280–283.

Wickner, S., Hoskins, J. and McKenney, K. (1991) Monomerization of RepA dimers by heat shock proteins activates binding to DNA replication origin. *Proc. Natl. Acad. Sci. USA* 88:7903–7907.

Wickner, S. Skowyra, D., Hoskins, J. and McKenney, K. (1992) DnaJ, DnaK, and GrpE heat shock proteins are required in oriP1 DNA replication solely at the RepA monomerization step. *Proc. Natl. Acad. Sci. USA* 89:10345–10349.

Wickner, S., Gottesman, S., Skowyra, D., Hoskins, J., McKenney, K. and Maurizi, M. (1994) A molecular chaperone, ClpA, functions like DnaK and DnaJ. *Proc. Natl. Acad. Sci. USA* 91:12218–12222.

Wojtkowiak, D., Georgopoulos, C. and Zylicz, M. (1994) Isolation and characterization of ClpX, a new ATP-dependent specificity component of the Clp protease of *Escherichia coli*. *J. Biol. Chem.* 268:22609–22617.

Wild, J., Altman, E., Yura, T. and Gross, C. (1992) DnaK and dnaJ heat shock proteins participate in protein export in *E. coli*. *Genes Dev.* 6:1165–1172.

Yaglom, J., Linskens, M., Sadis, S., Rubin, D., Futcher, B. and Finley, D. (1995) p34[Cdc28] mediated control of Cln3 cyclin degradation. *Mol. Cell Biol.* 15:731–741.

Yaglom, J., Goldberg, A.L., Finley, D. and Sherman, M.Y. (1996) The Molecular chaperone Ydj1 is required for p34[cdc28]-dependent phosphorylation of G1 cyclin, Clu3, that signals its degradation. *Mol. Cell. Biol.* 16:3679–3684.

Yura, T., Kawasaki, Y., Kusukawa, N., Nagai, H., Wada, C. and Yano, R. (1990) Roles and regulation of the heat shock sigma factor sigma 32 in *Escherichia coli*. *Anton. Leeuwenhoek Int. J. Gen. M.* 58:187–190.

Stress-Inducible Cellular Responses
ed. by U. Feige, R.I. Morimoto, I. Yahara and B. Polla
© 1996 Birkhäuser Verlag Basel/Switzerland

Molecular chaperoning of steroid hormone receptors

W. B. Pratt[1], U. Gehring[2] and D. O. Toft[3]

[1] *Department of Pharmocology, University of Michigan, Medical Science Research Building III, Ann Arbor, MI 48109-0626, USA*
[2] *Institut für Biologische Chemie, Universität Heidelberg, Im Neuenheimer Feld 501, D-69120 Heidelberg, Germany*
[3] *Department of Biochemistry and Molecular Biology, Mayo Clinic, Rochester, MN 55905, USA*

Summary. The study of the large, unactivated form of steroid receptors has led to the discovery of an hsp90/hsp70-based multicomponent protein folding system(s). For steroid receptors, the hsp90 chaperone system determines both repression of transcriptional activity in the absence of hormone and the proper folding of the hormone binding domain to produce the steroid binding conformation. Like steroid receptors, a number of other regulators of transcription and some protein kinases are now known to be associated with hsp90. Given the abundance of the proteins comprising the hsp90 chaperone system and the apparent ubiquity of the system in the animal and plant kingdoms, this system is thought to serve a fundamental role for protein folding, function and possibly trafficking within the cytoplasm and nucleus. In this chapter, we discuss the work on steroid receptor heterocomplex composition that has led to the discovery of new chaperone proteins and we summarize the mechanistic information developed in cell-free studies of receptor heterocomplex assembly.

Introduction

The steroid receptors are direct signal transducers in that they receive the signal by binding the hormone, an event that triggers their transformation from an inactive to an active state, and in their hormone-activated state they bind to response elements in the genome where they alter the transcription rates of specific genes. Separate domains of the receptors are responsible for signal reception and DNA binding. Nevertheless, the hormone binding domain (HBD) and the DNA binding domain cooperate intimately in bringing about the hormonal response in target cells, and deletion of the HBD yields receptors that are constitutive transcriptional activators (Evans, 1988). Thus, the HBD regulates the transcriptional activating activity of the rest of the receptor, and to do so, it can be joined either covalently, as in the natural situation, or noncovalently via a c-Jun/c-Fos leucine zipper introduced into separate constructs (Spanjaard and Chin, 1993). The regulatory function is transposable in that the HBD can regulate the activity of nonreceptor proteins. Picard et al. (1988) first showed that the HBD acts as a regulatory unit that can confer hormone responsiveness onto other proteins.

The HBD also determines the tight association of the receptor with the hsp90 component of a chaperone system that is responsible for its proper folding and function (Pratt, 1993; Smith and Toft, 1993). In the case of chimaeras, the HBD confers hormone-regulated binding to hsp90 onto the fusion protein (Scherrer et al., 1993). In the absence of hormone, a steroid receptor or a fusion protein remains "docked" via the HBD to hsp90. Steroid binding relaxes the tight association with hsp90, permitting the receptor to proceed to appropriate response elements, dimerize and bind to DNA (not necessarily in that order). The overall model, then, is that steroid receptors have evolved a tight interaction with the hsp90 component of a general protein folding system and the resulting multiprotein receptor heterocomplex probably represents a normal transition state in the folding process. Thus, the first step in the signal transduction pathway likely reflects an evolutionary process in which procession from this folding intermediate has been brought under hormonal control.

Our understanding of steroid receptor-chaperone protein interactions evolved from the study of cytosolic receptors. Rupturing of cells under low salt conditions and in the cold yields receptor structures with molecular weights in the 300 to 400 kDa range that are greatly in excess of those of the respective receptor polypeptides. These large complexes bind steroid but they do not bind to DNA, and they are called "unactivated" or "untransformed" receptors. Cytosolic receptors can be "activated" or "transformed" to the DNA binding state by a variety of treatments, most notably by warming or by exposure to conditions of high ionic strength. The molecular weights of the activated receptors (60–110 kDa) are similar to those of the respective receptor polypeptides themselves, suggesting that activation in cytosol results in dissociation of complexes to yield the receptor polypeptides with no other components associated.

These large, unactivated forms of the receptors are now known to be multiprotein receptor heterocomplexes with hsp90 (Smith and Toft, 1993; Pratt, 1993). This chapter focuses on these heteromeric receptor structures and the way they are formed by an hsp90/hsp70-based protein chaperone system. Other important aspects of steroid receptor action, such as nucleocytoplasmic shuttling, receptor dimerization and binding to DNA, and receptor interactions with the transcriptional apparatus of the cell are not considered here.

Structure of the unactivated receptors

Two lines of evidence led to the realization that the large, unactivated receptor forms were heterocomplexes. First, affinity purification of molybdate-stabilized progesterone and glucocorticoid receptors resulted in purification of a 90 kDa protein that was subsequently shown by site-specific affinity labeling to be different from the steroid binding proteins themselves (Dougherty et al., 1984; Renoir et al., 1984; Housley et al., 1985). An antibody raised against this 90 kDa protein was shown to shift the 9S untransformed form of several steroid receptors to a more rapidly sedi-menting peak on sucrose gradient centrifugation (Joab et al., 1984). Second, it was shown that the unactivated glucocorticoid receptor complex contains only one molecule of receptor polypeptide (Gehring and Arndt, 1985; Okret et al., 1985). The 90 kDa receptor-associated protein was identified as hsp90 (Catelli et al., 1985; Sanchez et al., 1985; Schuh et al., 1985), and there is now a general consensus that unactivated receptors con-tain a dimer of hsp90.

There is, however, a significant difference in molecular weights between hetero-trimeric structures of two molecules of hsp90 in association with one receptor polypeptide and experimentally determined values obtained for unactivated receptors. For example, the molecular weights of molyb-date-stabilized glucocorticoid and estrogen receptor complexes are roughly 330 and 300 kDa, respectively, as determined from Stokes' radii and sedimentation coefficients (Gehring et al., 1987; Segnitz and Gehring, 1995), while those of the respective trimeric structures would amount to only about 280 and 245 kDa. This difference is accounted for by the pre-sence of yet other proteins in the receptor heterocomplexes (Smith and Toft, 1993; Pratt, 1993).

Chemical cross-linking has proven to be a particularly useful approach in examining the composition and stoichiometry of the receptor hetero-complexes. Bifunctional reagents have been used to stabilize the high molecular weight forms of glucocorticoid, progesterone and estrogen receptors (Aranyi et al., 1988; Rexin et al., 1988a; Rehberger et al., 1992; Segnitz and Gehring, 1995). The molecular weights of the cross-linked receptor heterocomplexes were identical with those of molybdate-stabilized receptors, and subsequent cleavage of the cross-links yielded a receptor that could be dissociated with generation of DNA binding activity. Pro-gressive cross-linking of the receptor subunits following affinity labeling of glucocorticoid, progesterone and estrogen receptors revealed a series of labeled bands on denaturing gel electrophoresis with molecular weights gradually shifting from those of the receptor polypeptides them-selves to fully cross-linked hetero-tetramers of one receptor polypeptide in association with two molecules of hsp90 and one additional subunit of roughly 60 kDa. This additional cross-linked subunit is FKBP52, an immunophilin of the FK506-binding class that is discussed below.

Another heat shock protein that is easily detectable in various amounts in purified receptor preparations is hsp70 (Kost et al., 1989; Sanchez et al., 1990b; Rexin et al., 1991; Onate et al., 1991). It is not surprising that hsp70 should be present in some heterocomplexes because, as discussed below, it is a required component of the chaperone system that forms the complex. Like hsp90, hsp70 binds to the receptor HBD (Schowalter et al., 1991; Scherrer et al., 1993), but in contrast to hsp90, hsp70 does not complete-ly dissociate upon receptor activation. Hsp70 does not become cross-linked to the receptor (Rexin et al., 1991; Segnitz and Gehring, 1995), the

stoichiometry is unknown, and it is likely that only a portion of the untransformed receptors purified from cytosol is bound to hsp70.

An important point can be made here regarding differences in unactivated receptor structure. Cross-linking studies reveal a common heterotetrameric (1 receptor·2 hsp90·1 FKBP52) structure for glucocorticoid
(Rexin et al., 1991), progesterone (Rehberger et al., 1992) and estrogen
(Segnitz and Gehring, 1995) receptors, but that unit can be regarded as a
"core" structure and differences in heterocomplex structure may exist. For
example, the unactivated mouse glucocorticoid receptor of L cell is not
recovered in association with hsp70, yet the mouse receptor overexpressed
in Chinese hamster ovary cells is found in association with hsp70 (Sanchez
et al., 1990b). Even the core hetero-tetramer varies with respect to the
immunophilin component. For example, in extracts of human HeLa cells,
two forms of these non-activated glucocorticoid receptor were detected,
one with and one without FKBP52 (Alexis et al., 1992). As reviewed below,
receptor heterocomplex assembly can be viewed as a dynamic process
(Smith, 1993) and some components like hsp70 become tightly bound to
the receptor and then dissociate during the assembly process, generating
additional transient heterocomplex states (Smith et al., 1992).

The cross-linking technique has been used to demonstrate the existence
of heterotetrameric structures in intact cells (Rexin et al., 1988b; Rexin
et al., 1992; Segnitz and Gehring, 1995). These structures have the same
molecular weights and composition detected for molybdate-stabilized
unactivated glucocorticoid and estrogen receptors in cytosol preparations.
Taken together, the structural studies support the concept that the heterocomplex is the form of the receptor in the target cell that receives the
hormonal signal, and that binding of steroid to this structure is the first step
in hormone action.

Significance of hsp90 binding for receptor functions

Hsp90 is an abundant, essential cytosolic protein in eukaryotic cells, and it
appears to be a general molecular chaperone involved in folding/unfolding
of various proteins (Jakob and Buchner, 1994). The essential role of
hsp90 for receptor function in the cell was made evident by experiments in
yeast cells transfected with glucocorticoid or estrogen receptors (Picard
et al., 1990). Reduction of intracellular hsp90 levels resulted in impaired
hormonal responsiveness.

The interaction between receptors and hsp90 has been investigated by
mutational analysis of both components. A rather large portion of the HBD
participates in binding (Pratt, 1993; Cadepond et al., 1991; Simons, 1994).
Since no specific binding motif has been detected, it appears that multiple
sites of contact are involved. On the part of hsp90, several point mutations and deletions have also been studied (Bohen and Yamamoto, 1993;

Cadepond et al., 1993; Sullivan and Toft, 1993). While the carboxyl terminal portion of hsp90 is clearly involved in forming receptor complexes, some other parts of the molecule may also participate. Moreover, temperature-sensitive point mutations in the hsp90 gene of yeast were found to affect glucocorticoid receptor function (Nathan and Lindquist, 1995). Interestingly, different point mutations affect different receptors to different extents. These studies also show that association of hsp90 *per se* is not sufficient for eliciting hormone responsiveness (Bohen and Yamamoto, 1993).

The HSPs contained in nonactivated receptors, most notably hsp90, appear to play a dual role in modulating steroid receptor function. One function is to maintain the receptor in the unactivated form in the absence of steroidal ligands and a second function is to maintain the HBD in the proper conformation for high affinity steroid binding activity. This latter function has been demonstrated for glucocorticoid (Bresnick et al., 1989; Nemoto et al., 1990) and mineralocorticoid receptors (Caamaño et al., 1993), as well as for the dioxin receptor (Antonsson et al., 1995), which is in a different nuclear receptor class. It would seem that the hsp90/hsp70-based chaperone system is required to *fold* the receptor HBD to the appropriate steroid binding conformation and that the steroid binding sites of the glucocorticoid and mineralocorticoid receptors immediately *collapse* to a nonsteroid-binding conformation when separated from the hsp90 component of the chaperone system.

Essentially three models have been advanced to explain maintenance of the unactivated receptor. The most commonly acknowledged model can be called a *steric interference* model where occlusion of the DNA binding domain may be caused by direct protein-protein contacts involving hsp90. Because the receptor HBD carries an inactivation function that operates on activities of structurally different proteins (some without DNA binding activity), a model of conferrable hormone regulation based solely on steric interference of receptor function by hsp90 is probably not adequate.

Two models based on a role for hsp90 in receptor folding do accommodate a range of regulated structures and activities. Picard et al. (1988) proposed that binding of hsp90 to the HBD within a protein chimaera causes the polypeptide as a whole to assume an "unfolded" (i.e., partially unfolded) conformation that is reversed on hormone binding and receptor activation. This model assumes, then, that the function of the chimaera is inactivated because hsp90 determines an unfolded conformation in the region that is regulated as well as in the HBD that is conferring the regulation. However, Spanjaard and Chin (1993) have demonstrated reconstitution of hormone-mediated glucocorticoid receptor activity by expressing as individual proteins a fragment of the glucocorticoid receptor containing the *trans*-activation and DNA binding domains and a fragment containing the HBD, with each fragment being fused to either a c-Jun or c-Fos leucine zipper. As each fragment was translated and folded independently, this observation argues strongly against a model in which the HBD causes the polypeptide as a whole to assume an "unfolded" conformation.

The third model for maintenance of an inactive state in the absence of ligand has been called the *docking* model (Pratt, 1992) because it proposes that receptors are bound to the heat shock protein heterocomplex, to which they remain "docked" until released by steroid-mediated reversal of their unfolded conformation. In the docking model, hsp90 has to determine an unfolded conformation of only the HBD. In this case, the protein that is brought under control may have a properly folded active site but is biologically inactive until it is released from the heat shock protein structure and allowed to proceed to its ultimate site of action.

Immunophilin components of receptor heterocomplexes

FKBP52 is one of three immunophilins that have been identified in unactivated steroid receptor heterocomplexes. The immunophilins are ubiquitous and conserved proteins that bind immunosuppressant drugs, such as FK506, rapamycin and cyclosporin A (Walsh et al., 1992). All members of the immunophilin protein family have peptidylprolyl isomerase (PPIase) activity, suggesting that they may play a role in protein folding in the cell (Schmid, 1993).

FKBP52 (also called p59 or hsp56) was discovered when an antibody developed against the partially purified, untransformed progesterone receptor complex was found to react with a 59 kDa rabbit protein, p59 (Tai and Faber, 1985), that was subsequently shown to be associated with unactivated estrogen, progestin, androgen and glucocorticoid receptors (Tai et al., 1986). The human protein was found at increased levels under conditions of chemical and thermal stress (Sanchez, 1990), hence the term hsp56.

FKBP52 has been shown to have PPIase activity that is inhibited by FK506 (Peattie et al., 1992). When cytosol is passed through an FK506 affinity matrix, the unactivated glucocorticoid receptor and hsp90 are coisolated with FKBP52 (Tai et al., 1992), suggesting that the PPIase site is accessible within the receptor holostructure. FKBP52 binds directly to hsp90 (Czar et al., 1994b) via repetitive sequence motifs of 34 amino acids called tetratricopeptide repeat domains or TPR domains (Radanyi et al., 1994). However, cross-linking experiments, suggest that FKBP52 lies in close proximity to both the receptor and hsp90. Another FK506 binding protein, FKBP54, has been recently shown to be closely related to FKBP52 and to exist in progesterone receptor complexes (Smith et al., 1993a).

In addition to FKBP52 and the newly recognized FKBP54, a 40 kDa cyclosporin A-binding immunophilin has also been identified in steroid receptor heterocomplexes. This protein was found copurifying with the bovine estrogen receptor (Ratajczak et al., 1993), and it was found to be the same as cyclophilin 40 (CyP-40) cloned from a human library (Kieffer et al., 1993). CyP-40 has also been found in native progesterone and glucocorticoid receptor heterocomplexes (Milad et al., 1995; Renoir et al., 1995; Owens-Grillo et al., 1995). CyP-40 and FKBP52 bind to the same site on hsp90 (Owens-Grillo et al., 1995) and this site appears to be a general TPR domain binding site.

At this time, the role of the immunophilins in steroid receptor function is not defined. The prolyl isomerase activity of FKBPs and CyP-40 certainly points to a role in protein folding, but neither FK506 nor cyclosporin A affect assembly of the glucocorticoid receptor heterocomplex and conversion of the receptor to the steroid binding form (Hutchison et al., 1993; Owens-Grillo et al., 1995). We should note that the binding of FKBP54 and CyP-40 to receptor complexes has not yet been demonstrated with crosslinking techniques and further studies are needed.

Pratt et al. (1993) have suggested that FKBP52 may be involved in the intracellular transport of steroid hormone receptors through its interaction with nuclear localization signals via a conserved electrostatically complementary negative sequence. The majority of FKBP52 in the cell is located in the nucleus, where by double labeling and confocal imaging in WCL2 cells its localization is identical to that of the glucocorticoid receptor (Czar et al., 1994a). Such a localization might be consistent with a role in nuclear trafficking. Also consistent with a trafficking role is the observation that injection into L cells of an antibody raised against the conserved negative sequence of FKBP52 impedes the subsequent dexamethasone-mediated shift of the glucocorticoid receptor into the nucleus (Czar et al., 1995). Both FK506 and cyclosporin A have been found to potentiate steroid receptor-mediated gene transcription (Ning and Sanchez, 1993; Renoir et al., 1995; Tai et al., 1994; Milad et al., 1995). However, it has not yet been demonstrated that potentiation results from immunosuppressant drug binding to receptor-associated immunophilins. In some systems potentiation of receptor-mediated transcription has not been seen (Hutchison et al., 1993; Gehring, 1995), and the precise role of immunophilin chaperones in steroid receptor function remains cryptic.

Assembly of Receptor Complexes

Earlier attempts to bind receptors to hsp90 *in vitro* were not successful because the binding process was not a simple interaction between the two components. This was demonstrated with the progesterone receptor and the glucocorticoid receptor where assembly could be accomplished in rabbit reticulocyte lysate (Smith et al., 1990b; Scherrer et al., 1990). In this system, the receptor was isolated from target cells by adsorption to antibody resin, stripped of its associated proteins, and incubated in reticulocyte lysate. Assembly was shown to require Mg^{2+}, K^+, a continuous supply of

ATP and an incubation at elevated temperature (30–60 min. at 30 °C) (Hutchison et al., 1992a; Smith et al., 1992). ATP hydrolysis was necessary since non-metabolizable analogs would not support assembly. These results were very gratifying since they indicated the need for a specific bio-chemical process. This offered an explanation for the earlier lack of success and also provided more evidence that such complexes were bio-logically significant and not simply fortuitous interactions of sticky proteins.

Two complementary end-points have been used in these *in vitro* studies. The composition of the reconstituted complex was analyzed by SDS-PAGE and, under optimal conditions, this composition closely resembled that of receptor complexes from target cells. Thus, they contained receptor bound to hsp90, some hsp70, and the immunophilin FKBP52. Some additional receptor-associated proteins were more clearly revealed using this *in vitro* system such as FKBP54, CyP-40 and a 23 kDa protein, p23. p23 is not a heat shock protein or an immunophilin. It binds specifically to hsp90, but its function is unknown (Johnson and Toft, 1994). Two additional proteins, p60 and p48, have been observed to bind transiently during early stages of receptor complex formation. A second important parameter is the hormone binding activity of the receptor. For some receptors, this activity decays rapidly following the removal of hsp90 as noted above, but the activity is restored upon assembly of the complex *in vitro*.

At this stage, there are only hints as to the mechanism of the assem-bly process. Several components appear to participate in a cooperative manner. For example, the removal of hsp70 from reticulocyte lysate (Hutchison et al., 1994b) or its neutralization with an antibody (Smith et al., 1992) eliminates the binding of other proteins to the receptor. The removal of p23 also greatly reduces complex formation (Johnson and Toft, 1994).

While these studies are still incomplete, encouraging efforts have been made to isolate the proteins from reticulocyte lysate that are needed for receptor complex assembly. The focus has been on hsp90 which appears to exist in both a free state and also in complexes with other proteins. The immune adsorption of hsp90 from reticulocyte lysate and other cell extracts copurifies several proteins including hsp70 and some immuno-philins (Sanchez et al., 1990a; Perdew and Whitelaw, 1991; Smith et al., 1993b). Using this technique, Hutchison et al. (1994a) have isolated an hsp90 complex which has the ability to bind glucocorticoid receptor and restore its hormone binding activity in an ATP-dependent manner. Thus, the necessary components for assembly appear to exist in a performed complex with hsp90. All of the components of this com-plex, called a *foldosome*, have not yet been identified, but they include hsp70, p60, FKBP52 and p23. p23 is easily removed by extensive wash-ing and must be added back for successful assembly (Hutchison et al., 1995).

Using a different selection of antibodies, two types of hsp90 complexes have been identified which are both likely to exist in the foldosome complex cited above. An antibody to p23 has been used to isolate complexes that contain hsp90, the three immunophilins, FKBP52, FKBP54, and CyP-40, and minor amounts of hsp70 (Johnson and Toft, 1994). An interesting feature of this complex is that its assembly and maintenance requires ATP/Mg^{++}. However, unlike receptor complexes, ATP hydrolysis is not required since AMPPNP will support complex formation. Thus, one or more components of the p23 complex must contain bound ATP. Since this complex contains all of the major proteins of the receptor complex, it has been suggested that the final step in the assembly of the receptor complex is its association with the p23 complex accompanied by ATP hydrolysis (Johnson and Toft, 1995).

Another abundant hsp90 complex has been isolated that contains hsp90, hsp70, and p60 (Smith et al., 1993b). p60 was originally identified by Honoré et al. (1992) as a protein that is up-regulated by viral transformation and was given the code name IEF SSP 3521. Its homolog in yeast, Sti1, is a non-essential heat shock protein (Nicolet and Craig, 1989). p60 is of further interest because it and a less characterized protein p48 have been observed in progesterone receptor complexes at early stages of assembly (Smith, 1993) or when ATP is limiting (Smith et al., 1992). Thus, these proteins may be intermediates in the assembly process. One possibility is that initial recognition of receptor involves the hsp90, hsp70, p60 complex and this is replaced with or converted to the p23 type of complex containing immunophilins and p23, but little hsp70 or p60. It should be noted that all of these complexes are not static, but appear to be continuously dissociating and re-associating (Smith, 1993). Thus, it is possible that individual components could be gained or lost during the course of receptor complex assembly.

A perplexing feature of this system is the large number of proteins that has been implicated as components of receptor complexes (hsp90, hsp70, FKBP52, FKBP54, CyP-40, and p23) or as intermediates in complex formation (p60, p48). The importance of the components may vary from one receptor to another, some may represent alternate pathways to complex formation, and some components may not be required for the assembly process but may have a role subsequent to assembly, such as in nuclear transport or receptor processing. Additional studies are needed to identify those constituents that are required for receptor complex assembly, to describe the roles of individual components and the order of steps in the assembly process. It should be noted that although all of the published studies of cell-free receptor heterocomplex assembly have been carried out with rabbit reticulocyte lysate, the assembly system appears to be ubiquitous in that concentrated cell lysates from other mammalian, insect, and even plant cells fold the glucocorticoid receptor into a functional heterocomplex with hsp90 (Stancato et al., 1996a). Also, it has been shown that purified plant hsp70 interacts with mammalian chaperone proteins (Stancato et al., 1996a) and purified human p23 interacts with plant chaperone proteins (Hutchison et al., 1995), suggesting there exists a high conservation of the protein-protein interactions and of chaperoning mechanism between the animal and plant kingdoms.

Hsp90 as a component of other heteromeric protein structures

Although steroid receptor complexes have been a major focus for studies on the function of hsp90, several other classifications of proteins are now known to associate with hsp90. The first to be described was the transient association of the oncogenic protein kinase pp60$^{v\text{-}src}$ with hsp90 during its posttranslational processing (Brugge, 1981, Opperman et al., 1981). pp60^{src}·hsp90 complexes can also be reconstituted in reticulocyte lysate by an ATP/Mg^{2+}-dependent and K$^+$-dependent process that appears similar to that assembling steroid receptor·hsp90 complexes (Hutchison et al., 1992b). The Src-associated mitogenic signaling protein Raf is also present in a native complex with hsp90 that can be reconstituted in reticulocyte lysate (Stancato et al., 1993; Wartmann and Davis, 1994). However, the native steroid receptor·hsp90 and the protein kinase·hsp90 complexes are not identical. The Src and Raf complexes contain a unique 50 kDa protein, p50, that is not present in steroid receptor heterocomplexes, and they do not contain the FKBPs (FKBP52 and FKBP54) present in steroid receptor complexes. However, the Raf·hsp90 complex does bind [^{3}H]FK506 and therefore contains a novel immunophilin, which would be either p50 or an as yet unidentified protein (Stancato et al., 1994).

Table 1 lists several regulatory proteins that have been reported to associate with hsp90. The list is not exhaustive, but includes a number of proteins that are either regulators of transcription or protein kinases. Except for the glucocorticoid and progesterone receptors, little is known about the assembly process for these complexes and their actual composition. These may all reveal a general scheme of complex formation using the same basic machinery, but any differences that are observed should be quite instructive.

A key question is whether these complexes vary in purpose or fit a common theme of function. With steroid receptors, hsp90 complexes are inhibitory in maintaining the receptors in an unactivated state, but they have an additional purpose in managing the conformation of the receptor so that it can bind and respond to hormone. Thus, there is a negative and a positive aspect to their association. This is also the case for dioxin receptor complexes (Antonsson et al., 1995). pp60^{src} has been shown to be inactive as a protein kinase when in complex with hsp90 (Brugge, 1981), but studies in yeast indicate that hsp90 is also needed for activity, either for the

Table 1. hsp90-associated proteins. The table does not include immunophilins and other chaperones that bind to hsp90.

Associated protein	Reference
Transcription factors	
Glucocorticoid receptor	Sanchez et al., 1985
Progesterone receptor	Schuh et al., 1985; Catelli et al., 1985
Estrogen receptor	Redeuilh et al., 1987
Androgen receptor	Veldscholte et al., 1992
Mineralocorticoid receptor	Rafestin-Oblin et al., 1989
Ah receptor	Perdew, 1988; Denis et al., 1988
Myo D1	Shue and Kohtz, 1994
Heat shock factor	Nadeau et al., 1993
CN-tumor promotor-specific binding protein	Hashimoto and Shudo, 1991
Protein Kinases	
Tyrosine kinases	
v-src, c-src	Brugge et al., 1981; Oppermann et al., 1981; Hutchison et al., 1992b
v-fps	Lipsich et al., 1982
v-yes	Lipsich et al., 1982
v-fes	Ziemiecki et al., 1986
v-frg, c-frg	Ziemiecki et al., 1986; Hartson and Matts, 1994
lck	Hartson and Matts, 1994
Heme-regulated eIF-2α kinase	Rose et al., 1987; Matts and Hurst, 1989
eEF-2 kinase	Palmquist et al., 1994
Casein kinase II	Miyata and Yahara, 1992
v-Raf, c-Raf, Gag-Mil	Stancato et al., 1993; Wartmann and Davis, 1994; Lovric et al., 1994
MAP kinase kinase (Mek)	Stancato et al. (1996b)

production of active pp60^{src} or for a downstream effect in the mitogenic signaling pathway (Xu and Lindquist, 1993).

Further indication of the chaperoning role of hsp90 complexes was provided recently using the drug geldanamycin (Whitesell et al., 1994). This drug reverses the transformation phenotype of cells infected with v-Src and it appears to do this by binding to hsp90 and disrupting its complex with pp60^{v-src}. Geldanamycin has also been found to disrupt p23·hsp90 complexes and to dissociate some proteins from steroid receptor complexes (Johnson and Toft, 1994; Whitesell and Smith, personal communication). Thus, it will be important to further characterize the mechanism of geldanamycin and to test its effects in a variety of systems where hsp90 interactions are thought to occur.

Acknowledgements
The studies of the authors reported herein were supported by NIH grants DK31573 and CA28010 to W.B.P. and HD09140 and DK46249 to D.O.T. and by grants from the Deutsche Forschungsgemeinschaft to U.G.

References

Alexis, M.N., Mavridou, I. and Mitsiou, D.J. (1992) Subunit composition of the untransformed glucocorticoid receptor in the cytosol and in the cell. *Eur. J. Biochem.* 204:75–84.

Antonsson, C., Whitelaw, M.L., McGuire, J., Gustafsson, J.A. and Poellinger, L. (1995) Distinct roles of the molecular chaperone hsp90 in modulating dioxin receptor function via the basic helix-loop-helix and PAS domain. *Mol. Cell. Biol.* 15:756–765.

Arànyi, P., Radanyi, C., Renoir, M., Devin, J. and Baulieu E.-E. (1988) Covalent stabilization of the nontransformed chick oviduct cytosol progesterone receptor by chemical cross-linking. *Biochemistry* 27:1330–1336.

Bohen, S.P. and Yamamoto, K.R. (1993) Isolation of hsp90 mutants by screening for decreased steroid receptor function. *Proc. Natl. Acad. Sci. USA* 90:11424–11428.

Bresnic, E.H., Dalman, F.C., Sanchez, E.R. and Pratt, W.B. (1989) Evidence that the 90-kDa heat shock protein is necessary for the steroid binding conformation of the L cell glucocorticoid receptor. *J. Biol. Chem.* 264:4992–4997.

Brugge, J.S., Erikson, E. and Erikson, R.L. (1981) The specific interaction of the Rous sarcoma virus transforming protein, pp60^{src}, with two cellular proteins. *Cell* 25:363–372.

Caamaño, C.A., Morano, M.I., Patel, P.D., Watson, S.J. and Akil, H. (1993) A bacterially expressed mineralocorticoid receptor is associated *in vitro* with the 90-kilodalton heat shock protein and shows typical hormone- and DNA-binding characteristics. *Biochemistry* 32: 8589–8595.

Cadepond, F., Schweizer-Groyer, G., Segard-Maurel, I., Jibard, N., Hollenberg, S.M., Giguere, V., Evans, R.M. and Baulieu, E.-E. (1991) Heat shock protein 90 as a critical factor in maintaining glucocorticoid receptor in a nonfunctional state. *J. Biol. Chem.* 266:5834–5841.

Cadepond, F., Binart, N., Chambraud, B., Jibard, N., Schweizer-Groyer, G., Segard-Maruel, I. and Baulieu, E.-E. (1993) Interaction of glucocorticoid receptor and wild-type or mutated 90-kDa heat shock protein coexpressed in baculovirus-infected Sf9 cells. *Proc. Natl. Acad. Sci. USA* 90:10434–10438.

Catelli, M.G., Binart, N., Jung-Testas, I., Renoir, J.M., Baulieu, E.-E., Feramisco, J.R. and Welch, W.J. (1985) The common 90-kd protein component of non-transformed "8S"steroid receptors is a heat-shock protein. *EMBO J.* 4:3131–3135.

Czar, M.J., Lyons, R.H., Welsh, M.J., Renoir, J.-M. and Pratt, W.B. (1995) Evidence that the FK506-binding immunophilin hsp56 is required for trafficking of the glucocorticoid receptor from the cytoplasm to the nucleus. *Mol. Endo.* 9:1549–1560.

Czar, M.J., Owens-Grillo, J.K., Yem, A.W., Leach, K.L., Deibel, M.R., Welsh, M.J. and Pratt, W.B. (1994a) The hsp56 immunophilin component of untransformed steroid receptor complexes is localized both to microtubules in the cytoplasm and to the same nonrandom regions within the nucleus as the steroid receptor. *Mol. Endo.* 8:1731–1741.

Czar, M.J., Owens-Grillo, J.K., Dittmar, K.D., Hutchison, K.A., Zacharek, A.M., Leach, K.L., Deibel, M.R. and Pratt, W.B. (1994b) Characterization of the protein-protein interactions determining the heat shock protein (hsp90·hsp70·hsp56) heterocomplex. *J. Biol. Chem.* 269:11155–11161.

Denis, M., Cuthill, S., Wikstrom, A.-C., Poellinger, L. and Gustafsson, J.-A. (1988) Association of the dioxin receptor with the M$_r$ 90,000 heat shock protein: A structural kinship with the glucocorticoid receptor. *Biochem. Biophys. Res. Commun.* 155:801–807.

Dougherty, J.J., Puri, R.K. and Toft, D.O. (1984) Polypeptide components of two 8S forms of chicken oviduct progesterone receptor. *J. Biol. Chem.* 259:8004–8009.

Evans, R.M. (1988) The steroid and thyroid hormone receptor superfamily. *Science* 240: 389–395.

Gehring, U. and Arndt, H. (1985) Heteromeric nature of glucocorticoid receptors. *FEBS Lett.* 179:138–142.

Gehring, U., Mugele, K., Arndt, H. and Busch, W. (1987) Subunit dissociation and activation of wild-type and mutant glucocorticoid receptors. *Mol. Cell. Endocrinol.* 53: 33–44.

Gehring, U. (1995) Structure of the nonacitvated glucocorticoid receptor and activation to DNA binding. *In:* B. Gametchu (ed.): *Glucocorticoid Receptor Structure and Leukemic Cell Responses.* Molecular Biology Intelligence Unit, R.G. Landes Company, Austin, pp 33–56.

Hartson, S.D. and Matts, R.L. (1994) Association of hsp90 with cellular *Src*-family kinases in a cell-free system correlates with altered kinase structure and function. *Biochemistry* 33: 8912–8920.

Hashimoto, Y. and Shudo, K. (1991) Cytosolic-nuclear tumor promotor-specific binding protein: Association with the 90 kDa heat shock protein and translocation into nuclei by treatment with 12-O-tetradecanoylphorbol 13-acetate. *Jpn. J. Cancer Res.* 82:665–675.

Honoré, B., Leffers, H., Madsen, P., Rasmussen, H.H., Vandekerckhove, J. and Celis, J.E. (1992) Molecular cloning and expression of a transforming-sensitive human protein containing the TPR motif and sharing identity to the stress-inducible yeast protein STI1. *J. Biol. Chem.* 267:8485–8491.

Housley, P.R., Sanchez, E.R., Westphal, H.M., Beato, M. and Pratt, W.B. (1985) The molybdate-stabilized L cell glucocorticoid receptor isolated by affinity chromatography or with a monoclonal antibody is associated with a 90–92 kDa non-steroid-binding phosphoprotein. *J. Biol. Chem.* 260:13810–13817.

Hutchison, K.A. Czar, M.J., Scherrer, L.C. and Pratt, W.B. (1992a) Monovalent cation selectivity for ATP-dependent association of the glucocorticoid receptor with hsp70 and hsp90. *J. Biol. Chem.* 267:14047–14053.

Hutchison, K.A., Brott, B.K., DeLeon, J.H., Perdew, G.H., Jove, R. and Pratt, W.B. (1992b) Reconstitution of the multiprotein complex of pp60^{src}, hsp90 and p50 in a cell-free system. *J. Biol. Chem.* 267:2902–2908.

Hutchison, K.A., Scherrer, L.C., Czar, M.J., Ning, Y., Sanchez, E.R., Leach, K.L., Deibel, M.R. and Pratt, W.B. (1993) FK506 binding to the 56-kilodalton immunophilin (hsp56) in the glucocorticoid receptor heterocomplex has no effect on receptor folding or function. *Biochemistry* 32:3953–3957.

Hutchison, K.A., Dittmar, K.D. and Pratt, W.B. (1994a) All of the factors required for assembly of the glucocorticoid receptor into a functional heterocomplex with heat shock protein 90 are preassociated in a self-sufficient protein folding structure, a "foldosome". *J. Biol. Chem.* 269:27894–27899.

Hutchison, K.A., Dittmar, K.D., Czar, M.J. and Pratt, W.B. (1994b) Proof that hsp70 is required for assembly of the glucocorticoid receptor into a heterocomplex with hsp90. *J. Biol. Chem.* 269:5043–5049.

Hutchison, K.A., Stancato, L.F., Owens-Grillo, J.K., Johnson, J.L., Krishna, P., Toft, D.O. and Pratt, W.B. (1995) The 23 kDa acidic protein in reticulocyte lysate is the weakly bound component of the hsp foldosome that is required for assembly of the glucocorticoid receptor into a functional heterocomplex with hsp90. *J. Biol. Chem.* 270:18841–18847.

Jakob, U. and Buchner, J. (1994) Assisting spontaneity: the role of hsp90 and small hsps as molecular chaperones. *Trends Biochem. Sci.* 19:205–211.

Joab, I., Radanyi, C., Renoir, J.-M., Buchou, T., Catelli, M.-G., Binart, N., Mester, J. and Baulieu, E.-E. (1984) Common non-hormone binding component in non-transformed chick oviduct receptors for four steroid hormones. *Nature* 308:850–853.

Johnson, J.L. and Toft, D.O. (1994) A novel chaperone complex for steroid receptors involving heat shock proteins, immunophilins, and p23. *J. Biol. Chem.* 269:24989–24993.

Johnson, J.L. and Toft, D.O. (1995) Binding of p23 and hsp90 during assembly with the progesterone receptor. *Mol. Endo.* 9:670–678.

Kieffer, L.J., Seng, T.W., Li, W., Osterman, D.G., Handschumacher, R.E. and Bayney, R.M. (1993) Cyclophilin-40, a protein with homology to the p59 component of the steroid receptor complex: Cloning of the cDNA and further characterization. *J. Biol. Chem.* 268:12303–12310.

Kost, S.L., Smith, D.F., Sullivan, W.P., Welch, W.J. and Toft, D.O. (1989) Binding of heat shock proteins to the avian progesterone receptor. *Mol. Cell. Biol.* 9:3829–3838.

Lipsich, L.A., Cutt, J. and Brugge, J.S. (1982) Association of the transforming proteins of Rous, Fujinami and Y73 avian sarcoma viruses with the same two cellular proteins. *Mol. Cell. Biol.* 2:875–880.

Lovric, J., Bishof, O. and Moelling, K. (1994) Cell cycle-dependent association of Gag-Mil and hsp90. *FEBS* Lett. 343:15–21.

Matts, R.L. and Hurst, R. (1989) Evidence for the association of the heme-regulated eIF-2α kinase with the 90-kDa heat shock protein in rabbit reticulocyte lysate *in situ*. *J. Biol. Chem.* 264:15542–15547.

Milad, M., Sullivan, W.P., Diehl, E., Altmann, M., Nordeen, S., Edwards, D.P. and Toft. D.O. (1995) Interaction of the progesterone receptor with binding proteins for FK506 and cyclosporin A. *Mol. Endo.* 9:838–847.

Miyata, Y. and Yahara, I. (1992) The 90-kDa heat shock protein, hsp90, binds and protects casein kinase II from self-aggregation and enhances its kinase activity. *J. Biol. Chem.* 267: 7042–7047.

Nadeau, K., Das, A. and Walsh, C.T. (1993) Hsp90 chaperonins possess ATPase activity and bind heat shock transcription factors and peptidyl prolyl isomerases. *J. Biol. Chem.* 268: 1479–1487.

Nathan, D.F. and Lindquist, S. (1995) Mutational analysis of hsp90 function: interaction with a steroid receptor and a protein kinase. *Mol. Cell. Biol.* 15:3917–3925.

Nemoto, T.Y., Ohara-Nemoto, M. and Gustafsson, J.-A. (1990) The transformed glucocorticoid receptor has a lower steroid binding affinity than the nontransformed receptor. *Biochemistry* 29:1880–1886.

Nicolet, C.M. and Craig, E.A. (1989) Isolation and characterization of STI1, a stress-inducible gene from *Saccharomyces cerevisiae*. *Mol. Cell. Biol.* 9:3638–3646.

Ning, Y.-M. and Sanchez, E.R. (1993) Potentiation of glucocorticoid receptor-mediated gene expression by the immunophilin ligands FK506 and rapamycin, *J. Biol. Chem.* 268:6073–6076.

Okret, S., Wikstrom, A.-C. and Gustafsson, J.-A. (1985) Molybdate-stabilized glucocorticoid receptor: Evidence for a receptor heteromer. *Biochemistry* 24:6581–6586.

Oñate, S.A. Estes, P.A., Welch, W.J., Nordeen, S.K. and Edwards, D.P. (1991) Evidence that heat shock protein 70 associated with progesterone receptors is not involved in receptor-DNA binding. *Mol. Endo.* 5:1993–2004.

Opperman, H., Levinson, W. and Bishop, J.M. (1981) A cellular protein that associates with the transforming protein of Rous sarcoma virus is also a heat shock protein. *Proc. Natl. Acad. Sci. USA* 78:1067–1071.

Owens-Grillo, J.K., Hoffmann, K., Hutchison, K.A., Yem, A.W., Deibel, M.R., Handschumacher, R.E. and Pratt, W.B. (1995) The cyclosporin A-binding immunophilin CyP-40 and the FK506-binding immunophilin hsp56 (FKBP52) bind to a common site on hsp90 and exist in independent cytosolic heterocomplexes with the untransformed glucocorticoid receptor. *J. Biol. Chem.* 270:20479–20484.

Palmquist, K., Riis, B., Nilsson, A. and Nygard, O. (1994) Interaction of the calcium and calmodulin regulated eEF-2 kinase with heat shock protein 90. *FEBS Lett.* 349:239–242.

Peattie, D.A., Harding, M.W., Fleming, M.A., DeCenzo, M.T., Lippke, J.A., Livingston, D.J. and Benasutti, M. (1992) Expression and characterization of human FKBP52, an immunophilin that associates with the 90-kDa heat shock protein and is a component of steroid receptor complexes. *Proc. Natl. Acad. Sci. USA* 89:10974–10978.

Perdew, G.H. (1988) Association of the Ah receptor with the 90-kDa heat shock protein. *J. Biol. Chem.* 263:13802–13085.

Perdew, G.H. and Whitelaw, M.L. (1991) Evidence that the 90-kDa heat shock protein (hsp90) exists in cytosol in heteromeric complexes containing hsp70 and three other proteins with M_r of 63,000, 56,000, and 50,000. *J. Biol. Chem.* 266:6708–6713.

Picard, D., Khursheed, B., Garabedian, M.J., Fortin, M.G., Lindquist, S. and Yamamoto, K.R. (1990) Reduced levels of hsp90 compromise steroid receptor action *in vivo*. *Nature* 348: 166–168.

Picard, D., Salser, S.J. and Yamamoto, K.R. (1988) A moveable and regulable inactivation function within the steroid binding domain of the glucocorticoid receptor. *Cell* 54:1073–1080.

Pratt, W.B. (1992) Control of steroid receptor function and cytoplasmic-nuclear transport by heat shock proteins. *BioEssays* 14:841–848.

Pratt, W.B. (1993) The role of heat shock proteins in regulating the function, folding, and trafficking of the glucocorticoid receptor. *J. Biol. Chem.* 268:21455–21458.

Pratt, W.B., Czar, M.J., Stancato, L.F. and Owens, J.K. (1993) The hsp56 immunophilin component of steroid receptor heterocomplexes: Could this be the elusive nuclear localization signal-binding protein? *J. Steroid. Biochem. Molec. Biol.* 46:269–279.

Radanyi, C., Chambraud, B. and Baulieu, E.-E. (1994) The ability of the immunophilin FKBP59-HBI to interact with the 90-kDa heat shock protein is encoded by its tetratricopeptide repeat domain. *Proc. Natl. Acad. Sci. USA* 91:11197–11201.

Rafestin-Oblin, M.-E., Couette, B., Radanyi, C., Lombes, M. and Baulieu, E.-E. (1989) Mineralocorticoid receptor of the chick intestine. *J. Biol. Chem.* 264:9304–9309.

Ratajczak, T., Carrello, A., Mark, P.J., Warner, B.J., Simpson, R.J., Moritz, R.L. and House, A.K. (1993) The cyclophilin component of the unactivated estrogen receptor contains a tetra-tricopeptide repeat domain and shares identity with p59 (FKBP59) *J. Biol. Chem.* 268:13187–13192.

Redeuilh, G., Moncharmont, B., Secco, C. and Baulieu, E.-E. (1987) Subunit composition of the molybdate-stabilized "8-9S" nontransformed estradiol receptor purified from calf uterus. *J. Biol. Chem.* 262:6969–6979.

Rehberger, P., Rexin, M. and Gehring, U. (1992) Heterotetrameric structure of the human progesterone receptor. *Proc. Natl. Acad. Sci. USA* 89:8001–8005.

Renoir, J.-M., Buchou, T., Mester, J., Radanyi, C. and Baulieu, E.-E. (1984) Oligomeric structure of molybdate-stabilized, nontransformed 8S progesterone receptor form chicken oviduct cytosol. *Biochemistry* 23:6016–6023.

Renoir, J.-M., Mercier-Bodard, C., Hoffmann, K., Bihan, S.L., Ning, Y.-M., Sanchez, E.R., Handschumacher, R.E. and Baulieu, E.-E. (1995) Cyclosporin A potentiates the dexamethasone-induced MMTV-CAT activity in LMCAT cells: A possible role for different heat shock protein binding immunophilins (HBIs) in glucocorticoid receptor-mediated gene expression. *Proc. Natl. Acad. Sci. USA.* 92:4977–4981.

Rexin, M., Busch, W. and Gehring, U. (1988a) Chemical cross-linking of heteromeric glucocorticoid receptors. *Biochemistry* 27:5593–5601.

Rexin, M., Busch, W., Segnitz, B. and Gehring, U. (1988b) Tetrameric structure of the nonactivated glucocorticoid receptor in cell extracts and intact cells. *FEBS Lett.* 241:234–238.

Rexin, M., Busch, W. and Gehring, U. (1991) Protein components of the nonactivated glucocorticoid receptor. *J. Biol. Chem.* 266:24601–124605.

Rexin, M., Busch, W., Segnitz, B. and Gehring U. (1992) Structure of the glucocorticoid receptor in intact cells in the absence of hormone. *J. Biol. Chem.* 267:9619–9621.

Rose, D.W., Wettenhall, R.E.H., Kudlicki, W., Kramer, G. and Hardesty, B. (1987) The 90-kilodalton peptide of the heme-regulated eIF-2α kinase has sequence similarity with the 90-kilodalton heat shock protein. *Biochemistry* 26:6583–6587.

Sanchez, E.R., Toft, D.O., Schlesinger, M.J. and Pratt, W.B. (1985) Evidence that the 90-kDa phosphoprotein associated with the untransformed L-cell glucocorticoid receptor is a murine heat shock protein. *J. Biol. Chem.* 260:12398–12401.

Sanchez, E.R. (1990) Hsp56: A novel heat shock protein associated with untransformed steroid receptor complexes. *J. Biol. Chem.* 265:22067–22070.

Sanchez, E.R., Faber, L.E., Henzel, W.J. and Pratt, W.B. (1990a) The 56-59-kilodalton protein identified in untransformed steroid receptor complexes is a unique protein that exists in cytosol in a complex with both the 80- and 90-kilodalton heat shock proteins. *Biochemistry* 29:5145–5152.

Sanchez, E.R., Hirst, M., Scherrer, L.C., Tang, H.-Y., Welsh, M.J., Harmon, J.M., Simons, S.S., Ringold, G.M. and Pratt, W.B. (1990b) Hormone-free glucocorticoid receptors overexpressed in Chinese hamster ovary cells are localized to the nucleus and are associated with both hsp70 and hsp90. *J. Biol. Chem.* 265:20123–20130.

Scherrer, L.C., Dalman, F.C., Massa, E., Meshinchi, S. and Pratt, W.B. (1990) Structural and functional reconstitution of the glucocorticoid receptor-hsp90 complex. *J. Biol. Chem.* 265:21397–21400.

Scherrer, L.C., Picard, D., Massa, E., Harmon, J.M., Simons, S.S., Yamamoto, K.R. and Pratt, W.B. (1993) Evidence that the hormone binding domain of steroid receptors confers hormonal control on chimeric proteins by determining their hormone-regulated binding to heat shock protein 90. *Biochemistry* 32:5381–5386.

Schmid, F.X. (1993) Prolyl isomerase: Enzymatic catalysis of slow protein folding reactions. *Annu. Rev. Biomol. Struct.* 22:123–143.

Schowalter, D.B., Sullivan, W.P., Maihle, N.J., Dobson, A.D.W., Coneely, O.M., O'Malley, B.W. and Toft, D.O. (1991) Characterization of progesterone receptor binding to the 90- and 70 kDa heat shock proteins. *J. Biol. Chem.* 266:21165–121173.

Schuh, S., Yonemoto, W., Brugge, J., Bauer, V.J., Riehl, R.M., Sullivan, W.P. and Toft, D.O. (1985) A 90,000-dalton binding protein common to both steroid receptors and the Rous sarcoma virus transforming protein, pp 60^{v-src}. *J. Biol. Chem.* 260:14292–14296.

Segnitz, B. and Gehring, U. (1995) Subunit structure of the nonactivated human estrogen receptor. *Proc. Natl. Acad. Sci. USA* 92:2179–2183.

Shue, G. and Kohtz, D.S. (1994) Structural and functional aspects of basic helix-loop-helix protein folding by heat shock protein 90. *J. Biol. Chem.* 269:2707–2711.

Simons, S.S., Jr. (1994) Function/activity of specific amino acids in glucocorticoid receptors. *In:* G. Litwack (ed.): *Vitamins and Hormones*: Academic Press, New York, Vol. 49, pp 49–130.

Smith, D.F., Faber, L.E. and Toft, D.O. (1990a) Purification of unactivated progesterone receptor and identification of novel receptor-associated proteins. *J. Biol. Chem.* 265:3996–4003.

Smith, D.F., Schowalter, D.B., Kost, S.L. and Toft, D.O. (1990b) Reconstitution of progesterone receptor with heat shock proteins. *Mol. Endo.* 4:1704–1711.

Smith, D.F., Stensgard, B.A., Welch, W.J. and Toft, D.O. (1992) Assembly of progesterone receptor with heat shock proteins and receptor activation are ATP mediated events. *J. Biol. Chem.* 267:1350–1356.

Smith, D.F. (1993) Dynamics of heat shock protein 90-progesterone receptor binding and the disactivation loop model for steroid receptor complexes. *Mol. Endo.* 7:1418–1429.

Smith, D.F. and Toft, D.O. (1993) Steroid receptors and their associated proteins. *Mol. Endo.* 7:4–11.

Smith, D.F., Albers, M.W., Schreiber, S.L., Leach, K.L. and Deibel, M.R. (1993a) FKBP54, a novel FK506-binding protein in avian progesterone receptor complexes and HeLa extracts. *J. Biol. Chem.* 268:24270–24273.

Smith, D.F., Sullivan, W.P., Marion, T.N., Zaitsu, K., Madden, B., McCormick, D.J. and Toft, D.O. (1993b) Identification of a 60-kilodalton stress-related protein, p60, which interacts with hsp90 and hsp70. *Mol. Cell. Biol.* 13:869–876.

Spanjaard, R.A. and Chin, W.W. (1993) Reconstitution of ligand-mediated glucocorticoid receptor activity by *trans*-acting functional domains. *Mol. Endo.* 7:12–16.

Stancato, L.F., Chow, Y.-H., Hutchison, K.A., Perdew, G.H., Jove, R. and Pratt, W.B. (1993) Raf exists in a native heterocomplex with hsp90 and p50 that can be reconstituted in a cell-free system. *J. Biol. Chem.* 268:21711–21716.

Stancato, L.F., Chow, Y.-H., Owens-Grillo, J.K., Yem, A.W., Deibel, M.R., Jove, R. and Pratt, W.B. (1994) The native v-Raf·hsp90·p50 heterocomplex contains a novel immunophilin of the FK506 binding class. *J. Biol. Chem.* 269:22157–22161.

Stancato, L.F., Hutchison, K.A., Krishna, P. and Pratt, W.B. (1996a) Animal and plant cell lysates share a conserved chaperone system that assembles the glucocorticoid receptor into a functional heterocomplex with hsp90. *Biochemistry* 35:554–567.

Stancato, L.F., Owens-Grillo, J.K., Chow, Y.-H., Silverstein, A.M., Jove, R. and Pratt, W.B. (1996b) Formation of a multiprotein signalosome complex containing the mitogenic signaling proteins, Ras, Raf and Erk using a baculovirus expression system; *submitted.*

Sullivan, W.P. and Toft, D.O. (1993) Mutational analysis of hsp90 binding to the progesterone receptor. *J. Biol. Chem.* 268:20373–20379.

Tai, P.-K.K. and Faber, L.E. (1985) Isolation of dissimilar components of the 8.5S nonactivated uterine progestin receptor. *Can. J. Biochem. Cell. Biol.* 63:41–49.

Tai, P.-K.K., Maeda, Y., Nakao, K., Wakim, N.G., Duhring, J.L. and Faber, L.E. (1986) A 59-kilodalton protein associated with progestin, estrogen, androgen and glucocorticoid receptors. *Biochemistry* 25:5269–5275.

Tai, P.-K.K., Albers, M.W., Chang, H., Faber, L.E. and Schreiber, S.L. (1992) Association of a 59-kilodalton immunophilin with the glucocorticoid receptor complex. *Science* 256:1315–1318.

Tai, P.-K.K., Albers, M.W., McDonnell, D.P., Chang, H., Schreiber, S.L. and Faber, L.E. (1994) Potentiation of progesterone receptor-mediated transcription by the immunosuppressant FK506. *Biochemistry* 33:10666–10671.

Veldscholte, J., Berrevoets, C.A., Brinkmann, A.O., Grootegoed, J.A. and Mulder, E. (1992) Antiandrogens and the mutated androgen receptor of LNCaP cells: Differential effects on binding affinity, heat-shock protein interaction, and transcription activation. *Biochemistry* 31:2393–2399.

Walsh, C.T., Zydowsky, L.D. and McKeon, F.D. (1992) Cyclosporin A, the cyclophilin class of peptidylprolyl isomerases, and blockade of T cell signal transduction. *J. Biol. Chem.* 267:13115–13118.

Wartmann, M. and Davis, R.J. (1994) The native structure of the activated Raf protein kinase is a membrane-bound multi-subunit complex. *J. Biol. Chem.* 269:6695–6701.

Whitesell, L., Mimnaugh, E.G., DeCosta, B., Myers, C.E. and Neckers, L.M. (1994) Inhibition of heat shock protein hsp90-pp60$^{v\text{-}src}$ heteroprotein complex formation by benzoquinone ansamycins: Essential role for stress proteins in oncogenic transformation. *Proc. Natl. Acad. Sci. USA* 91:8324–8328.

Xu, Y. and Lindquist, S. (1993) Heat-shock protein hsp90 governs the activity of pp60$^{v\text{-}src}$ kinase. *Proc. Natl. Acad. Sci. USA* 90:7074–7078.

Ziemiecki, A., Catelli, M.-G., Joab, I. and Moncharmont, B. (1986) Association of the heat shock protein hsp90 with steroid hormone receptors and tyrosine kinase oncogene products. *Biochem. Biophys. Res. Commun.* 138:1298–1307.

Stress-Inducible Cellular Responses
ed. by U. Feige, R.I. Morimoto, I. Yahara and B. Polla
© 1996 Birkhäuser Verlag Basel/Switzerland

Protein disulfide isomerase: A multifunctional protein of the endoplasmic reticulum

J. M. Luz and W. J. Lennarz

Department of Biochemistry and Cell Biology, State University of New York at Stony Brook, Stony Brook, NY 11794-5215, USA

Summary. Protein disulfide isomerase (PDI) is a resident enzyme of the endoplasmic reticulum (ER) that was discovered over three decades ago. Contemporary biochemical and molecular biology techniques have revealed that it is present in all eukaryotic cells studied and retained in the ER via a -KDEL or -HDEL sequence at its C-terminus. However, evidence is accumulating that, in certain cell types, PDI can be found in other subcellular compartments, despite possessing an intact retention sequence. A wide range of studies has established that in presence of a redox pair, PDI acts catalytically to both form and reduce disulfide bonds, therefore acting as a disulfide isomerase. Recent studies have focused on the mechanism of the isomerization process and the precise role of the two active site sequences (-CGHC-) in the process. In addition, prokaryotes have been shown to possess a set of proteins that function in a similar fashion, being able to generate disulfide bonds on polypeptides translocated into the periplasmic space. Following the recent discovery that PDI binds peptides, coupled with earlier findings that PDI is a subunit of at least two enzymatic complexes (prolyl 4-hydroxylase and microsomal triglyceride transfer protein), it seems that it may serve functions other than merely that of a disulfide isomerase. In fact, it is now clear that PDI can facilitate protein folding independently of its disulfide isomerase activity. A major challenge for the future is to define mechanistically how it accomplishes isomerization and the relationship between this process and the protein folding steps that culminate in the final, fully mature protein.

Properties of PDI

Protein disulfide isomerase (PDI) was initially identified (Goldberger et al., 1964) based on its ability to catalyze the renaturation of reduced denatured RNAse (Epstein et al., 1963). It was later found to be an ER lumenal enzyme that catalyses the thiol-disulfide interchange reaction both *in vivo* and *in vitro* (reviewed in Freedman, 1991; Noiva and Lennarz, 1992; Freedman et al., 1994). This review will focus on recent findings on this protein.

PDI purified from bovine liver was found to be a homodimer with a mass of 57 kDa (Carmichael et al., 1977; Lambert and Freedman, 1983). The sequence of rat PDI, deduced from a cDNA clone (Edman et al., 1985), identified a protein which had two specific domains (CGHC) with homology to the bacterial protein thioredoxin (CGPC). The CXYC motif was later demonstrated to be the thiol-disulfide interchange catalytic domain (reviewed in Freedman, 1994; Bardwell, 1994).

The basic structural features of PDI (see Fig. 1) can be summarized as follows: (1) the presence of a putative signal sequence at the N-terminus

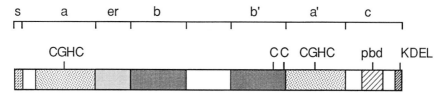

Figure 1. Schematic representation of human PDI (adapted from Kivirikko et al., 1992). Symbols: s signal sequence, a, a′ internally homologous regions with sequence homology to thioredoxin, b, b′ internally homologous regions, c highly acidic C-terminal region of the polypeptide, er region with homology to the estrogen receptor. The active site sequences (CGHC), additional cysteine residues (C), the peptide binding domain (pbd) and the C-terminal ER retention signal (KDEL) are also indicated.

of the protein consistent with its translocation into the lumen of the ER; (2) a KDEL or HDEL C-terminus tetrapeptide ER retention signal; (3) two thioredoxin-like domains (CGHC); (4) a molecular mass between 55 and 70 kDa depending on the extent of post-translational modifications; (5) the existence in a homodimeric form; (6) the existence of two pairs of internal repeats (a and a′, b and b′) (Freedman et al., 1989; Kivirikko et al., 1992); (7) a 43 amino acid segment exhibiting homology to a segment of the estrogen-binding domain of the estrogen receptor (Tsibris et al., 1989) and to a lower degree to the corresponding segments of the progesterone and glucocorticoid receptors (Fliegel et al., 1990), and (8) a peptide binding site found between residues 451–476 in the C-terminal region of bovine PDI (Noiva et al., 1993). The yeast protein differs from that of the higher eukaryotes in that it has 5 N-linked oligosaccharide chains. The PDI from trypanosomes is the only other N-glycosylated PDI thus far described (Hsu et al., 1989).

Subcellular localization

Several findings indicate that PDI is an ER lumenal protein: PDI co-localizes with ER markers; it is insensitive to attack by proteases added to isolated microsomes; and PDI activity is released by the rupture of the microsomal membranes (reviewed in Freedman et al., 1989 and in Noiva and Lennarz, 1992).

It is of interest that PDI has been found in other cellular compartments, although its origin is not always clear. In glial cells, approximately 25% of the protein was found in the cytoplasm (Safran et al., 1992). In rat exocrine pancreatic cells, 5–12% of total activity of PDI was found at the plasma membrane, in transit in the secretory pathway and in the acinar lumen (Yoshimori et al., 1990). A membrane associated form of PDI has also been described in the sinusoidal rat liver plasma membrane (Honscha et al., 1993) and in CHO (Chinese hamster ovary) cells (Mandel et al., 1993).

Significant amounts of PDI have been detected at the surface of human corneal stromal fibroblasts (Ahmad and Church, 1991). Most of the cell systems in which PDI occurs at subcellular compartments other than the ER are highly secretory. Secretion of PDI was detectable when it was over-expressed in mammalian cells (Mazzarella et al., 1990; Dorner et al., 1990) and in yeast (D. Natalia, J.M. Luz and M.F. Tuite, unpublished observations). In addition to PDI, two other major lumenal proteins, BiP (glucose regulated protein 78) and endoplasmin, were also secreted from rat pancreatic cells (Takemoto et al., 1992), suggesting that saturation of the ER retention system (Pelham, 1991) is a primary reason for existence of secreted PDI. However, the mechanism whereby the secreted PDI becomes associated with the cell surface has not been elucidated.

PDI as a catalyst of disulfide bond formation

PDI acts on a wide variety of substrates implying that it is a *bona fide* catalyst with a biological role *in vivo*. In this respect it is significant that the cellular level of PDI is increased in cells actively engaged in the production of disulfide-bonded proteins (Freedman et al., 1991; Kivirikko et al., 1992).

PDI (EC 5.3.4.1.) can be considered to be involved in three reactions: (1) the oxidation of cysteine sulfhydryl groups to produce disulfides, (2) the reduction of disulfide bonds, and (3) the isomerization of disulfide bonds, which is a composite of the reduction step followed by the oxidation step (Freedman, 1989; Hawkins et al., 1991a, b). The key chemical reaction is thiol-disulfide interchange, i.e., the attack of a thiol group (as the thiolate ion, RS^-) on a disulfide bond; Freedman et al., 1989; Hawkins and Freedman, 1991). The basic reaction sequence catalyzed by PDI in generation of disulfide bonds in a catalytic manner is shown in Figure 2.

In thioredoxin the N-terminal cysteine in the active site sequence is the nucleophilic group (Brandes et al., 1993). Similarly, in PDI and DsbA, the N-terminal cysteine in the active site of both proteins is the initial nucleophile (Hawkins and Freedman, 1991; Hwang et al., 1992). However, recent data suggest that, in addition to the differences between the cysteines within each active site, the N-terminal and C-terminal active sites are not equivalent (Lyles and Gilbert, 1994). Mutation of the active sites showed that, at a saturating concentration of substrate, a protein whose C-terminal active site was inactivated retained 95% of wild-type activity. In contrast, a mutant protein whose N-terminal, but not the C-terminal active site, has been inactivated exhibited only 32% of wild-type activity. According to these studies, the C-terminal active site contributes more to steady-state substrate binding whereas the N-terminal site is mainly involved in catalysis at saturating concentrations of substrate. These findings are not in contradiction with previous studies showing that each active site contributed

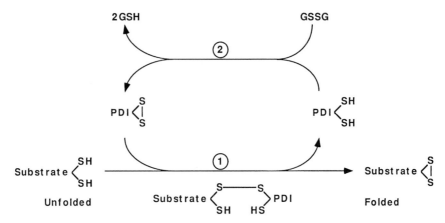

Figure 2. Catalytic activity of PDI. PDI, with its active sites in disulfide form, can oxidize the sulfhydryls of a substrate to form the disulfide bonded protein (reaction 1). Presumably in concert with this process the unfolded protein assumes its native state. As shown, an inter-mediate in the process is presumed to be a mixed disulfide between PDI and the substrate. Recycling of PDI requires its reoxidation by an oxidizing agent such as oxidized GSH (GSSG) as shown in reaction 2.

equally for activity (Hawkins et al., 1991a, b; Vuori et al., 1992a; LaMan-tia and Lennarz, 1993) in the sense that at concentrations of substrate below the Km value, it remains true that each active center contributes equally to activity.

Peptide binding by PDI

PDI binds synthetic peptides, with no specific sequence or amino acid compositional requirements, although cysteine containing peptides are preferred (Noiva et al., 1991; Morjana and Gilbert, 1991). Peptide binding to PDI occurred at sites distinct from the active site sequences, since che-mical modification of the thioredoxin-like sequences completely abolished disulfide isomerase activity but did not affect peptide binding (Noiva et al., 1993). The reverse was not true, since saturation of the peptide binding site, defined by the acidic region encompassing amino acids 451–476, which are located far away from the CXYC motifs, with a photoaffinity probe inhibited the disulfide isomerase activity of the protein (Noiva et al., 1993; Puig et al., 1994). Non-covalent binding of peptides to DsbA was also described as being necessary for catalysis (Darby and Creighton, 1995), but a specific region in the protein backbone has yet to be identified.

A role for PDI in quality control processes in the ER has been suggested due to its *in vivo* association with misfolded lysozyme in the ER lumen. This association indicates that PDI could transiently retain misfolded lyso-zyme molecules, until they refold and thereby become capable of moving to the Golgi apparatus (Otsu et al., 1994). A similar association has been described for PDI interacting with immunoglobulins, but it was ascribed to the mixed disulfide that PDI forms with its substrates (Roth and Pierce, 1987). Even in absence of cross-linker, depletion of GSH by diamide promoted the *in vivo* accumulation of disulfide-linked aggregates of immunoglobulins and PDI, suggesting that the normally transient association was indeed due to the catalytic action of PDI.

PDI as a multifunctional protein

In vivo, PDI catalyzes the generation and isomerization of disulfide bridges. However, other cellular roles have been ascribed to PDI. The elucidation of the sequence of a major cellular thyroid hormone binding protein (p55, Hasamura et al., 1986) demonstrated that this protein was identical to PDI (Cheng et al., 1987; Yamauchi et al., 1987; Fliegel et al., 1990) and that it possessed disulfide isomerase activity (Horiuchi et al., 1989).

In glial cells, the 25% of total PDI that was detected free in the cytoplasm relocalized to the cytoskeleton when the cells were grown in presence of thyroid hormone (Safran et al., 1992). It has been suggested that in this system PDI may play a role in the thyroid hormone-dependent regulation of actin polymerization and degradation of type II iodothyronine 5′-deiodinase.

PDI may also play a role in the transcriptional activation of interferon-inducible genes, probably via its redox properties, since it was shown to bind *in vitro* to complexes between nuclear proteins and regulatory domains of interferon-inducible genes (Johnson et al., 1992). PDI may also be involved in protection against halothane toxicity, by binding to one of its metabolites (Martin et al., 1991). Protection against oxidative damage was also suggested by the demonstration that an isoform of PDI is involved in the reversal of CCl_4 damage to the microsomal ATP-dependent calcium pump (Srivastava et al., 1991).

PDI was shown to have dehydroascorbarte reductase activity and could catalyze the conversion of dehydroascorbic acid to ascorbic acid (Wells et al., 1990). Thus, PDI may regenerate ascorbic acid in the ER lumen which could then be used as a co-factor in the prolyl hydroxylase reaction (see below). Additionally, PDI may provide a link between cellular ascorbic acid and glutathione metabolism.

PDI and cellular complexes

PDI is a component of the prolyl 4-hydroxylase (P4H, EC 1.14.11.2), an ER enzyme involved in the synthesis and maturation of collagens, that catalyzes the formation of 4-hydroxyproline residues in peptide linkages

(Kivirikko et al., 1989). The active enzyme is a $\alpha_2\beta_2$ tetramer (Chen-Kiang et al., 1977; Berg et al., 1980), and the β-subunits is PDI (Pihlajaniemi et al., 1987). Isolated PDI did not have P4H activity, and the P4H β-subunit isolated from the tetramer had the same amount of disulfide isomerase activity as PDI alone (Koivu et al., 1987). The active sites of PDI are not needed for the assembly of the $\alpha_2\beta_2$ P4H complex nor for P4H activity, since inactivation of one or both active sites, by substitution of the N-terminal cysteine by serine, resulted in a fully P4H active tetramer (Vuori et al., 1992b). The PDI activity of the $\alpha_2\beta_2$ P4H tetramer was almost exclusively due to the C-terminal CGHC sequence, explaining the fact that the PDI activity detected in the P4H tetramer was one-half the activity of PDI itself (Vuori et al., 1992a). Co-expression of human PDI and the α-subunit of P4H in insect cells (*Spodoptera frugiperda*) has demonstrated that the major function of the PDI subunit is to maintain the α-subunit in a catalytically active, non-aggregated conformation (Vuori et al., 1992b). The P4H α-subunit does not possess a ER retention signal (Kivirikko et al., 1992) and deletion of the C-terminal KDEL sequence of PDI resulted in considerable secretion of P4H (Vuori et al., 1992b), indicating that PDI also functions in the retention of prolyl hydroxylase in the ER lumen. Similar results were obtained when the α-subunit of P4H was synthesized in a cell-free system: in microsomes containing the usual complement lumenal proteins, the polypeptide chain was capable of being translocated and glycosylated. However, in microsomes depleted of lumenal proteins, the polypeptide precipitated after translocation, even under conditions that favored the formation of disulfide bonds (John et al., 1993). Recently, the gene encoding a second α-subunit isoform of P4H was isolated, and its product was able to form a $\alpha_2\beta_2$ complex with human PDI when co-expressed in the baculovirus system (Helaakoski et al., 1995). Human PDI was also able to form an active P4H complex when co-expressed in the same system with the P4H α-subunit of the nematode *Caenorhabditis elegans*, but it associated as a catalytically active $\alpha\beta$ P4H dimer instead of the usual tetramer (Veijola et al., 1994). This finding demonstrates that PDI can functionally associate with P4H α-subunits having marked differences in their amino-acid sequences.

Another case where PDI exists in a complex is the triglyceride transfer system. The microsomal triglyceride transfer protein complex is composed of two subunits: a 88 kDa polypeptide and PDI (Wetterau et al., 1990). When this complex was dissociated, the 88 kDa subunit aggregated and no triglyceride transfer activity could be recovered (Wetterau et al., 1991). Insulin negatively regulates the complex in $HepG_2$ cells, but its effects in transcription are almost limited to the 88 kDa subunit, with PDI showing only a modest decrease in mRNA synthesis (Lin et al., 1995). Mutations in the large subunit which result in the absence of triglyceride transfer activity have been shown to cause abetalipoproteinemia (Wetterau et al., 1992). An abetalipoproteinemic subject, in whom no microsomal tryglyceride transfer complex or activity could be detected, was shown to possess a truncated copy of the gene encoding the large subunit (88 kDa), predicting a protein whose C-terminal 30 amino acids were missing. Further, this segment was shown to be essential for assembly with PDI, since despite its high expression level in the baculovirus system, the truncated polypeptide failed to fold and assemble with PDI (Ricci et al., 1995).

A super family of PDI-like proteins

ERp72 is an ER lumenal protein that contains three CGHC motifs (Mazzarella et al., 1990). Although no disulfide isomerase activity could be detected for mouse ERp72 (Dorner et al., 1990), the rat homologue (CaBP2p) had significant activity (Nguyen-Van et al., 1993). Another rat PDI-like ER lumenal protein, CaBP1p, has two thioredoxin-like motifs (CGHC) and similar to CaBP2p, exhibited disulfide isomerase activity, albeit to a lower extent than PDI. Both proteins showed a significant synergistic effect in the refolding of reduced RNAse AIII (Rupp et al., 1994) and, like PDI, were shown to be substrates of thioredoxin reductase in a way consistent with a role in thiol-disulfide interchange reactions (Lundstrom-Ljung et al., 1995). Similar to PDI, an ATP- and Ca^{2+}-dependent chaperone activity has been suggested for ERp72 since it was found to be eluted by ATP (but not by a non-hydrolyzable analogue) after binding to immobilized denatured proteins (Nigam et al., 1994). Further support for a chaperone-like activity came from studies in the maturation of thyroglobulin in cultured thyroid cell lines. A complex consisting of ERp72, BiP, GRP94 and thyroglobulin could be imunoprecipitated by using an anti-thyroglobulin antibody, or antibodies directed to each of the individual proteins, providing evidence that these proteins play a role in the maturation of thyroglobulin in a chaperone-like manner (Kuznetsov et al., 1994). In addition, the disulfide isomerase inactive murine ERp72 can functionally replace the essential yeast PDI1 gene (Gunther et al., 1993).

The yeast *EUG1* gene encodes a non-essential lumenal protein whose cellular levels are increased in response to accumulation of native and unglycosylated proteins in the ER lumen (Tachibana and Stevens, 1992). *EUG1* has an unfolded protein responsive element in its promoter, leading to increased transcription under these conditions (Mori et al., 1992; Tachibana and Stevens, 1992). The Eug1p protein shows considerable homology to yeast PDI (43% identity at the amino acid level) (Tachibana and Stevens, 1992). Despite the C-terminal cysteine being replaced by serine in the thioredoxin-like sequences of the Eug1p (CLHS and CIHS, instead of CGHC), the *EUG1* gene could functionally replace the yeast *PDI1* gene, when overexpressed (Tachibana and Stevens, 1992). However, Eug1p did not rescue the phenotype of accumulation of unprocessed disulfide-bonded carboxlypeptidase Y (CPY) in the ER upon depletion of

PDI, indicating that restoration of disulfide isomerase activity is probably not the key function restored by overexpression of the *EUG1* gene product in *pdi1* null cells (LaMantia and Lennarz, 1993).

Disulfide isomerase activity has also been detected in the gonadotropic hormones, lutropin and folliotropin. Thioredoxin-like domains were found for these hormones, and it was proposed that this activity may function in receptor activation and signal transduction (Boniface and Reichert, 1990).

R-cognin is another protein highly homologous to PDI that mediates the adhesion and reaggregation *in vitro* of retina cells from chicken and rat embryos. This protein, which was found associated with the plasma membrane, contained the binding site for thyroid hormone found in PDI (encoded by exon 3 of human PDI cDNA). It has been proposed that R-cognin may be involved in the cellular response to hormonal signals (Krishna-Rao and Hausman, 1993) by mediation of hormone-receptor interactions.

Misidentification of a PDI isoform cDNA clone as a phospholipase Cα (Benett et al., 1988) initially led to the belief that the two proteins were identical (Srivastava et al., 1991; Haugejorden et al., 1991). This cDNA was subsequently shown to encode a minor isoform of PDI, less than 50% identical to the classical isoform of PDI (Srivastava et al., 1993). The murine homologue of this isoform is ERp60, which exhibited both thiol-disulfide isomerase activity and a CXYC motif-dependent protease activity (Urade et al., 1992; Urade and Kito, 1992). A recent study showed that ERp60 associated *in vivo* with misfolded human mutant lysozyme expressed in mouse L cells, but not with wild-type protein. In addition, denatured lysozyme was a substrate for the ERp60 cysteine protease activity, whereas native lysozyme was not. These data suggest that ERp60 could be a component of the proteolytic apparatus responsible for ER quality control (Otsu et al., 1995). The major isoform of mouse PDI did not exhibit a protease activity similar to ERp60 and ERp72 (Urade et al., 1993).

Bacterial PDI (the Dsb system of bacteria)

DsbA is a 21 kDa protein that facilitates disulfide bond formation and is required for correct folding and stability of a number of proteins exported into the periplasm of the bacterium *Escherichia coli* (Martin et al., 1993). The gene was initially identified in *E. coli* pleiotropic mutants defective in disulfide bond formation (Bardwell et al., 1991; Kamitani et al., 1992). The *dsbA* gene product was found to localize in the periplasm (the bacterial equivalent to the ER), had a thioredoxin-like sequence CPHC, and was capable of catalysis of disulfide bond formation *in vitro* (Akiyama et al., 1992) and *in vivo* (Wunderlich and Glockshuber, 1993). DsbA homologues have been isolated from *Vibrio cholera* (Peek and Taylor, 1992; Yu et al., 1992), *Haemophilus influenzae* (Tomb, 1992), *Erwinia chrysanthemi*, and

Legionella (Shevchik et al., 1994; Bardwell, 1994). Recently, DsbA was crystallized and shown to be very similar in structure to thioredoxin (Martin et al., 1993). The catalytic center is present at a domain interface and is surrounded by grooves and exposed hydrophobic residues suggesting that DsbA may bind the partially folded peptide before oxidation of its cysteine residues (Martin et al., 1993).

A second *E. coli* gene is also required for disulfide bond formation; this gene, *dsbB*, encodes a plasma membrane protein which is proposed to function in the transfer of redox equivalents from the cytoplasm to the periplasm to reduce DsbA so that it can act catalytically (Bardwell et al., 1993; Dailey and Berg, 1993; Missiakas et al., 1993). *In vivo*, overexpression of DsbA alone does not improve the yield of correctly folded proteins, unless reduced glutathione is added to the medium (Wunderlich and Glockshuber, 1993), which is consistent with the fact that the defect in *dsbB* mutants could be complemented by addition of GSSG or cystine to the medium (Bardwell et al., 1993). Fusions of DsbB with alkaline phosphatase allowed the determination of the membrane topology of the protein: the CXYC motif of DsbB faces the periplasm, together with two other cysteines all of which were shown to be essential for DsbB function (Jander et al., 1994). The DsbA-DsbB system is not involved in the formation of disulfide bonds in the intramembranous domains of proteins (Whitely and von Heijne, 1993).

A recent study provided valuable insights into the oxidation reduction pathway that DsbA follows after biosynthesis (Kishigami et al., 1995). In this study the levels of DsbA were manipulated in a *dsbB* mutant or wild type background. Overexpression of DsbA in wild type cells yielded considerable amounts of DsbA in the reduced state, which was interpreted as being due to saturation of DsbB. However, in a *dsbB⁻* background, normal levels of DsbA expression yielded the molecule exclusively in the reduced state, but overexpression lead to a considerable portion of DsbA in an oxidized state. Based on these findings the authors proposed that newly synthesized DsbA is rapidly oxidized by preexisting DsbA, while oxidation of mature (functional) DsbA to sustain a steady state requires DsbB.

A third gene involved in thiol-disulfide interchange of *E. coli* is *dsbC* (Missiakas et al., 1994; Shevchik et al., 1994). The protein has the characteristic CXYC motif, and was capable of complementing a *dsbA* null mutation. Replacement of either cysteine by another amino acid in the active site results in loss of activity (Missiakas et al., 1994). Double mutants (either *dsbA-dsbC* or *dsbB-dsbC*) showed more severe defects than mutations in either of the individual genes, and a mixture containing both DsbA and DsbC was much more active in folding reactions than either protein separately. DsbC seems to be more active in isomerization reactions (Fig. 3), whereas DsbA seems more active in catalyzing the net formation of disulfide bridges (Shevchik et al., 1994).

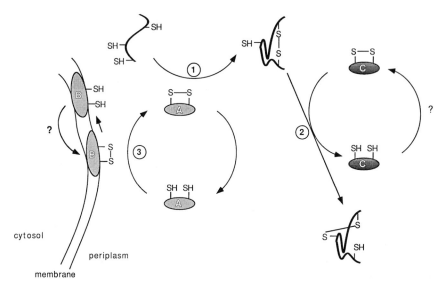

Figure 3. The Dsb system of bacteria. A newly translocated polypeptide chain in the periplasm is believed to be the substrate for oxidized DsbA (reaction 1), leading to the formation of disulfide bridges in the former. If these disulfides are not the "correct" ones found in the native protein, the DsbC protein is believed to act in the isomerization reaction (reaction 2). The mechanism for regeneration of oxidized DsbC is presently unknown. Reduced DsbA can be reoxidized by the integral membrane protein DsbB (reaction 3), which is believed to transduce oxidizing equivalents from the cytoplasm, by an unknown mechanism. In this diagram, only the generally accepted *in vivo* functions of DsbA (formation of disulfide bridges) and DsbC (isomerization of thiol-disulfides) are depicted, despite the fact that both enzymes can catalyze the net oxidation, reduction or isomerization thiol-disulfides, depending on the imposed *in vitro* redox potential. A, B, and C refer to DsbA, DsbB and DsbC, respectively.

Comparison of DsbA catalytic activity versus that of PDI

DsbA and PDI share the CXYC motif with thioredoxin, have similar redox potentials, and both are thought to facilitate net protein disulfide bond formation *in vivo* (Bardwell, 1994). Both proteins can catalyze sulfhydryl oxidation, as well as disulfide reduction and isomerization, depending on the *in vitro* conditions. PDI is much more active in isomerization reactions than DsbA (Zapun and Creighton, 1994), which may be the result of the involvement of two independent active sites in PDI versus only one in DsbA. However in both cases, transient intermolecular disulfide bonds are presumed to be formed between each of the enzymes and their substrates (Kanaya et al., 1994). Another factor that may explain a primary role for DsbA in the formation of disulfide bonds is that its disulfide is 1000-fold more reactive than a normal disulfide (Zapun et al., 1993), which is indicative of high instability of the folded, disulfide linked state. The high reactivity of the N-terminal thiol group (C30) of DsbA is explained by its pKa of approx. 3.5, which means that it exists as the reactive thiolate (S⁻)

species, instead of the unreactive thiol (SH), at physiological pH and even at pH values as low as 4 (Nelson and Creighton, 1994). Like other CXYC-carrying proteins, the pKa of the C-terminal thiol group (C33) is 9.5, resulting in a much less reactive group. Alkylation studies have also demonstrated that the cysteine residue(s) of thioredoxin and PDI equivalent of C33 in DsbA, is buried in the protein and therefore inaccessible to solvent (Kallis and Holmgreen, 1980; Hawkins and Freedman, 1991). In agreement with this data, Wunderlich et al. (1995) demonstrated that a mutant DsbA, in which the C33 was mutated to alanine, was 60% as active as the wild-type protein in catalysis of disulfide bond formation, while mutation of the C30 or both C30 and C33 to alanines, lead to loss of disulfide bond formation activity. A similar mutation (C33 to S33) yielded a DsbA protein capable of catalysis of the reduction of insulin by DTT, in contrast to the S30 and S30–S33 mutant DsbA, which were inactive (Zapun et al., 1994). However, in the DsbA from *V. cholera*, both of the cysteines in the sequence CPHC have been implicated in disulfide bond formation, since Cys→Ala mutation at either of the two cysteines abolishes activity and secretion of a disulfide-bonded protein (Yu et al., 1993). Whether these findings are due to characteristics of *V. cholera* DsbA, or whether it reflects the use of different activity assays remains to be established.

Dsb-like proteins

Like PDI, several DsbA-like proteins have been described. The *bdb* gene from *Bacillus brevis* complemented a *dsbA* null, when introduced in *E. coli*, but did not show overall homology to DsbA, besides the presence of the CXYC motif (Ishiara et al., 1995).

Other bacterial proteins that have a CXYC motif are HelX from *Rhodobacter*, TlpA and TlpB from *Bradyrhizobium* and DipZ from *E. coli* (Beckman and Kranz, 1993; Loferer et al., 1993; Thony-Meyer et al., 1994; Crooke and Cole, 1995). The membrane topology of DipZ is not yet known, but all of the other proteins have their CXYC motifs facing the periplasm (Crook and Cole, 1995), suggesting that one of their functions might be the catalysis of thiol-disulfide interchange.

Is PDI a molecular chaperone?

The yeast *PDI1* gene encodes an essential product (Farquhar et al., 1991; LaMantia et al., 1991; Gunther et al., 1991; Tachikawa et al., 1991). However, there are three observations that indicate that the disulfide isomerase activity of the protein is not essential for cell viability. First, Eug1p is a yeast protein homologous to PDI containing CXYZ, instead of the traditional CXYC, at the same relative positions in the molecule. Second when this protein is overexpressed, it is able to functionally rescue *pdi1* null mutants. Third, mutation of the C-terminal-most cysteines of both active sites of yeast PDI leads to loss of disulfide isomerase *in vitro*, but is not lethal (LaMantia and Lennarz, 1993). Based on these observations, and the fact that both yeast and mammalian PDI bind peptides, a chaperone-like activity was postulated for PDI. Consistent with this idea, it has been shown by competition experiments with a labeled peptide that unfolded proteins have a higher binding affinity for PDI than their native counterparts (R. Niva, personal communication). In contrast, Wunderlich et al. (1995) recently reported that a DsbA molecule whose C-terminal cysteine was mutated to alanine exhibits reduced, but still detectable catalysis of disulfide bond formation and refolding of hirudin. These authors argue that this result explains the fact that Eug1p (CXYZ) can rescue a yeast *pdi1* null, and that PDI with both active sites mutated to resemble those of Eug1p is still capable of rescuing the same *pdi1* null. Based on this they suggest that postulating a chaperone-like function for PDI is not necessary. However, the different results in enzymatic activity of mutated PDI and DsbA molecules could be ascribed to the different functions of PDI and DsbA *in vivo* (namely isomerization versus formation of disulfide bonds), or to the different assays employed (disulfide bond formation and refolding of BPTI versus refolding of hirudin). Clearly, further investigation is needed to clarify this issue.

The refolding of lysozyme in the presence of PDI was recently studied by Puig and Gilbert (1994a). They found that the yield of folded lysozyme showed a strong dependence of the concentration of both reduced lysozyme, and PDI. PDI is proposed to have both chaperone and anti-chaperone activity, with the yield of folded protein being dependent on the experimental conditions used. As pointed out by Freedman et al. (1994) the overall action of PDI seems to be that of a catalyst of thiol-disulfide interchange and the different amounts of folded product result from the complexity of the system used. Interestingly, the molecular chaperone BiP shows identical properties in a similar experimental system, (i.e., the yield of folded product depends on the concentrations of BiP and lysozyme) and BiP competes with PDI for binding of denatured lysozyme (Puig and Gilbert, 1994b).

Simpler refolding experiments have determined that the main function of PDI is as an isomerase. For example, Lilie et al. (1994) studied the refolding of a Fab fragment, and found that PDI facilitates the refolding of the denatured and reduced Fab fragment, but not the refolding of the Fab fragment containing intact disulfide bonds. Thus, Lilie et al. (1994) conclude that in this system, a chaperone-like activity of PDI was not detected. However, it is interesting that PDI can facilitate refolding of a protein that does not contain disulfide bonds (D-glyceraldehyde 3-phosphate dehydrogenase, GAPDH). GAPDH shows only a limited extent of spontaneous reactivation following denaturation. Inclusion of PDI in the reaction mixture markedly increased reactivation of GAPDH, and simultaneously decreased aggregation (Cai et al., 1994). Consequently, in this *in vitro* system a chaperone-like activity was postulated. Further support for this idea was provided by studies in which PDI inactivated with respect to isomerase activity (by reduction and alkylation) was shown to still facilitate folding of GAPDH (Quan et al., 1995). Moreover, a peptide that presumably binds to the peptide binding site inhibited isomerase activity in the native PDI and blocked the chaperone activity.

A chaperone activity for PDI was also postulated as a result of its selective binding to immobilized denatured proteins, and subsequent elution by ATP, (but not by a non-hydrolyzable analogue of ATP) (Nigam et al., 1994). It is of interest that although PDI has a peptide binding domain (see below), an ATP binding domain is not recognizable in the structure of PDI. Further experimentation is necessary to define whether these observations might be explained by the presence of complexes between PDI and other ER proteins containing ATPase activity.

Potential regulatory functions of PDI

Evidence is accumulating regarding a multitude of processes that occur at the cell surface and require a reductive activity of the plasma membrane. For instance, treatment of CHO cells with membrane-impermeant sulfhydryl reactive agents, reduced the cytotoxicity of diphtheria toxin (a disulfide linked heterodimer), thereby implicating cell-surface sulfhydryl compounds in the initial stages of entry of this toxin (Ryser et al., 1991). Further investigation demonstrated that entry of the toxin was also inhibited by bacitracin (an inhibitor of PDI, Mizunaga et al., 1990) and significantly, by anti-PDI antibodies (Mandel et al., 1993). The same group demonstrated later that HIV infection of human lymphoid cells could be inhibited by the same membrane-impermeant sulfhydryl blockers, or by bacitracin or anti-PDI antibodies (Ryser et al., 1994). These results, although indirect, suggest that PDI might serve to maintain reduced sulfhydryls at the cell surface. In the case of HIV, they imply that HIV and its target cell engage in a thiol-disulfide interchange mediated by PDI.

An isoform of PDI (p55) has been shown to be present in elevated levels in the myeloid cells of chronic myelogenous leukemia patients, and to regulate the complex formation between nuclear proteins and the regulatory domains of interferon-inducible genes (Johnson et al., 1992). The level of p55 is believed to be regulated by interferon α in these cells.

Autoimmune anti-PDI antibodies have been described in rats with liver injuries that had been induced by diethylmaleate with acetaminophen, diethylmaleate with carbon tetrachloride or D-galactosamine (Nagayama et al., 1994a). In addition, the same antibody could be found in the serum of patients with liver disease (alcohol hepatitis, liver cirrhosis and hepatoma) and chronic alcoholism (Nagayama et al., 1994b) suggesting that the occurrence of the anti-PDI antibody played a role in the progression of hepatic disorders. Occurrence of anti-PDI antibody has also been suggested to play an important immunological role in the progression and persistence of hepatitis leading to death in LEC rats (Yokoi et al., 1994).

PDI as a component of P4H has been implicated in the increased synthesis of collagen associated with severe lung injury (fibrosing alveolitis), in several lung cell types (Kasper et al., 1994). The PDI homologue in the parasitic nematode *Onchocerca volvulus* has been isolated and characterized (Wilson et al., 1994). In this organism a major role of PDI seems to be the biosynthesis of the parasite cuticle (rich in disulfide bridges), which is the first protective barrier against host defenses and therefore a valuable drug target. Recently, it has been reported that synthesis of PDI and several molecular chaperones in the pancreatic exocrine secretory pathway is coordinately regulated by hormones. After hormonal stimulus, expression of PDI and the molecular chaperones precedes the synthesis of secretory proteins (Hensel et al., 1994). This observation is in agreement with a regulated role for PDI in protein folding.

An integrated model for the mode of action of PDI

Studies on the steps in folding of hCG-β and of hCG subunit assembly, as measured by the order of disulfide bond formation, indicate that these processes are identical *in vivo* and *in vitro* (Huth et al., 1993). Despite this finding, confusion about the mode of action of PDI remains, probably as a result of the different assays used in *in vitro* to measure PDI activity. It is important to point out that under normal physiological conditions PDI is acting on newly synthesized polypeptide chains in the lumen of the ER. In fact, there is recent experimental evidence (Chen et al., 1995), supporting earlier studies, that PDI can act on nascent polypeptide chains. It is therefore reasonable to propose that the normal mode of action of PDI is to convert cysteine sulfhydryls to disulfide bonds. With a protein with only two cystine residues there is no chance of forming the incorrect disulfide bridge and this oxidation process would be the only necessary event with respect to sulfhydryls or disulfide bonds. But, of course, most secretory and membrane proteins contain more than just two cysteine residues, and therefore PDI almost certainly must also act as an isomerase when it acts on large proteins with many disulfide bonds. In these cases, it would seem essential that as the protein folds, the incorrect disulfide bridges are reduced and new ones are formed. Under these conditions one would expect that PDI would be covalently tethered via a mixed sulfide to the nascent chain until all the correct disulfide bonds had been formed. Indeed, evidence for such bonds has been reported (Roth and Pierce, 1987). It seems likely that during this tethered state PDI is acting as a chaperone, keeping the protein from being converted into insoluble folded forms that cannot readily be reversed. According to this model PDI acts like other chaperones, but does so while it is transiently covalently linked via a mixed disulfide to its protein substrate. Since its chaperone property is not lost when the active sites are alkylated or mutated, it still can function in protein folding provided that another oxidation-reduction system is available, although this system may not be as effective as PDI. This situation seems to occur in the processing and disulfide bond formation in the case of CPY in yeast (LaMantia and Lennarz, 1993). Thus, a yeast strain in which the active sites of PDI have been mutated survives, but the rate of formation of disulfide linked CPY is reduced. The idea that disulfide bond formation via a mixed disulfide between PDI and its substrate and the chaperone action of PDI are independent processes is consistent with its apparent role as a stabilizing component of prolyl hydroxylase and tryglyceride transfer protein complex (see above). Support for this hypothesis comes from the fact that PDI can assist folding of a protein containing no disulfide bonds (Cai et al., 1994).

In summary, the proposed model assumes that initially PDI oxidatively generates disulfides, but can also act to reduce and reform a disulfide bond between another pair of cysteine residues as the protein is progressing towards its native configuration; during this overall process PDI serves as a chaperone. A challenge for the future is to devise experiments to test the various aspects of this model.

Acknowledgements
The authors thank Elizabeth Anderson and Dr. Robert Noiva for reviewing the manuscript. We acknowledge Lorraine Conroy for her assistance with the preparation of this manuscript. The work reported from our laboratory was supported by NIH Grant GM 33184 to WJL.

References

Ahmad, M. and Church, R.L. (1991) Amino acid sequence analysis of proteins in the human corneal stromal membrane. *Curr. Eye Res.* 10:35–46.

Akiyama, Y., Kamitani, S., Kusukawa, N. and Iko, K. (1992) *In vitro* catalysis of oxidative folding of disulfide bonded protein by the *Escherichia coli* DsbA (*ppfA*) gene product. *J. Biol. Chem.* 267:22440–22445.

Bardwell, J.C.A., McGovern, K. and Beckwith, J. (1991) Identification of a protein required for disulfide bond formation *in vivo*. *Cell* 67:581–589.

Bardwell, J.C.A., Lee, J.-O., Jander, G., Martin, N., Belin, D. and Beckwith, J. (1993) A pathway for disulfide bond formation *in vivo*. *Proc. Natl. Acad. Sci. USA* 90:1038–1042.

Bardwell, J.C.A. (1994) Building bridges: disulfide bond formation in the cell. *Mol. Microbiol.* 14:199–205.

Beckman, D.L. and Kranz, R.G. (1993) Cytochrome *c* biogenesis in a photosynthetic bacterium requires a periplasmic thioredoxin-like protein. *Proc. Natl. Acad. Sci. USA* 90:2179–2183.

Bennett, C.F., Balcarek, J.M., Varrichio, A. and Crooke, S.T. (1988) Molecular cloning and complete sequence of form-I phosphoinositide-specific phospholipase C. *Nature* 334:268–270.

Berg, R.A., Kao, W.W.-Y. and Kedersha, N.L. (1980) The assembly of tetrameric prolyl hydroxylase in tendon fibroblasts form newly synthesized α-subunits and from preformed cross-reacting protein. *Biochem. J.* 189:491–499.

Boniface, J.J. and Reichert, L.E., Jr. (1990) Evidence for a novel thioredoxin-like catalytic property of gonadotropic hormones. *Science* 247:61–64.

Brandes, H.K., Larimer, F.W., Geck, M.K., Stringer, C.D., Schurmann, P. and Hartman, F.C. (1993) Direct identification of the primary nucleophile of thioredoxin *f. J. Biol. Chem.* 268: 18411–18414.

Cai, H., Wang, C.-C. and Tsou, C.-L. (1994) Chaperone-like activity of protein disulfide isomerase in the refolding of a protein with no disulfide bonds. *J. Biol. Chem.* 269:24550–24552.

Carmichael, D.F., Morin, J.E. and Dixon, J.E. (1977) Purification and characterization of a thiol-protein disulfide oxireductase from bovine liver. *J. Biol. Chem.* 252:7163–7167.

Chen, W., Helenius, J., Braakman, I. and Helenius, A. (1995) Co-translational folding and calnexin binding during glycoprotein synthesis. *Proc. Natl. Acad. Sci. USA* 92:6229–6233.

Chen-Kiang, S., Cardinale, G.J. and Udenfriend, S. (1977) Homology between a prolyl hydroxylase subunit and a tissue protein that cross-reacts immunologically with the enzyme. *Proc. Natl. Acad. Sci. USA* 74:4420–4424.

Cheng, S., Gong, Q., Parkinson, C., Robinson, E.A., Appella, E., Merlino, G.T. and Pastan, I. (1987) The nucleotide sequence of a human cellular thyroid hormone binding protein present in the endoplasmic reticulum. *J. Biol. Chem.* 262:11221–11227.

Crooke, H. and Cole, J. (1995) The biogenesis of *c*-type cytochromes in *Escherichia coli* requires a membrane bound protein, DipZ, with a protein disulfide isomerase-like domain. *Mol. Microbiol.* 15:1139–1150.

Dailey, F.E. and Berg, H.C. (1993) Mutants in disulfide bond formation that disrupt flagellar assembly in *Escherichia coli. Proc. Natl. Acad. Sci. USA* 90:1043–1047.

Darby, N.J. and Creighton, T.E. (1995) Catalytic mechanism of DsbA and its comparison with that of protein disulfide isomerase. *Biochemistry* 34:3576–3587.

Dorner, A.J., Wasley, L.C., Raney, P., Haugejorden, S., Green, M. and Kaufman, R.J. (1990) The stress response in chinese hamster ovary cells. *J. Biol. Chem.* 265:22029–22034.

Edman, J.E., Ellis, L., Blacker, R.W., Roth, R.A. and Ruthe, W.J. (1985) Sequence of protein disulfide isomerase and implications of its relationship to thioredoxin. *Nature* 317: 267–270.

Epstein, C.J., Goldberger, R.F. and Anfinsen, C.F. (1963) The genetic control of tertiary protein structure: studies with model systems. *Cold Spring Harbor Symp. Quant. Biol.* 28: 439–449.

Farquhar, R., Honey, N., Murant, S.J., Bossier, P., Schultz, L., Montgomery, D., Ellis, R.W., Freedman, R.B. and Tuite, M.F. (1991) PDI is essential for viability in *Saccharomyces cerevisiae. Gene* 108:81–89.

Fliegel, L., Newton, E., Burns, K. and Michalak, M. (1990) Molecular cloning of a cDNA encoding a 55 kDa multifunctional thyroid hormone binding protein of skeletal muscle sarcoplasmic reticulum. *J. Biol. Chem.* 265:15496–15502.

Freedman, R.B., Hirst, T.R. and Tuite, M.F. (1994) Protein disulfide isomerase: building bridges in protein folding. *Trends Biol. Sci.* 19:331–336.

Freedman, R.B. (1989) Protein disulfide isomerase: multiple roles in the modification of nascent secretory proteins. *Cell* 57:1069–1072.

Freedman, R.B., Bulleid, N.J., Hawkins, H. and Paver, J.L. (1989) Role of protein disulfide-isomerase in the expression of native proteins. *Biochemical Society Symposia* 55:167–192.

Freedman, R.B. (1991) Protein disulfide isomerase- an enzyme that catalyses protein folding in the test tube and in the cell. *In:* E.T. Nall and K.A. Dill (eds): *Conformations and Forces in Protein Folding.* AAAS, vol. 16, pp 204–216.

Freedman, R.B. (1994) Folding helpers and unhelpful folders. *Curr. Biol.* 4:933–935.

Goldberger, R.F., Epstein, C.J. and Anfinsen, C.B. (1964) Purification and properties of a microsomal enzyme system catalyzing the reactivation of reduced ribonuclease and lysozyme. *J. Biol. Chem.* 239:1406–1410.

Gunther, R., Brauer, C., Janetzky, B., Forster, H.-H., Elbrech, I.-M., Lehle, L. and Kuntzel, H. (1991) The *Saccharomyces cerevisiae TRG1* gene is essential for growth and encodes a lumenal endoplasmic reticulum glycoprotein involved in the maturation of vacuolar carboxypeptidase. *J. Biol. Chem.* 266:24557–24563.

Gunther, R., Srinivasan, M., Haugejorden, S., Green, M., Ehbrecht, I.-M. and Kuntzel, H. (1993) Functional replacement of the *Saccharomyces cerevisiae* Trg1/Pdi1 protein by members of the mammalian protein disulfide isomerase family. *J. Biol. Chem.* 268:7728–7732.

Hasumura, S., Kitagawa, S., Lovelace, E., Willingham, M.C., Pastan, I. and Cheng, S. (1986) Characterization of a membrane-associated 3,3′,5-triiodo-L-tyronine binding protein by use of monoclonal antibodies. *Biochemistry* 25:7881–7888.

Haugejorden, S., Srinivasan, M. and Green, M. (1991) Analysis of retention signals of two resident endoplasmic reticulum proteins by *in vitro* mutagenesis. *J. Biol. Chem.* 266: 6015–6018.

Hawkins, H.C. and Freedman, R.B. (1991) The reactivities and ionization properties of the active-site dithiol groups of mammalian protein disulfide-isomerase. *Biochem. J.* 275: 335–340.

Hawkins, H.C., DeNardi, M. and Freedman, R.B. (1991a) Redox properties and cross-linking of the dithiol/disulfide active sites of mammalian protein disulfide-isomerase. *Biochem. J.* 275:341–348.

Hawkins, H.C., Blackburn, E.C. and Freedman, R.B. (1991b) Comparison of the activities of protein disulfide-isomerase and thioredoxin in catalyzing disulfide isomerisation in a protein substrate. *Biochem. J.* 275:349–353.

Helaakoski, T., Annunen, P., Vuori, K., MacNeil, I.A., Pihlajaniemi, T. and Kivirikko, K.I. (1995) Cloning, baculovirus expression, and characterization of a second mouse prolyl 4-hydroxylase α-subunit isoform: formation of an $\alpha_2\beta_2$ tetramer with the protein disulfide isomerase/β-subunit. *Proc. Natl. Acad. Sci. USA* 92:442–4431.

Hensel, G., Abmann, V. and Kern, H.F. (1994) Hormonal regulation of protein disulfide isomerase and chaperone synthesis in the rat exocrine pancreas. *Europ. J. Cell Biol.* 63: 208–218.

Honscha, W., Ottallah, M., Kistner, A., Platte, H. and Petzinger, E. (1993) A membrane-bound form of protein disulfide isomerase (PDI) and the uptake of organic anions. *Biochim. Biophys. Acta* 1153:175–183.

Horiuchi, R., Yamauchi, K., Hayashi, H., Koya, S., Takeuchi, Y., Kato, K., Kobayashi, M. and Takikawa, H. (1989) Purification and characterization of 55-kDa protein with 3,5,3′-triiodo-L-thyronine-binding activity and protein disulfide isomerase from beef liver membrane. *Europ. J. Biochem.* 183:529–538.

Hsu, M.P., Muhich, M.L. and Boothroyd, J.C. (1989) A developmentally regulated gene of Trypanosomes encodes a homologue of rat protein disulfide isomerase and phosphoinositolphospholipase C. *Biochemistry* 28:6440–6446.

Huth, J.R., Perin, F., Lockridge, O. Bedows, E. and Ruddon, R.W. (1993) Protein folding and assembly *in vitro* parallel intracellular folding and assembly. *J. Biol. Chem.* 268:16472–16482.

Hwang, C., Sinskey, A.J. and Lodish, H.F. (1992) Oxidized redox state of glutathione in the ER. *Science* 257:1496–1502.

Ishihara, T., Tomita, H., Hasegawa, Y., Tsukagoshi, N., Yamagata, H. and Udaka, S. (1995) Cloning and characterization of the gene for a protein thiol-disulfide oxidoreductase in *Bacillus brevis. J. Bacteriol.* 177:745–749.

Jander, G., Martin, N.L. and Beckwith, J. (1994) Two cysteines in each periplasmic domain of the membrane protein DsbB are required for its function in protein disulfide bond formation. *EMBO J.* 13:5121–5127.

John, D.C.A., Grant, M.E. and Bulleid, N. (1993) Cell-free synthesis and assembly of prolyl 4-hydroxylase: the role of the β-subunit (PDI) in preventing misfolding and aggregation. *EMBO J.* 12:1578–1595.

Johnson, E., Henzel, W. and Deisseroth, A. (1992) An isoform of protein disulfide isomerase isolated from chronic myelogenous leukemia cells alters complex formation between nuclear proteins and regulatory regions of interferon-inducible genes. *J. Biol. Chem.* 267: 14412–14417.

Kallis, G.-B. and Holmgreen, A. (1980) Differential reactivity of the functional sulfhydryl groups of cysteine-32 and cysteine-35 present in the reduced form of thioredoxin from *Escherichia coli. J. Biol. Chem.* 255:10261–10265.

Kamitani, S., Akiyama, Y. and Ito, K. (1992) Identification and characterization of an *Escherichia coli* gene required for the formation of correctly folded alkaline phosphatase, a periplasmic enzyme. *Cell* 11:57–62.

Kanaya, E., Anaguchi, H. and Kikuchi, M. (1994) Involvement of the two sulfur atoms of protein disulfide isomerase and one sulfur atom of the DsbA/PpfA protein in the oxidation of mutant human lysozyme. *J. Biol. Chem.* 269:4273–4278.

Kasper, M., Schuh, D. and Muller, M. (1994) Immunohistochemical localization of the β-sub-unit of prolyl 4-hydroxylase in human alveolar epithelial cells. *Acta Histochem.* 96:309–313.

Kishigami, S., Akiyama, Y. and Ito, K. (1995) Redox states of DsbA in the periplasm of *Escherichia coli. FEBS Lett.* 364:55–68.

Kivirikko, K.I., Myllyla, R. and Philajaniemi, T. (1989) Protein hydroxylation: prolyl 4-hydro-xylase, an enzyme with four cosubstrates and a multifunctional subunit. *FASEB J.* 3:1609–1617.

Kivirikko, K.I., Myllyla, R. and Pihlajaniemi, T. (1992) Hydroxylation of proline and lysine residues in collagens and other animal and plant proteins. *In:* J.J. Harding and J.C. Crabbe (eds) *Post-translational Modifications of Proteins,* CRC Press, Boca Raton, pp 1–51.

Koivu, J., Myllyla, R., Helaakoski, T., Pihlajaniemi, T., Tasanen, K. and Kivirikko, K.I. (1987) A single polypeptide acts both as the β-subunit of prolyl 4-hydroxylase and as protein disulfide isomerase. *J. Biol. Chem.* 262:6447–6449.

Krishna-Rao, A.S.M. and Hausman, R.E. (1993) cDNA for R-cognin: homology with a multi-functional protein. *Proc. Natl. Acad. Sci. USA* 90:2950–2954.

Kuznetsov, G., Chen, L.B. and Nigam, S.K. 81994) Several endoplasmic reticulum stress proteins, including ERp72, interact with thyroglobulin during its maturation. *J. Biol. Chem.* 269:22990–22995.

LaMantia, M. and Lennarz, W.J. (1993) The essential function of protein disulfide isomerase does not reside in its isomerase activity. *Cell* 74:1–20.

LaMantia, M., Miura, T., Tachikawa, H., Kaplan, H., Lennarz, W.J. and Mizunaga, T. (1991) Glycosylation site binding protein and protein disulfide isomerase are identical and essential for cell viability in yeast. *Proc. Natl. Acad. Sci. USA* 88:4453–4457.

Lambert, N. and Freedman, R.B. (1983) Kinetics and specificity of homogenous protein disulfide-isomerase in protein disulfide isomerization and thiol-protein-disulfide oxido-reduction. *Biochem. J.* 213:235–243.

Lilie, H., McLaughlin, S., Freedman, R.B. and Buchner, J. (1994) Influence of protein disulfide isomerase on antibody folding *in vitro. J. Biol. Chem.* 269:14290–14296.

Lin, M.C., Gordon, D. and Wetterau, J.R. (1995) Microsomal triglyceride transfer protein (MTP) regulation in HepG2 cells: insulin negatively regulates MTP gene expression. *J. Lipid Res.* 36:1073–1081.

Loferer, H., Bott, M. and Hennecke, H. (1993) *Bradyrhizobium japonicum* TlpA, a novel mem-brane-anchored thioredoxin-like protein involved in the biogenesis of cytochrome *aa3* and development of symbiosis. *EMBO J.* 12:3373–3383.

Lundstrom-Ljung, J., Birnbach, U., Rupp, K., Soling, H.D. and Holmgreen, A. (1995) Two resident ER-proteins, CaBP1 and CaBP2, with thioredoxin domains, are substrates for thioredoxin reductase: comparison with protein disulfide isomerase. *FEBS Lett.* 357:305–308.

Lyles, M.M. and Gilbert, H.F. (1994) Mutations in the thioredoxin sites of protein disulfide iso-merase reveal functional nonequivalence of the N- and C-terminal domains. *J. Biol. Chem.* 269:30946–30952.

Mandel, R., Ryser, H.J.-P., Ghani, F., Wu, M. and Peak, D. (1993) Inhibition of a reductive function of the plasma membrane by bacitracin and antibodies against protein disulfide isomerase. *Proc. Natl. Acad. Sci. USA* 90L:4112–4116.

Martin, J.L., Pumford, N.R., Larosa, A.C., Martin, B.M., Gonzaga, H.M.S., Beaven, M.A. and Pohl, L.R. (1991) A metabolite of halothane covalently binds to an endoplasmic reticulum protein that is highly homologous to phosphatidylinositol-specific phospholipase Ca but has no activity. *Biochem.* 178:679–685.

Martin, J.L., Bardwell, J.C.A. and Kuriyan, J. (1993) Crystal structure of the DsbA protein required for disulfide bond formation *in vivo. Nature* 365:464–468.

Mazzarella, R.A., Srinivasan, M., Haugejorden, S.M. and Green, M. (1990) ERp72 an abundant lumenal endoplasmic reticulum protein, contains three copies of the active site sequences of protein disulfide isomerase. *J. Biol. Chem.* 265:1094–1101.

Missiakas, D., Georgopoulos, C. and Raina, S. (1993) Identification and characterization of the *Escherichia coli* gene *dsbB*, whose product is involved in the formation of disulfide bonds *in vivo. Proc. Natl. Acad. Sci. USA* 90:7084–7088.

Missiakas, D., Georgopoulos, C. and Raina, S. (1994) The *Escherichia coli dsbC (xprA)* gene encodes a periplasmic protein involved in disulfide bond formation. *EMBO J.* 13:2013–2020.

Mizunaga, T., Katakura, Y., Miura, T. and Maruyama, Y. (1990) Characterization and purification of yeast protein disulfide isomerase. *J. Biochem.* 108:846–851.

Mori, K., Sant, A., Kohno, K., Normington, K., Gething, M.-J. and Sambrook, J.F. (1992) A 22bp cis-acting element is necessary and sufficient for the induction of the yeast *KAR2* (BiP) gene by unfolded proteins. *EMBO J.* 11:2583–2593.

Morjana, N.A. and Gilbert, H.F. (1991) Effect of protein and peptides inhibitors on the activity of protein disulfide isomerase. *Biochemistry* 30:4985–4990.

Nagayama, S., Yokoi, T., Kawaguchi, Y. and Kamataki, T. (1994a) Occurrence of autoantibody to protein disulfide isomerase in rats with xenobiotic-induced hepatitis. *J. Toxicol. Sci.* 19: 1255–1261.

Nagayama, S., Yokoi, T., Tanaka, H., Kawaguchi, Y., Shirasaka, T. and Kamataki, T. (1994b) Occurrence of autoantibody to protein disulfide isomerase in patients with hepatic disorder. *J. Toxicol. Sci.* 19:163–169.

Nelson, J.W. and Creighton, T.E. (1994) Reactivity and ionization of the active site cysteine residues of DsbA, a protein required for disulfide bond formation *in vivo. Biochemistry* 33:5974–5983.

Nguyen-Van, P.N., Rupp, K., Lampen, A. and Soling, H.-D. (1993) CaBP2 is a rat homolog of ERp72 with protein disulfide isomerase activity. *Europ. J. Biochem.* 213:789–795.

Nigam, S.K., Goldberg, A.L., Ho, S., Rohde, M.F., Bush, K.T. and Sherman, M.Y. (1994) A set of endoplasmic reticulum proteins possessing properties of molecular chaperones includes Ca^{2+} binding proteins and members of the thioredoxin superfamily. *J. Biol. Chem.* 269: 1744–1749.

Noiva, R., Kimura, H., Roos, J. and Lennarz, W.J. (1991) Peptide binding by protein disulfide isomerase, a resident protein of the endoplasmic reticulum lumen. *J. Biol. Chem.* 266: 19645–19649.

Noiva, R. and Lennarz, W.J. (1992) Protein disulfide isomerase- a multifunctional protein resident in the lumen of the endoplasmic reticulum. *J. Biol. Chem.* 267:3553–3556.

Noiva, R., Freedman, R.B. and Lennarz, W.J. (1993) Peptide binding to protein disulfide isomerase occurs at a site distinct from the active sites. *J. Biol. Chem.* 268:19210–19217.

Otsu, M., Omura, F., Yoshimori, T. and Kikuchi, M. (1994) Protein disulfide isomerase associates with misfolded human lysozyme *in vivo. J. Biol. Chem.* 269:6874–6877.

Otsu, M., Urade, R., Kito, M., Omura, F. and Kikuchi, M. (1995) A possible role of ER-60 protease in the degradation of misfolded proteins in the endoplasmic reticulum. *J. Biol. Chem.* 270:14958–14961.

Peek, J.A. and Taylor, R.K. (1992) Characterization of a periplasmic thiol:disulfide interchange protein required for the functional maturation of secreted virulence factors secreted by *Vibrio cholerae. Proc. Natl. Acad. Sci. USA* 89:6210–5214.

Pelham, H.R.B. (1991) Recycling of proteins between the endoplasmic reticulum and the Golgi complex. *Curr. Opin. Cell Biol.* 3:585–591.

Phihlajaniemi, T., Helaakoski, T., Tasanen, K., Myllyla, R., Huhtala, M.-L., Koivu, J. and Kivirikko, K.I. (1987) Molecular cloning of the β-subunit of human prolyl 4-hydroxylase. This subunit and protein disulfide isomerase are products of the same gene. *EMBO J.* 6:643–649.

Puig, A. and Gilbert, H.F. (1994a) Protein disulfide isomerase exhibits chaperone and anti-chaperone activity in the oxidative refolding of lysozyme. *J. Biol. Chem.* 269:7764–7771.

Puig, A. and Gilbert, H.F. (1994b) Anti-chaperone behavior of BiP during protein disulfide isomerase-catalyzed refolding of reduced denatured lysozyme. *J. Biol. Chem.* 269:25889–25896.

Puig, A., Lyles, M.M., Noiva, R. and Gilbert, H.F. (1994) The role of thiol/disulfide centers and the peptide binding site in the chaperone and anti-chaperone activities of protein disulfide isomerase. *J. Biol. Chem.* 269:19128–19125.

Quan, H., Fan, G. and Wang, C-C. (1985) Independence of the chaperone activity of protein disulfide isomerase from its thioredoxin-like active site. *J. Biol. Chem.* 270:17078–17080.

Ricci, B., Sharp, D., O'Rourke, E., Kienzle, B., Blinderman, L., Gordon, D. Smith-Monroy, C., Robinson, G., Gregg, R.E., Rader, D.J. and Wetterau, J.R. (1995) A 30-amino acid truncation of the microsomal triglyceride transfer protein large subunit disrupts its interaction with protein disulfide isomerase and causes abetalipoproteinemia. *J. Biol. Chem.* 270: 14281–14285.

Roth R. and Pierce, S.B. (1987) *In vivo* cross-linking of protein disulfide isomerase to immunoglobulins. *Biochemistry* 26:4179–4182.

Rupp, K., Birnbach, U., Lundstrom, J., Nguyen-Van, P. and Soling H.-D. (1994) Effects of CaBP2, the rat analog of ERp72, and of CaBP1 on the refolding of denatured reduced proteins. *J. Biol. Chem.* 269:2501–2507.

Ryser, H.J.P., Mandel, R. and Ghani, F. (1991) Cell surface sulfhydryls are required for the cytotoxicity of diphtheria toxin but not of ricin in chinese hamster ovary cells. *J. Biol. Chem.* 266:18439–18442.

Ryser, H.J.P., Levy, E.M., Mandel, R. and DiSciullo, G.J. (1994) Inhibition of human immunodeficiency virus infection by agents that interfere with thiol-disulfide interchange upon virus-receptor interaction. *Proc. Natl. Acad. Sci. USA* 91:4559–4563.

Safran, M., Farwell, A.P. and Leonard, J.A. (1992) Thyroid hormone-dependent redistribution of the 55 kilodalton monomer of protein disulfide isomerase in cultured glial cells. *Endocrinology* 131:2413–2418.

Shevchik, V.E., Condemine, G. and Robert-Baudouy, J. (1994) Characterization of DsbC, a periplasmic protein of *Erwinia chrysanthemi* and *Escherichia coli* with disulfide isomerase activity. *EMBO J.* 13:2007–2012.

Srivastava, S.P., Chen, N.Q., Liu, Y.X. and Holtzman, J.L. (1991) Purification and characterization of a new isozyme of thiol:protein disulfide oxireductase from rat hepatic microsomes. *J. Biol. Chem.* 266:20337–20344.

Srivastava, S.P., Fuchs, J.A. and Holtzman, J.L. (1993) The reported cDNA sequence for phospholipase C_α encodes protein disulfide isomerase, isozyme Q2 and not phospholipase-C. *Biochem. Biophys. Res. Comm.* 193:971–978.

Tachibana, C. and Stevens, T.H. (1992) The yeast *EUG1* encodes an endoplasmic reticulum protein that is functionally related to protein disulfide isomerase. *Mol. Cell. Biol.* 12:4601–4611.

Tachikawa, H., Miura, T., Katakura, Y. and Mizunaga, T. (1991) Molecular structure of a yeast gene, *PDI1,* encoding protein disulfide isomerase that is essential for cell growth. *J. Biochem.* 110:306–313.

Takemoto, H., Yoshimori, T., Yamamoto, A., Miyata, Y., Yahara, I., Inoue, K. and Tashiro, Y. (1992) Heavy chain binding protein (BiP/GRP78) and endoplasmin are exported from the endoplasmic reticulum of rat exocrine pancreatic cells, similar to protein disulfide-isomerase. *Arch. Biochem. Biophys.* 296:129–136.

Thony-Meyer, L., Ritz, D. and Hennecke, H. (1994) Cytochrome *c* biogenesis in bacteria: a possible pathway begins to emerge. *Mol. Microbiol.* 12:1–9.

Tomb, J.-F. (1992) A periplasmic protein disulfide oxireductase is required for transformation of *Haemophilus influenzae* Rd. *Proc. Natl. Acad. Sci. USA* 89:10252–10256.

Tsibris, J.C.M., Hunt, L.T., Ballejo, G., Barker, W.C., Toney, L.J. and Spellacy, W.N. (1989) Selective inhibition of protein disulfide isomerase by estrogens. *J. Biol. Chem.* 264:13967–13970.

Urade, R. and Kito, M. (1992) Inhibition by acidic phospholipids of protein degradation by ER-60 a novel cystein protease of the endoplasmic reticulum. *FEBS Lett.* 312:83–86.

Urade, R., Nasu, M., Moriymana, T., Wada, K. and Kito, M. (1992) Protein degradation by the phosphoinositide-specific phospholipase $C\alpha$ family from the rat liver endoplasmic reticulum. *J. Biol. Chem.* 267:15152–15159.

Urade, R., Takenaka, Y. and Kito, M. (1993) Protein degradation by ERp72 from rat and mouse liver endoplasmic reticulum. *J. Biol. Chem.* 268:22004–22009.

Veijola, J., Koivunen, P., Annunen, P., Pihlajaniemi, T. and Kivirikko, K.I. (1994) Cloning, baculovirus expression, and characterization of the α-subunit of prolyl 4-hydroxylase from the nematode *Caenorhabditis elegans. J. Biol. Chem.* 269:26746–26753.

Vuori, K., Myllyla, R., Philajaniemi, T. and Kivirikko, K.I. (1992a) Expression and site-directed mutagenesis of human protein disulfide isomerase in *Escherichia coli. J. Biol. Chem.* 267:7211–7214.

Vuori, K., Pihlajaniemi, T., Myllya, R. and Kivirikko, K.I. (1992b) Site-directed mutagenesis on human protein disulfide isomerase: effect on assembly, activity and endoplasmic reticulum retention of human prolyl 4-hydroxylase in *Spodoptera frugiperda* insect cells. *EMBO J.* 11:4213–4217.

Wells, W.W., Xu, D.P., Yang, Y. and Rocque, P.A. (1990) Mammalian thioltransferase (glutaredoxin) and protein disulfide isomerase have dehydroascorbate reductase activity. *J. Biol. Chem.* 265:15361–15364.

Wetterau, J.R., Aggerbeck, L.P., Laplaud, P.M. and McLean, L.R. (1990) Protein disulfide isomerase is a component of the triglyceride transfer protein complex. *J. Biol. Chem.* 265:9800–9807.

Wetterau, J.R., Combs, K.A., McLean, L.R., Spinner, S.N. and Aggerbeck, L.P. (1991) Protein disulfide isomerase appears to be necessary to maintain the catalytically active structure of the microsomal triglyceride transfer protein. *Biochemistry* 30:9728–9735.

Wetterau, J.R., Aggerbeck, L.P., Bouma, M.E., Eisenberg, C., Munck, A., Hermier, M., Schmitz, J., Gay, G., Rader, D.J. and Gregg, R.E. (1992) Absence of microsomal tryglyceride transfer protein in individuals with abetalipoproteinemia. *Science* 258:999–1001.

Whitley, P. and von Heijne, G. (1993) The DsbA-DsbB system affects the formation of disulfide bonds in periplasmic but not in intramembranous protein domains. *FEBS Lett.* 332:49–51.

Wilson, W.R., Tuan, R.S., Shepley, K.J., Freedman, D.O., Greene, B.M., Awadzi, K. and Unnasch, T.R. (1994) The *Onchocerca volvulus* homologue of the multifunctional polypeptide protein disulfide isomerase. *Mol. Biochem. Parasit.* 68:103–107.

Wunderlich, M. and Glockshuber, R. (1993) *In vivo* control or redox potential during protein folding catalyzed by bacterial protein disulfide isomerase (DsbA) *J. Biol. Chem.* 268:24547–24550.

Wunderlich, M., Otto, A., Maskos, K., Mucke, M., Seckler, R. and Glockshuber, R. (1995) Efficient catalysis of disulfide bond formation during protein folding with a single active-site cysteine. *J. Mol. Biol.* 247:28–33.

Yamauchi, K., Yamamoto, T., Hayashi, H., Koya, S., Takikawa, H., Toyoshima, K. and Horiuchi, R. (1987) Sequence of membrane-associated thyroid hormone binding protein from bovine liver: its identity with protein disulfide isomerase. *Biochem. Biophys. Res. Comm.* 146:1485–1492.

Yokoi, T., Nagayama, S., Kajiwara, R., Kawaguchi, Y. and Kamataki, T. (1994) Effects of cyclosporin-A and D-penicillamine in the development of hepatitis and the production of antibody to protein disulfide isomerase in LEC rats. *Res. Comm. Mol. Pathol. Pharmacol.* 85:73–81.

Yoshimori, T., Semba, T., Takemoto, H., Akagi, S., Yamamoto, A. and Tashiro, Y. (1990) Protein disulfide-isomerase in rat exocrine pancreatic cells is exported form the endoplasmic reticulum despite possessing the retention signal. *J. Biol. Chem.* 265:15984–15990.

Yu, J., Webb, H. and Hirst, T.R. (1992) A homologue of *Escherichia coli* DsbA protein involved in disulfide bond formation is required for enterotoxin biogenesis in *Vibrio cholerae. Mol. Microbiol.* 6:1949–1958.

Yu, J., McLaughlin, S., Freedman, R.B. and Hirst, T.R. (1993) Cloning and active site mutagenesis of the *Vibrio cholera* DsbA, a periplasmic enzyme that catalyzes disulfide bond formation. *J. Biol. Chem.* 268:4326–4330.

Zapun, A., Bardwell, J.C.A. and Creighton, T.E. (1993) The reactive and destabilizing disulfide bond of DsbA, a protein required for protein disulfide bond formation *in vivo. Biochemistry* 32:5083–5092.

Zapun, A. and Creighton, T.E. (1994) Effects of DsbA on the disulfide folding of bovine pancreatic trypsin inhibitor and α-lactalbumin. *Biochemistry* 33:5202–5211.

Zapun, A., Cooper, L. and Creighton, T.E. (1994) Replacement of the active-site residues of DsbA, a protein required for disulfide bond formation *in vivo. Biochemistry* 33:1907–1914.

Part II
Regulation of inducible stress responses

Introduction

R. I. Morimoto

Department of Biochemistry, Northwestern University, Molecular and Cell Biology, 2153 Sheridan Road, Evanston, IL 60208-3500, USA

The response to environmental and physiological stress includes a highly ordered set of events that is often represented by rapid changes in gene expression followed by the synthesis of proteins involved in adaptation to the stress. Two examples presented in this section include the SOS response to DNA damaged the heat shock response to protein damage. The genes involved in the SOS response are under the regulation of recA and lexA gene products of which lexA acts to repress the transcription of genes involved in repair mechanisms. A consequence of these events is the reduced fidelity of DNA replication to allow error-prone synthesis, the accumulation of mutations and cell survival. The SOS response provides unique insights into cell survival mechanisms at the level of protection of the genome. By comparison, the heat shock response in prokaryotes utilizes the s^{32} and s^E proteins as positive activators of genes encoding heat shock proteins and molecular chaperones. The mechanism by which s^{32} levels increase following heat shock involves a novel post-translational process in which there is increased translation of s^{32} RNA and stabilization of s^{32} which involves interactions with dnaK, dnaJ, and grpE in a complex autoregulatory loop. In eukaryotes, the heat shock transcriptional response is mediated by members of the heat shock factor (HSF) family. The process by which the HSFs are regulated in response to stress and negatively regulated under conditions of normal cell growth reveals an understanding of the mechanisms by which the eukaryotic cell detects stress. Among the other well studied events of the stressed cell are changes in protein folding/protein unfolding and on protein modification such as phosphorylation. These "inducible" events may be integrated to diminish some of the deleterious effects of prolonged stress and as a means of signaling components of the replication, transcription, and translation machinery. Together, these examples provide a representative presentation of inducible stress response.

Stress-Inducible Cellular Responses
ed. by U. Feige, R.I. Morimoto, I. Yahara and B. Polla
© 1996 Birkhäuser Verlag Basel/Switzerland

Sensing stress and responding to stress

R. Voellmy

Department of Biochemistry and Molecular Biology, University of Miami School of Medicine, Miami, FL 33101, USA

Summary. Heat shock protein gene expression is enhanced by proteotoxic stress, i.e., by conditions favoring protein unfolding. This upregulation of heat shock protein genes is mediated by heat shock transcription factor HSF1. A mechanism, the details of which are still elusive, senses adverse conditions and causes HSF1 to oligomerize and to acquire DNA-binding ability. The DNA-binding form of HSF1 then undergoes further conformational changes that render it transcriptionally competent. The current model in which heat shock protein 70 acts both as sensor of stress and as negative regulator of HSF1 oligomerization as well as alternative models involving additional protein factors are discussed.

Introduction

When cells are exposed to temperatures that exceed their normal growth temperature, they respond by inducing the synthesis of several polypeptides referred to as heat shock proteins (HSPs). This phenomenon is seen in eukaryotic as well as prokaryotic organisms. The focus of this review is on eukaryotic cells, and readers interested in prokaryotes are referred to the chapter by Yura that deals with the latter organisms.

Increases in rates of *hsp* gene expression are generally proportional to the severity of the heat stress applied to cells (DiDomenico et al, 1982). However, the inducibility of different hsps can differ drastically. In mammalian cells, for example, hsp90, hsc70 and hsp60 synthesis is only slightly heat-induced during severe heat treatment, whereas that of hsp70 and, in some cell types, of a small hsp increases many fold (Welch, 1992). Differences in heat induction have even been noted between different *hsp70* genes. One human *hsp70* gene (*hsp70A*) was found to be upregulated 10–20 fold (Wu et al., 1985; Choi et al., 1991), whereas the expression of another (*hsp70B*) increased by several orders of magnitude (Voellmy et al., 1985; Dreano et al., 1986). Furthermore, as noted previously in studies involving *Drosophila* cell lines, different *hsp* genes appear to display different sensitivities to heat stress, with the *hsp90* gene being induced at lower temperatures than small *hsp* genes, and small *hsp* genes at lower temperatures than *hsp70* genes (Lindquist, 1980). Particularly intriguing is a phenomenon referred to as attenuation. As originally shown in *Drosophila* cells (DiDomenico et al., 1982) and more recently also in mammalian cells (Abravaya et al., 1991), cells exposed to extreme heat shock quickly initiate massive HSP RNA and protein synthesis which is maintained for the duration of the stress. When cells are subjected to a more moderate, sustained heat shock, they initially respond by drastically up-regulating HSP synthesis. However, rates of HSP synthesis soon decline to near pre-heat shock levels. Related to these observations is the finding by DiDomenico et al. (1982) that exposure of cells to a given level of heat stress programs them to synthesize a defined amount of HSPs, i.e., that subsequent to stress activated *hsp* genes are only shut down after a set amount of HSPs has been synthesized. Thus, it appears that *hsp* gene expression is controlled by some type of feedback mechanism, presumably involving one or more Hsps. While these experiments did not reveal the identity of the putative regulatory Hsp(s), Hsp70 was considered the prime candidate (see below).

Heat shock transcription factor and heat shock elements

Heat regulation of *hsp* genes is mediated by heat shock transcription factor (HSF) (Parker and Topol, 1984; Wu, 1984). HSF has been shown to recognize modular sequence elements within *hsp* gene promoters, referred to as heat shock elements (HSE), consisting of three or more 5-bp units (Amin et al., 1988; Xiao et al., 1988). Whereas yeasts and fruit flies appear to have only a single type of HSF, plants and vertebrates (i.e., chicken, mouse and man) express two or more different factors. In vertebrates only one of these factors (HSF1) is capable of mediating stress regulation of hsp genes (Baler et al., 1993; Sarge et al., 1993).

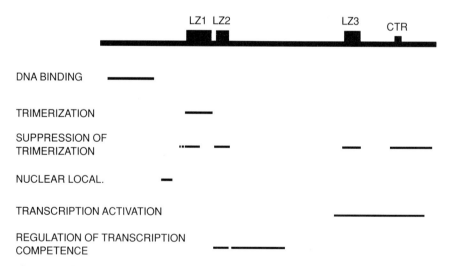

Figure 1. Schematic representation of the structure of human HSF1. The solid line represents the 529 amino acid-long protein-coding region. Below are regions identified by various mapping studies to play a role in different aspects of factor regulation.

HSFs from different organisms share a number of structural features. All have a conserved DNA-binding domain located near the amino terminus that includes a helix-turn-helix motif similar to that found in catabolite activator protein (Harrison et al., 1994; Vuister et al., 1994). Further inside the molecule are two closely spaced regions containing hydrophobic heptad repeats of the 4–3 type (Hodges et al., 1972; McLachlan and Stewart, 1975), referred to as leucine zippers 1 and 2 (LZ1 and 2). The central portion of LZ1 and LZ2 sequences is particularly well conserved. Following LZ2 is a 110-residue region that is well conserved in vertebrate HSF1. The latter region is separated by less well conserved sequence from a third 4–3 heptad repeat (LZ3) that appears to be present in all HSF species. Finally, near the carboxy terminus of vertebrate HSF is a short element referred to as carboxy-terminal repeat (CTR) (Rabindran et al., 1993). For the purposes of this review, the term HSF shall be used exclusively to describe HSF species involved in the stress regulation of *hsp* genes, including HSF1 in vertebrates.

The rate of expression of the HSF gene does not appear to be increased during stress (Rabindran et al., 1991). Thus, HSF is present in both unstressed and stressed cells. However, HSF in unstressed cells is in an inactive form, and stress is required to convert it into an active transcription factor. There appear to be basic differences in the regulation of HSF activity in budding yeasts and other eukaryotes. In yeasts, HSF is trimeric and capable of DNA binding even in the absence of stress (Sorger et al., 1987; Sorger and Nelson, 1989). Stress is required for the transcriptional competence of the factor. In other eukaryotes, HSF is subject to regulation at the level of oligomerization (Westwood et al., 1991) and, in at least some cases, also at the level of transcriptional competence (see references below). In extracts from unstressed cells, HSF is present either as a monomer (Westwood et al., 1991; Baler et al., 1993, Sarge et al., 1993) or as a heterodimer with an hsp70-type protein (Baler, 1992; Rabindran et al., 1994: Baler et al., 1995). As discussed below, it is now suspected that HSF monomers may associate with large multiprotein complexes. HSF monomers or heterodimers are essentially incapable of binding HSE DNA sequences. Upon heat treatment or other stressful events, HSF oligomerizes to form homotrimers (Westwood et al., 1991; Baler et al., 1993; Sarge et al., 1993). The acquisition of DNA-binding ability and the translocation of HSF to the nucleus are thought to be consequences of its oligomerization (Baler et al., 1993; Sarge et al., 1993).

However, oligomerization alone does not render vertebrate HSF transcriptionally competent (Zuo et al., 1995). A second event needs to occur to uncover the transcription-enhancing activity of the factor. This second event can be inferred from several, diverse observations. First, a number of situations has been described in which HSF undergoes oligomerization, but is not transcriptionally activated. For example, in certain human cell types, salicylate, menadione and hydrogen peroxide have been reported to induce HSF oligomerization and DNA-binding ability, without rendering the factor transcriptionally competent (Jurivich et al., 1992; Bruce et al., 1993). Second, a murine erythroleukemia cell line that is unable to increase hsp70 synthesis upon heat treatment is nevertheless capable of forming HSF oligomers (Hensold et al., 1990). Third, studies with permeabilized mouse cells have revealed an ATP-dependent step specifically required for the transcriptional activation of HSF (Price and Calderwood, 1991). Finally, although human HSF overexpressed in human HeLa cells is largely trimeric, DNA binding and nuclear-localized even in the absence of stress, stress is still required to render the factor transcriptionally active (Zuo et al., 1995). Thus, it appears that the activity of HSF in species other than budding yeasts is initially controlled at the level of DNA binding and, possibly, nuclear transport. Furthermore, in vertebrates there is a second level of control regulating transcriptional competence.

Several sequence elements in HSF that are important for regulation of its activity have been defined by mutagenesis experiments (Fig. 1). A first study, by Rabindran et al. (1993), overexpressed human HSF in human 293 cells. The authors reported that oligomerization and DNA binding of exogenous HSF is heat-regulated in this cell type, provided that the level of overexpression of the exogenous factor is carefully controlled. Using this system they demonstrated that interruption of LZ3 or deletion of sequences near the carboxy terminus caused HSF to become constitutively trimeric and DNA binding. Expressing human HSF in *Xenopus* oocytes, a system in which oligomerization and DNA-binding ability of the exogenous factor are tightly heat-regulated, Zuo et al. (1994) confirmed the importance of LZ3. Moreover, they showed that single amino acid substitutions in any of the three LZs cause deregulation of oligomerization and DNA binding. Thus, human HSF is maintained in a monomeric form by interactions involving all three LZs which may bind one another to form a triple-stranded coiled-coil structure. Whereas this coiled-coil structure of human HSF may be stable in *Xenopus* oocytes routinely maintained at 20 °C, in human cells it is thought to require further stabilization through interactions with other proteins (see below).

Transcription activation domains have been mapped to regions near the carboxy terminus of HSF (Nieto-Sotelo et al., 1990, Sorger, 1990; Green et al., 1995). In addition, *S. cerevisiae* HSF has an amino-terminal activation domain that precedes the DNA-binding region (Nieto-Sotelo et al., 1990; Sorger, 1990). Different approaches have been used to map the region(s) involved in the regulation of transcriptional competence of HSF. When consecutively longer amino-terminal regions in *S cerevisiae* HSF were replaced with a heterologous DNA-binding domain, transcriptional activity became deregulated as soon as sequences close to or including LZs 1 and 2 were removed (Nieto-Sotelo et al., 1990). Analogous observations were made with human HSF (Zuo et al., 1995). Experiments which systematically replaced sequences from the carboxy terminus in amino-terminally truncated *S. cerevisiae* HSF with a VP16 activation domain implicated an extended region including LZ1, LZ2 and sequences following these zippers in the regulation of the carboxy-terminal activation domains (Bonner et al., 1992). Chen et al. (1993) showed that deletion of LZ2 sequences alone deregulated *S. cerevisiae* HSF activity. Chimeras containing GAL4 DNA-binding, nuclear localization and dimerization domains linked to human HSF sequences at a position immediately after LZ2 were found to enhance the expression of a GAL4-responsive reporter gene several fold better in heat-treated than in untreated HeLa cells (Green et al., 1995). Further deletion produced a chimera containing the 110-residue region immediately following LZ2 and a 160-residue region from the carboxy terminus of HSF that still showed heat-stimulated transactivation ability. As noted before, transient overexpression of human HSF in HeLa cells results in trimeric and DNA binding but transcriptionally inactive factor. When deregulation of transcription ability by deletion and substitution mutants of HSF was examined in transfected HeLa cells, the same region adjacent to LZ2 was identified as being critical to transcriptional regulation as that revealed by the above experiments with GAL4-HSF chimeras (Zuo et al., 1995). In addition, these assays also showed that LZ2 sequences are essentially required for maintaining heat regulation of transcriptional activity. In fact, single substitutions in LZ2 caused substantial deregulation. Together, these data identify two types of elements that play important roles in the regulation of transcriptional competence. The first is an extended region adjacent to LZ2 that is alone capable of conferring at least some degree of heat regulation on HSF activation domains, and the second is a short region including LZ2 whose presence is essential for heat-regulated transcriptional competence of the intact HSF molecule.

Note that the majority of studies on regulation of HSF in organisms other than yeasts has been concerned with regulation at the level of oligomerization and DNA binding. The following discussion therefore focuses on this aspect of HSF control. Although oligomerization alone does not render HSF transcriptionally competent (see before), the complete absence of information, including from the yeast system, regarding the nature of the event(s) regulating transcriptional competence precludes a meaningful discussion at this time.

Inducers of hsp gene expression and the abnormal protein mechanism of regulation

In addition to heat, a bewildering number of agents and treatments stimulates the expression of *hsp* genes. For an abbreviated listing the reader is referred to Ananthan et al. (1986). While some inducers do not fit neatly in any category (for example, prostaglandins and their precursors), many can be assigned to two groups, one of which may be further divided into a small number of subgroups. The first group includes amino acid analogues, puromycin and ethanol. Compounds in this group cause the synthesis of non-native proteins either through incorporation of amino acid residues with altered side chains, premature termination of nascent polypeptides or errors in translation. The second group includes agents and treatments that are likely to alter the conformation of preformed native proteins. One subgroup is comprised of heat treatment which can be expected to increase the rate of protein unfolding. A second subgroup includes certain heavy metal ions, arsenite, certain chelators and other thiol-reactive compounds such as p-mercuribenzoate, iodoacetamide, diamide etc. Compounds in this subgroup may interact either directly or indirectly, via changes in the redox state, e.g., in the ratios of oxidized to reduced forms of glutathione or other redox pairs, with exposed thiols, causing protein thiolation and protein-protein disulfide formation (Grimm et al. 1985; Collison et al, 1986; see also Kosower and Kosower, 1978; Brigelius, 1985). Both types of interactions may result in conformational changes in target proteins. A third subgroup includes recovery from anoxia, hydrogen peroxide, superoxide ions and other free radicals. Also placed in this subgroup may be a number of hepatotoxicants such as bromobenzene, carbon tetrachloride, chloroform, acetaminophen, cocaine, etc., that induce hsp gene expression in the intact animal or in a hepatocyte-derived cell lines (S. Roberts et al., unpublished observations). Redox cyclers such as naphtho- and benzoquione-containing molecules may also be grouped here. While free radicals formed may interact directly with target proteins and may cause their fragmentation, there is also the possibility that they may act indirectly by changing the redox state of the cell which in turn may lead to covalent modification of target proteins (see before). A final subgroup includes amytal, antimycin, azide, dinitrophenol, rotenone, etc. These compounds cause inhibition of oxidative phosphorylation, leading to changes in the ATP level and the redox state. Conformational changes in proteins may result from covalent modification or may be induced by the absence of ATP ligand.

Thus, it appears that both groups of inducers cause, although by different mechanism, the formation of non-native proteins. These considerations led to the hypothesis that *hsp* gene activation may be triggered by the accumulation of non-native proteins. The first attempt to verify this hypothesis was undertaken by Goff and Goldberg (1985). The design of their experiment was based on the observation that *E. coli* is incapable of producing secreted eukaryotic proteins having proper disulfide bridges and thus a native conformation. The authors showed that *E. coli* overexpressing tissue plasminogen activator or bovine serum albumin also expressed high levels of HSPs. This up-regulation of HSP synthesis was dependent on σ^{32}, the factor controlling the heat shock response in bacteria. The hypothesis was tested directly by Ananthan et al. (1986). Their experiment relied on the previous observation that a bacterial β-galactosidase gene under the control of a eukaryotic *hsp* gene promoter injected into the nucleus of *Xenopus laevis* oocytes is silent, but can be activated by heat treatment of the oocytes (Voellmy et al., 1985). The authors chose several purified proteins that were irreversibly denatured by reductive carboxymethylation. Co-injection of these chemically denatured proteins caused up-regulation of the β-galactosidase reporter gene in the absence of heat treatment. When the native forms of the same proteins were co-injected, the reporter gene remained silent. Similar experiments using reporter genes with consecutively larger deletions in the linked hsp promoter showed that activation by non-native proteins was dependent on the presence of HSE sequences. These results indicate that the system responsible for stress regulation of *hsp* genes senses and reacts to the presence of non-native proteins.

Over time, a variety of mechanisms has been suggested to explain how *hsp* genes are induced by stress. One proposition was that the regulatory mechanism may be based on redox changes in the cell. Interest in this idea has been rekindled by a recent study by Huang et al. (1994; see also Freeman et al., 1995). These authors have shown that preincubation of human HeLa cells with reducing agents such as dithiothreitol or β-mercaptoethanol prevents induction of HSF oligomerization and DNA-binding activity by mild but not severe heat shock, oxygen deprivation, exposure to hydrogen peroxide or dinitrophenol. These observations have been interpreted as providing strong support for a "redox" mechanism of regulation of *hsp* gene expression.

It is important to note that aspects of the postulated "redox" mechanism are integral parts of the abnormal protein mechanism of induction discussed before. Oxygen deprivation and hydrogen peroxide exposure were envisaged to cause conformational changes by producing a more oxidized environment, leading to protein thiolation and formation of protein-protein disulfides. Dinitrophenol was expected to act by a similar mechanism and, in addition, to cause unfolding of particular proteins requiring ATP binding. Protein thiolation and subsequent formation of protein-protein disulfides is expected to be prevented by excess reducing agent. In contrast to the abnormal protein mechanism outlined before, a pure redox mechanism is not well suited to explain HSF activation by heavy metals, since it is unlikely that heavy metal ions are reduced by the intracellular environment. The abnormal protein mechanism presumes that heavy metal ions interact directly with susceptible proteins, causing conformational changes resulting in a non-native structure. This notion is based on the observation that the main heavy metal ions that induce *hsp* gene expression are Cd^{2+}, Cu^{2+}, Hg^{2+}, and Zn^{2+} (Levinson et al., 1980). These metal ions are known to interact with thiols. The finding reported by Huang et al. (1994) that dithiothreitol and β-mercaptoethanol prevent induction by these metal ions may be explained simply as being due to the capture of the metal ions by the thiol-containing reducing agents. The only finding that was not predicted by the abnormal protein mechanism as outlined before is that induction of *hsp* genes by a mild heat shock can also be preempted by exposure to reducing agents. This effect may be explained by the formation during heat shock of reactive oxygen intermediates (Mitchell and Russo, 1983). Such an explanation would also be compatible with the observations that oxidation of glutathione occurs during heat stress (Skibba et al., 1989) and that depletion of glutathione renders *hsp* gene expression more sensitive to temperature stress (Freeman et al., 1993). Thus, some of the protein unfolding caused by heat shock may be due to thiolation and protein-protein disulfide formation and/or radical-induced damage rather than to direct effects of high temperature as envisaged previously. In summary, although it was not presented in this light, the study by Huang et al. (1994) provided substantive support for some aspects of the abnormal protein mechanism. It confirmed and expanded our notion of how certain inducers of the heat shock response may cause protein unfolding.

Hsp70 as the sensor of proteotoxic stress

As alluded to before, a number of laboratories including our own became enamored with the idea that hsp70-type proteins may feedback-regulate the stress-inducible *hsp* genes. The demonstration that accumulation of non-native proteins elicits the response provided considerable conceptual support for this idea. It required that a molecule be identified that is capable of sensing elevated concentrations of non-native proteins and of transmitting this information to HSF. Hsp70 proteins are a class of proteins whose major functions appear to be directly related to their ability to generically recognize non-native conformations in proteins (Bole et al., 1986; Beckmann et al., 1990, Nelson et al, 1992; Palleros et al., 1993: Hightower et al., 1994). They, therefore, would be ideally suited to play the postulated regulatory role. Observations that HSF overexpressed in several systems accumulates as a trimer (Clos et al., 1990; Rabindran et al., 1991; Zuo et al., 1995) suggested that the postulated regulatory factor should act negatively. This then gave rise to the following simple model of regulation of HSF oligomerization and DNA-binding ability (Fig. 2A). In nonstressed cells, HSF oligomerization is prevented by the formation of HSF-hsp70 heterodimers or higher order complexes containing HSF (see below). When cells are stressed, they accumulate non-native proteins that compete with HSF for hsp70 binding. As a consequence of this competition, the fraction of unbound, monomeric HSF increases, and with it the probability of HSF trimer formation. Once HSF has trimerized and acquired DNA-binding ability, a further reaction(s) is required to render the factor transcription-competent. As discussed before, the model does not concern itself with the latter reaction. Activated HSF enhances HSP synthesis. During recovery from stress HSPs continue to be synthesized, and non-native proteins either are returned to their native conformation or are degraded by the cell's proteolytic systems. With time the concentration of hsp70 available for binding HSF increases to the pre-stress level, HSF trimers are disassembled, and HSF-hsp70 heterodimers are formed again.

There is considerable indirect support for aspects of this model. First, manipulations that transiently alter the level of hsp70 available for binding to HSF produce the expected regulatory outcome. When the level of hsp70 in human HeLa cells was increased by a pre-heat treatment (in HeLa cells the only hsps whose levels drastically increase following heat shock are of the hsp70 type), HSF became considerably more refractory to oligomer- ization and induction of DNA-binding ability during a subsequent heat treatment (Baler et al., 1992). Cells contain two types of non-native proteins, proteins partially unfolded as a consequence of either stress or normal wear and tear, and nascent polypeptides that have not yet acquired their final, stable conformation. Hsp70 proteins are known to interact with both types of non-native proteins (see references above). It could, therefore, be predicted that a reduction in the level of nascent polypeptides would result in an increased level of hsp70 available for HSF binding. It was indeed found that human HeLa or K562 cells exposed to protein synthesis inhibitors that deplete the pool of nascent polypeptides become refractory to induction of HSF trimerization and DNA-binding activity by a moder- ate heat shock (Baler et al., 1992). As also expected, this effect is abolished by exposure of the cells to a more severe heat stress capable of compensat- ing for the absence of nascent polypeptides by causing unfolding of a suf- ficiently large quantity of native protein. Furthermore, when rodent cells were microinjected with anti-hsc70 antibody, decreasing the pool of func- tional hsp70-type proteins, the rate of inducible hsp70 synthesis was increased (W. Welch, personal communication). Finally, Baler (1992) observed that he could transiently raise, by several fold, the level of hsp70 proteins in COS monkey cells through transfection with an RSV promoter- driven human hsp70A gene. When he co-transfected the cells with the hsp70A construct and a β-galactosidase reporter gene under control of the highly heat-inducible human hsp70B promoter, he found that the reporter gene was no longer heat-inducible. When he used a mutant hsp70A con- struct with a stop codon early in the hsp70-coding sequence in the co-trans- fection assay, the heat inducibility of the reporter gene was unaffected. Second, Mosser et al. (1993) isolated human cell lines overexpressing the human hsp70A protein and found that HSF in these cells was refractory to heat activation. Finally, Abravaya et al. (1992) observed that an hsp70 pro- tein inhibited the *in vitro* heat activation of HSF.

Recently, Rabindran et al. (1994) reported results from experiments involving overexpression of hsp70 in rat cells suggesting that hsp70 levels and HSF regulation are not correlated in these cells. Given the substantial body of evidence supporting a role of hsp70-type proteins in HSF regula- tion (see before), one wonders whether the negative results of these authors may be related to the observation that inducible *hsp* genes are more stringently repressed in the absence of stress in rodent cells than in human cells, the latter having been used in most of the experiments cited above. It is interesting in this regard to take note again of the experiment by the Welch group showing that depletion of hsc70 by antibody injection activates inducible *hsp* genes in rodent cells.

Obviously, the prediction most basic to the above model is that hsp70 proteins should be able to bind HSF. Abravaya et al. (1992) and Baler et al. (1992) were able to demonstrate that a fraction of activated human HSF formed complexes with hsp70 proteins. Baler at al. also used purified components to establish that hsp70 did bind directly to oligomeric HSF. More importantly, hsp70 was recently also found to interact with inactive monomeric HSF. Limiting pore size gel electrophoresis experiments revealed that a fraction of inactive human HSF exists as a HSF-hsp70 heterodimer (Baler, 1992; Baler et al., 1995). Rabindran et al. (1994) showed by immunoprecipitation assays that inactive *Drosophila* HSF also forms complexes containing hsp70. Thus, the most critical prediction of the model that hsp70 should be capable of interacting with HSF could be verified. However, because they discovered hsp70 binding to both inactive and active forms of HSF, these experiments did not provide independent evidence for a regulatory role of hsp70. Note that the observation of hsp70 binding to activated HSF is consistent with the model which predicts reassociation of hsp70 with HSF during recovery from stress.

Other protein-protein interactions involving HSF

While the data presented above generally support the model that hsp70 proteins serve as negative regulators of HSF, they do not exclude the possibility that additional proteins, particularly other chaperones, may interact with HSF and also play a role in its regulation. Early support for this possibility came from studies with an hsp90 affinity column which showed that chromatography of cell extract on this affinity matrix enriches HSF (Nadeau et al., 1993) That complexes of HSF with proteins other than hsp70 have not been discovered in cell extracts does not argue against their existence in the cell since such associations may be relatively unstable and/or transient. Paradigms for regulatory molecules forming multiprotein complexes including chaperones are the steroid receptors that were found to interact with hsp90, hsp70, p60, p48, p23 and immunophilins (Pratt, 1992; Smith and Toft, 1993). Intriguingly, these steroid receptor complexes are unstable and have been notoriously difficult to isolate prior to the discovery that they can be stabilized by molybdate. Reconstitution experiments suggested that steroid receptor can form different types of complexes containing one or more of the above proteins (Smith et al., 1990; Scherrer et al., 1990; Smith et al., 1992). Note that hsp70 has been found to be essential for the formation of these complexes (Smith and Toft, 1993; Hutchison et al., 1994). At least under *in vitro* conditions steroid receptor complexes are relatively short-lived, and receptor may cycle relatively rapidly and in an orderly fashion through different complexes (Smith, 1993). Hormonal activation interrupts the cycling process, presumably by preventing the assembly of the earliest complex.

Using the above reconstitution assay, Smith's laboratory (Nair, S. and Smith, D., unpublished observations; confirmed by the author's laboratory in collaboration with D. Smith) has recently been able to show that human HSF assembles into analogous complexes containing most, if not all, of the proteins that were previously found to associate with steroid receptors (for a representation of complexes see Fig. 2B). Presumably, although this has not yet been proven, these complexes form on inactive monomeric HSF.

Although the relevance of the above multiprotein complexes including HSF is still unclear, their identification should prompt us to revisit the model presented before in which hsp70 negatively regulates HSF oligomerization. At this time, two comments may be made. First, the existence of multiprotein complexes containing HSF is compatible with the hsp70 regulation model. It seems likely that the function of these complexes is to prevent oligomerization of HSF. If binding of hsp70 to HSF is a prerequisite for the assembly of these complexes on HSF, then hsp70 remains the regulator of HSF oligomerization. Indeed, as discussed before, there is a precedent for such a role of hsp70: assembly of multiprotein complexes on steroid receptor is critically dependent on hsp70 binding to receptor. Second, the observation of multiprotein complexes containing HSF opens up the possibility that proteins in addition to hsp70 may be targets for regulatory events that cause complex disruption and activation of HSF. Of particular interest in this regard may be hsp90. The benzoquinone ansamycins, herbimycin and geldanamycin were recently found to induce HSF oligomerization in HeLa cells (Hedge et al., 1995). These same compounds had been shown previously to have a strong affinity for hsp90 (Whitesell et al., 1994). Furthermore, the compounds prevent activation of glucocorticoid receptor *in vivo* (Whitesell et al., 1995) as well as block the formation of an hsp90- and p23-containing receptor complex *in vitro* by interfering with the recruitment of p23 into the complex (Johnson and Toft, 1995; Smith et al., 1995). Thus, the finding that both compounds cause activation of HSF in human cells supports the notion derived from *in vitro* assembly experiments that HSF enters into similar multiprotein complexes as the steroid receptors and that interruption of these complexes can cause HSF oligomerization.

Direct versus representative regulation of HSF activation by non-native proteins

While the hsp70 mechanism discussed before is based on a native regulatory molecule sensing bulk non-native protein, alternative mechanisms

A

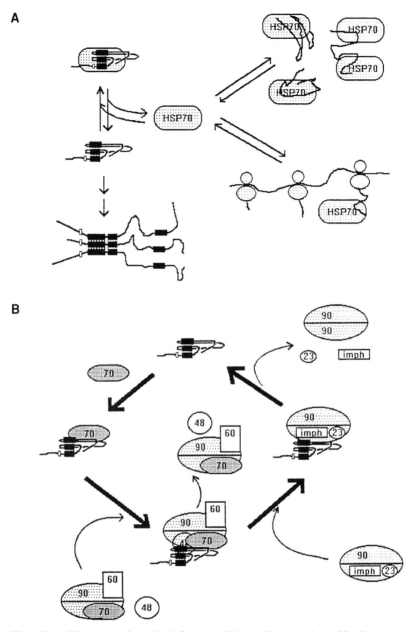

B

Figure 2. **A** The cartoon shows basic features of the hsp70 mechanism of feedback regulation of HSF. Hsp70 is shown to interact with monomeric HSF having intramolecularly complexed LZs (top left), unfolded cellular proteins (top right) and nascent polypeptides (bottom right). HSF monomers not bound by hsp70 unfold and undergo trimerization. **B** Model showing composition of HSF-containing multiprotein complexes. HSF is shown as a structure containing intramolecularly bound LZs. In analogy to steroid receptor, HSF is suggested to cycle through these complexes. Note that disruption of any of these complexes would remove HSF from the cycle which may cause it to undergo trimerization.

may function without a need for a sensing mechanism. They could be dependent on a regulatory protein, presumably one of the components of the multiprotein complexes containing HSF, which include hsp70, that may serve as a representative rather than as a sensor of bulk non-native protein. To illustrate this concept by way of example, assume that hsp90 is such a regulatory factor. Hsp90 may have one or more accessible thiols on the surface that interacts with HSF or other essential components of an HSF-containing multiprotein complex. In the absence of inducer these thiol groups are free and do not hinder binding of hsp90 to HSF or other components of a HSF multiprotein complex. When cells are exposed to inducers of group 2, the hsp90 thiols would be complexed, thiolated or otherwise derivatized. Obviously, many other cellular proteins would suffer the same fate, i.e., the above reactions may be identical with the proteotoxic events discussed before that cause generalized conformational change in proteins. Thus, hsp90 behaves as a representative of a great many proteins affected in a similar fashion. In its altered form hsp90 would be unable to interact with HSF or other proteins contained in an HSF multiprotein complex. Furthermore, if HSF multiprotein complexes have a short half-life similar to that of steroid receptor complexes, HSF multiprotein complexes will rapidly decay, and free HSF will be forced into the trimerization reaction.

As alluded to previously, a mechanism of this kind, if it indeed exists, may not be the only means by which HSF oligomerization may be regulated. The hsp70 mechanism outlined before may well operate in concert with the latter mechanism. Obviously, it will not be a trivial matter to find out whether such alternative mechanisms play a role in the regulation of HSF. It is hoped that the *in vitro* reconstitution system will provide a useful tool for defining such mechanisms. Ultimate proof of the importance of any alternative mechanism will involve the identification of a critical regulatory molecule(s), the demonstration that its conformation/binding properties are affected by multiple inducers of the heat shock response as well as showing that overexpressed mutant regulator alters stress-induced oligomerization of HSF.

Concluding remarks

The molecular analysis of the regulation of the stress response is a fairly recent area of research. Consequently, answers obtained to date are both preliminary and incomplete. In addition, there are important issues that are only now beginning to be addressed such as the mechanism that regulates the transcriptional competence of HSF, including the role that stress-induced phosphorylation of HSF plays in its regulation (reviewed in Voellmy, 1994). Whereas experiments with inhibitors of protein kinases suggested that transcriptional activity of *hsp* genes is regulated by phosphorylation/

dephosphorylation events, it is unknown whether it is phosphorylation of HSF or of a downstream transcription-associated protein that is assayed in these inhibitor experiments. The fact that potentially phosphorylatable residues abound in the HSF sequence has so far proved a formidable obstacle to the identification and mutagenesis of target sites for kinases. No doubt, this information, although difficult to obtain, will become available with time and will define the relationship between phosphorylation and transcriptional competence and/or rate of return of HSF to the inactive form, a role that has been suggested for certain phosphorylation events in yeast HSF (Hoj and Jakobsen, 1994).

Acknowledgment
The author's work on the stress response has been supported by NIH grant GM31125. S. Roberts and N. Spector have been kind enough to critically read this manuscript.

References

Abravaya, K., Philips, B. and Morimoto, R.I. (1991) Attenuation of the heat shock response in HeLa cells is mediated by the release of bound heat shock transcription factor and is modulated by changes in growth and in heat shock temperature. *Genes Dev.* 5: 2117–2127.

Abravaya, K.A., Myers, M., Murphy, S.P. and Morimoto, R.I. (1992) The human heat shock protein hsp70 interacts with HSF, the transcription factor that regulates heat shock gene expression. *Genes Dev.* 6:1153–1164.

Amin, J., Ananthan, J. and Voellmy, R. (1988) Key features of heat shock regulatory elements. *Mol. Cell. Biol.* 8:3761–3769.

Ananthan, J., Goldberg, A.L. and Voellmy, R. (1986) Abnormal proteins serve as eukaryotic stress signals and trigger the activation of heat shock genes. *Science* 232:522–524.

Baler, R. (1992) Ph. D. Thesis, University of Miami, Miami, FL.

Baler, R., Welch, W.J. and Voellmy, R. (1992) Heat shock gene regulation by nascent polypeptides and denatured proteins: hsp70 as a potential autoregulatory factor. *J. Cell. Biol.* 117:1151–1159.

Baler, R., Dahl, G. and Voellmy, R. (1993) Activation of human heat shock genes is accompanied by oligomerization, modification, and rapid translocation of heat shock transcription factor HSF1. *Mol. Cell. Biol.* 13:2486–2496.

Baler, R., Guettouche, T. and Voellmy, R. (1995) On the model of feedback regulation of heat shock gene expression by the shock proteins: Demonstration of heat shock protein 70-containing complexes of unactivated heat shock transcription factor 1. *In:* W.J. Whelan et al. (eds.): *Miami Bio/Technology Short Reports,* Vol 6. IRL Press at Oxford University Press, Oxford, p 35.

Beckmann, R.P., Mizzen, L. and Welch, W.J. (1990) Interaction of hsp70 with newly synthesized proteins: implications for protein folding and assembly. *Science* 248:850–854.

Bole, D.G., Hendershot, L.M. and Kearney, J.F. (1986) Post-translational associations of immunoglobulin heavy chain binding protein with nascent heavy chains in non-secreting and secreting hybridomas. *J. Cell. Biol.* 102:1558–1566.

Bonner, J.J., Heyward, S. and Fackenthal, D.L. (1992) Temperature-dependent regulation of a heterologous transcriptional activation domain fused to yeast heat shock transcription factor. *Mol. Cell. Biol.* 12:1021–1030.

Brigelius, R. (1985) Mixed disulfides: biological functions and increase in oxidative stress. *In: Oxidative Stress.* Academic Press, Inc. London, pp 243–272.

Bruce, J.L., Price, B.D., Coleman, C.N. and Calderwood, S.K. (1993) Oxidative injury rapidly activates the heat shock transcription factor but fails to increase levels of heat shock proteins. *Cancer Res.* 53:12–15.

Chen, Y., Barlev, N.A., Westergaard, O. and Jakobsen, B.K. (1993) Identification of the C-terminal activator domain in yeast heat shock factor: independent control of transient and sustained transcriptional activity. *EMBO J.* 12:5007–5018.

Choi, H.-S., Li, B., Lin, Z., Huang, L.E. and Liu, A.Y.-C. (1991) cAMP and cAMP-dependent protein kinase regulate the human heat shock protein 70 gene promoter activity. *J. Biol. Chem.* 266:11858–11865.

Clos, J., Westwood, T., Becker, B., Wilson, S., Lambert, K. and Wu, C (1990) Molecular cloning and expression of a hexameric *Drosophila* heat shock factor subject to negative regulation. *Cell* 63:1085–1097.

Collison, M.W., Beidler, D., Grimm, L.M. and Thomas, J.A. (1986) A comparison of protein S-thiolation (protein mixed-disulfide formation) in heart cells treated with t-butyl hydroperoxide or diamide. *Biochim. Biophys. Acta* 885:58–67.

DiDomenico, B.J., Bugaisky, G.E. and Lindquist, S. (1982) The heat shock response is self-regulated at both the transcriptional and posttranscriptional levels. *Cell* 31:593–603.

Dreano, M., Brochot, J., Myers, A., Cheng-Meyer, C., Rungger, D., Voellmy, R. and Bromley, P. (1986) High-level, heat-regulated synthesis of proteins in eukaryotic cells. *Gene* 49:1–8.

Freeman, M.L., Sierra-Rivera, E, Voorhees, G.J., Eisert, D.R. and Meredith, M.J. (1993) Synthesis of hsp70 is enhanced in glutathione-depleted Hep G2 cells. *Radiat. Res.* 135:387–393.

Freeman, M.L., Sierra-Rivera, E., Meredith, M.J., Borelli, M.J. and Lepock, J.R. (1995) Formation of protein disulfides, generated by oxidative stress, represent a signal for induction of the stress response. *J. Cell. Biochem.* 19B:197.

Goff, S.A. and Goldberg, A.L. (1985) Production of abnormal proteins in *E. coli* stimulates transcription of *lon* and other heat shock genes. *Cell* 41:587–595.

Green, M., Schuetz, T.J., Sullivan, E.K. and Kingston, R.E. (1995) A heat shock-responsive domain of human HSF1 that regulates transcription activation domain function. *Mol. Cell. biol.* 15:3354–3362.

Grimm, L.M., Collison, M.W., Fisher, R.A. and Thomas, J.A. (1985) Protein mixed-disulfides in cardiac cells. S-Thiolation of soluble proteins in response to diamide. *Biophys. Biochem. Acta* 844:50–54.

Harrison, C.J., Bohm, A.A. and Nelson, H.C.M. (1994) Crystal structure of the DNA binding domain of the heat shock transcription factor. *Science* 263:224–227.

Hedge, R.S., Zuo, J., Voellmy, R. and Welch, W.J. (1995) Short-circuiting stress protein expression via a tyrosine kinase inhibitor, herbimycin A. *J. Cell. Physiol.* 165:186–200.

Hensold, J.O., Hunt, C.R., Calderwood, S.K., Housman, D.E. and Kingston, R.E. (1990) DNA binding of heat shock factor to the heat shock element is insufficient for transcriptional activation in murine erythroleukemia cells. *Mol. Cell. Biol.* 10:1600–1608.

Hightower, L.E., Sadis, S.E. and Takenaka, J.M. (1994) Interactions of vertebrate hsc70 and hsp70 with unfolded proteins and peptides. *In:* R.I. Morimoto, A. Tissieres and C. Georgopoulos (eds): *The Biology of Heat Shock Proteins and Molecular Chaperones.* Cold Spring Harbor Laboratory Press, Cold Spring Harbor, N.Y., pp 179–207.

Hodges, R.S., Sodeck, J., Smillie, L.B. and Jurasek, L. (1972) Tropomyosin: amino acid sequence and coiled-coil structure. *Cold Spring Harb. Symp. Quant. Biol.* 37:299–310.

Hoj, A. and Jakobsen, B.K. (1994) A short element required for turning off heat shock transcription factor: evidence that phosphorylation enhances deactivation. *EMBO J.* 13:2617–2624.

Huang, L.E., Zhang, H., Bae, S.W. and Liu, A.Y.-C. (1994) Thiol reducing reagents inhibit the heat shock response. *J. Biol. Chem.* 269:30718–30725.

Hutchison, K.A., Dittmar, K.D., Czar, M.J. and Pratt, W.B. (1994) Proof that hsp70 is required for assembly of the glucocorticoid receptor into a heterocomplex with hsp90. *J. Biol. Chem.* 269:5043–5049.

Johnson, J.L. and Toft, D.O. (1995) binding of p23 and hsp90 during assembly with the progesterone receptor. *Mol. Endocrin.* 9:670–678.

Jurivich, D.A., Sistonen, L., Kroes, R.A. and Morimoto, R.I. (1992) Effect of sodium salicylate on the human heat shock response. *Science* 255:1243–1245.

Kosower, N.S. and Kosower, E.M. (1978) The glutathione status of cells. *Internat. Rev. Cytol.* 54:109–160.

Levinson, W., Oppermann, H. and Jackson, J. (1980) Transition series metals and sulfhydryl reagents induce the synthesis of four proteins in eukaryotic cells. *Biochim. Biophys. Acta* 606:170–180.

Lindquist, S. (1980) Varying patterns of protein synthesis in *Drosophila* during heat shock: implications for regulation. *Dev. Biol.* 77:463–479.

McLachlan, A.D. and Stewart, M. (1975) Tropomyosin coiled-coil interactions: evidence for an unstaggered structure. *J. Mol. Biol.* 98:293–304.

Mitchell, J.B. and Russo, A. (1983) Thiols, thiol depletion, and thermosensitivity. *Radiat. Res.* 95:471–485.

Mosser, D.D., Duchaine, J. and Massie, B. (1993) The DNA-binding activity of the human heat shock transcription factor is regulated *in vivo* by hsp70. *Mol. Cell. Biol.* 13:5427–5438.

Nadeau, K., Das, A. and Walsh, C.T. (1993) Hsp90 chaperonins possess ATPase activity and bind heat shock transcription factor and peptidyl prolyl isomerases. *J. Biol. Chem.* 268:1479–1487.

Nelson, R.J., Ziegelhofer, T., Nicolet, C., Werner-Washburne, M. and Craig, E.A. (1992) The translation machinery and 70 kd heat shock protein cooperate in protein synthesis. *Cell* 71:97–105.

Nieto-Sotelo, J., Wiederrecht, G., Okuda, A. and Parker, C.S. (1990) The yeast heat shock transcription factor contains a transcriptional activation domain whose activity is repressed under nonshock conditions. *Cell* 62:807–817.

Palleros, D.R., Reid, K.L., Shi, L., Welch, W.J. and Fink, A.L. (1993) ATP-induced protein-hsp70 complex dissociation requires K⁺ and does not involve ATP hydrolysis. *Nature* 365:664–666.

Parker, C.S. and Topol, J. (1984) A *Drosophila* RNA polymerase II transcription factor binds to the regulatory site of an *hsp70* gene. *Cell* 37:273–283.

Pratt, W.B. (1992) Control of steroid receptor function and cytoplasmic-nuclear transport by heat shock proteins. *BioEssays* 14:841–848.

Price, B.D. and Calderwood, S.K. (1991) Ca²⁺ is essential for multistep activation of the heat shock factor in permeabilized cells. *Mol. Cell. Biol.* 11:3365–3368.

Rabindran, S.K., Giorgi, G., Clos, J. and Wu, C. (1991) Molecular cloning and expression of a human heat shock transcription factor, HSF1. *Proc. Natl. Acad. Sci. USA* 88:6906–6910.

Rabindran, S.K., Haroun, R.I., Clos, J., Wisniewski, J. and Wu, C. (1993) Regulation of heat shock factor trimerization: role of a conserved leucine zipper. *Science* 259:230–234.

Rabindran, S.K., Wisniewski, J., Li, L., Li, G.C. and Wu, C. (1994) Interaction between heat shock factor and hsp70 is insufficient to suppress induction of DNA-binding activity *in vivo*. *Mol. Cell. Biol.* 14:6552–6560.

Sarge, K.D., Murphy, S.P. and Morimoto, R.I. (1993) Activation of heat shock gene transcription by heat shock factor 1 involves oligomerization, acquisition of DNA-binding activity, and nuclear localization and can occur in the absence of stress. *Mol. Cell. Biol.* 13:1392–1407.

Scherrer, L.C., Dalman, F.C., Massa, E., Meshinchi, S. and Pratt, W.B. (1990) Structural and functional reconstitution of the glucocorticoid-hsp90 complex. *J. Biol. Chem.* 265:21397–21400.

Skibba, J.L., Stadnick, A., Kalbfleisch, J.H. and Powers, R.H. (1989) Effects of hyperthermia on xantine oxidase activity and glutathione levels in the perfused rat liver. *J. Biochem. Toxicol.* 4:119–125.

Smith, D.F., Schowalter, D.B., Kost, S.L. and Toft, D.O. (1990) Reconstitution of progesterone receptor with heat shock proteins. *Mol. Endocrinol.* 4:1704–1711.

Smith, D.F., Stensgard, B.A., Welch, W.J. and Toft, D.O. (1992) Assembly of progesterone receptor with heat shock proteins and receptor activation are ATP-mediated events. *J. Biol. Chem.* 267:1350–1356.

Smith, D. (1993) Dynamics of heat shock protein 90 – progesterone receptor binding and the disactivation loop model for steroid receptor complexes. *Mol. Endocrinol.* 7:1418–1429.

Smith, D.F. and Toft, D.O. (1993) Steroid receptors and their associated proteins. *Mol. Endocrinol.* 7:4–11.

Smith, D.F., Whitesell, L., Nair, S.C., Chen, S., Prapapanich, V. and Rimerman, R.A. (1995) Progesterone receptor structure and function altered by geldanamycin, an hsp90-binding agent; submitted.

Sorger, P.K., Lewis, M.J. and Pelham, H.R.B. (1987) Heat shock factor is regulated differently in yeast and HeLa cells. *Nature* 329:81–84.

Sorger, P.K. and Nelson, H.C.M. (1989) Trimerization of a yeast transcriptional activator via a coiled-coil motif. *Cell* 59:807–813.

Sorger, P. (1990) Yeast heat shock factor contains separable transient and sustained response transcriptional activators. *Cell* 62:793–805.

Voellmy, R., Ahmed, A., Schiller, P., Bromley, P. and Rungger, D. (1985) Isolation and functional analysis of a human 70,000 dalton heat shock protein gene segment. *Proc. Natl. Acad. Sci. USA* 82:4949–4953.

Voellmy, R. (1994) Transduction of the stress signal and mechanisms of transcriptional regulation of heat shock/stress protein gene expression in higher eukaryotes. *Crit. Rev. Eukaryotic Gene Expression* 4:357–401.

Vuister, G.W., Kim, S.-J., Orosz, A., Marquardt, J., Wu, C. and Bax, A. (1994) Solution structure of the DNA-binding domain of *Drosophila* heat shock transcription factor. *Structural Biology* 1:605–614.

Welch, W.J. (1992) Mammalian stress response: cell physiology, structure/function of stress proteins, and implications for medicine and disease. *Physiol. Rev.* 72:1063–1081.

Westwood, J.T., Clos, J. and Wu, C. (1991) Stress-induced oligomerization and chromosomal relocalization of heat-shock factor. *Nature* 353:822–827.

Whitesell, L., Mimnaugh, E.G., DeCosta, B., Myers, C.E. and Neckers, L.M. (1994) Inhibition of heat shock protein hsp90–pp60[v-src] heteroprotein complex formation by benzoquinone ansamycins: essential role for stress proteins in oncogenic transformation. *Proc. Natl. Acad. Sci. USA* 91:8324–8328.

Whitesell, L., Cook, P.H. and Bagatell, R. (1995) Stable and specific binding of hsp90 by the benzoquinone ansamycins inhibits glucocorticoid receptor function. *J. Cell. Biochem.* 19B:206.

Wu, B.J., Hunt, C. and Morimoto, R.I. (1985) Structure and expression of the human gene encoding major heat shock protein hsp70. *Mol. Cell. Biol.* 5:330–343.

Wu, C. (1984) Two protein-binding sites in chromatin implicated in the activation of heat shock genes. *Nature* 309:229–234.

Xiao, H. and Lis, J.T. (1988) Germline transformation used to define key features of heat shock response elements. *Science* 239:1139–1142.

Zuo, J., Baler, R., Dahl, G. and Voellmy, R. (1994) Activation of the DNA-binding ability of human heat shock transcription factor 1 may involve the transition from an intramolecular to an intermolecular triple-stranded coiled-coil structure. *Mol. Cell. Biol.* 14:7557–7568.

Zuo, J., Rungger, D. and Voellmy, R. (1995) Multiple layers of regulation of human heat shock transcription factor 1. *Mol. Cell. Biol.* 15:4319–4330.

Stress-Inducible Cellular Responses
ed. by U. Feige, R.I. Morimoto, I. Yahara and B. Polla
© 1996 Birkhäuser Verlag Basel/Switzerland

The transcriptional regulation of heat shock genes: A plethora of heat shock factors and regulatory conditions

R. I. Morimoto, P. E. Kroeger and J. J. Cotto

Department of Biochemistry, Molecular Biology and Cell Biology Northwestern University, 2153 Sheridan Road, Evanston, IL 60208, USA

Summary. The inducible regulation of heat shock gene transcription is mediated by a family of heat shock factors (HSF) that respond to diverse forms of physiological and environmental stress including elevated temperature, amino acid analogs, heavy metals, oxidative stress, anti-inflammatory drugs, arachidonic acid, and a number of pathophysiological disease states. The vertebrate genome encodes a family of HSFs which are expressed ubiquitously, yet the DNA binding properties of each factor are negatively regulated and activated in response to specific conditions. This chapter will discuss the regulation of the HSF multi-gene family and the role of these transcriptional activators in the inducible expression of genes encoding heat shock proteins and molecular chaperones.

Introduction

The heat shock (HS) response has afforded numerous insights into the molecular mechanisms by which a cell detects and responds to diverse forms of chemical, environmental, and physiological stress. The hallmark of the HS response is the inducible expression of a highly conserved family of proteins known as the heat shock proteins (HSPs) or molecular chaperones which confers a safeguard mechanism to ensure survival of the stressed cell against normally lethal forms of stress (Lindquist, 1986; Lindquist and Craig, 1988; Morimoto et al, 1994). Among the most dramatic and extensively investigated aspects of the HS response is at the level of HS-inducible transcription (also reviewed by Voellmy, this Volume). Within seconds of temperature elevation, the heat shock transcription factor (HSF) is activated and results in the rapid increase in transcription of genes encoding HSPs (Morimoto, 1993; Lis and Wu, 1993; Morimoto et al., 1994; Wu et al., 1994). The transcriptional regulation of HS genes is complex as exemplified by the diverse array of conditions identified to date.

Conditions that regulate heat shock gene expression

An understanding of the HS response requires an appreciation of the myriad of conditions that lead to the elevated expression of HSPs. Initial

Cellular Stress Response

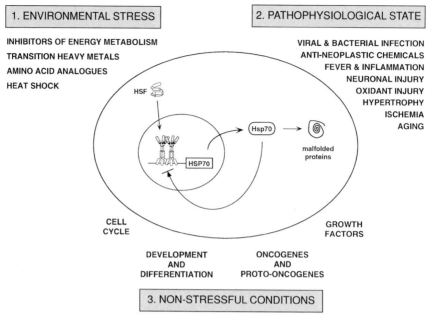

Figure 1. Conditions that induce heat shock gene expression. Three classes of cellular stress are indicated: (1) environmental stress; (2) pathophysiological state; and (3) non stressful conditions. Representative examples of inducers of HS gene expression in each class are listed. The diagram in the center of the figure corresponds to the stress-dependent activation of HSF which occurs in response to environmental and pathophysiological stress and results in elevated expression of hsp70.

studies on the heat shock response in *Drosophila* identified conditions such as elevated temperature, salicylate, dinitrophenol, ethanol, and anoxia, all of which induced the appearance of specific chromosomal puffs that corresponded to heat shock gene loci (Ritossa, 1962; Ashburner, 1970; Tissieres et al., 1974; Lindquist-McKenzie et al., 1975). Subsequent studies carried out in diverse plants and animals have revealed that the expression of HSPs is inducibly regulated by an ever growing list of conditions that fall into three general classes: environmental stress, physiological stress, and nonstressful conditions (Fig. 1).

Environmental stress
Environmental stress is the prototypical condition that has been employed in the study of the HS response and includes exposure of cells to heat shock, heavy metals, and chemicals. Although it remains to be demonstrated whether the HS response has a common denominator, many of these conditions result in increased levels of malfolded proteins which directly or indirectly lead to the elevated synthesis of HSPs including HSP70 (Goff

and Goldberg, 1985; Ananthan et al., 1986; Beckmann et al., 1990; Baler et al., 1992). Whereas elevated temperature is thought to have an immediate consequence on the folding of nascent polypeptides, induction of the HS response by amino acid analogs requires their incorporation into nascent polypeptides which are improperly folded (DiDomenico et al., 1982; Mosser et al., 1988).

Numerous observations have indicated that the magnitude of the heat shock-induced transcriptional response reflects the severity of the stress (DiDomenico et al., 1982; Mosser et al., 1988; Abravaya et al., 1991b). Exposure of HeLa cells grown at 37 °C to a heat shock of 42 °C ($\Delta T = 5$ °C) leads to the transient activation of HSF 1 DNA binding and transcription of HS genes. The inducible response at 42 °C is rapid and transient with attenuation occurring within $2-3$ h even though the cells remain at the heat shock temperature. The negative regulation of the heat shock transcriptional response at 42 °C correlates with the elevated synthesis and accumulation of HSPs. This response is dependent on ongoing protein synthesis which suggests that a component of the stress signaling mechanism is linked to the synthesis of nascent polypeptides (Amici et al., 1992). These results, together with genetic studies on the regulation of the heat shock response in yeast, have led to the proposal that the transcription of heat shock genes involves an auto regulatory mechanism (Stone et al, 1990). This is further supported by experiments with amino acid analogue treatment of cells which have demonstrated that it is the appearance of malfolded proteins which correlates with the activation of the response (DiDomenico et al., 1982; Mosser et al., 1988). The lack of attenuation in the latter case is thought to be due to the synthesis of malfolded heat shock proteins (due to incorporation of the amino acid analogue) which are ineffective in binding and refolding the accumulated malfolded proteins (Beckmann et al., 1990).

The temperature at which the HS response within cells of a particular organism occurs reflects an increment above the ambient optimal growth temperature consequently the heat shock response in mammalian cells (42 °C) is typically activated at $\approx 5-6$ °C above this optimal ambient temperature of 37 °C. The temperature at which the HS response is induced is not an absolute condition. Growth of HeLa cells at the suboptimal temperature of 35 °C resulted in activation of HSF1 at $40-41$ °C, a temperature which is normally insufficient to induce a HS response in cells grown at 37 °C (Abravaya et al., 1991b). Additional evidence that the temperature optimum for heat shock factor can be altered follows from studies in which human HSF1 was expressed in *Drosophila* cells. Although the normal activation temperature for human HSF1 is 42 °C, when expressed in *Drosophila*, human HSF1 was activated at the intermediate temperature of 32 °C (Clos et al., 1993). Conversely, when *Drosophila* HSF was expressed in human cells, it acquired constitutive activity at the ambient temperature of 37 °C. Taken together, these data suggest that the temperature at which a HSF is activated is not strictly intrinsic to the factor, but can be reprogrammed according to the cellular environment in which the HSF is expressed. Moreover, it seems likely that there are intersecting pathways for the detection and response to environmental stress since exposure to multiple stresses has synergistic effects on the activation of the HS response (Mosser et al., 1988; Lee et al., 1995).

Pathophysiological stress
The list of conditions (Fig. 1) collectively known as pathophysiological stress has grown and includes the cellular response to bacterial and viral infection (Santoro, this volume), neuronal trauma and ischemia (Nagata, this volume). The cell and tissue damage from ischemia results from the blockage of blood flow and is accompanied by anoxia (Marber et al., 1993). An early response to tissue damage and oxygen and nutrient deprivation is the induction of HSPs which has been shown to be essential for cell and tissue viability. The hypoxic component of this stress, alone, induces HSP70; furthermore overexpression of HSP70 by transient transfection or by a priming ischemic event has been shown to have protective effects (Benjamin et al., 1992; Williams and Benjamin, 1993). An additional bout of stress due to oxidative damage has been shown to occur upon reperfusion (Currie and White, 1993). Similar events have been described for the ischemic response in the brain. Unlike studies on cells from other somatic tissues, certain populations of neuronal cells have been shown to be more sensitive to the effects of stress (Nowak and Abe, 1994). In general, the brain responds to global and isolated ischemic events by the elevated expression of heat shock proteins; furthermore, preischemic tolerance has been shown to have protective effects against subsequent neuronal damage (Pulsinelli et al., 1982; Blake et al., 1990a; Liu et al., 1992; Aoki et al., 1993). Certain neuronal cell lines exhibit a selective inability to activate HSP70 expression which, in part, appears to be duet o chromatin inaccessibility of transcription factors (Mathur et al., 1994). Although HSF1 is activated upon heat shock in Y79 retinoblastoma cells, the endogenous HSP70 gene was induced poorly which correlated with the inability of HSF1 to bind *in vivo* to the HSE in the HSP70 promoter. In contrast, heat shock of Y79 cells led to transcription induction of the HSP90 gene and *in vivo* association of HSF1 to the respective HSEs in the HSP90 promoter. The absence of a heat shock response has also been observed for explanted embryonic rat hippocampal cells in culture which appears to be due to the absence of HSF1 (Marcucilli et al., 1996). Taken together, these results indicate the potential for tissue specific differences in the stress response and gene-specific differences within a cell type.

The role of neuro-hormonal stress in activation of the HS response provides another interesting example of tissue-specific regulation (Blake et al., 1990b; Blake et al., 1991; Udelsman et al., 1993). The selective expression of HSP70 in the aorta and adrenals of a rodent is induced by physical restraint (Blake et al., 1991; Udelsman et al., 1993). This effect is mediated by adrenocorticotropic hormone (ACTH) in the pituitary thus leading to the induction of HSF1 and the synthesis of HSP70 in the inner cortex of the adrenal glands (Blake et al., 1991). The induction of HSP70 in the aorta was not affected by ACTH but was blocked by an a_1-adrenergic blocking agent, prazosin (Udelsman et al., 1993). These results reveal that the concept of physiological stress extends to induce hormonal regulations of neuroendocrine stress and that these effectors result in a tissue specific HS response.

Nonstressful inducers

The transcription of HS genes is modulated during progression through the cell cycle, by various growth factors, during development and differentiation, and by certain oncogenes (Fig. 1) (Wu et al., 1986a, b; Morimoto, 1991; Phillips and Morimoto, 1991; Sistonen et al., 1992). The human HSP70 gene exhibits features of a delayed growth responsive gene. Transcription is regulated in response to serum stimulation of growth arrested HeLa cells and E1a transformed human embryonic kidney 293 cells (Wu et al., 1985) and upon IL-2 stimulation of resting T lymphocytes (Ferris et al., 1988). Transcription of the HSP70 gene has also been shown to be modulated during the cell cycle leading to a burst of HSP70 message at the G1/S boundary (Milarski and Morimoto, 1986).

The transcriptional activation of the HSP70 gene by serum, growth factors, and the cell cycle is mediated through the basal promoter of the HSP70 gene (Fig. 2) whose function is independent of the HS response. The identity and role of the cis-acting elements in the HSP70 basal promoter have been determined by deletion and mutagenesis studies (Wu et al., 1986a; Greene et al., 1987; Williams et al., 1989; Greene and Kingston, 1990; Williams and Morimoto, 1990). The CCAAT box located at -74 bp is important for the function of the basal promoter and interacts with two factors, CBF and CTF (Morgan et al., 1987; Lum et al., 1990). CBF activity can be regulated through direct interaction with p53 (Lum et al., 1990; Agoff et al., 1993). Additional binding sites for several other ubiquitous factors including Sp1 can also function as a positive activator of HSP70 gene transcription (Morgan, 1989; Williams et al., 1989). The basal promoter is also the target for the trans-activating effects of the adenovirus E1a protein during viral infection (Kao et al., 1983; Wu et al., 1986b; Williams et al., 1989; Phillips and Morimoto, 1991). *In vivo* footprinting studies have shown that the basal promoter sites are constitutively occupied (Abravaya et al., 1991a). Taken together, the complex structure of the HSP70 promoter and the many inducers of the heat shock response demonstrate the diversity of conditions that can modulate and activate HSP70 transcription.

HSP70 Promoter

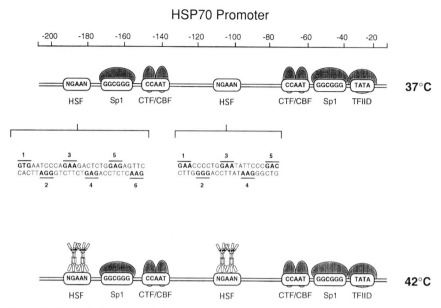

Figure 2. Transcription factor binding sites located in the promoter of the human HSP70 gene. The cis-acting elements are indicated as labeled. The binding sites include the TATA box, Sp1 sites, CCAAT box, and proximal and distal HSEs and are indicated relative to the +1 site of transcription. A combination of *in vitro* footprinting with purified factors and *in vivo* genomic footprinting of the human hsp70 promoter was used to demonstrate that the basal promoter elements are occupied at 37 °C, whereas upon heat shock at 42 °C, the basal elements remain occupied and the proximal and distal HSEs are inducibly bound by HSF.

Stress-induced transcription via direct interaction of heat shock transcription factor and the heat shock element

Certain sequence motifs perhaps involved in the regulation of HS genes were initially identified by alignment of 5′ flanking regions of various cloned *Drosophila* HS genes (Holmgren et al., 1981). Among these homologous sequences was the dyad symmetrical element (5′-CnnGAAnnTTCnnG-3′) which was subsequently referred to as the heat shock element or HSE (Pelham, 1982). The HSE, when linked to the promoter of a heterologous reporter gene, was capable of rendering the reporter heat inducible, thus providing definitive evidence of a unique regulatory element (Corces et al., 1981; Mirault et al., 1982; Pelham and Bienz, 1982; Bienz and Pelham, 1986). The HSE can be more accurately described as a series of pentameric units arranged as inverted adjacent arrays of the sequence 5′-nGAAn-3′ (Amin et al., 1988; Xiao and Lis, 1988; Perisic et al., 1989). A functional HSE is composed of a minimum of three pentamers with additional reiteration of the pentamer resulting in higher affinity interactions between HSF and the HSE (Xiao et al., 1991; Kroeger et al., 1993). Comparison of the nucleotide composition of the HSE indicates that the guanine residue in the second position of each pentamer (5′-nGAAn-3′) is absolutely conserved. The adjacent adenine residues are also highly conserved, but may be substituted by other bases. In the mammalian HSP70 and HSP90 promoters (Fig. 2) the HSE is composed of multiple pentameric units, five and six respectively, in close proximity to the basal promoter elements (Williams and Morimoto, 1990). The activity and interdependence of the HSE in the context of the basal promoter elements have been examined (Greene and Kingston, 1990; Williams and Morimoto, 1990). These studies demonstrated that the HSE was rotationally independent of the basal promoter; however, the degree of induction by heat was dependent on the distance between the HSE and the basal promoter and cooperation with basal factors.

An important step in the understanding of HS gene regulation was provided by the *in vivo* analysis of chromatin structure as revealed by DNaseI hypersensitive sites in the 5′ flanking region of the *Drosophila* HSP70 gene (Wu, 1980). Subsequent experiments using exonuclease III footprinting demonstrated the presence of constitutively bound factors to the TATA box before and after heat stress and that a HSE-binding protein was only detected upon temperature elevation (Wu, 1984). The subsequent biochemical identification, purification, and cloning of the HSE DNA binding activity was first accomplished for *S. cerevisiae* and *D. melanogaster* HSFs (Sorger et al., 1987; Sorger et al., 1988; Wu et al., 1987). The native heat shock factors (HSFs) differed in size (*S. cerevisiae* = 130 kDa; *D. melanogaster* = 110 kDa) and activity. Whereas the yeast HSF bound constitutively to DNA, the *Drosophila* HSF was strictly negatively regulated and heat shock responsive (Sorger et al., 1987; Jakobsen and Pelham, 1988). Even though *S. cerevisiae* HSF was bound to DNA constitutively, heat shock was required to activate transcription which correlated with the inducible phosphorylation of HSF (Sorger and Pelham, 1988). Not all yeasts exhibited this type of HSF regulation as the fission yeast *Schizosaccharomyces pombe* required heat shock to activate both HSF DNA-binding and phosphorylation (Gallo et al., 1991). A related heat induced HSE binding activity was also detected in crude extracts of mammalian cells using a gel shift assay (Kingston et al., 1987). HSF from *Drosophila* and mammalian cells was shown to have latent activity which was activated for DNA binding and transcriptional activity upon heat shock (Zimarino and Wu, 1987; Larson et al., 1988; Mosser et al., 1988). The latent HSF could be activated *in vitro* by a variety of treatments that affect protein structure including *in vitro* heat shock, detergent, or mild denaturant treatment (Larson et al., 1988; Mosser et al., 1990). These results demonstrated that the transcription factor responsible for the transcriptional activation of HS genes was functionally conserved and regulated posttranslationally.

The cloning of the *S. cerevisiae* and *D. melanogaster* HSF molecules revealed a number of important features regarding HSF structure and provided the tools for the subsequent cloning of homologues from a variety of organisms (Wiederrecht et al., 1988; Clos et al., 1990). Alignment of the cloned HSFs from yeasts, *Drosophila*, tomato, chicken, mouse, and humans identified a conserved ≈100 amino acid DNA-binding domain located in the amino-terminus. Immediately adjacent to the DNA-binding domain was a second conserved region that contained three leucine zipper repeats responsible for trimerization of HSF (Sorger and Nelson, 1989; Peteranderl and Nelson, 1992). The HSFs in *D. melanogaster* and larger eukaryotes contain an additional leucine zipper in the carboxyl-terminus which has been suggested to have a role in the negative regulation of HSF activity (Rabindran et al., 1993; Nakai and Morimoto, 1993).

A multi-gene family of heat shock transcription factors

An unexpected finding during the cloning the HSF genes from plants and vertebrates was the identification of a HSF multi-gene family. At least three HSFs have been isolated from the human, mouse, chicken, and tomato genomes and an additional factor, HSF4 has been identified in humans (Scharf et al., 1990; Rabindran et al., 1991; Sarge et al., 1991; Schuetz et al., 1991; Nakai and Morimoto, 1993; Nakai et al., in preparation). Comparison of the structure and sequences of the vertebrate HSFs (Fig. 3) reveals that within a single species the HSFs are ≈40% related in amino acid sequences (i.e., mouse HSF1-HSF2 or chicken HSF1-HSF2-HSF3) which is primarily due to the extensive identity of the DNA binding and oligomerization domains (Clos et al., 1990; Scharf et al., 1990; Rabindran et al., 1991; Sarge et al., 1991; Schuetz et al., 1991; Nakai and Morimoto, 1993). Interspecies comparisons, i.e. between human, mouse, and chicken HSF1 indicate 85–95% conservation in sequence. Comparison of the three chicken HSFs with other cloned HSFs suggested that there was a common ancestor from which they diverged (Nakai and Morimoto, 1993). By comparison, the tomato HSFs (L. p. HSF8, HSF24, and HSF30), although similar in structure, were isolated independently through binding site screening; furthermore, the plant HSFs do not share the same family relationships as established for the vertebrate HSFs (Scharf et al., 1990).

What is the role of multiple HSFs? One possibility is that larger more complex organisms may utilize multiple HSFs to respond to an increasingly diverse array of developmental and environmental cues. Another consideration is that *S. cerevisiae* HSF, which is an essential gene, has at least two transcriptional activation domains that respond to sustained or transient heat shock (Nieto-Sotelo et al., 1990; Sorger, 1990; Chen et al., 1993). Perhaps the duplication of HSFs in the larger eukaryotes was a response to evolutionary events that required the co-regulation of the heat shock genes in response to distinct signals. An answer to some of these questions was provided by the demonstration that HSF1 corresponded to the rapidly activated stress-responsive factor whereas the co-expressed HSF2 was activated in response to distinct developmental cues (Fig. 4; Sistonen

HSF Gene Family and Structural Domains

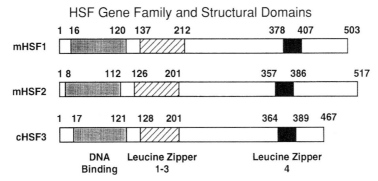

DNA   Leucine Zipper   Leucine Zipper
Binding   1-3   4

Figure 3. HSF gene family and structural domains. General features of mouse HSF1, mHSF2 and chicken HSF3 are schematically presented. The DNA binding domain of approximately 100 amino acids is localized to the amino terminus (shaded box). Adjacent is the conserved leucine zipper motifs responsible for HSF trimerization (cross-hatched box). The carboxyl terminus contains an additional leucine zipper involved in negative regulation (black box). The relative location (in amino acids relative to the amino-terminus) of each conserved region of HSF1, 2, and 3 is indicated.

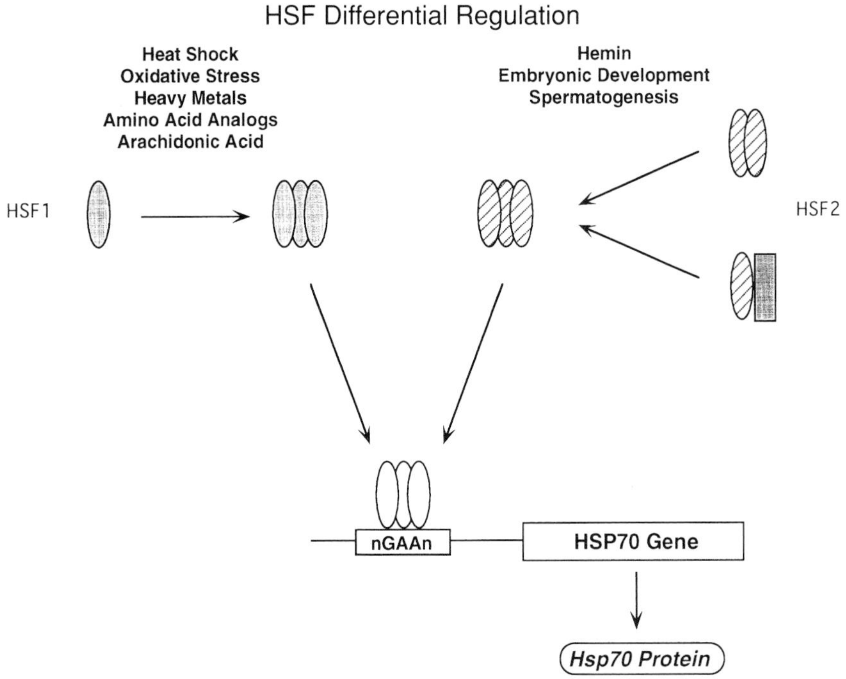

HSF Differential Regulation

Figure 4. Differential regulation of HSF1 and HSF2. Vertebrate cells express multiple HSFs including HSF1 and HSF2. HSF1 is maintained in the control non-DNA binding state as a monomer that becomes stress-activated by heat shock, oxidative stress, heavy metals and exposure to amino acid analogs to a homotrimer that binds to the promoter of the HSP70 gene. HSF2 is maintained in the control state as a dimer and becomes activated to the trimeric state during hemin treatment of human K562 erythroleukemia cells, Activation of either HSF1 or HSF2 leads to elevated transcription of the HSP70 gene and other genes that contain HSEs in the promoter region. HSF2 is constitutively active as a trimer during early embryonic development and spermatogenesis.

et al., 1992; Sarge et al., 1993). During hemin-treatment of K562 erythro-leukemia cells, which leads to non-terminal erythroid differentiation, HSF2 is converted from an inert dimer to an active trimer (Theodorakis et al., 1989; Sistonen et al., 1994). An important distinction, however, is that HSF1 undergoes rapid (within seconds) activation in response to heat and attenuates rapidly, whereas hemin induced HSF2 is activated over a period of time requiring 16–24 h and remains activated through 72 h (Sistonen et al., 1994). Simultaneous activation of HSF1 and HSF2 in K562 cells by exposing hemin-treated cells to heat shock led to a synergis-tic transcriptional activation of the HSP70 gene suggesting that both HSF1 and HSF2 can co-exist in their respective activated states (Sistonen et al., 1994). Other studies on mouse tissues revealed that HSF2 mRNA was developmentally regulated and accumulated to high levels in the testis at the spermatocyte and round spermatid stages of development (Fiorenza et al., 1995; Sarge et al., 1994). Additionally, several embryonic cell lines (F9, MEL) contain a constitutive HSF2 DNA binding activity, however, there was no apparent activation of HS genes in these cells in the absence of stress (Murphy et al., 1994).

Recent studies on chicken HSF3 reveal additional diversity in the pathways for activation of heat shock factors (Nakai and Morimoto, 1993; Nakai et al., 1995). HSF3 is expressed ubiquitously, however, its activity is induced upon heat shock in chicken erythroblastic cells (HD6) and not in cells from other lineages (Nakai et al., 1995). HSF3 has many characteris-tics similar to that of HSF1 such as sequence-specific binding to the heat shock element, negative regulation, and activation to a trimer, however, unlike HSF1, HSF3 is an inert dimer. As for HSF1, deletion of the leucine zipper 4 resulted in constitutive DNA binding activity which suggests a role for zipper 4 in the negative regulation of HSF3 DNA-binding activity (Nakai and Morimoto, 1993). HSF3 has the activity of a positive activator as demonstrated by co-transfection with a reporter gene. These data suggest that HSF1 and HSF3 could enhance the ability of the cell to tightly regulate the heat shock response in a cell type-specific manner (Nakai et al., 1995).

A reexamination of the HSE in the context of multiple HSFs

The DNA-binding activities of HSF1 and HSF2 in human erythroleukemia K562 cells can be inducibly regulated in response to distinct signals. This has made possible a direct *in vivo* comparison of the DNA binding pro-perties and transcription activities of both HSFs within the same cell. Al-though HSF1 and HSF2 exhibit similar DNA binding properties *in vitro*, HSF1 is a more potent transcriptional activator (Sistonen et al., 1992). Comparison of the *in vivo* occupancy of the hsp70 HSE by HSF1 and HSF2 reveals that HSF1 interacts with all five pentameric repeats of the HSE, whereas hermin-induced HSF2 recognized a subset of the binding sites (Sistonen et al., 1992). These observations were further supported by *in vitro* DNase1 footprinting analyses which established that the preference exhibited by HSF1 could be explained by cooperative interactions between HSF1 trimers (Kroeger et al., 1993; Kroeger and Morimoto, 1994) and that HSF1 was a more potent activator in a nuclear extract transcription system than HSF2 (Kroeger et al., 1993). Perhaps these differences in occupancy of the HSE offers an explanation for the more potent transcriptional activity of HSF1.

To further explore the significance of these DNA binding studies on HSF1 and HSF2, experiments were performed to establish the DNA binding preference of HSF1 and HSF2 in an unbiased manner through the selection of new binding sites from a pool of random oligonucleotides (Kroeger and Morimoto, 1994). This analysis revealed that both HSFs recognized nearly identical consensus inverted pentamers and that the nucleotides at the "n" positions were not random. The dimeric consensus sites determined for HSF1 and HSF2 were 5'-aGAA^c/tgTTCg-3' and 5'-^{a}/gGAAnnTTC^g/t-3', respectively. Notably, in the HSF1 consensus sequence, adenine and cytosine residues were alternately preferred at the first position of each nGAAn repeat, and pyrimidines were preferred at the fifth position. The similarity in consensus sequences for both factors suggested that the differential recognition of the HSP70 HSE was likely due to other factors such as differences in cooperativety between trimers. In this regard, HSF1 selected binding sites that contained more pentameric repeats per binding site (4−5) than HSF2 (2−3), which suggests that the binding of multiple HSF1 trimers to a selected oligonucleotide resulted in increased stability of the DNA-protein complex. Perhaps this lack of stable cooperative interactions between HSF2 trimers is in part responsible for the lower level of transcriptional activation observed *in vitro* and *in vivo*. In support of this hypothesis, there is evidence from the *S. cerevisiae* system that the ability to bind multiple HSF trimers to an HSE correlates with increased levels of transcription (Bonner et al., 1994).

Analysis of HSF1 and HSF2: Regulation of the oligomeric state

A prominent regulatory feature of HSF is oligomerization to an active trimeric DNA binding state. *Drosophila* and the vertebrate HSFs are maintained in a latent monomeric form and converted into a DNA binding competent trimer upon heat shock (Westwood et al., 1991). The biochemical events associated with oligomerization have been studied through the use of a variety of biochemical methods including chemical crosslinking reagents and hydrodynamic studies. The latent HSF molecule in *D. melanogaster* is a very elongated monomer with a frictional ratio (f/f_0) of ≈1.9 (Westwood and Wu, 1993), whereas studies with human HSF1 and

HSF1 Domains and Activation

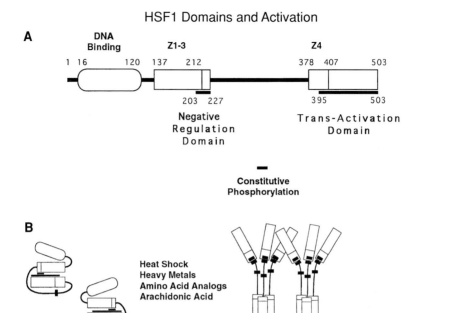

Figure 5. Analysis of HSF1 structural and functional domains: A model for the activation mechanism of HSF1. **A** Schematic representation of HSF1 structural motifs (DNA binding, leucine zippers 1–3 and leucine zipper 4) and functional domains (negative regulation domain, activation domain, constitutive phosphorylation, and inducible phosphorylation). The location of each structural motif or regulatory domain is indicated by the amino acid position **B** Model for HSF1 regulation. In control cells, intramolecular interactions between leucine zippers 1–3 and leucine zipper 4 are important for negative regulation. Additional interactions between the negative regulatory domain and activation domain are also indicated. During heat shock and other stresses, HSF1 activation is mediated by disruption of the intermolecular interactions between leucine zipper 1–3 and 4 and the negative regulatory domain with the activation domain in factor of new intermolecular interactions between zippers 1–3 leading to HSF1 trimerization and subsequent acquisition of DNA binding activity.

HSF2 have demonstrated that the f/f_0 for monomeric HSF ≈ 1.7 and for HSF2 is also ≈ 1.7 (Sistonen et al., 1994). Upon conversion to the trimeric form during heat shock the frictional ratio for *Drosophila* HSF increases ($f/f_0 = 2.6$) suggesting an unfolding event in conjunction with the trimerization, whereas the vertebrate HSF1 is essentially unaffected. The lack of a significant change in the axial ratio of the vertebrate HSF may be due to its smaller size as human HSF1 and HSF2 are 54 and 58 kDa, respectively, whereas *Drosophila* HSF is 110 kDa.

Investigation of the physical parameters for HSF2 from K562 cells revealed, surprisingly, that it was a dimer in the latent form and converted upon hemin treatment into an active DNA-binding trimer (Sistonen et al.,

1994). The difference in the oligomeric state of latent HSF1 and HSF2 corroborates the observed differences in activation in response to different signaling pathways.

A central question has been to understand how HSF is maintained in the latent state. Studies on *Drosophila* HSF have suggested a role for intra-molecular interactions between leucine zipper 4, at the carboxyl-terminus of the protein, with the leucine zippers of the oligomerization domain with the result to form the inert monomer (Rabindran et al., 1993). In these studies, mutation of leucine zipper 4 resulted in constitutive DNA-binding activity. These results were corroborated by an analysis of HSF3 which demonstrated that deletion of leucine zipper 4 also resulted in constitutive DNA-binding activity (Nakai and Morimoto, 1993). Taken together, these studies suggest that intramolecular interactions between the amino- and carboxyl-terminal leucine zippers regulate HSF1 activity and are consistent with the observed unfolding events during HS activation (Fig. 5).

The role of stress-induced phosphorylation of HSF1

The activation of transcription by *S. cerevisiae* HSF has been associated with heat induced serine phosphorylation which results in an altered mobility on SDS-PAGE (Sorger and Pelham, 1988). Recent studies in yeast suggest that hyperphosphorylation of residues adjacent to a conserved element in the carboxy-terminus of HSF may also be important in the deactivation of the factor (Hoj and Jakobsen, 1994). In mammalian cells, the role of HSF1 phosphorylation is not well understood. In mouse and humans, HSF1 is constitutively and inducibly phosphorylated at serine residues (Sarge et al., 1993; M. Kline, unpublished observation; Cotto et al., 1996). Treatment of mammalian cells with salicylate or azetidine led to activation of HSF1 trimerization and DNA-binding *in vivo*, however, this form of HSF1 was not inducibly phosphorylated and this correlated with a decreased levels of HSP70 transcription (Jurivich et al., 1992; Sarge et al., 1993). Anti-inflammatory drugs have proven useful to uncouple HSF1 activity and in determining the role of inducible phosphorylation in the activation of HSF1. The salicylate-induced form of HSF1 can function as a substrate for inducible phosphorylation upon exposure of drug-treated cells to heat shock. This indicates that an intermediate in HSF1 activation is the constitutively phosphorylated, DNA binding trimer which lacks transactivation function. Comparison of the phosphopeptide maps of the heat shock-induced form of HSF1 with the salicylate and heat shock-induced form of HSF1 demonstrates that the patterns of HSF1 inducible phosphorylation are indistinguishable; furthermore, this form of HSF1 is now transcriptionally active (Cotto et al., 1996). However, lack of phosphorylation does not completely abolish transcriptional activation, in fact, it seems that the requirements for phosphorylation vary depending on the type of

HSF. For example, hemin treatment of K562 cells resulted in activation of HSF2, but did not induce HSF2 phosphorylation (Sistonen et al., 1994). Whether the observed differences in HSF1 and HSF2 phosphorylation are related to their relative strength as transcription factors remains to be determined.

Transcription activation domains of HSF1

Trimerization of HSF is essential to its function and provides the high affinity DNA binding required for interaction with the HSE. When bound to the DNA through the amino-terminal DNA binding domain, HSF is likely to affect transcriptional activation via sequences in the carboxyl-terminus. The activation domains of the yeast HSFs (*S. cerevisiae* and *K. lactis*) and the tomato HSFs have been dissected and reveal differences in the nature of the respective activation domains (Chen et al., 1993; Treuter et al., 1993). Yeast HSF has multiple transcriptional activation domains that mediate the response to sustained and transient heat shock. The amino-terminal activator of *S. cerevisiae* is responsible for activation of transcription during sustained stress (Nieto-Sotelo et al., 1990; Sorger, 1990), whereas the carboxyl-terminal activator responds to transient stress. The *S. cerevisiae* carboxyl-terminal activation domain is ≈ 180 amino acids long and appears to be bipartite with one domain responsible for activation (aa 595–713) and the adjacent sequences (aa 713–780) having a supporting function for the activator. In contrast, the corresponding region in *K. lactis* HSF is a 30 amino acid (aa 592–623) sequence with some α-helical character and acidic quality. The tomato HSFs apparently have a third type of activation motif located in the carboxyl-terminus which is composed of acidic sequences that contain a central tryptophan residue.

The activation domains of mammalian HSF (human and mouse) were characterized through the use of chimeric factors in which the heterologous GAL4 DNA binding domain (DBD) was fused to various segments of HSF1 to assess the activity of respective HSF sequences. Such studies have identified two activation domains, referred to as AD1 and AD2, both of which are localized at the extreme carboxyl-terminus of HSF1 (Green et al., 1995; Zuo et al., 1995; Shi et al., 1995). Fusion of the GAL4 DNA binding domain to residues 124–503 of HSF1 results in a chimeric factor that binds DNA yet lacks any transcriptional activity. In order to detect transcription activity from GAL4-HSF1, it was necessary to delete an array of leucine zippers (zippers 1–3) positioned between residues 1224 and 227 of HSF1. The minimum negative regulatory domain was shown to require residues between amino acids 188–227. The minimal region for transcription activation was mapped to the extreme carboxyl-terminal 108 amino acids, corresponding to a region rich in acidic and hydrophobic residues. The carboxyl-terminal trans-activation region, corresponding to AD2 is likely to represent a large domain a loss of residues 395–425 or 451–503 located at either end of this activation domain severely diminished activity. The minimal activation domain of HSF1 is also sufficient to generate an enhanced (four-fold) transcriptional response to heat shock or cadmium treatment. These results demonstrate that the transcriptional activation domain of HSF1 is maintained in a negatively regulated state, presumably through intramolecular interactions and that the signal for stress induction is also mediated by the carboxyl-terminal activator region of HSF1.

The cell biology of HSF1 activation

Studies on the subcellular localization of HSF1 have provided additional insights into the mechanism of regulation upon heat shock. In *Drosophila*, HSF1 is localized to the nucleus both before and after heat shock (West-wood et al., 1991). However, in mammalian cells the subcellular localization of HSF1 is regulated (Sarge et al., 1993). In the absence of stress, HSF1 is distributed in a diffuse pattern throughout the cytoplasm and nucleus of 3T3 and HeLa cells and upon exposure to heat shock, HSF1 is localized to the nucleus. Detailed studies on the kinetics of HSF1 nuclear relocalization have demonstrated that the inactive HSF1 resides mostly in the cytoplasm of the cell, within minutes of activation ($\approx$ 2 min), HSF1 rapidly translocates to the soluble nuclear fraction and shortly thereafter becomes associated with the insoluble nuclear fraction, perhaps reflecting binding to chromosomal target sequences (Baler et al., 1993, Sarge et al., 1993, Cotto, unpublished observation). From these observations it can be concluded that HSF1 activation is not only negatively regulated through intramolecular interactions, but additionally by compartmentalization in the cytosol. Comparison of different rodent and human cell lines indicates that the subcellular locale of HSF1 can vary such that certain cell lines exhibit cytosolic and nuclear translocation and others exhibit strictly nuclear localization before and after heat shock.

Comparison of the immunofluorescence staining patterns for HSF1 in heat shocked 3T3 and HeLa cells has revealed distinct, brightly staining granules in heat shocked HeLa cells but not in 3T3 cells (Fig. 6; Sarge et al., 1993). Confocal analysis of human cells stained with HSF1 specific antisera revealed that there are an average of five to seven granules ranging in size from 4 to 6 μm in the nucleus of a heat shocked HeLa cell. The number and size of the HSF1 granules is variable and highly dependent on the magnitude or type of stress to which the cell is exposed. The appearance of HSF1 granules seems to be specific to human cells and has not been detected in cell lines form any other organism (Cotto et al., in preparation). The presence of these nuclear HSF1 granules correlates closely with acquisition of DNA binding activity and transcription. Furthermore, the granules disappear during attenuation of the heat shock response (Fig. 7). Although these results

37°C   **42°C**

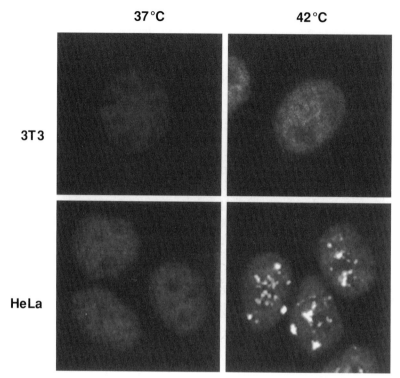

Figure 6. Intracellular localization of HSF1 in unstressed and heat shocked 3T3 and HeLa cells; Evidence that HSF1 granules are specific to human cells. Immunofluorescence analysis of unstressed and heat shocked 3T3 or HeLa cells using a mab to HSF1. Distinct, brightly staining granules are visible only in the nucleus of the heat shocked HeLa cells but not in the nucleus of 3T3 cells.

suggest a correlation between activation of HSF1 and the appearance of granules in the stressed cell, the role of these granules remains uncertain. One possibility is that the HSF1 granules provide a means to locally increase the concentration of HSF1 to ensure rapid kinetics of HSF1 DNA binding and elevated transcription of heat shock genes. At least one observation seems to support this hypothesis; the number of HSF1 granules has been determined for different human lines in which the DNA content is known; results from these experiments suggest that the number of granules varies according to chromosome ploidy (Cotto et al., in preparation).

Activation of heat shock factor by anti-inflammatory drugs

Anti-inflammatory drugs have provided key insights into the multistep pathway of HSF1 activation. Studies with salicylate, a classical inducer of chromosome puffing in *Drosophila* salivary gland chromosomes, have

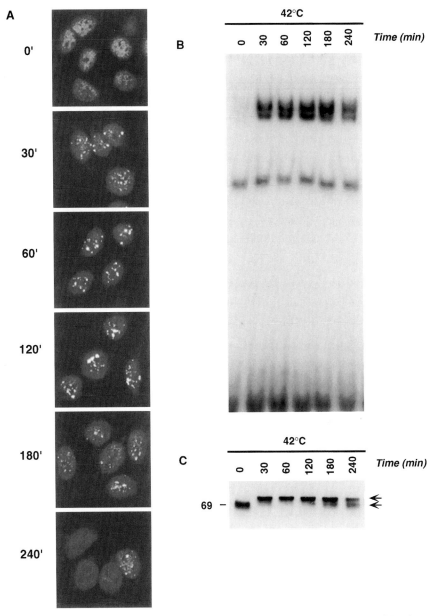

Figure 7. Kinetics of HSF1 granule formation, DNA binding activity and inducible phosphorylation in heat shocked HeLa cells. **A** Immunofluorescence analysis in HeLa cells prior to heat shock (0′) or after incubation at 42 °C for 60, 120, 180 and 240 min using HSF1 specific monoclonal antibodies. **B** Kinetics of HSF1 DNA binding activity by gel-shift analysis, and **C** Western blot analysis of whole cell extracts from samples indicated in panel B. The slower mobility of HSF1 in heat shocked cells is due to inducible serine phosphorylation.

shown that HSF1 can undergo trimerization and translocation to the nucleus without the subsequent event of HSP70 transcription (Jurivich et al., 1992). Exposure of the salicylate induced-HSF1 to heat shock results in the rapid transcription of heat shock genes, thus the inert trimer is likely to represent an intermediate which can be converted to a functional state. Furthermore, pre-treatment of mammalian cells with salicylate or indomethacin decreases the threshold for heat shock induction and leads to increased cell survival under sub-optimal conditions (Lee et al., 1995). Therefore, it is conceivable that the ability of anti-inflammatory drugs to activate the DNA binding activity of HSF1 could be an important aspect in the pharmacological effectiveness of these drugs.

Among the well-known effects of salicylates and indomethacin are their actions as potent inhibitors of arachidonic acid metabolism. Arachidonic acid itself has been shown to activate HSF (Jurivich and Morimoto, 1994). Treatment of cultured cells with lower concentrations of arachidonic acid was shown to have a synergistic effect with mild heat stress, suggesting that temperatures that are physiologically approached during fever or inflammatory processes might activate the heat shock response. The activation of HSF1 has also been observed in cells treated with specific prostaglandins (PGA$_2$), additional products of arachidonic acid metabolism which have anti-proliferative properties (Amici et al., 1992; Holbrook et al., 1992).

Another example of a pharmacologically active compound that influences the HS response is the flavenoid compound quercetin which interferes with the activation of HSF1 DNA binding activity by heat shock (Hosokawa et al., 1992). Quercetin was shown to inhibit heat induced phosphorylation of HSF1, which suggests a role for inducible phosphorylation in the activation of HSF1.

A multi-step model for HSF1 activation

The kinetics of the heat shock response is proportional to the synthesis of HSPs such that HS transcription attenuates after a period of several hours at the elevated temperature. The process of transcriptional attenuation of the HS response involves the release of HSF1 from the endogenous HSE and conversion of the trimeric HSF1 back to the latent non-DNA-binding monomer. An estimate for the *in vivo* equilibrium dissociation rate of HSF1 from the HSE of the endogenous hsp70 gene was shown to be ≈10-fold faster than occurs *in vitro*, suggesting an active process that facilitates the release of active HSF1 from DNA (Abravaya et al., 1991b). The attenuation and release of HSF1 correlated with increased levels of HSP70. Evidence for the association of HSP70 with HSF was obtained by analysis of HSF1-containing complexes using gel mobility shift experiments (Abravaya et al., 1992; Baler et al., 1992). These observations revealed that HSP70 was in a stable complex with HSF and that this complex was

HSF1 Stress Cycle

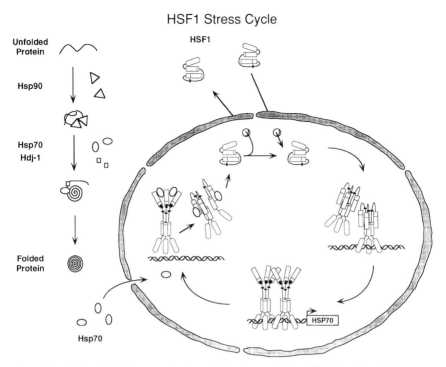

Figure 8. The heat shock factor cycle: A model for the regulation of HSF1 activity in the stressed cell. The activation by HS of latent cytoplasmic HSF1 is shown with the accompanying translocation to the nuclear compartment and oligomerization into an active trimer. The interaction of two HSF1 trimers with the HSE of the HSP70 gene promoter, follows by the subsequent events of HSF1 phosphorylation, and activation of HSP70 transcription are shown. We speculate that the dissociation of HSF1 from the HSE is facilitated by events initiated by formation of the HSF1:HSP70 complex. HSF1 remains associated with HSP70 until it is converted back into a latent monomer in the nucleus or cytoplasm. These events are presumed to be related to the levels of HSP70 and other chaperones during and following exposure to heat shock.

competent to bind DNA. Notably, the HSP70:HSF complex was dissociated with upon addition of ATP, a cofactor known to mediate the release of bound substrates from HSP70 (Hartl et al., 1992; Hendrick and Hartl, 1993; Freeman et al., 1995). The association between HSP70 and HSF was most readily observed during the attenuation phase of the heat shock response which suggested a role for HSP70 in disassembly of the HSF trimer. It is unlikely that HSP70 acts alone to accomplish this process as there are known co-chaperones that act in concert with HSP70 to facilitate protein folding (Hartl et al., 1992; Craig, 1993; Craig et al., 1993).

We propose a scenario for the activation and attenuation of mammalian HSF1 activity (Fig. 8). (1) In the initial phase, heat shock results in increased levels of unfolded protein substrates that require HSP70 and other chaperones for their correct refolding. The sequestration of HSP70 and other chaperones releases HSF1 from its negatively regulated state and, consequently, the HSF monomer translocates to the nucleus. (2) Upon entry to the nucleus, HSF1 undergoes conversion to a trimer and binds to the HSE (step 3) in the promoters of HS genes. This form of HSF1 is DNA binding competent yet remains transcriptionally inert. (4) HS induced HSF1 is subsequently stress-inducibly phosphorylated which correlates with elevated transcription of HSP70 and other HS genes. This results in the elevated synthesis of heat shock proteins. (5) As heat or stress-induced protein damage is repaired, the released HSP70 protein transiently reassociates with HSF and facilitates the release of HSF from the HSE. (6) HSF1 remains complexed with HSP70 and HSF1 trimers are then converted to monomeric state. The maintenance of HSF1 in the latent state is most likely due to the continued transient association of HSP70.

References

Abravaya, K., Myers, M.P., Murphy, S.P. and Morimoto, R.I. (1992) The human heat shock protein hsp70 interacts with HSF, the transcription factor that regulates heat shock gene transcription. *Genes and Dev.* 6:1153–1164.

Abravaya, K., Phillips, B. and Morimoto, R.I. (1991a) Heat shock-induced interactions of heat shock transcription factor and the human hsp70 promoter examined by *in vivo* footprinting. *Mol. Cell. Biol.* 11:586–592.

Abravaya, K., Phillips, B. and Morimoto, R.I. (1991b) Attenuation of the heat shock response in HeLa cells is mediated by the release of bound heat shock transcription factor and is modulated by changes in growth and in heat shock temperatures. *Genes and Dev.* 5:2117–2127.

Agoff, S.N., Hou, J., Linzer, D.I. and Wu, B. (1993) Regulation of the human hsp70 promoter by p53. *Science* 259:84–87.

Amici, C, Sistonen, L., Santoro, M.G. and Morimoto, R.I. (1992) Antiproliferative prostaglandins activate heat shock transcription factor. *Proc. Natl. Acad. Sci. USA.* 89:6227–6231.

Amin, J., Ananthan, J. and Voellmy, R. (1988) Key features of heat shock regulatory elements. *Mol. Cell. Biol.* 8:3761–3769.

Ananthan, T., Goldberg, A.L. and Voellmy, R. (1986) Abnormal proteins serve as eukaryotic stress signals and trigger the activation of heat shock genes. *Science* 232:522–524.

Aoki, M., Abe, K., Kawagoe, J.I., Sato, S., Nakamura, S. and Kogure, K. (1993) Temporal profile of the induction of heat shock protein 70 and heat shock cognate protein 70 mRNAs after transient ischemia in gerbil brain. *Brain Res.* 601:185–192.

Ashburner, M. (1970) Pattern of puffing activity in the salivary gland chromosomes of *Drosophila*. V. Response to environmental treatments. *Chromosoma* 31:356–376.

Baler, R., Welch, W.J. and Voellmy, R. (1992) Heat shock gene regulation by nascent polypeptides and denatured proteins: hsp70 as a potential autoregulatory factor. *J. Cell. Biol.* 117:1151–1159.

Baler, R., Dahl, G. and Voellmy, R. (1993) Activation of human heat shock genes is accompanied by oligomerization, modification, and repaid translocation of heat shock transcription factor HSF1. *Mol. Cell. Biol.* 13(4):2486–2496.

Beckmann, R.P., Mizzen, L.E. and Welch, W.J. (1990) Interaction of Hsp70 with newly synthesized proteins: implications for protein folding and assembly. *Science* 248:850–854.

Benjamin, I.J., Kroger, B. and Williams, R.S. (1992) Induction of stress proteins in cultured myogenic cells: Molecular signals for the activation of heat shock transcription factor during ischemia. *J. Clin. Invest.* 89:1685–1689.

Bienz, M. and Pelham, H.R.B. (1986) Heat shock regulatory elements function as an inducible enhancer in the *Xenopus hsp70* gene and when linked to a heterologous promoter. *Cell* 45:753–760.

Blake, M.J., Gershon, D., Fargnoli, J. and Holbrook, N.J. (1990a) Discordant expression of heat shock protein mRNAs in tissues of heat-stressed rats. *J. Biol. Chem.* 265:15275–15279.

Blake, M.J., Nowak, T.S., Jr. and Holbrook, N.J. (1990b) *In vivo* hyperthermia induces expression of HSP70 mRNA in brain regions controlling the neuroendocrine response to stress. *Mol. Brain Res.* 8:89–92.

Blake, M.J., Udelsman, R., Feulner, G.J., Norton, D.D. and Holbrook, N.J. (1991) Stress-induced heat shock protein 70 expression in adrenal cortex: an adrenocorticotropic hormone-sensitive, age-dependent response. *Proc. Natl. Acad. Sci. USA* 88: 9873–9877.

Bonner, J.J., Ballou, C. and Fackenthal, D.L. (1994) Interactions between DNA-bound trimers of the yeast heat shock factor. *Mol. Cell. Biol.* 14:501–508.

Chen, Y., Barlev, N.A., Westergaard, O. and Jakobsen, B.K. (1993) Identification of the C-terminal activator domain in yeast heat shock factor: independent control of transient and sustained transcriptional activity. *EMBO J.* 12:5007–5018.

Clos, J., Rabindra, S., Wisniewski, J. and Wu, C. (1993) Induction temperature of human heat shock factor is reprogrammed in a *Drosophila* cell environment. *Nature* 364:252–255.

Clos, J., Westwood, J.T., Becker, P.B., Wilson, S., Lambert, K. and Wu, C. (1990) Molecular cloning and expression of a hexameric *Drosophila* heat shock factor subject to negative regulation. *Cell* 63:1085–1097.

Corces, V., Pellicer, A., Axel, R. and Meselson, M. (1981) Integration, transcription, and control of a *Drosophila* heat shock gene in mouse cells. *Proc. Natl. Acad. Sci. USA* 78: 7038–7042.

Cotto, J.J., Kline, M. and Morimoto, R.I. (1996) Activation of heat shock factor 1 DNA binding precedes stress-induced serine phosphorylation. *J. Biol. Chem.* 271:3355–3358.

Craig, E.A. (1993) Chaperones: Helpers along the pathway to protein folding. *Science* 260: 1902–1903.

Craig, E.A., Gambill, B.D. and Nelson, R.J. (1993) Heat shock proteins: molecular chaperones of protein biogenesis. *Microbiol. Rev.* 57:402–414.

Currie, R.W. and White, F.P. (1993) Heat shock and limitation of tissue necrosis during occlusion/reperfusion in rabbit hearts. *Circulation* 87:963–971.

DiDomenico, B.J., Bugaisky, G.E. and Lindquist S. (1982) The heat shock response is regulated at both the transcriptional and posttranscriptional levels. *Cell* 31:593–603.

Ferris, D.K., Harel-Bellan, A., Morimoto, R.I., Welch, W.J. and Farrar, W.L. (1988) Mitogen and lymphokine stimulation of heat shock proteins in T lymphocytes. *Proc. Natl. Acad. Sci. USA* 85:3850–3854.

Fiorenza, M.T., Farkas, T., Dissing, M., Kolding, D. and Zimarino, V. (1995) Complex expression of murine heat shock transcription factors. *Nucl. Acid. Res.* 23(3):467–474.

Freeman, B.C., Myers, M.P., Schumacher, R. and Morimoto, R.I. (1995) Identification of a regulatory motif in Hsp70 that affects ATPase activity, substrate binding and interaction with HDJ-1. *EMBO J.* 14(10):2281–2292.

Gallo, G.J., Schuetz, T.J. and Kingston, R.E. (1991) Regulation of heat shock factor in *Schizosaccharomyces pombe* more closely resembles regulation in mammals than in *Saccharomyces cerevisiae*. *Mol. Cell. Biol.* 11:281–288.

Goff, S.A. and Goldberg, A.L. (1985) Production of abnormal proteins in *E. coli* stimulates transcription of *lon* and other heat shock genes. *Cell* 41:587–595.

Green, M., Schuetz, T.J., Sullivan, E.K. and Kingston, R.E. (1995) A heat shock-responsive domain of human HSF1 that regulates transcription activation domain function. *Mol. Cell. Biol.* 15(6):3354–3362.

Greene, J.M. and Kingston, R.E. (1990) TATA-dependent and TATA-independent function of the basal and heat shock elements of a human hsp70 promoter. *Mol. Cell. Biol.* 10(4): 1319–1328.

Greene, J.M., Larin, Z., Taylor, I.C., Prentice, H., Gwinn, K.A. and Kingston, R.E. (1987) Multiple basal elements of a human hsp70 promoter function differently in human and rodent cell lines. *Mol. Cell. Biol.* 7(10):3646–3655.

Hartl, F.U., Martin, J. and Neupert, W. (1992) Protein folding in the cell: the role of molecular chaperones Hsp70 and Hsp60. *Annu. Rev. Biophys. Biomol. Struct.* 21:292–322.

Hendrick, J.P. and Hartl F.-U. (1993) Molecular chaperone functions of heat-shock proteins. *Annu. Rev. Biochem.* 62:349–384.

Hoj, A. and Jakobsen, B.K. (1994) A short element required for turning off heat shock transcription factor: evidence that phosphorylation enhances deactivation. *EMBO J.* 13: 2617–2624.

Holbrook, N.J., Carlson, S.G., Choi, A.M.K. and Fargnoli, J. (1992) Induction of HSP70 gene expression by the antiproliferative prostaglandin PGA$_2$: a growth-dependent response mediated by activation of heat shock transcription factor. *Mol. Cell. Biol.* 12:1528–1534.

Holmgren, R., Corces, V., Morimoto, R., Blackman, R. and Meselson, M. (1981) Sequence homologies in the 5' regions of four *Drosophila* heat-shock genes. *Proc. Natl. Acad. Sci. USA* 78:3775–3778.

Hosokawa, N., Hirayoshi, K., Kudo, H., Takechi, H., Aoike, A., Kawai, K. and Nagata, K. (1992) Inhibition of activation of heat shock factor *in vivo* and *in vitro* by flavanoids. *Mol. Cell. Biol.* 12:3490–3498.

Jakobsen, B.K. and Pelham, H.R. (1988) Constitutive binding of yeast heat shock factor to DNA *in vivo. Mol. Cell. Biol.* 8:5040–5042.

Jurivich, D. and Morimoto, R.I. (1994) Arachidonate is a potent modulator of human heat shock gene transcription. *Proc. Natl. Acad. Sci. USA* 91:2280–2284.

Jurivich, D.A., Sistonen, L., Kroes, R.A. and Morimoto, R.I. (1992) Effect of sodium salicylate on the human heat shock response. *Science* 255:1243–1245.

Kao, H.T. and Nevins, J.R. (1983) Transcriptional activation and subsequent control of the human heat shock gene during adenovirus infection. *Mol. Cell. Biol.* 3(11):2058–2065.

Kingston, R.E., Schuetz, T.J. and Larin, Z. (1987) Heat-inducible human factor that binds to a human hsp70 promoter. *Mol. Cell. Biol.* 7:1530–1534.

Kroeger, P.E., Sarge, K.D. and Morimoto, R.I. (1993) Mouse heat shock transcription factors 1 and 2 prefer a trimeric binding site but interact differently with the HSP70 heat shock element. *Mol. Cell. Biol.* 13:3370–3383.

Kroeger, P.E. and Morimoto, R.I. (1994) Selection of new HSF1 and HSF2 DNA-binding sites reveals difference in trimer cooperatively. *Mol. Cell. Biol.* 14(11):7592–7603.

Larson, J.S., Schuetz, T.J. and Kingston, R.E. (1988) Activation *in vitro* of sequence-specific DNA binding by a human regulatory factor. *Nature* 335:372–375.

Lee, B.S., Chen, J., Angelidis, C., Jurivich, D.A. and Morimoto, R.I. (1995) Pharmacological modulation of Heat Shock Factor 1 by anti-inflammatory drugs results in protection against stress-induced cellular damage. *Proc. Natl. Acad. Sci. USA* 92:7307–7211.

Lindquist, S. (1986) The Heat-Shock response. *Ann. Rev. Biochem.* 55:1151–1191.

Lindquist, S. and Craig E.A. (1988) The heat shock proteins. *Annu. Rev. Genet.* 22:631–677.

Lindquist-McKenzie, S.L., Henikoff, S. and Meselson, M. (1975) Localization of RNA from heat-induced polysomes at puff sites in *Drosophila melanogaster. Proc. Natl. Acad. Sci. USA* 72:1117–1121.

Lis, J. and Wu, C. (1993) Protein traffic on the heat shock promoter: parking, stalling, and trucking along. *Cell* 74:1–4.

Liu, Y., Kato, H., Nakata, N. and Kogure, K. (1992) Protection of rat hippocampus against ischemic neuronal damage by pretreatment with sublethal ischemia. *Brain Res.* 586:121–124.

Lum, L.S.Y., Sultzman, L.A., Kaufman, R.J., Linzer, D.I.H. and Wu, B. (1990) A cloned human CCAAT-box-binding factor stimulates transcription from the human hsp70 promoter. *Mol. Cell. Biol.* 10:6709–6717.

Marber, M.S., Latchman, D.S., Walker, J.M. and Yellon, D.M. (1993) Cardiac stress protein elevation 24 hours after brief ischemia or heat stress is associated with resistance to myocardial infarction. *Circulation* 88:1264–1272.

Marcuccilli, C.J., Mathur, S.K., Morimoto, R.I. and Miller, R.J. (1996) Regulatory differences in the stress response of hippocampal neurons and glial cells after heat shock. *J. Neurosci.* 16(2):478–485.

Mathur, S., Sistonen, L., Brown, I.B., Murphy, S.P., Sarge, K.D. and Morimoto, R.I. (1994) Deficient induction of human HSP70 gene transcription in Y79 retinoblastoma cells despite activation of HSF1. *Proc. Natl. Acad. Sci. USA* 91(18):8695–8699.

Milarski, K.L. and Morimoto, R.I. (1986) Expression of human HSP70 during the synthetic phase of the cell cycle. *Proc. Natl. Acad. Sci. USA* 83:9517–9521.

Mirault, M.-E., Southgate, R. and Delwart, E. (1982) Regulation of heat shock genes: a DNA sequence upstream of *Drosophila* hsp70 genes is essential for their induction in monkey cells. *EMBO J.* 1:1279–1285.

Morgan, W.D. (1989) Transcription factor Sp1 binds to and activates a human HSP70 gene promoter. *Mol. Cell. Biol.* 9:4099–4104.

Morgan, W.D., Williams, G.T., Morimoto, R.I., Greene, J., Kingston, R.E. and Tjian, R. (1987) Two transcriptional activators, CCAAT-box binding transcription factor and heat shock transcription factor, interact with a human HSP70 gene promoter. *Mol. Cell. Biol.* 7: 1129–1138.

Morimoto, R.I., Jurivich, D.A., Kroeger, P.E., Mathur, S.K., Murphy, S.P., Nakai, A., Sarge, K., Abravaya, K. and Sistonen, L. (1994) The regulation of heat shock gene expression by a family of heat shock factors. *In:* R.I. Morimoto, A. Tissieres and C. Georgopoulos (eds): *Biology of Heat Shock Proteins and Molecular Chaperones*. Cold Spring Harbor Laboratory Press, Cold Spring Harbor, NY, pp 417–455.

Morimoto, R.I. (1993) Chaperoning the nascent polypeptide chain. *Curr. Biol.* 3: 101–102.

Morimoto, R.I. (1991) Heat shock: the role of transient inducible responses in cell damage, transformation, and differentiation. *Cancer Cells* 3: 297–301.

Mosser, D.D., Kotzbauer, P.T., Sarge, K.D. and Morimoto, R.I. (1990) *In vitro* activation of heat shock transcription factor DNA-binding by calcium and biochemical conditions that affect protein conformation. *Proc. Natl. Acad. Sci. USA* 87: 3748–3752.

Mosser, D.D., Theodorakis, N.G. and Morimoto, R.I. (1988) Coordinate changes in heat shock element-binding activity and hsp70 gene transcription rates in human cells. *Mol. Cell. Biol.* 8: 4736–4744.

Murphy, S.P., Gorzowski, J.J., Sarge, K.D. and Phillips, B. (1994) Characterization of constitutive HSF2 DNA-binding activity in mouse embryonal carcinoma cells. *Mol. Cell. Biol.* 14(8): 5309–5317.

Nakai, A., Kawazoe, Y., Tanabe, M., Nagata, K. and Morimoto, R.I. (1995) The DNA-binding properties to two heat shock factors, HSF1 and HSF3, are induced in the avian erythroblast cell line HD6. *Mol. Cell. Biol.* 15(10): 5268–5278.

Nakai, A. and Morimoto, R.I. (1993) Characterization of a novel chicken heat shock transcription factor, HSF3, suggests a new regulatory pathway. *Mol. Cell. Biol.* 13: 1983–1997.

Nieto-Sotelo, J., Wiederrecht, G., Okuda, A. and Parker, C.S. (1990) The yeast heat shock transcription factor contains a transcriptional activation domain whose activity is repressed under nonshock conditions. *Cell* 62: 807–817.

Nowak, T.S. and Abe, H. (1994) The postischemic stress response in brain. *In:* R.I. Morimoto, A. Tissieres and C. Georgopoulos (eds): *The Biology of Heat Shock Proteins and Molecular Chaperones*. Cold Spring Harbor Laboratory Press, Cold Spring Harbor, NY, pp 553–575.

Pelham, H.R.B. (1982) A regulatory upstream promoter element in the *Drosophila* hsp70 heat-shock gene. *Cell* 30: 517–528.

Pelham, H.R.B. and Bienz, M. (1982) A synthetic heat-shock promoter element confers heat-inducibility on the herpes simplex virus thymidine kinase gene. *EMBO J.* 1: 1473–1477.

Perisic, O., Xiao, H. and Lis, J.T. (1989) Stable binding of *Drosophila* heat shock factor to head-to-head and tail-to-tail repeats of a conserved 5 bp recognition unit. *Cell* 59: 797–806.

Peteranderl, R. and Nelson, H.C.M. (1992) Trimerization of the heat shock transcription factor by a triple-stranded a helical coiled-coil. *Biochemistry* 31: 12272–12276.

Phillips, B. and Morimoto, R.I. (1991) Transcriptional regulation of human hsp70 genes: relationship between cell growth, differentiation, virus infection, and the stress response. *In:* L.E. Hightower and L. Nover (eds): *Heat Shock and Development*. Springer Verlag, Berlin, Heidelberg, pp 167–187.

Pulsinelli, W.A., Brierley, J.B. and Plum, F. (1982) Temporal profile of neuronal damage in a model of transient forebrain ischemia. *Ann. Neurol.* 11: 491–498.

Rabindran, S.K., Haroun, R.I., Clos, J., Wisniewski, J. and Wu, C. (1993) Regulation of heat shock factor trimer formation: role of a conserved leucine zipper. *Science* 259: 230–234.

Rabindran, S.K., Giorgi, G., Clos, J. and Wu, C. (1991) Molecular cloning and expression of a human heat shock factor, HSF1. *Proc. Natl. Acad. Sci. USA* 88: 6906–6910.

Ritossa, F.M. (1962) A new puffing pattern induced by a temperature shock and DNP in *Drosophila*. *Experientia* 18: 571–573.

Sarge, K.D., Park-Sarge, O.Y., Kirby, J.D., Mayo, K. and Morimoto, R.I. (1994) Expression of heat shock factor 2 in mouse testis: potential role as a regulator of heat-shock protein gene expression during spermatogenesis. *Biol. Reprod.* 50(6): 1334–1343.

Sarge, K.D., Murphy, S.P. and Morimoto, R.I. (1993) Activation of heat shock gene transcription by HSF1 involves oligomerization, acquisition of DNA binding activity, and nuclear localization and can occur in the absence of stress. *Mol. Cell. Biol.* 13:1392–1407.

Sarge, K.D., Zimarino, Holm, K., Wu, C. and Morimoto, R.I. (1991) Cloning and characterization of two mouse heat shock factors with distinct inducible and constitutive DNA-binding ability. *Genes and Dev.* 5:1902–1911.

Scharf, K.-D., Rose, S., Zott, W., Schoff, F. and Nover, L. (1990) Three tomato genes code for heat stress transcription factors with a remarkable degree of homology to the DNA-binding domain of the yeast HSF. *EMBO J.* 9:4495–4501.

Schuetz, T.J., Gallo, G.J., Sheldon, L., Tempst, P. and Kingston, R.E. (1991) Isolation of a cDNA for HSF2: evidence for two heat shock factor genes in humans. *Proc. Natl. Acad. Sci. USA* 88:6910–6915.

Shi, Y., Kroeger, P.E. and Morimoto, R. (1995) The carboxyl-terminal transactivation domain of heat shock factor 1 is negatively regulated and stress responsive. *Mol. Cell. Biol.* 15(8): 4309–4318.

Sistonen, L., Sarge, K.D. and Morimoto, R. (1994) Human heat shock factors 1 and 2 are differentially activated and can synergistically induce hsp70 gene transcription. *Mol. Cell. Biol.* 14(3):2087–2099.

Sistonen, L., Sarge, K.D., Phillips, B., Abravaya, K. and Morimoto, R. (1992) Activation of heat shock factor 2 during hemin-induced differentiation of human erythroleukemia cells. *Mol. Cell. Biol.* 12(9):4104–4111.

Sorger, P.K. (1990) Yeast heat shock factor contains separable transient and sustained response transcriptional activators. *Cell* 62:793–805.

Sorger, P.K. and Nelson, H.C.M. (1989) Trimerization of a yeast transcriptional activator via a coiled-coil motif. *Cell* 59:807–813.

Sorger P.K. and Pelham, H.R.B. (1988) yeast heat shock factor is an essential DNA-binding protein that exhibits temperature-dependent phosphorylation. *Cell* 54:855–864.

Sorger, P.K., Lewis, M.J. and Pelham, H.R.B. (1987) Heat shock factor is regulated differently in yeast and HeLa cells. *Nature* 329:81–84.

Stone, D.E. and Craig, E.A. (1990) Self-regulation of 70-kilodalton heat shock proteins in *Saccharomyces cerevisiae*. *Mol. Cell. Biol.* 10(4):1622–1632.

Theodorakis, N.G., Zand, D.J., Kotzbauer, P.T., Williams, G.T. and Morimoto, R.I. (1989) Hemin-induced transcriptional activation of the HSP70 gene during erythroid maturation in K562 cells is due to a heat shock factor-mediated stress response. *Mol. Cell. Biol.* 9(8): 3166–3173.

Tissieres, A., Mitchell, H.K. and Tracy, V.M. (1974) Protein synthesis in salivary glands of *Drosophila melanogaster*: Relation to chromosome puffs. *J. Mol. Biol.* 84:389–398.

Treuter, E., Nover, L., Ohme, K. and Scharf, K.-D. (1993) Promoter specificity and deletion analysis of three tomato heat stress transcription factors. *Mol. Gen. Genet.* 240:113–125.

Udelsman, R., Blake, M.J., Stagg, C.A., Li, D., Putney, D.J. and Holbrook, N.J. (1993) Vascular heat shock protein expression in response to stress. *J. Clin. Invest.* 91:465–473.

Westwood, J.T. and Wu, C. (1993) Activation of *Drosophila* heat shock factor: conformational change associated with a monomer-to-trimer transition. *Mol. Cell. Biol.* 13:3481–3486.

Westwood, J.T., Clos, J. and Wu, C. (1991) Stress-induced oligomerization and chromosomal relocalization of heat-shock factor. *Nature* 353:822–827.

Wiederrecht, G., Seto, D. and Parker, C.S. (1988) Isolation of the gene encoding the *S. cerevisiae* heat shock transcription factor. *Cell* 54:841–853.

Williams, G.T. and Morimoto, R.I. (1990) Maximal stress-induced transcription from the human hsp70 promoter requires interactions with the basal promoter elements independent of rotational alignment. *Mol. Cell. Biol.* 10:3125–3136.

Williams, G.T., McClanahan, T.K. and Morimoto, R.I. (1989) E1a transactivation of the human HSP70 promoter is mediated through the basal transcriptional complex. *Mol. Cell. Biol.* 9:2574–2587.

Williams, R.S. and Benjamin, I.J. (1993) Human HSP70 protects murine cells from injury during metabolic stress. *J. Clin. Invest.* 92:503–508.

Wu, C. (1980) The 5′ ends of *Drosophila* heat shock genes in chromatin are hypersensitive to DNase I. *Nature* 286:854–880.

Wu, C. (1984) Two protein-binding sites in chromatin implicated in the activation of heat-shock genes. *Nature* 309:229–234.

Wu, B., Hunt, C. and Morimoto, R.I. (1985) Structure and expression of the human gene encoding major heat shock protein HSP70. *Mol. Cell. Biol.* 5:330–341.

Wu, B.J., Kingston, R.E. and Morimoto, R.I. (1986a) Human HSP70 promoter contains at least two distinct regulatory domains. *Proc. Natl. Acad. Sci. USA* 83:629–633.

Wu, B., Hurst, H., Jones, N. and Morimoto, R. (1986b) The E1a 13S product of adenovirus 5 activates transcription of the cellular human HSP70 gene. *Mol. Cell. Biol.* 6:2994–2999.

Wu, C., Wilson, S., Walker, B., Dawid, I., Paisley, T., Zimarino, V. and Ueda, H. (1987) Purification and properties of *Drosophila* heat shock activator protein. *Science* 238:1247–1253.

Wu, C., Clos, J., Giorgi, G., Haroun, R.I., Kim, S.J., Rabindran, S.K., Westwood, J.T., Wisniewski, J. and Yim, G. (1994) Structure and regulation of heat shock transcription factor. *In:* R.I. Morimoto, A. Tissieres and C. Georgopoulos (eds): *The Biology of Heat Shock Proteins and Molecular Chaperones.* Cold Spring Harbor Laboratory Press, Cold Spring Harbor, NY, pp 395–416.

Xiao, H. and Lis, J.T. (1988) Germline transformation used to define key features of the heat shock response element. *Science* 239:1139–1142.

Xiao, H., Perisic, O. and Lis, J.T. (1991) Cooperative binding of *Drosophila* heat shock factor to arrays of a conserved 5 bp unit. *Cell* 64:585–593.

Zimarino, V. and Wu, C. (1987) Induction of sequence-specific binding of *Drosophila* heat shock activator protein without protein synthesis. *Nature* 327:727–730.

Zuo, J., Rungger, D. and Voellmy, R. (1995) Multiple layers of regulation of human heat shock transcription factor 1. *Mol. Cell. Biol.* 15(8):4319–4330.

Stress-Inducible Cellular Responses
ed. by U. Feige, R.I. Morimoto, I. Yahara and B. Polla
© 1996 Birkhäuser Verlag Basel/Switzerland

Transcriptional regulation of stress-inducible genes in procaryotes

T. Yura, K. Nakahigashi and M. Kanemori

HSP Research Institute, Kyoto Research Park, Shimogyo-ku, Kyoto 600, Japan

Summary. In procaryotes such as *Escherichia coli,* transcriptional activation of heat shock genes in response to elevated temperature is caused primarily by transient increase in the amount of σ^{32} (*rpoH* gene product) specifically required for transcription from the heat shock promoters. The increase in σ^{32} level results from increased translation of *rpoH* mRNA and from stabilization of σ^{32} which is ordinarily very unstable. Some of the factors and cis-acting elements that constitute the complex regulatory circuits have been identified and characterized, but detailed mechanisms as well as nature of sensors and signals remain to be elucidated. Whereas this "classical" heat shock regulon (σ^{32} regulon) provides major protective functions against thermal stress, a second heat shock regulon mediated by σ^{E} (σ^{24}) encodes functions apparently required under more extreme conditions, and is activated by responding to extracytoplasmic signals. These regulons mediated by minor σ factors (σ^{32} in particular) appear to be conserved in most gram-negative bacteria, but not in gram-positive bacteria.

Introduction

When cells of any bacteria are exposed to high temperature, the synthesis of a set of heat-shock proteins (HSPs) is rapidly and transiently induced. Like in most eucaryotes, transient induction of HSPs represents a major cellular response to cope with increased damage in proteins caused by heat shock and other stress. However, the role of HSPs is not limited to stress conditions: many HSPs are synthesized at low levels under nonstress conditions during normal growth or development. Some of the major HSPs are essential for cell growth at all temperatures, others are essential at a certain range of temperature, and still others are apparently dispensable or required only under special circumstances (Georgopoulos et al., 1990, 1994). The nature of differential, complementary, and overlapping functions of bacterial HSPs in assisting protein folding, assembly, and other processes is beginning to emerge by extensive work using diverse experimental systems and approaches. Evidently, a fairly large set of HSPs including those to be discovered play fundamental roles in normal growth as well as under stress conditions, particularly as molecular chaperones or ATP-dependent proteases.

Consistent with the conserved structure and function of HSPs, the rapid and tight adjustment of HSPs to meet physiological and environmental stresses appears to be universal and found even in bacteria living under extreme conditions. The major strategy employed is transcriptional regulation, particularly suited for procaryotes that are likely to be exposed to most versatile environments. One or more transcription factors controlling expression of a specific set of heat shock genes are involved in this important task. The amount or activity of such transcription factors is in turn modulated by a variety of mechanisms working at many different levels, perhaps reflecting various requirements of ecological niches for a given species. We shall first summarize our current understanding of the heat shock regulation in *E. coli* where most extensive work has been done, and then discuss some of the recent developments with other bacterial systems. For more comprehensive or historical treatment of the subject, readers should consult earlier reviews (e.g. Gross et al., 1990; Bukau, 1993; Yura et al., 1993; Georgopoulos et al. 1994).

E. coli heat shock regulons defined by the minor σ factors

Two major heat shock regulons whose transcription depend on a specific minor σ factor have been studied in *E. coli*: the "classical" σ^{32} regulon that encodes a well-known set of HSPs such as DnaK, GroE, Lon and Clp, and the newly discovered σ^E regulon whose members include *htrA (degP)* and *rpoH*. The σ^{32} regulon plays a major role in coping with heat shock stress, whereas the σ^E regulon appears to provide certain auxiliary functions important for survival under more extreme conditions. The fact that the latter regulon includes *rpoH* encoding σ^{32} illustrates the intimate functional link between the two regulons. It should be noted however that there are a number of other heat-inducible genes that appear to be under control of other regulatory factors (e.g., Weiner et al., 1991). Thus, a regulatory net work system may serve to coordinate or integrate the cellular responses to heat shock and related stress in *E. coli*.

The σ^{32} regulon: major protections against thermal stress

When *E. coli* cells grown at 30 °C are shifted to 42 °C, a set of at least 30 heat shock genes, whose transcription depends on RNA polymerase containing σ^{32}, is rapidly and transiently induced. About 20 of these HSPs were initially identified by examining synthesis rates of individual proteins by one- or two-dimensional gel electrophoresis (Lemaux et al., 1978; Herendeen et al., 1979; Yamamori et al., 1978, 1980). Additional HSPs were subsequently identified particularly by hybridizing cDNA (obtained with mRNA from heat-shocked cells) with membrane filters containing an ordered *E. coli* genomic library (Chuang and Blattner, 1993). Identification of genes that belong to this regulon depends on the demonstration that synthesis of a given protein can be transiently enhanced upon temperature shift (e.g. from 30 to 42 °C) in a wild-type strain but not in an isogenic *rpoH*

(= *htpR, hin*) mutant defective in the heat shock response (Neidhardt and VanBogelen, 1981; Yamamori and Yura, 1982) due to a deficiency in σ^{32} (Grossman et al., 1984). Such observations are usually followed by analyses of transcripts *in vivo* and *in vitro*.

The overall function of σ^{32} regulon was assessed by studying properties of the *rpoH* mutants as well as individual heat shock gene mutants. When the cellular level of *rpoH* gene product (σ^{32}) was varied quantitatively, more σ^{32} resulted in more HSPs and permitted cell growth at higher temperatures (Yamamori and Yura, 1982). At least some of the HSPs encoded by the σ^{32} regulon are essential for growth at high temperature, because the $\Delta rpoH$ mutant lacking σ^{32} cannot grow at above 20 °C (Zhou et al., 1988). This mutant was used to determine which of the HSPs are most critical for cell growth at higher temperature. The results showed that overexpression of GroEL-GroES chaperones in the $\Delta rpoH$ mutant permit cell growth up to 40 °C, and that simultaneous overproduction of the GroEL-GroES and DnaK-DnaJ chaperones support growth up to 42°C (Kusukawa and Yura, 1988). Massive accumulation of protein aggregates was found in the $\Delta rpoH$ mutant, but many less aggregates were seen when these chaperones were overexpressed (Gragerov et al., 1992). These results suggest that the two major chaperone teams are functionally among the most important members of the σ^{32} regulon. This is consistent with other genetic and biochemical evidence indicating that these HSPs play major roles in assisting protein folding by preventing misfolding and aggregation of proteins at normal growth temperatures (Georgopoulos and Welch, 1993; Hendrick and Hartl, 1993). Interestingly, these chaperones are also important in modulating energy-dependent protein degradation (Straus et al., 1988).

Besides those major chaperones that appear to be primarily involved in protein folding (and turnover), a number of ATP-dependent proteases including some of the putative regulatory subunits are encoded by the σ^{32} regulon. In addition to the well known proteases Lon and ClpP, some proteins that belong to the ClpA family (regulatory subunits of Clp proteases that exhibit ATPase activity) are members of this regulon (see Georgopoulos et al., 1994). The most recent addition to this list is HflB (= FtsH; Herman et al., 1995), which is a membrane-bound metalloprotease that can degrade σ^{32} *in vitro* (Tomoyasu et al., 1995), implying its potential regulatory role in the heat shock response.

The σ^E regulon: protections against extracytoplasmic damage

The existence of the second heat shock regulon under control of a distinct σ factor was first suggested from transcription studies of the *rpoH* gene encoding σ^{32}. Among the four promoters involved in *rpoH* transcription, three (P1, P4 and P5) are recognized by the primary σ factor (σ^{70}), whereas one (P3) is recognized by another σ factor called σ^E or σ^{24} (Erickson and

Gross, 1989; Wang and Kaguni, 1989; see Fig. 2 A). The latter promoter is only slightly active at $30-37\,^{\circ}$C, but quite active at $50\,^{\circ}$C where most transcription is arrested. Members of this regulon include *htrA* (*degP*) encoding a periplasmic protease, and *rpoE* (encoding σ^E) itself, besides *rpoH* (see Raina et al., 1995; Rouvière et al., 1995). In addition, synthesis of at least 10 proteins is enhanced when σ^E is overproduced, suggesting that some of these proteins may belong to this regulon. As expected, the mutant lacking σ^E is sensitive to high temperature, and grows normally at $37\,^{\circ}$C, but not at $42\,^{\circ}$C. The σ^E regulon appears to provide essential functions, particularly related to cellular surfaces, that are important for growth and survival at high temperatures.

Regulatory mechanisms of the σ^{32} regulon

Transcription of the σ^{32} regulon is markedly affected by either upshift or downshift of temperature. Upon temperature upshift, the rate of transcription of heat shock genes rapidly increases, with larger increases observed at higher temperatures. For example, after shift from 30° to $42\,^{\circ}$C, the rate of HSP synthesis increases 10-fold within 5 min (Yamamori et al., 1978; Lemaux et al., 1978). Under conditions of maximum induction, HSPs represent a substantial fraction ($> 20\%$) of total proteins synthesized. This induction phase is shortly followed by an adaptation phase in which the HSP synthesis decreases to a new steady-state level. By $20-30$ min, the amounts of HSPs increase 2- to 3-fold over the preshift level. Upon exposure to more extreme temperatures (e.g. $50\,^{\circ}$C), synthesis of most proteins is arrested, while that of HSPs is enhanced and continues to be made as long as cells can produce proteins.

Upon temperature downshift, on the other hand, the rate of transcription of heat shock genes decreases abruptly. For example, after shift from 42° to $30\,^{\circ}$C, the rate of transcription decreases about 10-fold within 5 min, followed by gradual decrease and then slow recovery to a new steady-state level (Straus et al., 1989; Taura et al., 1989). The rapid and dramatic change in the transcription rate of heat shock genes following temperature shift represents the mechanism by which the cell tries to attain a new steady-state level of HSPs as rapidly as possible. Since the difference in HSP levels between cells growing at different temperatures is generally small (two- to three-fold between 30° and $42\,^{\circ}$C), the above observations indicate that the level of HSPs needs to be regulated very tightly to assure maximum cell growth and survival. Various lines of experiments summarized below amply support this general conclusion.

Besides changes in temperature, sublethal concentrations of ethanol or production of abnormal or unfolded proteins cause less rapid but prolonged induction of HSPs to appreciable extents. Other agents can also induce at least some HSPs though the induction is less striking and the underlying mechanisms remain unresolved. The latter includes starvation for carbon source or amino acids, exposure to DNA-damaging agents, antibiotics or heavy metals, oxidation stress, and phage infection (see Neidhardt et al., 1987).

Regulation of the level and activity of σ^{32}

Following temperature upshift, the cellular level of σ^{32} increases rapidly but transiently. Time-course experiments employing immunoprecipitation with σ^{32} antiserum revealed that the concentration of σ^{32} increases 15- to 20-fold within 5 min after shift from 30° to 42°C, and then declines to a new steady-state level which is two- to three-fold higher than the preshift level. Both the time-course and the increase in σ^{32} level are sufficient to account for the increased transcription of heat shock genes (Grossman et al., 1987; Straus et al., 1987). The amount of σ^{32} during steady-state growth is very low, only 10 to 30 molecules per cell. This is because σ^{32} is very unstable (half-life = ca. 1 min), and its synthesis is largely restricted at the level of translation (Straus et al., 1987; Tilly et al., 1989; Nagai et al., 1991).

The unusual instability and translational repression provide effective means of maintaining σ^{32} concentration to the minimum during steady-state growth where only low levels of HSPs are needed. Upon temperature upshift, the levels of σ^{32} and of HSPs are rapidly increased to meet the requirements for growth at higher temperature. The increase in σ^{32} level results from changes in both synthesis and stability of σ^{32} (Straus et al., 1987). In the case of temperature downshift, the rapid drop in heat shock gene transcription results from a decrease in σ^{32} activity rather than from a decrease in σ^{32} level (Straus et al., 1989; Taura et al., 1989). Within 5 min after shift from 42° to 30°C, the σ^{32} activity drops 10-fold with only less than two-fold decrease in the σ^{32} level. Thus, transcription of the heat shock genes can be controlled at three levels: synthesis, degradation, and activity of σ^{32} (Fig. 1). Our ultimate understanding of transcriptional regulation of the σ^{32} regulon will therefore depend on understanding of the mechanisms of each of these processes.

Transcriptional control of σ^{32} synthesis

Four promoters are known to participate in transcription of the *rpoH* gene; three (P1, P4 and P5) are recognized by RNA polymerase containing primary σ factor σ^{70}, while one (P3) is recognized by RNA polymerase containing σ^E (Fig. 2A). The P1 and P4 promoters are responsible for *rpoH* transcription under most conditions (Erickson et al., 1987; Fujita and Ishihama, 1987). Whereas P1 does not seem to be highly regulated, P4

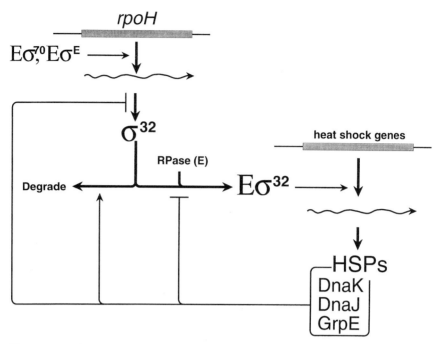

Figure 1. Regulatory circuits of the σ^{32} regulon in *E. coli*. The central regulatory factor, σ^{32}, confers to RNA polymerase (E) the specificity to transcribe heat shock genes. Induction of HSPs upon heat shock is achieved by an increase in the cellular concentration of σ^{32}, primarily through increased translation of *rpoH* mRNA and stabilization of ordinarily very unstable σ^{32}. During adaptation phase, the chaperone team of DnaK, DnaJ and GrpE heat shock proteins acts as negative feedback modulators by repressing *rpoH* mRNA translation, by accelerating σ^{32} degradation, and by inhibiting σ^{32} activity. Transcription of *rpoH* depends on Eσ^{70} (σ^{70} is the primary σ factor) under most conditions, but on Eσ^{E} at lethal temperature (50°C).

activity is enhanced two- to three-fold upon temperature upshift. P5 is a weak promoter, stimulated by the lack of glucose or by ethanol, and requires cAMP and cAMP-receptor protein *in vitro* (Nagai et al., 1990). The P3 promoter transcribed by σ^{E}-RNA polymerase is particularly interesting, because it is hardly active under normal conditions, but becomes quite active at extremely high temperatures (45–50°C) where σ^{70} is virtually inactivated. In addition, *rpoH* transcription appears to be negatively modulated by DnaA protein which is known to control the initiation of DNA replication.

In spite of the complex organization and potential regulatory importance of the various promoters, the bulk of transient increase in σ^{32} synthesis observed upon mild heat shock (e.g., a shift from 30° to 42°C) results from increased translation rather than increased transcription. However, transcriptional control of σ^{32} synthesis is clearly important not only in determining the basal level of expression under a variety of steady-state

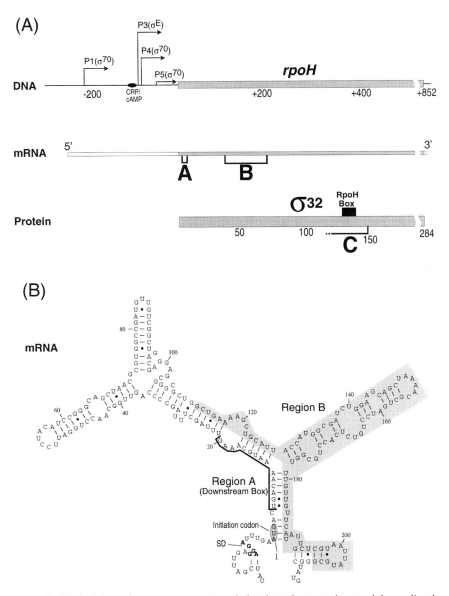

Figure 2. *Escherichia coli rpoH* promoters, translational regulatory regions, and the predicted secondary structure of 5′-portion of mRNA. The numbering of nucleotide is in reference to the coding region (begins at +1 and ends at 852). (A) Promoters (P1, P2, P3 and P5) for *rpoH* transcription, regions A and B (on mRNA) involved in thermoregulation of *rpoH* translation, and region C (on σ^{32} polypeptide) involved in chaperone-mediated negative feedback regulation (translational repression and σ^{32} degradation) are indicated. See Nakahigashi et al. (1995) for "RpoH box". (B) Computer-predicted secondary structure of *rpoH* mRNA involving base pairings between region A and part of region B (shaded). The structure shown differs slightly from that of Nagai et al. (1991) due to correction of two nucleotides, 67G and 68C.

conditions, but also in maintaining critical levels of *rpoH* mRNA when cells are exposed to extreme conditions such as lethal temperature. For further discussion on this subject, see earlier reviews (Gross et al., 1990; Bukau, 1993; Yura et al., 1993; Georgopoulos et al., 1994).

Translational control of σ^{32} synthesis

Some evidence has suggested that increased synthesis of σ^{32} observed upon heat shock (shift from 30° to 42 °C) occurs at the translational level (Straus et al., 1987; Nagai et al., 1991). First, *lacZ* expression from *rpoH-lacZ* gene (or protein) fusions but not operon fusions exhibits increased synthesis upon temperature upshift, suggesting that the translational initiation region rather than the promoter region is important for regulation. Indeed, heat-induced expression can be observed with a gene fusion construct in which transcription is initiated from a heterologous promoter and not the authentic *rpoH* promoters. Second, the increase of σ^{32} (or fusion protein) synthesis precedes that of *rpoH* mRNA level following temperature upshift. Finally, rifampicin that strongly inhibits new RNA synthesis does not inhibit heat induction of the gene fusion significantly.

Translational induction of σ^{32} is transient and consists of two phases: induction and repression. These phases are closely correlated in time with the induction and adaptation phases of the heat shock response, resulting in transient induction of HSPs. The fact that transcription of *rpoH* is relatively high but translation is poor initially suggested that temperature upshift relieves repression of σ^{32} synthesis at the translational level (Straus et al., 1987). The analysis of cis-acting elements required for temperature regulation of *rpoH* translation agrees well with this expectation (see below). Besides, both cis-acting element and trans-acting factors specifically required for translational repression observed during adaptation phase have been identified, suggesting that two separate though interrelated mechanisms may be involved.

Thermoinduction of rpoH translation

Extensive analyses of 3′- and 5′-deletion derivatives of *rpoH-lacZ* gene fusion led to identify two 5′-proximal regions of *rpoH* mRNA involved in thermoregulation (Kamath-Loeb and Gross, 1991; Nagai et al., 1991). One is a stretch of 15-nucleotide "downstream box" (Sprengart et al., 1990) which is located downstream of the translational start site (nucleotide 6–20) and complementary to 16S rRNA (region A in Fig. 2). Deletion or alteration of this region results in very low non-inducible synthesis of fusion protein. Thus, it is a positive element acting as a translational enhancer. The other (region B in Fig. 2) is an internal negative element of about 100 nucleotides (110–210) whose deletion from either 3′ or 5′-end causes high constitutive synthesis. Consistent with the results of deletion analysis, random point mutations causing high constitutive expression were found to be localized within these two regions (Yuzawa et al., 1993). It was therefore suspected that a secondary structure formed between the two mRNA regions might explain limited translation under nonstress conditions.

Computer prediction revealed that a fairly stable secondary structure ($G_0 = -61$ Kcal/mole) may be formed within the 5'-proximal region of mRNA defined above: this includes base parings of the initiation codon and region A with part of the internal region B (Nagai et al., 1991; Fig. 2B). Analyses of mutations that are expected to disrupt a specific base pairing or compensatory mutations that are expected to recover the base pairing indicated that some of the base pairings (e.g., 15G::124C and 16C::123G) are important for maintaining low level expression under nonstress conditions. However, the recovery of base pairing is not necessarily sufficient for regulation, and base sequence for part of the stem-loop structure involving region A (15–20 and 120–124) seems also to be critical for normal heat inducibility (Yuzawa et al., 1993 and unpublished results). It is most likely that a transacting factor binds to this region and modulates mRNA secondary structure and its translatability. A hypothetical protein factor may either repress translation under nonstress conditions or activate translation upon exposure to heat shock. However, σ^{32} itself or the DnaK, DnaJ, and GrpE chaperones do not seem to participate in this regulation.

Translational repression
Almost immediately after attaining the maximum induction, usually within several minutes after temperature upshift, *rpoH* translation is rapidly shut off. This translational repression requires some additional cis-acting and trans-acting factors. It has been shown that a subset of HSPs (DnaK, DnaJ and GrpE) exerts a negative feedback control on heat shock induction of HSPs and of σ^{32}, because induction occurs normally but hardly shuts off in mutants defective in any of these HSPs (Tilly et al., 1983, 1989; Straus et al., 1987, 1990). Neither synthesis of σ^{32} during steady-state growth nor induction itself is affected, however.

Recent analysis of 3'- or 5'-deletion derivatives or *rpoH-lacZ* gene fusion revealed that a specific segment of σ^{32} (around residues 122–144; region C in Fig. 2A) is required for normal shutoff of σ^{32} synthesis following heat induction (Nagai et al., 1994; Mori et al., unpublished). The amino acid sequence (protein structure) rather than the nucleotide sequence is important in this case, because a frameshift mutation specifically affecting this segment (alteration of 23 residues) abolishes repression. In view of the recent demonstration of physical interaction between these HSPs and σ^{32} *in vivo* and *in vitro* (Gamer et al., 1992; Liberek et al., 1992, 1993), the above finding raises the interesting possibility that the DnaK/DnaJ chaperones directly interact with this segment of either mature or nascent σ^{32}, causing translational repression during adaptation phase. Consistent with such an expectation, 13-aa peptides that contain part of region C bind to DnaK (but not DnaJ) with the highest affinity among all the peptides tested spanning the entire σ^{32} polypeptide (J. McCarty and B. Bukau, personal communication).

Regulation of σ^{32} stability

One of the earliest events that occurs in response to heat-shock stress may be the transient stabilization of otherwise unstable σ^{32}. The rapid stabilization of σ^{32} probably explains an almost instantaneous increase (within 1 min) in the rates of HSP synthesis observed. Thus, the instability of σ^{32} is probably a key element in this regulatory system. When cells are shifted from 30° to 42 °C, σ^{32} is markedly stabilized for the first 4–5 min and is quickly followed by resumed instability during transition from the induction to the adaptation phase. The mechanisms underlying the extreme instability as well as transient stabilization are not known, but mutations in several genes have been shown to stabilize σ^{32}. Among them is a temperature-sensitive *ftsH (hflB)* mutation, recently shown to stabilize σ^{32} at the restrictive temperature (Herman et al., 1995; Tomoyasu et al., 1995). Moreover, purified FtsH protein can degrade histidine-tagged σ^{32} in an ATP-dependent manner (Tomoyasu et al., 1995), suggesting that this membrane-bound ATP-dependent metalloprotease may play a major role in the rapid turnover of σ^{32} *in vivo*. At present, FtsH is the only ATP-dependent protease, encoded by the σ^{32} regulon, that can degrade σ^{32} *in vitro*: Lon and Clp proteases are virtually inactive. However, the possibility that protease(s) other than FtsH is also involved in *in vivo* turnover of σ^{32} cannot be excluded.

Besides the *ftsH* mutant with a deficiency in protease, mutations affecting any of the DnaK, DnaJ and GrpE chaperones stabilize σ^{32} (Tilly et al., 1989; Straus et al., 1990). Thus, these chaperones also contribute to the characteristic instability of σ^{32}, perhaps by interacting with the substrate or a protease, facilitating proteolysis. In this connection, the deletions or the frameshift mutation that prevent translational repression (see above) also stabilize σ^{32}-β-galactosidase fusion protein, suggesting that this segment of σ^{32} (region C) is involved in protein instability as well as translational repression (Nagai et al., 1994).

Earlier work indicated that production of abnormal, incomplete, or heterologous proteins in *E. coli* can induce HSP synthesis without heat shock treatment (Goff and Goldberg, 1985; Ito et al., 1986). Synthesis of unfolded proteins rather than protein degradation appears to be important in enhancing HSP synthesis (Parsell and Sauer, 1989). Accumulation of secretory protein precursors, abnormal proteins made in the presence of azetidine, or a human protein all result in induction of HSPs by stabilizing σ^{32} (Wild et al., 1993; Kanemori et al., 1994) and not by inducing synthesis of σ^{32} (Kanemori et al., 1994).

Regulation of σ^{32} activity

Whereas increased level rather than increased activity of σ^{32} primarily accounts for the HSP induction following temperature upshift, inactivation of σ^{32} partially contributes to the rapid shutoff of HSP synthesis early during adaptation phase (Straus et al., 1987; see Yura et al., 1993). Furthermore inactivation (reversible) of σ^{32} is a predominant regulatory event that occurs following temperature downshift or under other conditions where HSPs are present in excess (Straus et al., 1989; Taura et al., 1989). The DnaK, DnaJ and GrpE chaperones are expected to participate in this process, because mutants affected in any of these genes seem to be defective in inactivation (Kusukawa and Yura, unpublished; Gross, personal communication). These chaperones probably inactivate σ^{32} either by preventing formation or functioning of RNA polymerase holoenzyme containing σ^{32}.

Recent studies *in vivo* and *in vitro* indicate that DnaK and DnaJ can bind independently to σ^{32} (Gamer et al., 1992; Liberek et al., 1992, 1993). Whereas DnaK binds weakly to σ^{32} and is dissociated by ATP, DnaJ binds strongly to σ^{32} and is not dissociated by ATP. A σ^{32}-DnaK-DnaJ ternary complex can also be formed in an ATP-dependent manner and is supposed to be involved in sequestering or inactivating σ^{32}: inactivation probably occurs when σ^{32} is associated with DnaK and DnaJ and not when associated with core RNA polymerase forming holoenzyme. However, the question of whether the chaperones can associate only with free σ^{32} or also with RNA polymerase holoenzyme containing σ^{32} dissociating them from each other remains to be established.

Regulatory signals and feedback circuits

Although the nature of cellular sensors and signals for the heat shock response in bacteria have been the subject of much discussion, little experimental data are currently at hand. Since the DnaK-DnaJ-GrpE chaperones negatively modulate the expression of the σ^{32} regulon at three different levels (see above; Fig. 1), and since they are supposed to interact with many unfolded or incomplete proteins under normal and heat shock conditions, they are most likely to serve as central mediators in the heat shock regulation. Such a negative feedback control mediated by chaperones, particularly by members of the HSP70 family, appears to be important not only in procaryotes but in eucaryotes as well (see R.I. Morimoto et al., this volume).

Even a mild heat shock supposedly increases partially unfolded proteins that tend to bind and titrate free DnaK and/or DnaJ in the cell, and such titration presumably stabilizes σ^{32} by shifting the equilibrium from the chaperone-bound form to RNA polymerase holoenzyme ($E\sigma^{32}$). Specifically, the pool of free DnaK (Craig and Gross, 1991) and or DnaJ (Bukau, 1993) has been proposed to act as the cellular thermometer which would sense the state of unfolded proteins and adjust the amount of active σ^{32} and expression of the regulon accordingly. On the other hand, translational induction of σ^{32} is probably mediated by a separate signal because it involves transient disruption of the mRNA secondary structure which is apparently independent of DnaK/DnaJ activities (Straus et al., 1990; Nagai et al., 1991). Furthermore, various abnormal proteins can markedly stabilize σ^{32} but do not induce *rpoH* translation significantly (Wild et al., 1993; Kanemori et al., 1994). Thus at least two distinct signals may be involved in enhancing the σ^{32} level and assuring the rapid and transient induction of heat shock genes upon temperature upshift.

Among the chaperone-mediated feedback control of the σ^{32} level that occurs at three levels, translational repression and destabilization involve the same or overlapping segment (region C) of σ^{32}, suggesting a possible link between the two processes (Nagai et al., 1994; Yura et al., 1993). One speculative model is to assume that this segment of nascent σ^{32} interacts with DnaK (or DnaJ) and facilitates both proteolytic cleavage and translational repression (attenuation) of incomplete polypeptides even before leaving the ribosome. On the other hand, inactivation of σ^{32} observed upon temperature downshift or early during adaptation phase of the heat shock response involves interaction of the DnaK/DnaJ chaperones with mature σ^{32} protein or RNA polymerase containing σ^{32}. The nature of cellular signals for the chaperone-mediated feedback controls at these different levels remains obscure at present.

As noted above, the thermoregulation and feedback control of *rpoH* translation are likely to represent separate mechanisms that are complementary to each other. Further support for such bipartite controls of σ^{32} synthesis came from the recent isolation and sequence analyses of *rpoH* homologues encoding σ^{32}-like proteins from a variety of gram-negative bacteria (Nakahigashi et al., 1995). Whereas the cis-acting element required for the DnaK/DnaJ-mediated feedback control (region C) is highly conserved among all *rpoH* homologues examined (eight divergent species), those involved in thermoregulation (regions A and essential features of mRNA secondary structure) are conserved only among the species that are more closely related to *E. coli* (see below).

Regulatory mechanisms of the σ^{E} regulon

In search for mutations and factors affecting expression of the σ^{E} regulon, Mecsas et al. (1993) discovered that the rate of synthesis of outer membrane proteins (OMP) modulates the functioning of σ^{E} (but not σ^{32})-

regulon: the higher or lower OMP levels enhanced or reduced σ^E activity, respectively. In contrast, accumulation of cytoplasmic precursors of OMP decreased rather than increased σ^E activity. Furthermore, overproduction of mutant OMP that are incapable of being inserted into outer membrane can stimulate σ^E activity, suggesting that the signals are generated before membrane insertion of OMP. These results suggest that extracytoplasmic accumulation of immature or misfolded precursors of OMP generate signals that can enhance σ^E activity, although a detailed mechanism of activation is not known. In any event, the σ^E regulon appears to provide functions that are complementary to those encoded by the σ^{32} regulon, and the two regulons respond to signals generated in different cellular compartments.

The *rpoE* gene encoding σ^E was recently isolated and characterized in several laboratories (Raina et al., 1995; Rouvière et al., 1995; Hiratsu et al., 1995). The mechanisms that regulate *rpoE* expression (σ^E synthesis) seem to be very different from those known for *rpoH* regulation. Transcription from a major promoter of *rpoE* is mediated by σ^E-RNA polymerase and thus autoregulated. This suggests that σ^E synthesis is at least in part under transcriptional control. However, increased transcription of *rpoE* following temperature upshift is not sufficient to account for the rapid increase of σ^E activity observed. Besides, the positive autoregulatory circuit of σ^E is likely to be counteracted by a negative regulatory circuit, possibly at the level of activity by means of anti-sigma factor.

Heat shock regulation in other bacteria

Although major heat shock genes such as *groE* and *dna*K have been isolated and characterized from numerous bacteria over the past years, relatively little is known about the regulatory mechanisms of their expression. Recent work from several laboratories resulted in isolation of *rpoH* homologues encoding σ^{32}-like proteins in a number of gram-negative bacteria including γ-proteobacterium (*Citrobacter freundii, Enterobacter cloacae, Serratia marcescens, Proteus mirabilis, Pseudomonas aeruginosa*) and α-proteobacterium (*Agrobacterium tumefaciens, Zymomonas mobilis*) (Naczynski et al., 1995; Benvenisti et al., 1995; Nakahigashi et al., 1995). Each of the homologues can support growth of the *E. coli* mutant lacking σ^{32} at 30 °C or higher temperatures, and permit transcription of heat shock genes to various extents, thus functionally substituting for *E. coli rpoH* at least partially. Consistent with this finding, promoters similar to "σ^{32}-consensus" have been found in some of the heat shock genes from these bacteria.

Comparative sequence analyses revealed a stretch of nine residues absolutely and uniquely conserved among all the RpoH homologues tested (Nakahigashi et al., 1995). Interestingly, this region called "RpoH box", Q(R/K)(K/R)LFFNLR, is located between generally conserved regions 2.4

and 3.1 of bacterial σ factors, and overlaps with region C thought to be involved in the DnaK/DnaJ-mediated feedback control in *E. coli* (Fig. 2 A). On the other hand, other putative regulatory elements such as downstream box, mRNA secondary structure, and σ^E-promoter are well conserved among γ subgroup, but not α subgroup. These results suggest that the RpoH box together with its flanking sequences was evolved to permit translational and posttranslational control of σ^{32} as a basic regulatory strategy. Furthermore, additional regulatory elements were evolved in γ- but not α-proteobacterium perhaps as further refinement with regulatory advantage. Alternatively, α subgroup might have lost these latter elements during evolution. In any event, the heat shock response mediated by σ^{32} homologue appears to be widely distributed among gram-negative bacteria.

In contrast, regulatory mechanisms in gram-positive bacteria seem to be quite different from those known in *E. coli*. No evidence for σ^{32} like transcription factors has been reported despite much efforts to find them. Moreover, transcription of major heat shock genes *groE* and *dnaK* appears to start from vegetative promoters in both *B. subtilis* and *Clostridium acetobutilicum* (e.g. Narberhaus and Bahl, 1992; Wetzstein et al., 1992). Although σ^E-like factors have been found in gram-positive bacteria (see Rouvière et al., 1995), their roles in the heat shock or other stress response remain unknown. In this connection, transcription of a separate group of genes encoding "general stress proteins" appears to be under control of σ^B, one of the minor σ factors in *B. subtilis* (Boylan et al., 1993; Völker et al., 1994).

Finally, extensive analyses of HSP genes revealed the presence of a highly conserved 9-nucleotide inverted repeat sequence separated by nine nucleotides, presumably a negative regulatory element, in front of major chaperone genes *groE* and *dnaK(dnaJ)* of *B. subtilis, C. acetobutilicum, Lactococcus lactis* and many others (e.g. Narberhaus and Bahl, 1992; Van Asseldonk et al., 1993; Zuber and Schumann, 1994). Special interest is attached to the regulatory roles of these sequences because of their wide distribution among eubacteria (gram-positive, -negative, and cyanobacterium), although they seem to be limited to some of the chaperone genes. In view of the results on σ^{32} homologues discussed above, it would be interesting to assess the differential and perhaps complementary roles of the inverted repeat and σ^{32} homologues in the heat shock response of bacteria such as *Agrobacterium tumefaciens* that appear to contain both these regulatory systems.

Acknowledgements
We thank our colleagues for critical reading of the manuscript. Work from our laboratory at Institute for Virus Research, Kyoto University was supported by grants form Ministry of Education, Science and Culture, Japan. More recent work done at HSP Research Institute was supported in part by grants from Japan Health Sciences Foundation, Tokyo.

References

Benvenisti, L., Koby, S., Rutman, A., Giladi, H., Yura, T. and Oppenheim, A.B. (1995) Cloning and primary sequence of the *rpoH* gene from *Pseudomonas aeruginosa*. *Gene* 155:73–76.

Bukau, B. (1993) Regulation of the *Escherichia coli* heat-shock response. *Mol. Microbiol.* 9:671–680.

Boylan, S.A., Redfield, A.R., Brody, M.S. and Price, C.W. (1993) *J. Bacteriol.* 175:7931–7937.

Chuang, S.-E. and Blattner, F.R. (1993) Characterization of twenty-six new heat shock genes of *Escherichia coli*. *J. Bacteriol.* 175:5242–5252.

Craig, E.A. and Gross, C.A. (1991) Is hsp70 the cellular thermometer? *Trends Biochem. Sci.* 16:135–140.

Erickson, J.W., Vaughn, V., Walter, W.A., Neidhardt, F.C. and Gross, C.A. (1987) Regulation of the promoters and transcripts of *rpoH*, the *Escherichia coli* heat shock regulatory gene. *Genes Dev.* 1:419–432.

Erickson, J.W. and Gross, C.A. (1989) Identification of the σ^E subunit of *Escherichia coli* RNA polymerase: a second alternate σ factor involved in high-temperature gene expression. *Genes Dev.* 3:1462–1471.

Fujita, N., Nomura, T. and Ishihama, A. (1987) Promoter selectivity of *Escherichia coli* RNA polymerase: purification and properties of holoenzyme containing the heat-shock σ subunit. *J. Biol. Chem.* 262:1855–1859.

Gamer, J., Bujard, H. and Bukau, B. (1992) Physical interaction between heat shock proteins DnaK, DnaJ, and grpE and the bacterial heat shock transcription factor σ^{32}. *Cell* 69:833–842.

Georgopoulos, C., Ang, D., Liberek, K. and Zylicz, M. (1990) Properties of the *Escherichia coli* heat shock proteins and their role in bacteriophage λ growth. *In:* R. Morimoto, A. Tissières and C. Georgopoulos (eds): *Stress Proteins in Biology and Medicine,* Cold Spring Harbor Lab. Press, Cold Spring Harbor, NY, pp 191–221.

Georgopoulos, C. and Welch, W.J. (1993) Role of the major heat shock proteins as molecular chaperones. *Annu. Rev. Cell Biol.* 9:601–634.

Georgopoulos, C., Liberek, K., Zylicz, M. and Ang, D. (1994) Properties of the heat shock proteins of *Escherichia coli* and the autoregulation of the heat shock response. *In:* R.I. Morimoto, A. Tissières and C. Georgopoulos (eds): *The Biology of Heat Shock Proteins and Molecular Chaperones.* Cold Spring Harbor Lab. Press, Cold Spring Harbor, NY, pp 209–249.

Goff, S.A. and Goldberg, A.L. (1985) Production of abnormal proteins in *E. coli* stimulates transcription of *lon* and other heat shock genes. *Cell* 41:589–595.

Gragerov, A., Nudler, E., Komissarova, N., Gaitanaris, G.A., Gottesman, M.E. and Nikiforov, V. (1992) Cooperation of GroEL/GroES and DnaK/DnaJ heat shock proteins in preventing protein misfolding in *Escherichia coli*. *Proc. Natl. Acad. Sci. USA* 89:10341–10344.

Gross, C.A., Straus, D.B., Erickson, J.W. and Yura, T. (1990) The function and regulation of heat shock proteins in *Escherichia coli*. *In:* R. Morimoto, A. Tissières and C. Georgopoulos (eds): *Stress Proteins in Biology and Medicine,* Cold Spring Harbor Lab. Press, Cold Spring Harbor, NY, pp 167–189.

Grossman, A.D., Erickson, J.W. and Gross, C. (1984) The *htpR* gene product of *E. coli* is a sigma factor for heat-shock promoters. *Cell* 38:383–390.

Grossman, A.D., Straus, D.B., Walter, W.A. and Gross, C.A. (1987) σ^{32} synthesis can regulate the synthesis of heat shock proteins in *Escherichia coli*. *Genes Dev.* 1:179–184.

Hendrick, J.P. and Hartl, F.-U. (1993) Molecular chaperone functions of heat-shock proteins. *Annu. Rev. Biochem.* 62:349–384.

Herendeen, S.L., VanBogelen, R.A. and Neidhardt, F.C. (1979) Levels of major proteins of *Escherichia coli* during growth at different temperatures. *J. Bacteriol.* 139:185–194.

Herman, C., Thevenet, D., D'Ari, R. and Bouloc, P. (1995) Degradation of σ^{32}, the heat shock regulator in *Escherichia coli*, is governed by HflB. *Proc. Natl. Acad. Sci. USA* 92:3516–3520.

Hiratsu, K., Amemura, M., Nashimoto, H., Shinagawa, H. and Makino, K. (1995) The *rpoE* gene of *Escherichia coli,* which encodes σ^E, is essential for bacterial growth at high temperature. *J. Bacteriol.* 177:2918–2922.

Ito, K., Akiyama, Y., Yura, T and Shiba, K. (1986) Diverse effects of the MalE-LacZ hybrid protein on *Escherichia coli* cell physiology. *J. Bacteriol.* 167:201–204.

Kamath-Loeb, A.S. and Gross, C.A. (1991) Translational regulation of σ^{32} synthesis: requirement for an internal control element. *J. Bacteriol.* 173:3904–3906.

Kanemori, M., Mori, H. and Yura, T. (1994) Induction of heat shock proteins by abnormal proteins results form stabilization and not increased synthesis of σ^{32} in *Escherichia coli*. *J. Bacteriol.* 176:5648–5653.

Kusukawa, N. and Yura, T. (1988) Heat shock protein GroE of *Escherichia coli*: key protective roles against thermal stress. *Genes Dev.* 2:874–882.

Lemaux, P.G., Herendeen, S.L., Bloch, P.L. and Neidhardt, F.C. (1978) Transient rates of synthesis of individual polypeptides in *E. coli* following temperature shifts. *Cell* 13:427–434.

Liberek, K., Galitski, T.P., Zylicz, M. and Georgopoulos, C. (1992) The DnaK chaperone modulates the heat shock response of *Escherichia coli* by binding to the σ^{32} transcription factor. *Proc. Natl. Acad. Sci. USA* 89:3516–3520.

Liberek, K. and Georgopoulos, C. (1993) Autoregulation of the *Escherichia coli* heat shock response by the DnaK and DnaJ heat shock proteins. *Proc. Natl. Acad. Sci. USA* 90: 11019–11023.

Mecsas, J., Rouvière, P.E., Erickson, J.W., Donohue, T. and Gross, C.A. (1993) The activity of σ^E, an *Escherichia coli* heat-inducible σ-factor, is modulated by expression of outer membrane proteins. *Genes Dev.* 7:2619–2628.

Naczynski, Z.M., Mueller, C. and Kropinski, A.M. (1995) Cloning the gene for the heat shock response positive regulator (σ^{32} homolog) from *Pseudomonas aeruginosa. Can. J. Microbiol.* 41:75–87.

Nagai, H., Yano, R., Erickson, J.W. and Yura, T. (1990) Transcriptional regulation of the heat shock regulatory gene *rpoH* in *Escherichia coli*: involvement of a novel catabolite-sensitive promoter. *J. Bacteriol.* 172:2710–2715.

Nagai, H., Yuzawa, H. and Yura, T. (1991) Interplay of two cis-acting mRNA regions in translational control of σ^{32} synthesis during the heat shock response of *Escherichia coli. Proc. Natl. Acad. Sci. USA* 88:10515–10519.

Nagai, H., Yuzawa, H., Kanemori, M. and Yura, T. (1994) Distinct segment of σ^{32} polypeptide is involved in the DnaK-mediated negative control of heat shock response in *Escherichia coli. Proc. Natl. Acad. Sci. USA* 91:10280–10284.

Nakahigashi, K., Yanagi, H. and Yura, T. (1995) Isolation and sequence analysis of *rpoH* genes encoding σ^{32} homologues from gram-negative bacteria: conserved mRNA and protein segments for heat shock regulation. *Nucleic Acids Res.* 23:4383–4390.

Narberhaus, F. and Bahl, H. (1992) Cloning, sequencing, and molecular analysis of the *groESL* operon of *Clostridium acetobutylicum. J. Bacteriol.* 174:3282–3289.

Neidhardt, F.C. and VanBogelen, R.A. (1981) Positive regulatory gene for temperature-controlled proteins in *Escherichia coli. Biochem. Biophys. Res. Commun.* 100:894–900.

Neidhardt, F.C. and VanBogelen, R.A. (1987) Heat shock response. *In:* F.C. Neidhardt, J.L. Ingraham, K.B. Low, B. Magasanik, M. Schaechter and H.E. Umbarger (eds): *Escherichia coli and Salmonella typhimurium: Cellular and Molecular Biology,* Am. Soc. Microbiol., Washington, DC., pp 1334–1345.

Parsell, D.A. and Sauer, R.T. (1989) Induction of a heat shock-like response by unfolded protein in *Escherichia coli*: dependence on protein level not protein degradation. *Genes Dev.* 3: 1226–1232.

Raina, S., Missiakas, D. and Georgopoulos, C. (1995) The *rpoE* gene encoding the σ^E (σ^{24}) heat shock sigma factor of *Escherichia coli. EMBO J.* 14:1043–1055.

Rouvière, P.E., Peñas, A. De Las, Mecsas, J., Lu, C.Z., Rudd, K.E. and Gross, C.A. (1995) *rpoE*, the gene encoding the second heat-shock sigma factor, σ^E, in *Escherichia coli. EMBO J.* 14:1032–1042.

Sprengart, M.L., Fatscher, H.P. and Fuchs, E. (1990) The initiation of translation in *E. coli*: apparent base pairing between the 16S rRNA and downstream sequences of the mRNA. *Nuclic Acids Res.* 18:1719–1723.

Straus, D.B., Walter, W.A. and Gross, C.A. (1987) The heat shock response of *E. coli* is regulated by changes in the concentration of σ^{32}. *Nature* 329:348–351.

Straus, D.B., Walter, W.A. and Gross, C.A. (1988) *Escherichia coli* heat shock gene mutants are defective in proteolysis. *Genes Dev.* 2:1851–1858.

Straus, D.B., Walter, W.A. and Gross, C.A. (1989) The activity of σ^{32} is reduced under conditions of excess heat shock protein production in *Escherichia coli. Genes Dev.* 3:2003–2010.

Straus, D., Walter, W. and Gross, C.A. (1990) DnaK, DnaJ, and GrpE heat shock proteins negatively regulate heat shock gene expression by controlling the synthesis and stability of σ^{32}. *Genes Dev.* 4:2202–2209.

Taura, T., Kusukawa, N., Yura, T. and Ito, K. (1989) Transient shutoff of *Escherichia coli* heat shock protein synthesis upon temperature shift down. *Biochem. Biophys. Res. Commun.* 163:438–443.

Tilly, K., McKittrick, N., Zylicz, M. and Georgopoulos, C. (1983) The *dnaK* protein modulates the heat-shock response of *Escherichia coli. Cell* 34:641–646.

Tilly, K., Spence, J. and Georgopoulos, C. (1989) Modulation of stability of the *Escherichia coli* heat shock regulatory factor σ^{32}. *J. Bacteriol.* 171:1585–1589.

Tomoyasu, T., Gamer, J., Bukau, B., Kanemori, M., Mori, H., Rutman, A.J., Oppenheim, A.B., Yura, T., Yamanaka, K., Niki, H., Hiraga, S. and Ogura, T. (1995) *Escherichia coli* FtsH is a membrane-bound, ATP-dependent protease which degrades the heat-shock transcription factor σ^{32}. *EMBO J.* 14:2551–2560.

Van Asseldonk, M., Simons, A., Visser, H., De Vos, W.M. and Simons, G. (1993) Cloning, nucleotide sequence, and regulatory analysis of the *Lactococcus lactis dnaJ* gene. *J. Bacteriol.* 175:1637–1644.

Völker, U., Engelmann, S., Maul, B., Riethdorf, S., Völker, A., Schmid, R., Mach, H. and Hecker, M. (1994) Analysis of the induction of general stress proteins of *Bacillus subtilis. Microbiology* 140:741–752.

Wang, Q. and Kaguni, J.M. (1989) A novel sigma factor is involved in expression of the *rpoH* gene of *Escherichia coli. J. Bacteriol.* 171:4248–4283.

Weiner, L., Brissette, J.L. and Model, P. (1991) Stress-induced expression of the *Escherichia coli* phage shock protein operon is dependent on σ^{54} and modulated by positive and negative feedback mechanisms. *Genes Dev.* 5:1912–1923.

Wetzstein, M., Völker, U., Dedio, J., Löbau, S., Zuber, U., Schiesswohl, M., Herget, C., Hecker, M. and Schumann, W. (1992) Cloning, sequencing, and molecular analysis of the *dnaK* locus from *Bacillus subtils. J. Bacteriol.* 174:3300–3310.

Wild, J., Walter, W.A., Gross, C.A. and Altman, E. (1993) Accumulation of secretory protein precursors in *Escherichia coli* induces the heat shock response. *J. Bacteriol.* 175:3992–3997.

Yamamori, T., Ito, K., Nakamura, Y. and Yura, T. (1978) Transient regulation of protein synthesis in *Escherichia coli* upon shift-up of growth temperature. *J. Bacteriol.* 134:1133–1140.

Yamamori, T. and Yura, T. (1980) Temperature-induced synthesis of specific proteins in *Escherichia coli*: evidence for transcriptional control. *J. Bacteriol.* 142:843–851.

Yamamori, T. and Yura, T. (1982) Genetic control of heat-shock protein synthesis and its bearing on growth and thermal resistance in *Escherichia coli* K-12. *Proc. Natl. Acad. Sci. USA* 79:860–864.

Yura, T., Nagai, H. and Mori, H. (1993) Regulation of the heat shock response in bacteria. *Annu. Rev. Microbiol.* 47:321–350.

Yuzawa, H., Nagai, H., Mori, H. and Yura, T. (1993) Heat induction of σ^{32} synthesis mediated by mRNA secondary structure: a primary step of the heat shock response in *Escherichia coli. Nucleic Acids Res.* 21:5449–5455.

Zhou, Y.-N., Kusukawa, N., Erickson, J.W., Gross, C.A. and Yura, T. (1988) Isolation and characterization of *Escherichia coli* mutants that lack the heat shock sigma factor σ^{32}. *J. Bacteriol.* 170:3640–3649.

Zuber, U. and Schumann, W. (1994) CIRCE, a novel heat shock element involved in regulation of heat shock operon *dnaK* of *Bacillus subtilis. J. Bacteriol.* 176:1359–1363.

Stress-Inducible Cellular Responses
ed. by U. Feige, R.I. Morimoto, I. Yahara and B. Polla
© 1996 Birkhäuser Verlag Basel/Switzerland

The impact of oxidative stress on eukaryotic iron metabolism

T. A. Rouault and R. D. Klausner

Cell Biology and Metabolism Branch, National Institute of Child Health and Human Development, Bethesda, MD 20892, USA

Summary. The processes of iron uptake and distribution are highly regulated in mammalian cells. Expression of the transferrin receptor is increased when cells are iron-depleted, while expression of the iron sequestration protein ferritin is increased in cells that are iron-replete. Regulation of expression of proteins of iron uptake (transferrin receptor) and iron sequestration (ferritin) presumably ensures that levels of reactive free iron are not high in cells. Formation of reactive oxygen species occurs when free iron reacts with oxygen, and tight regulation of iron metabolism may enable cells to avoid engaging in destructive chemical reactions. Levels of intracellular iron are directly sensed by two iron sensing proteins. Iron regulatory protein 1 (IRP1) is a bifunctional protein; in cells that are iron-replete, IRP1 contains an iron-sulfur cluster and functions as cytosolic aconitase. In cells that are iron-depleted, IRP1 binds stem-loop structures in RNA transcripts known as iron responsive elements (IREs). Iron regulatory protein 2 (IRP2) binds similar stem-loop structures, but the mode of regulation of IRP2 is different in that IRP2 is rapidly degraded in iron-replete cells. The post-transcriptional regulation of genes of iron metabolism in mammalian cells ensures that cells have an adequate supply of iron, and also ensures that cells do not generate excess reactive oxygen species through the interaction of free iron and oxygen.

Introduction

Iron is the most abundant transition metal in the earth's crust. The facility with which iron can change its oxidation state from Fe^{3+} to Fe^{2+} and its availability make it an important cofactor for numerous cellular proteins, including proteins important in the transport and storage of dioxygen, electron transfer, and nitrogen fixation, to name but a few. Because of the fundamental role of iron in numerous metabolic pathways, an adequate supply of iron is required for maintenance of homeostasis and growth of cells. However, the very reactivity which is so useful in a metabolic cofactor has a significant drawback; iron is highly reactive with oxygen, and the by-products of the interaction of iron and oxygen can be damaging to cellular constituents. Generation of reactive oxygen species is promoted by the interaction of reduced iron with dioxygen-$Fe^{2+}+O_2-Fe^{3+}+O_2^-$. The superoxide anion O_2^- generated by this reaction can undergo dismutation mediated by superoxide dismutase to yield hydrogen peroxide in the reaction $2O_2^-+2H^+-H_2O_2+O_2$. Hydrogen peroxide thus generated can react with reduced iron, Fe^{2+}, in a reaction known as the Fenton reaction, to yield hydroxyl radical as follows-$Fe^{2+}+H_2O_2-Fe^{3+}+OH^·+OH^-$. The superoxide anion and the hydroxyl radical are highly reactive species which can cause oxidative damage to lipids, nucleic acids, and proteins. These reactions between iron and oxygen species account for the intimate relationship between oxygen and iron toxicity.

Because of the reactivity of iron and oxygen, regulatory systems have developed to ensure that levels of iron and oxygen byproducts are regulated. In this chapter, we will consider the mechanisms that have evolved in mammalian cells that assure a sufficient iron supply to support cellular metabolism, along with those which protect against potential iron toxicity. The role of the transferrin receptor in iron uptake, ferritin in iron sequestration, and iron regulatory proteins in regulation of expression of genes of iron metabolism will be reviewed. The mechanism of iron sensing will be discussed in greater detail, and the impact of oxidative stress on iron regulatory proteins and iron-sulfur cluster proteins will be reviewed.

Iron uptake in mammalian cells

The insolubility of iron is a major obstacle which must be overcome by mammalian cells. While iron is the single most abundant metal in the earth's crust, it is most frequently found in the oxidized Fe^{3+} form, a form which is highly insoluble at neutral pH. A complex iron uptake system has evolved to deal with the issue of solubility. The solubilization of ferric iron is accomplished in mammalian cells through binding to the iron carrier protein transferrin. Each transferrin has the capacity to bind two ferric ions and to transport the bound iron through the circulation. Transferrin binds to the transferrin receptor, a 90 kD homodimer expressed on the plasma membrane. Each transferrin receptor can bind two transferrins, and the resulting complex is internalized in an endocytic vesicle formed by invagination of the plasma membrane. Acidification of the endosome results in the release of iron from transferrin within the endosome. The released free iron then must cross the membrane of the endosome in order to be available to the cell (reviewed in Klausner et al., 1993). Though the mechanism of transport across the plasma membrane of the endosome is not presently understood, it is likely that a transport mechanism analogous to that which has been described in yeast is involved (Klausner and Dancis, 1994). Once the iron crosses the membrane, it is available for use in metabolism. When the amount of iron absorbed exceeds the metabolic needs of the cell, the iron is detoxified by sequestration in a cytosolic protein known as ferritin.

Iron sequestration and storage

The main iron sequestration protein in mammalian cells is ferritin, a 24 subunit protein in which the subunits assemble to form a hollow sphere.

The subunits are 18–20 kD in size, and there are two types, known as H (heavy or derived from heart) and L (light or derived from liver). Iron enters the ferritin sphere and is deposited within as an insoluble precipitate. The capacity of each ferritin to store iron is as high as 4500 atoms of iron per protein complex. Ferritin serves to protect cells from iron toxicity by sequestering iron, thereby preventing reactions with O_2 and damage to cellular proteins and lipids (reviewed in Theil, 1987).

Regulation of iron uptake and sequestration

Regulation of the expression of ferritin, the transferrin receptor, and the erythrocyte form of aminolevulinic acid synthetase (eALAS), the rate-limiting step in heme biosynthesis, is accomplished by a post-transcriptional regulatory system involving cis acting RNA elements and trans acting iron-sensing proteins. RNA stem-loops were initially identified in the 5'UTR of ferritin transcripts through comparisons of 5'UTR sequences between ferritin H and L chains and comparisons between species (Aziz and Munro, 1987: Hentze et al., 1987).

Though there was considerable variation in the sequences of the 5' non-coding regions, there was conservation of a 32 nucleotide sequence which had the potential to assume the conformation of a stem-loop. Deletional analysis demonstrated that this 32 nucleotide sequence was required for regulation of ferritin biosynthesis. When this sequence element was placed in the 5'UTR of reporter genes, translation of the reporter genes was high when cells were iron replete and impeded when iron levels were low (Aziz and Munro, 1987; Hentze et al., 1987). Because of its function in determining the regulatory response to changes in iron levels, the RNA stem-loop sequence was named the iron responsive element (IRE).

Expression of the transferrin receptor (TfR) is also regulated post-transcriptionally by iron levels. Levels of TfR mRNA decrease in cells that are iron-replete, and sequences in the 3' UTR of the mRNA mediate this regulation (Mullner et al., 1989; Casey et al., 1989). The region of the mRNA responsible for this regulation contains five IREs and an endonucleolytic cleavage site. When iron levels are high, an endonucleolytic cleavage occurs in a single stranded region flanked by two IREs in the mRNA regulatory region, yielding a 3' cleavage product and a truncated 5' product which is then rapidly degraded (Binder et al., 1994). When cells are iron-starved, the TfR mRNA is protected from endonucleolytic cleavage and levels of TfR mRNA increase, resulting in increased biosynthesis of the TfR and increased iron uptake in iron-depleted cells.

Identification of a cis acting RNA sequence element permitted development of an assay which could reveal the presence of a putative trans acting factor. An RNA gel retardation assay was developed which made use of principles similar to those which had been previously used to identify DNA

binding proteins; the radiolabeled target nucleic acid sequence was incubated with cellular extracts and subsequently electrophoresed on a non-
denaturing acrylamide gel (Leibold and Munro, 1988). Proteins that bound
to the target sequence created an RNA-protein complex with electrophoretic characteristics different from those of the unbound RNA sequence.
Using this assay, it became clear that lysates containened a protein(s) which
bound with high affinity to the RNA stem-loop sequence when isolated
from cells that were iron depleted, and the protein was therefore called the
IRE binding protein, or IRE-BP (Rouault et al., 1988). A similar assay,
based on *in vitro* translation of ferritin transcripts revealed that in the
lysates of reticulocytes, but not wheat germ, there was a protein which
could repress translation of ferritin, and which was therefore named the
ferritin repressor protein, or FRP (Walden et al., 1988).

   These protein(s) were subsequently purified using either RNA affinity
chromatography (Rouault et al., 1989), traditional chromatographic methods (Patino and Walden, 1992; Yu et al., 1992), or a combination of the
two approaches (Hirling et al., 1991). Using peptide sequence from the
purified IRE binding protein and degenerate oligonucleotides to identify
candidate clones, two related genes were identified. The encoded proteins
were highly homologous, with amino acid homology of 58%. One of the
two proteins was clearly the protein which had been identified by peptide
sequencing (Rouault et al., 1990) and the second related protein was later
shown to also bind IREs (Rouault et al., 1992). Sequence comparisons between the newly cloned genes and the data base revealed that IRE-BP1 was
remarkably similar in sequence to the mitochondrial aconitase of pig and
yeast (Rouault et al., 1991; Hentze and Argos, 1991), and this similarity led
to immediate insights into the mode of regulation of IRE-BP1 which are
discussed below. IRE-BP1 has subsequently been renamed iron regulatory
protein 1 (IRP1) to clarify nomenclature, and this nomenclature will be
followed in the remainder of this chapter.

IRP1 is an iron-sulfur cluster protein

Prior to discovery of the relationship between mitochondrial aconitase and
IRP1, a considerable body of work had accumulated on mitochondrial
aconitase which included extensive physico-chemical characterization,
solution of the crystal structure, and site-directed mutagenesis. Mitochondrial aconitase, a Krebs' cycle enzyme that catalyzes the conversion of
citrate to isocitrate, ligates an iron-sulfur cluster to the third domain of the
four domain protein. The cubane cluster contains four irons and four sul-
furs; three of the irons are ligated to cysteines and to inorganic sulfur atoms
of the cluster, whereas the fourth lacks a cysteinyl ligand and is bound
solely by the inorganic sulfur atoms of the cluster. The fourth iron binds
the hydroxyl and carboxyl groups of the substrates isocitrate and citrate

(Beinert and Kennedy, 1989; Beinert, 1990; Lauble et al., 1992). The fourth iron, which is required for enzymatic activity, is relatively labile and is readily lost during aerobic purifications, leaving a $[3Fe-4S]^+$ cluster in place. Aconitase activity is easily restored by treatment of the protein with iron and reducing agents (Beinert and Kennedy, 1989).

Though the sequence similarity between IRP1 and mitochondrial aconitase was less than 30% overall, the sequence homology extended throughout all four domains of the protein. Even more impressive, the 23 active site residues identified in mitochondrial aconitase were conserved in IRP1, and IRP1 was shown to function as an aconitase under iron-replete conditions (Haile et al., 1992a). It had long been known that an active aconitase was present in the cytosol, and the cytosolic aconitase was purified from beef liver. Chemical and spectroscopic characterization demonstrated that the cytosolic aconitase, like mitochondrial aconitase, ligates a $[4Fe-4S]^{2+}$ cluster. Peptide sequencing of the beef cytosolic aconitase revealed that there was a single mammalian cytosolic aconitase with 98% homology to the human sequence in the six peptides and 89 amino acids sequenced, supporting the conclusion that IRP1 is identical to cytosolic aconitase (Kennedy et al., 1992).

IRP1 performs two mutually exclusive functions depending on the status of the iron-sulfur cluster

When cells are iron-replete, IRP1 is almost entirely in the form of cytosolic aconitase, whereas when cells are iron-depleted, IRP1 no longer functions as an aconitase and it becomes a high affinity RNA binding protein, capable of repressing translation of mRNAs of ferritin and eALAS (Cox et al., 1991; Dandekar et al., 1991), and prolonging the half-life of the TfR mRNA. The transition in function takes place without change in the levels of the protein (Tang et al., 1992; Pantopoulos and Hentze, 1995). The major definable change is in the status of the associated [4Fe-4S] cluster. Since mitochondrial aconitase was known to easily lose the fourth iron from the cluster during aerobic purification, an attractive hypothesis was that the fourth iron was the determinant of whether the protein would function as an RNA binder (Hentze and Argos. 1991; Rouault et al., 1991). In order to determine the status of the iron-sulfur cluster in the RNA binding form of the protein, the RNA binding properties of IRP containing a $[3Fe-4S]^+$ cluster and chemically assessed apoprotein were evaluated. The RNA binding form of the protein derived from cells treated with desferrioxamine, an iron chelator, was determined to be most similar in its behavior to chemically characterized apoprotein, devoid of cluster, while the IRP containing a $[3Fe-4S]^+$ cluster did not bind RNA (Haile et al., 1992b). Numerous studies have supported the conclusion that the RNA binding form of the protein is apoprotein (Emery-Goodman et al., 1993; Gray et al.,

1993; Basilion et al., 1994a). Chemically characterized apoprotein binds with picomolar affinity to IREs (Basilion et al., 1994a); these measurements are comparable to the affinities derived from IRP in lysates of cells (Haile et al., 1989).

Much evidence indicates that the IRE binding form of IRP1 has a different conformation from the form that functions as an aconitase. Studies using *uv* cross-linking to identify a contact point between IRP1 and the IRE have indicated that the peptide to which the IRE is cross-linked is derived from sequences that would be expected, based on analogy to the crystal structure of mitochondrial aconitase, to be within the active site cleft between domains 1–3 and domain 4 of aconitase (Basilion et at., 1994a). Modeling using known dimensions of an RNA stem-loop comparable in size to the IRE and the dimensions of aconitase derived from solution of the crystal structure supports the conclusion that domain 4 must move with respect to domains 1–3 in the apoprotein in order for identified RNA protein contact to be made. If in fact the structure of cytosolic aconitase is comparable, which we expect, then one could postulate that the cleft remains closed when the iron-sulfur cluster is intact and substrate is bound, but that the cleft could open when cluster is disassembled and substrate is not present to bridge the two sides of the cleft (Klausner and Rouault, 1993).

Because the results of *uv* cross-linking implicated the active site cleft in RNA binding, residues of the enzymatic active site were evaluated for a possible role in ligation of IREs. Since arginines have been shown to play an important role in other RNA binding proteins, the contribution to IRE binding of arginines in the active site was studied. Site directed mutagenesis demonstrated that arginines from the enzymatic active site were indispensable to the RNA binding affinity, further supporting the view that regions of the active site cleft serve as the IRE binding site in the apoprotein (Philpott et al., 1994). In this model, the binding site for RNA would be accessible only in the apoprotein, thus ensuring that the enzymatic activity and RNA binding activity are mutually exclusive properties of the protein (Basilion et al., 1994a). (See Fig. 1).

Though the assembled data is most consistent with the conclusion that the form of the protein from cells treated with iron chelators does not contain iron, these data do not address whether inorganic sulfur is bound by the protein. Considerable work has been done on *in vitro* assembly of the iron-sulfur cluster of mitochondrial aconitase (reviewed in Kennedy and Beinert, 1989). The ease with which the iron-sulfur cluster of mitochondrial aconitase is reassembled depends on how the apoprotein has been generated. Oxidative destruction of the cluster with ferricyanide generates protein that contains entrapped sulfur, and reassembly can be accomplished with addition of DTT and an iron source only (Kennedy and Beinert, 1988). When IRP1 is isolated from cells that have been treated prior to lysis with the iron chelator desferrioxamine, the cluster can be

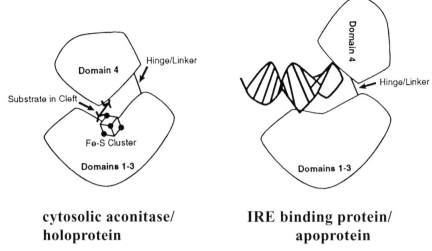

cytosolic aconitase/   IRE binding protein/
holoprotein   apoprotein

Figure 1. The bifunctional iron regulatory protein functions as cytosolic aconitase when a [4Fe-4S] cluster is bound (left) or as an IRE binding protein when the iron-sulfur cluster is absent (right). A conformational change is postulated to occur upon loss of the cluster. Contact points for IRE binding are within the active site cleft that was sterically unavailable for interaction with RNA in the holoprotein.

reassembled *in vitro* using DTT and an iron source only, suggesting that the apoprotein from cells also contains inorganic sulfur. When the protein is over-expressed in insect cells, the procedure used to rebuild the cluster is unsuccessful unless a source of reduced sulfur is added, and even then the procedure is inefficient (Basilion et al., 1994b). Thus, the apoprotein from cells that regulate IRP expression can be distinguished from the apoprotein that is over-expressed non-physiologically. The implication of these observations is that the endogenous IRP in cells treated with desferrioxamine may contain sufficient inorganic sulfur to permit assembly of an iron sulfur cluster with the addition of a reducing agent and iron only.

Little is known about iron-sulfur cluster assembly in eukaryotic cells, but the study of prokaryotes is yielding important clues about the process. Two gene products involved in nitrogen fixation, Nif S and Nif U, are critical to assembly of the iron-sulfur cluster of the nitrogenase subunits of Azotobacter vinelandii (Zheng et al., 1993; Fu et al., 1994). Recent work has established that Nif S protein serves as a source of sulfur in iron-sulfur cluster assembly (Zheng et al., 1994). Interestingly, Nif S protein, over-expressed and purified from bacteria, will catalyze *in vitro* formation of iron-sulfur clusters in IRP1 (Basilion et al., 1994b), thus supporting the viewpoint that cluster assembly may require participation of iron-sulfur cluster assembly enzymes.

Iron-sulfur clusters are often destabilized by oxidative stress

The mechanism by which the iron-sulfur cluster is degraded in IRP1 is not completely understood, but several pieces of information are relevant. Notably, the iron-sulfur cluster of mitochondrial aconitase readily loses the fourth iron which is not bound to a cysteinyl ligand in an aerobic atmosphere. Though the fourth iron of purified cytosolic aconitase/IRP1 is less labile than that of mitochondrial aconitase when exposed to air (Kennedy et al., 1992), experiments have shown that cytosolic aconitase activity is lost when cells are exposed to nitric oxide (Drapier et al., 1993; Weiss et al., 1993) and to hydrogen peroxide (Gardner et al., 1995; Martins et al., 1995; Pantopoulos and Hentze, 1995a). Nitric oxide is thought to cause complete disassembly of the iron-sulfur cluster, as evidenced by observations that IRP1 acquires IRE binding activity after exposure to nitric oxide, since it is known that the form of protein that contains a stable partially disassembled $[3Fe-4S]^+$ cluster does not bind RNA (Haile et al., 1992b). Nitric oxide is proposed to cause cluster disassembly either through direct binding to the cluster with formation of a iron-thiol-nitrosyl adduct (Kennedy and Beinert, 1994) or through reaction with hydrogen peroxide to produce peroxynitrite which oxidizes and destabilizes the cluster (Castro et al., 1994; Hausladen and Fridovich, 1994). These experiments suggest that the iron-sulfur cluster is partially or completely disassembled upon exposure to oxidizing reagents. Both nitric oxide and hydrogen peroxide are produced by cells in the course of normal metabolism, and it is therefore possible that the iron-sulfur cluster of cytosolic aconitase is routinely subjected to oxidative stresses sufficient to cause disassembly of the cluster.

Several recent studies provide support for a model of oxidation induced disassembly of the iron-sulfur cluster of IRP1. Treatment of cells with H_2O_2 has been shown to result in loss of cytosolic aconitase activity, and the recovery of aconitase activity is significantly slower when cells treated with hydrogen peroxide are simultaneously treated with desferrioxamine (Gardner et al., 1995). Since a complete [4Fe-4S] cluster is required for aconitase activity, the loss of aconitase activity does not distinguish between loss of the labile fourth iron of the cluster and more extensive disassembly of the cluster. The assay of RNA binding activity is more revealing about the status of the iron-sulfur cluster since evidence (reviewed above) supports the conclusion that the RNA binding form of the protein is apoprotein. Cells treated with H_2O_2 rapidly activate the IRE binding activity of IRP, though treatment of IRP with H_2O_2 *in vitro* does not induce RNA binding activity, suggesting that disassembly is mediated by cellular species that arise from treatment with H_2O_2 (Martins et al., 1995; Pantopoulos and Hentze, 1995b). The nature of the reactive species is unclear, since scavengers of hydroxyl radical do not interfere with induction of RNA binding activity (Martins et al., 1995), and treatment with paraquat to produce superoxide anion is inefficient in inducing RNA binding activity

(Pantopoulos and Hentze, 1995 b). Though the exact cause of cluster disassembly is not yet defined, it appears that treatment of cells with hydrogen peroxide results in disassembly of the iron-sulfur cluster of IRP1. Furthermore, activation of the IRE binding activity of IRP1 by treatment with hydrogen peroxide is associated with repression of ferritin biosynthesis and increased biosynthesis of the transferrin receptor (Pantopoulos and Hentze, 1995 b).

A comparison of the time-course of induction of RNA binding activity by treatment of cells with H_2O_2 revealed that H_2O_2 induces IRE binding activity much more rapidly than treatment with desferrioxamine alone (Pantopoulos and Hentze, 1995 b). These results are consistent with several interpretations. While it is possible that there are two different induction pathways, another possibility is that the desferrioxiamine effect cannot be seen in the absence of oxidative disassembly of the iron-sulfur cluster, thus explaining the difference in timing of induction. If degradation of the iron-sulfur cluster of aconitase were to occur routinely as a result of oxidative stress associated with normal aerobic growth, then IRP1 would be poised to function as a sensitive indicator of iron levels. The protein would remain as apoprotein and a high affinity binder of IREs only when iron was scarce. Otherwise the cluster would be resynthesized efficiently, iron uptake would remain low, and sequestration would remain high. The presence of constant low level oxidative stress in cells would thus serve to create a sensitive monitor of iron levels in cells, based on the lability of iron-sulfur clusters.

The example of dihydroxyacid dehydratase of E. coli is relevant to the cluster assembly disassembly model which has just been described. Dihydroxyacid dehydratase is an iron-sulfur cluster protein that functions in the branched chain amino acid biosynthetic pathway and was identified many years ago as the enzyme of E. coli most sensitive to treatment with hyperbaric O_2. Exposure to hyperbaric oxygen results in complete disassembly of the [4Fe-4S] cluster, but the protein backbone remains stable. When the hyperbaric O_2 stress is discontinued, cluster is rapidly reassembled (Flint et al., 1993). It has also been shown that E. coli fumarase A, fumarase B, and aconitase (Gardner and Fridovich, 1992) are inactivated in cells treated with hyperbaric O_2; all of these enzymes ligate a $[4Fe-4S]^{2+}$ cluster, and each is a member of the hydro-lyase class of enzymes (Flint et al., 1993). Though cluster disassembly in these examples does not result in execution of a specific recognized regulatory event related to oxidative stress, the lability of such clusters to oxidative stress is apparent and forms the basis for a potential intersection of oxidative stress pathways with iron.

Speculations on the evolution of bifunctionality of IRP1

One could ask from an evolutionary standpoint how a complex regulatory system such as that which has been described here could arise. Some clues may exist in nature. In *E. coli*, there is only one aconitase, and its sole apparent function is to interconvert citrate and isocitrate in the Krebs' cycle, and thus contribute to carbon fixation and energy capture. The *E. coli* aconitase does not bind IREs to a measurable degree, and it is not implicated at this time in any post-transcriptional gene regulation pathway. The cytosolic aconitase of mammalian cells, IRP1, is much more similar to *E. coli* aconitase than to mitochondrial aconitase from mammalian cells. It seems possible therefore, that the direct evolutionary antecedent of cytosolic aconitase is the aconitase of *E. coli*, rather than the mitochondrial aconitase of present day mammalian cells (Prodromou et al., 1992).

How then has IRP1 acquired the capacity to bind IREs in appropriate gene targets, and appropriately influence mammalian iron homeostasis when cells are depleted of iron? The possibility that a conformational change in the apoprotein allows presentation of specific protein binding surface unique to the iron-depleted state raises the possibility that RNA transcripts evolved specifically to take advantage of the capability of IRP to sense iron levels and oxidative stress. This process would be analogous to the SELEX procedure, and *in vitro* selection procedure which postulates that a high affinity RNA ligand can be found for any protein because RNA can fold into a wide variety of complex complementary structures (Irvine et al., 1991). In the case of IRP1, those mRNAs that contained a sequence that could be bound by the apoprotein and for which binding produced a desirable adjustment to iron scarcity could confer an advantage to cells in which they were expressed. Because excess ferritin could potentially compete for iron in iron-depleted cells, decreasing ferritin synthesis in iron-depleted cells could be desirable and similarly, stabilization of the mRNA of TfR in iron-depleted cells would also be expected to yield an improvement in overall iron homeostasis. Thus, we postulate that the role of IRP1 in post-transcriptional gene regulation evolved from a cytosolic enzyme which contained a labile oxygen sensitive iron-sulfur cluster; cluster disassembly permitted a conformational change that revealed a binding surface to which specific RNA stem-loops could bind. When RNA stem-loops arose in mRNAs involved in cellular iron metabolism, a desirable post-transcriptional regulatory response could be produced and retained.

A second IRP, IRP2, adds complexity to the system

As was mentioned earlier, a second IRP, IRP2, is present in mammalian cells which is highly related to IRP1. IRP2 binds consensus IREs with an affinity and specificity comparable to that of IRP1 (Rouault et al., 1992; Guo et al., 1994; Samaniego et al., 1994). However, IRP2 differs from IRP1 in several important respects. One is that it does not have the capacity to function as an aconitase. Two of the 23 active site residues identified for IRP1 are different in IRP2, including a serine to glutamine residue change.

This residue change is important because the serine in question is postulated to function as the catalytic base in the aconitase enzymatic reaction. In the majority of cell types examined, IRP1 is found in substantially greater quantities than is IRP2 (Henderson et al., 1993; Guo et al., 1994; Samaniego et al., 1994). There is no evidence as yet that IRP2 binds an iron-sulfur cluster. However, IRP2 binding activity is regulated by intracellular iron levels. When cells are iron-replete, IRP2 is rapidly degraded in some cell types (Samaniego et al., 1994). This differs from the regulation of IRP1, which depends on modifications to a prosthetic group against the background of a relatively stable protein. The degradation is largely dependent on the presence of an extra 73 amino acids present in IRP2 relative to IRP1. This sequence is encoded by a separate exon and it appears to contain the information required for iron-dependent degradation (Iwai et al., (1995).

When the 73 amino acid sequence is deleted from IRP2, the deletion mutant retains its ability to bind IREs with high affinity and specificity, but the iron-induced rapid degration is no longer seen. When the sequence is ligated to IRP1, the chimeric IRP1/IRP2-specific exon becomes rapidly degraded by iron. Thus it appears that the 73 amino acid sequence contains the sequences required for sensing of high iron levels and conversion of the information into a signal for rapid degradation. It is quite possible that the information is conveyed by direct binding of iron to cysteines in the IRP2 specific exon. Site directed mutagenesis of cysteines within the exon prevents the iron induced degradation phenomenon (Iwai et al., 1995).

Direct binding of iron could facilitate degradation by several possible scenarios. Assembly of an iron sulfur cluster could produce a conformational change in IRP2 which would reveal an epitope that targets the protein for degradation. There are a number of examples in which changes in status of an iron-sulfur cluster result in changes in the structure and stability of the associated protein. The bacterial enzyme, glutamine amidophosphoribosyl (PRPP) amidotransferase requires an intact [4Fe-4S] cluster for maintenance of tertiary structure, and oxidative damage to the cluster results in denaturation of the protein and unmasking of degradation signals in the protein backbone (Smith et al., 1994). The iron-sulfur cluster of endonuclease III stabilizes a DNA binding motif (Kuo et al., 1992), and the transcription factor FNR requires an intact iron sulfur cluster to maintain the active dimeric form of the protein (Khoroshilova et al., 1995). In the case of bacterial PRPP amidotransferase, sensitivity of the iron-sulfur cluster to oxidative damage is proposed to play a regulatory role; nutrient starvation is thought to lead to a decrease in oxygen consumption resulting in an elevation in intracellular oxygen concentrations and oxygen-mediated inactivation of PRPP amidotransferase. Thus the cell would be protected from wasteful consumption of substrates in purine biosynthesis when cells are nutrient deprived (Switzer, 1989). Similarly, the oxidative destruction of the iron-sulfur cluster of FNR performs a regulatory function in that the transcription factor activates transcription of genes required in anaerobic growth and is degraded when cells are oxygen replete, a situation in which expression of genes of anaerobic metabolism is not required.

Another possible means by which direct binding of metal can lead to protein degradation has been described in which iron bound to substrate leads to localized oxidative damage to proteins and subsequent degradation of these damaged proteins by the proteasome. It is proposed that iron bound to protein reacts with hydrogen peroxide to produce free radicals in the Fenton reaction, and the localized generation of free radicals results in degradation of the amino groups of amino acids adjacent to the metal binding site. Oxidatively modified proteins bearing carbonyl groups are thought to then be degraded by the proteasome (Stadtman and Oliver, 1991; Grune et al., 1995). The *E. coli* glutamine synthetase is the best characterized example of a protein in which the protein levels are regulated by metal catalyzed oxidation and degradation (Stadtman, 1993):

Our work indicates that proteasome function is required for the degradation of IRP2 (Iwai et al., 1995). However, further work will be required to fully characterize the cis acting signal for degradation along with the trans acting factors required for degradation.

Conclusions

Iron is both indispensable and highly reactive in eukaryotic cells. A complex post-transcriptional regulatory system has developed that ensures that iron uptake and sequestration will be appropriate to the needs of the cell. Iron sensing may in fact be quite dependent on the reactivity of iron with oxygen species. Oxidative stress may result in constant turnover of iron-sulfur clusters, including the labile iron-sulfur cluster of IRP1; IRP1 may therefore be capable of registering changes in overall iron status, because reassembly of an iron sulfur cluster can proceed only when sufficient quantities of bioavailable iron and sulfur are present. In addition, metal catalyzed oxidation reactions could be important in the turnover of IRP2, though there is as yet no proof that IRP2 directly binds iron. However, the fact that cysteines are indispensable to the degradation signal makes attractive a hypothesis that iron is bound. Examples exist in nature in which iron-sulfur clusters are used to sense changes in oxygen status, through changing the activity or stability of the associated protein.

Larger questions about the scope of this regulatory system and the interplay between the two IRPs remain to be answered. An exciting possibility is that more targets of IRP binding will be identified, and that these additional targets will include transcripts which encode proteins involved in the response to oxidative stress in mammalian cells.

References

Aziz, N. and Munro, H.N. (1987) Iron regulates ferritin mRNA translation through a segment of its 5′ untranslated region. *Proc. Natl. Acad. Sci. USA* 84:8478–8482.

Basilion, J.P., Rouault, T.A., Massinople, C.M., Klausner, R.D. and Burgess, W.H. (1994a) The iron-responsive element-binding protein: Localization of the RNA binding site to the aconitase active-site cleft. *Proc. Natl. Acad. Sci. USA* 91:574–578.

Basilion, J.P., Kennedy, M.C., Beinert, H., Massinople, C.M., Klausner, R.D. and Rouault, T.A. (1994b) Overexpression of iron-responsive element binding protein and its analytical characterization as the RNA binding form, devoid of an iron-sulfur cluster. *Arch. Biochem. and Biophys.* 311:517–522.

Beinert, H. and Kennedy, M.C. (1989) 19th Sir Hans Krebs lecture. Engineering of protein bound iron-sulfur clusters. A tool for the study of protein and cluster chemistry and mechanism of iron-sulfur enzymes. *Eur. J. Biochem.* 186:5–15.

Beinert, H. (1990) Recent developments in the field of iron-sulfur proteins. *FASEB J.* 4:2483–2491.

Binder, R., Horowitz, J.A., Basilion, J.P., Koeller, D.M., Klausner, R.D. and Harford, J.B. (1994) Evidence that the pathway of transferrin receptor mRNA degradation involves an endonucleolytic cleavage within the 3′UTR and does not involve poly(A) tail shortening. *EMBO J.* 13:1969–1980.

Castro, L., Rodriguez, M. and Radi, R. (1994) Aconitase is readily inactivated by peroxynitrite, but not by its precursor, nitric oxide. *J. Biol. Chem.* 269:29409–29415.

Cox, T.C., Bawden, M.J., Martin, A. and May, B.K. (1991) Human erythroid 5-aminolevulinate synthase: promoter analysis and identification of an iron-responsive element in the mRNA. *EMBO J.* 10:1891–1902.

Dandekar, T., Stripecke, R., Gray, N.K., Goossen, B., Constable, A., Johansson, H.E. and Hentze, M.W. (1991) Identification of a novel iron-responsive element in murine and human erythroid delta-aminolevulinic acid synthase mRNA. *EMBO J.* 10:1903–1909.

Drapier, J.C., Hirling, H., Wietzerbin, J., Kaldy, P. and Kuhn, L.C. (1993) Biosynthesis of nitric oxide activates iron regulatory factor in macrophages. *EMBO J.* 12:3643–3649.

Emery-Goodman, A., Hirling, H., Scarpellino, L., Henderson, B. and Kuhn, L. (1993) Iron regulatory factor expressed from recombinant baculovirus: conversion between the RNA-binding apoprotein and Fe-S cluster containing aconitase. *Nuc. Acids Res.* 21:1457–1461.

Flint, D.H., Smyk-Randall, E., Tuminello, J.F., Draczynska-Lusiak, B. and Brown, O.R. (1993) The inactivation of dihydroxy-acid dehydratase in *Escherichia coli* treated with hyperbaric oxygen occurs because of the destruction of its Fe-S cluster, but the enzyme remains in the cell in a form that can be reactivated. *J. Biol. Chem.* 268:25547–25552.

Fu, W., Jack, R.F., Morgan, T.V., Dean, D.R. and Johnson, M.K. (1994) nifU gene product from Azotobacter vinelandii is a homodimer that contains two identical [2Fe-2S] clusters. *Biochemistry* 33:13455–13463.

Gardner, P.R. and Fridovich, I. (1992) Superoxide sensitivity of the *Escherichia coli* aconitase. *J. Biol. Chem.* 266:19328–19333.

Gardner, P.R., Raineri, I., Epstein, L.B. and White, C.W. (1995) Superoxide radical and iron modulate aconitase activity in mammalian cells. *J. Biol. Chem.* 270:13399–13405.

Gray, N.K., Quick, S., Goossen, B., Consable, A., Hirling, H., Kuhn, L.C. and Hentze, M.W. (1993) Recombinant iron-regulatory factor functions as an iron-responsive-element-binding protein, a translational repressor and an aconitase. A functional assay for translational repression and direct demonstration of the iron switch. *Eur. J. Biochem.* 218:657–667.

Grune, T., Reinheckel, T., Joshi, M. and Davies, K.J. (1995) Proteolysis in cultured liver epithelial cells during oxidative stress. Role of the mulitcatalytic proteinase complex proteasome. *J. Biol. Chem.* 270:2344–2351.

Guo, B., Yu, Y. and Leibold, E.A. (1994) Iron regulates cytoplasmic levels of a novel iron-responsive element-binding protein without aconitase activity. *J. Biol. Chem.* 268:24252–24260.

Haile, D.J., Hentze, M.W., Rouault, T.A., Harford, J.B. and Klausner, R.D. (1989) Regulation of interaction of the iron-responsive element binding protein with iron-responsive RNA elements. *Mol. Cell. Biol.* 9:5055–5061.

Haile, D.J., Rouault, T.A., Tang, C.K., Chin, J., Harford, J.B. and Klausner, R.D. (1992a) Reciprocal control of RNA binding and aconitase activity in the regulation of the iron responsive element binding protein: Role of the iron-sulfur cluster. *Proc. Natl. Acad. Sci. USA* 89: 7536–7540.

Haile, D.J., Rouault, T.A., Harford, J.B., Kennedy, M.C., Blondin, G.A., Beinert, H. and Klausner, R.D. (1992b) Cellular regulation of the iron-responsive element binding protein: Disassembly of the cubane iron-sulfur cluster results in high affinity RNA binding. *Proc. Natl. Acad. Sci. USA* 89:11735–11739.

Hausladen, A. and Fridovich, I. (1994) Superoxide and peroxynitrite inactivate aconitases, but nitric oxide does not. *J. Biol. Chem.* 269:290405–29408.

Hentze, M.W., Caughman, S.W., Rouault, T.A., Barriocanal, J.G., Dancis, A., Harford, J.B. and Klausner, R.D. (1987) Identification of the iron-responsive element for the translational regulation of human ferritin mRNA. *Science* 238:1570–1573.

Hentze, M.W. and Argos, P. (1991) Homology between IRE-BP, a regulatory RNA-binding protein, aconitase, and isopropylmalate isomerase. *Nuc. Acids. Res.* 19:1739–1740.

Hirling H., Emery-Goodman, A., Thompson, N., Neupert, B., Seiser, C. and Kuhn, L.C. (1992) Expression of active iron regulatory factor from a full-length human cDNA by *in vitro* transcription/translation. *Nuc. Acids Res.* 20:33–39.

Irvine, D., Tuerk, C. and Gold, L. (1991) SELEXION. Systematic evolution of ligands by exponential enrichment with integrated optimization by non-linear analysis. *J. Mol. Biol.* 222:739–761.

Iwai, K., Klausner, R.D., Rouault, T.A., (1995) Requirements for iron regulated degradation of the RNA binding protein, Iron Regulatory Protein 2 (IRP2), *EMBO J.* 14:5350–5357.

Kennedy, M.C. and Beinert, H. (1988) The state of cluster SH and S2- of aconitase during cluster interconversions and removal. *J. Biol. Chem.* 263:8194–8198.

Kennedy, M.C., Mende-Müller, L., Blondin, G.A. and Beinert, H. (1992) Purification and characterization of cytosolic aconitase from beef liver and its relationship to the iron-responsive element binding protein (IRE-BP). *Proc. Natl. Acad. Sci. USA* 89:11730–11734.

Kennedy, C. and Beinert, H. (1994) *In vitro* studies on the disassembly of the Fe-S cluster of cytosolic and mitochondreal aconitases on reaction with nitric oxide. *J. Inorg. Biochem.* 56: L13.

Khoroshilova, N., Beinert, H. and Kiley, P.J. (1995) Association of a polynuclear iron-sulfur center with a mutant FNR protein enhances DNA binding. *Proc. Natl. Acad. Sci. USA* 92:2499–2503.

Klausner, R.D. and Rouault, T.A. (1993) A double life: Cytosolic aconitase as a regulatory RNA binding protein. *Mol. Biol. of Cell* 4:1–5.

Klausner, R.D., Rouault, T.A. and Harford, J.B. (1993) Regulating the fate of mRNA: the control of cellular iron metabolism. *Cell* 72:19–28.

Klausner, R.D. and Dancis, A. (1994) A genetic approach to elucidating eukaryotic iron metabolism. *FEBS Lett.* 355:109–113.

Lauble, H., Kennedy, M.C., Beinert, H. and Stout, C.D. (1992) Crystal structures of aconitase with isocitrate and nitroisocitrate bound. *Biochemistry* 31:2735–2748.

Leibold, E.A. and Munro, H.N. (1988) Cytoplasmic protein binds *in vitro* to a highly conserved sequence in the 5′ untranslated region of ferritin heavy- and light-subunit mRNAs. *Proc. Natl. Acad. Sci. USA* 85:2171–2175.

Martins, E.A., Robalinho, R.L. and Meneghini, R. (1995) Oxidative stress induces activation of a cytosolic protein responsible for control of iron uptake. *Arch. Biochem. Biophys.* 316:128–134.

Mullner, E.W., Neupert, B. and Kuhn, L.C. (1989) A specific mRNA binding factor regulates the iron-dependent stability of cytoplasmic transferrin receptor mRNA. *Cell* 58:-373–382.

Pantopoulos, K. and Hentze, M.W. (1995a) Nitric oxide signaling to iron-regulatory protein: direct control of ferritin mRNA translation and transferrin receptor mRNA stability in transfected fibroblasts. *Proc. Natl. Acad. Sci. USA* 92:1267–1271.

Pantopoulos, K. and Hentze, M.W. (1995b) Rapid responses to oxidative stress mediated by iron regulatory protein. *EMBO J.* 14:2917–2924.

Patino, M.M. and Walden, W.E. (1992) Cloning of a functional cDNA for the rabbit ferritin mRNA repressor protein: Demonstration of a tissue specific pattern of expression. *J. Biol. Chem.* 267:19011–19016.

Philpott, C.C., Klausner, R.D. and Rouault, T.A. (1994) The bifunctional iron-responsive element binding protein/cytosolic aconitase: The role of active-site residues in ligand binding and regulation. *Proc. Natl. Acad. Sci. USA* 91:7321–7325.

Prodromou, C., Artymiuk, P.J. and Guest, J.R. (1992) The aconitase of *E. Coli. Eur. J. Biochem.* 204:599–609.

Rouault, T.A., Hentze, M.W., Caughman, S.W., Harford, J.B. and Klausner, R.D. (1988) Binding of a cytosolic protein to the iron-responsive element of human ferritin messenger RNA. *Science* 241:1207–1210.

Rouault, T.A., Hentze, M.W., Haile, D.J., Harford, J.B. and Klausner, R.D. (1989) The iron-responsive element binding protein: a method for the affinity purification of a regulatory RNA-binding protein. *Proc. Natl. Acad. Sci. USA* 86:5768–5772.

Rouault, T.A., Tang, C.K., Kaptain, S., Burgess, W.H., Haile, D.J., Samaniego, F., McBride, O.W., Harford, J.B. and Klausner, R.D. (1990) Cloning of the cDNA encoding an RNA regulatroy protein – the human iron-responsive element-binding protein. *Proc. Natl. Acad. Sci. USA* 87:7958–7962.

Rouault, T.A., Stout, C.D., Kaptain, S., Harford, J.B. and Klausner, R.D. (1991) Structural relationship between an iron-regulated RNA-binding protein (IRE-BP) and aconitase: functional implications. *Cell* 64:881–882.

Rouault, T.A., Haile, D.H., Downey, W.E., Philpott, C.C., Tang, C., Samaniego, F., Chin, J., Paul, I., Orloff, D., Harford, J.B. and Klausner, R.D. (1992) An iron-sulfur cluster plays a novel regulatory role in the iron-responsive element binding protein. *BioMetals* 5:131–140.

Samaniego, F., Chin, J., Iwai, K., Rouault, T.A. and Klausner, R.D. (1994) Molecular characterization of a second iron responsive element binding protein, iron regulatroy protein 2 (IRP2): Structure, function and post-translational regulation. *J. Biol. Chem.* 269:30904–30910.

Stadtman, E.R. and Oliver, C.N. (1991) Metal-catalyzed oxidation of proteins. *J. Biol. Chem.* 266:2005–2008.

Stadtman, E.R. (1993) Oxidation of free amino acids and amino acid residues in proteins by radiolysis and by metal-catalyzed reactions. *Annu. Rev. Biochem.* 62:797–821.

Switzer, R.L. (1989) Non-redox roles for iron-sulfur clusters in proteins. *BioFactors* 2:77–86.

Theil, E.C. (1989) Ferritins. *Annu. Rev. Biochem.* 56:289–315.

Walden, W.E., Daniels-McQueen, S., Brown, P.H., Gaffield, L., Russell, D.A., Bielser, D., Bailey, L.C. and Thach, R.E. (1988). Translational repression in eukaryotes: Partial purification and characterization of a repressor of ferritin mRNA translation. *Proc. Natl. Acad. Sci. USA* 85:9503–9507.

Weiss, G., Goossen, B., Doppler, W., Fuchs, D., Pantopoulos, K., Werner-Felmayer, G., Wachter, H. and Hentze, M.W. (1993) Translational regulation via iron-reponsive elements by the nitric oxide/NO-synthase pathway. *EMBO J.* 12:3651–3657.

Yu, Y., Radisky, E. and Leibold, E.A. (1992) The iron-responsive element binding protein: Purification, cloning and regulation in rat liver. *J. Biol. Chem.* 267:19005–19010.

Zheng, L., White, R.H., Cash, V.L., Jack, R.F. and Dean, D.R. (1993) Cysteine desulfurase activity indicates a role for NIFS in metallocluster biosynthesis. *Proc. Natl. Acad. Sci. USA* 90:2754–2758.

Zheng, L. Dean, D.R. (1994) Catalytic formation of a nitrogenase iron-sulfur cluster. *J. Biol Chem.* 269:18723–18726.

Stress-Inducible Cellular Responses
ed. by U. Feige, R.I. Morimoto, I. Yahara and B. Polla
© 1996 Birkhäuser Verlag Basel/Switzerland

Heat-shock induced protein modifications and modulation of enzyme activities

O. Bensaude, S. Bellier, M.-F. Dubois, F. Giannoni and V.T. Nguyen

Génétique Moléculaire, Ecole Normale Supérieure, 46 rue d'Ulm F-75230 Paris Cedex 05, France

Summary. Upon heat stress, the cell physiology is profoundly altered. The extent of the alterations depends on the severity of the stress and may lead to cell death. The heat shock response is an array of metabolic changes characterized by the impairment of major cellular functions and by an adaptive reprogramming of the cell metabolism. The enhanced synthesis of the HSPs is a spectacular manifestation of this reprogramming.

Numerous post translational modifications of proteins occur in response to heat stress and can be related to altered cellular functions. Some proteins are heat-denatured and temporarily inactivated. Heat-denaturation is reversible, chaperones may contribute to the repair. The extent of heat-denaturation depends on the cell metabolism: (a) it is attenuated in thermotolerant cells or in cells overexpressing the appropriate chaperones (b) it is enhanced in energy-deprived cells. Covalent modifications may also rapidly alter protein function. Changes in protein glycosylation, methylation, acetylation, farnesylation, ubiquitination have been found to occur during stress. But protein phosphorylation is the most studied modification. Several protein kinase cascades are activated, among which the various mitogen activated protein kinase (MAP kinase) cascades which are also triggered by a wide range of stimuli. As a possible consequence, stress modifies the phosphorylation status and the activity of components from the transcriptional and translational apparatuses. The same kinases also target key enzymes of the cellular metabolism.

Protein denaturation results in constitutive hsp titration, this titration is a signal to trigger the heat-shock gene transcription and to activate some of the protein kinase cascades.

The cell physiology is altered when cells are submitted to heat stress (reviewed in (Laszlo, 1992a; Nover, 1991; Welch, 1992). For example, RNA splicing and protein synthesis are impaired, the cytoskeleton collapses, and an extensive reprogramming of transcription occurs. The extent of the alterations depends on the severity of the stress and may lead to cell death. These alterations constitute the heat shock response as an array of metabolic changes characterized by the impairment of major cellular functions and by an adaptive reprogramming of the cell metabolism. The enhanced synthesis of the HSPs is a spectacular manifestation of this reprogramming.

This chapter will review post translational modifications of proteins reported to occur in response to heat stress and which can be related to altered cellular functions.

Protein denaturation

The involvement of protein denaturation as a cause of the noxious effects of heat on living organisms is a trivial hypothesis since the end of last century. However a renewed interest in this idea has been triggered by the recent identification of the HSPs as being molecular chaperones capable of avoiding sticky situations and repairing unfolded proteins (Georgopoulos and Welch, 1993: Hartl et al., 1994). There are several evidences that protein denaturation is critical for cell survival from heat-stress: (i) solvents such as monoalcohols decrease the survival following heat stress (Dewey, 1989); (ii) addition of protein heat-protectors, such as glycerol and deuterium oxide to the culture medium, protects cells against heat stress (Edington et al., 1989); (iii) there is a positive correlation between the temperature adaptation of a species and the interspecific differences in protein thermal stability *in vitro* (Jaenicke, 1991; Somero, 1995). Thus, the intrinsic properties of the protein components would determine the cellular resistance of cells to temperature.

In vivo *denaturation of individualized proteins*

Evidences for occurrence of protein denaturation within heat shocked cells were obtained using differential scanning calorimetry, loss of protein solubility and activity in lysates from stressed cells (reviewed in Bensaude et al., 1990; Kampinga, 1993). A small proportion of the proteins which are soluble in non denaturing lysates remains aggregated to the large cellular debris from heat-shocked cells. Not all proteins aggregate, some proteins are more thermosensitive than others. The constitutively expressed heat-shock proteins are the most abundant insolubilized proteins. Cytoplasmic proteins such as the dsRNA-dependent protein kinase (PKR) are insolub-ilized and inactivated (Dubois et al, 1991). PKR contributes in signaling the antiviral effects of interferon and its inactivation corresponds to a partial loss in these effects (Dubois et al., 1989). Recombinant proteins expressed in transfected cells behave like endogenous proteins. Thus, β-galactosidase (*E. coli*), firefly and bacterial luciferases are used as models to investigate protein denaturation both *ex vivo* within the living cells or *in vitro* (Nguyen et al., 1989; Parsell et al., 1994; Schröder et al., 1993).

   Many nuclear proteins have been found to aggregate. The protein content of nuclei isolated from heat-shocked cells is markedly increased most like-ly as a result of protein aggregation (Kampinga, 1993; Laszlo et al., 1992). Are nuclear proteins particularly thermosensitive or is the intranuclear environment destabilizing? To address this question, a mutated cyto-plasmic luciferase was targeted to the nucleus after fusing a nuclear local-ization sequence (nls) (Michels et al., 1995) (firefly luciferase is normally localized in the peroxisomes but a point mutation abolishes this localiza-tion). The luciferase heat-denatures more rapidly in the nucleus than in the cytoplasm. This effect could, in part, be explained by a higher local protein concentration.

Repair of the heat-denatured proteins

The damaged proteins might be targeted for degradation or renaturation (reviewed in Parsell and Lindquist, 1993, and Goldberg, this volume). Upon recovery from stress, both the solubility and the activity of many aggregated proteins can partially be recovered (Dubois et al., 1991; Parsell et al., 1994; Pinto et al., 1991; Siverio et al., 1993). This repair does not require *de novo* protein synthesis. Extensive studies suggest that the molecular chaperones catalyze protein folding and hence should be involved in protein renaturation (Georgopoulos and Welch, 1993; Hartl et al., 1994); here, we restrictively mention studies with heat-denatured proteins. In bacteria, the chaperones dnaK, dnaJ, grpE are involved *ex vivo* in the renaturation of RNA polymerase (Ziemienowicz et al., 1993) and firefly luciferase (Schröder et al., 1993). In yeast, hsp104 is involved in the renaturation of a bacterial luciferase (Parsell et al., 1994). Purified mammalian hsp90, hsp70 and a 60 kDa protein homologous to the yeast STI1 protein cooperate to renature thermally denatured firefly luciferase (Schumacher et al., 1994).

Increased protein stability in thermoresistant cells

A priming non lethal heat-shock promotes a transient increased (adaptative) resistance to a second challenging stress (G. Li, this volume). Acquisition of this thermotolerance correlates with the accumulation of HSPs. Overexpression of these proteins in mammalian cells increase their thermal resistance. Conversely, a decreased expression is associated with an enhanced heat sensitivity.

The increased HSP/chaperone concentration might contribute to thermotolerance through two mechanisms: either an accelerated damage repair or a protection against initial damage (Laszlo, 1992 b). The first interpretation is supported by the known involvement of chaperones in protein folding. Overexpression of chaperones increases the thermal stability of proteins and hence supports the protection against initial damage interpretation. Overexpression of the *groE* operon suppresses some thermosensitive mutations in bacteria (Van Dyk et al., 1989). Aggregation/inactivation of proteins followed by differential scanning calorimetry or by activity and solubility is attenuated in thermoresistant cells (Ciavarra et al., 1992; Dubois et al., 1991; Konstantinova et al., 1994; Lepock et al., 1990; Nguyen et al., 1989). The heat-induced intranuclear protein aggregation is attenuated in hsp70 overexpressing cells (Stege et al., 1994). The HSPs might protect proteins from heat-induced aggregation by capturing unfolded intermediates (Höll-Neugebauer et al., 1991; Martin et al., 1992; Taguchi and Yoshida, 1993). DNA polymerases are heat-inactivated (Kampinga et al., 1985; Mivechi and Dewey, 1985). Calf thymus hsc70 protects and reactivates

DNA polymerases in an ATP-dependent manner (Ziemienowicz et al.,
1995). Heat-stress induces hsc70/nuclear topoisomerase I complex formation *ex vivo* and *in vitro*, hsc70 mediates reactivation of this enzyme
(Ciavarra et al., 1994). Interestingly, topoisomerase I reactivation mediated
by hsc70 does not require ATP hydrolysis. This finding might explain why
overexpression of an hsp70 deletion mutant lacking a functional ATP binding domain is efficient in increasing the thermoresistance of mammalian
cells and preventing protein aggregation (Li et al., 1992; Stege et al., 1994).
Prevention of protein aggregation by chaperones might occur only above
a specific temperature threshold. In example, α-crystallin exposes hydrophobic surfaces at temperatures above 30 °C and prevents insulin B chain
aggregation above that temperature (Raman et al., 1995). An increasing
efficiency is obtained at higher temperatures up to 41 °C.

Changing the point of view, denatured proteins capture and titrate the
preexisting chaperones (Bensaude et al., 1990; Craig and Gross, 1991).
Titration of the preexisting 70 kDa hsps (hsp70 and hsc70, the dnaK homologs) might be a signal which triggers the expression of the heat-shock
genes (Hightower and Voellmy, this volume).

Decreased protein stability in energy deprived cells

Energy deprived cells are more heat-sensitive than controls (Haveman and
Hahn, 1981). Aggregation of heat-sensitive proteins including the HSPs is
enhanced in energy-deprived cells (Kabakov and Gabai, 1994; Nguyen and
Bensaude, 1994). Similarly, the thermal aggregation of a recombinant
dihydrofolate reductase (DHFR) was enhanced in ATP-depleted mitochondria (Martin et al., 1992). Many (but not all) chaperone-captured denaturation intermediates require energy (from ATP hydrolysis) to be reactivated.
In vitro, the heat denaturation of recombinant DHFR is enhanced in the
presence of groEL and in the absence of ATP, however, when both groEL
and ATP are present, the heat denaturation of DHFR is attenuated (Martin
et al., 1992). In conditions of ATP-depletion, the conformational flexibility
of thermosensitive proteins will lead to the capture and, as a consequence,
to the titration of the constitutive heat shock proteins (Beckmann et al.,
1992). This titration might explain the original finding of Ritossa: heat and
dinitrophenol induce the same (heat-shock) puffs on polytene chromosomes (Ritossa, 1962). Thermotolerant cells expressing high amounts of
HSPs are more resistant to ATP deprivation (Kabakov and Gabai, 1994;
Kabakov and Gabai, 1995). The ATP depletion-promoted aggregation of
reporter proteins is prevented in thermotolerant cells (Kabakov and
Bensaude, unpublished). This observation could be in agreement with the
above-mentioned protection brought about by ATP binding-deficient
hsp70. It may also suggest that factors which do not hydrolyze ATP are
involved. Some chaperone/protein interactions require ATP to be disrupted,
others not (Jakob and Buchner, 1994).

Heat protectors

Inhibition of protein synthesis by cycloheximide increases the heat-resistance of cells (Widelitz et al, 1986). This protection is associated with an increase in thermostability of proteins as detected by differential scanning calorimetry (Borelli et al., 1991). In growing cells, the chaperones such as hsp70 bind to the newly synthesized polypeptide chains, hence the inhibition of protein synthesis is expected to increase the pool of free available chaperones.

The thermal stability of proteins might be increased by small metabolites which are accumulated by cells in the acquired thermotolerant state. Yeast cells accumulate glycerol after activation of the HOG kinase cascade (see below) (Brewster et al., 1993). Trehalose is a disaccharide accumulated by yeast exposed to high osmolarity medium and heat stress, the trehalose accumulation contributes to the establishment of adaptive thermotolerance (De Virgilio et al., 1994). Prevention of protein denaturation is likely to be involved since physiological concentration of trehalose increases the thermal stability of proteins *in vitro* (Hottiger et al., 1994). Despite the transcription of several genes coding for enzymes of trehalose metabolism is regulated by the HOG kinase cascade (Gounalaki and Thireos, 1994). Accumulation of trehalose also results from an activation of the enzymes of trehalose metabolism which does not require protein synthesis. In bacteria, the enzymes of the trehalose metabolism are involved in stationary phase but not in adaptive thermoresistance (Hengge-Aronis et al., 1991).

The thermal stability of membrane proteins is dependent upon the fatty acid composition of the membrane. Increasing the cholesterol content protects the calcium adenosine triphosphatase from thermal inactivation (Cheng et al., 1987). Hence, membrane fluidity affects the thermal sensitivity of cells (reviewed in Hazel, 1995) and modulates the heat shock gene expression (Maresca and Kobayashi, 1993): Membrane fluidity has also been suggested to be involved in protein kinase cascade activation (see below).

Covalent modifications of proteins

*Glycosylation, ubiquitination, methylation, farnesylation
and phosphorylation*

Heat-shock has been shown to promote numerous covalent modifications of proteins such as methylation, glycosylation, ubiquitination, acetylation (previously reviewed in Nover, 1991). Calreticulin, a calcium-binding protein present in the sarcoplasmic reticulum, the endoplasmic reticulum, the nucleus and the nuclear envelope is promptly glycosylated upon heat shock (Jethmalani et al., 1994). The glycosylation of phosphoglucomutase is modulated by heat stress and nutrient deprivation (Dey et al., 1994). Mannosylation increases whereas glycosylation decreases and these changes may affect the membrane association of this protein. An enhanced glycosyltransferase activity parallels the development of adaptive thermo-tolerance (Henle et al., 1990). But inhibition of glycosylation with tunic-amycin did not prevent thermotolerance development (Lee et al., 1991).

Ubiquitin is a 6 kDa protein which can be covalently linked to proteins and such a link can be a signal to target for degradation via the proteasome or the lysosomal pathways (reviewed in Jennissen, 1995; Jentsch, 1992). Ubiquitin synthesis is heat shock-inducible. During stress marked changes occur, histone 2A is deubiquitinated, whereas new conjugates are formed. Stress resistance is strongly associated with multiubiquitination by the yeast ubc4 and ubc5 conjugating enzymes (Arnason and Ellison, 1994). Ubiquitination might be required to target degradation of the stress-damaged proteins, but some ubiquitinated proteins are stable. Indeed, ubiquitin also functions as a covalent chaperone (Finley et al., 1989; Witt-liff et al., 1990).

An increased protein methylation on aspartyl residues is found in stress-ed cells (Ladino and O'Connor, 1992; Tanguay and Desrosiers, 1990). Heat shock induces an increase in histone H2B and a decrease in histone H3 methylation (Arrigo, 1983; Desrosiers and Tanguay, 1988). These effects might be associated with disruption of the chromatin structure and are also observed upon transcriptional inhibition (Tanguay and Desrosiers, 1990). The intracellular concentrations in HSPs increases spectacularly after stress. But stress also promotes several post translational modifications of the preexisting HSPs. Methylation was first described (Wang et al., 1981) and a protein methyltransferase is involved in heat-shock resistance in *E. coli* (Li and Clarke, 1992).

An increased farnesylation of ydj1p (a yeast dnaJ homolog) occurs and is required for function at elevated temperatures (Caplan et al., 1992).

However, protein phosphorylation is by far the most extensively studied covalent modification which modulates protein activity. The phosphoryla-tion state of numerous proteins is modified by heat stress. For example, the intermediate filament protein, vimentin, is increasingly phosphorylated (Cheng and Lai, 1994), whereas the retinoblastoma protein Rb is dephos-phorylated (Khandjian, 1995). The following paragraphs will review the phosphorylation of the HSPs and several protein kinase cascades activated by heat stress.

Phosphorylation of the HSPs in response to stress

Heat-shock increases the proportion of phosphorylated groEL and dnaK in *E. coli* (Sherman and Goldberg, 1994; Sherman and Goldberg, 1993). Phosphorylation increases the affinity of these proteins towards their peptide substrates. The mammalian hsp90s are phosphoproteins (Csermely and Kahn, 1991), their phosphate turnover is increased during heat stress (Legagneux et al., 1991). The "small" HSPs (23 to 28 kd) become phosphorylated and aggregate upon a wide range of stimuli including heat stress (reviewed in Arrigo and Landry, 1994). Conflicting reports have discussed the role of small HSP phosphorylation during stress. Phosphorylation is concomitant with the formation of large aggregates of small HSP (Mehlen and Arrigo, 1994). Phosphorylation of the human hsp27 might contribute to the dissociation of the aggregates (Kato et al., 1994; Lavoie et al., 1995). Depolymerization of the actin microfilaments occurs in response to heat stress. *In vitro*, the unphosphorylated chicken small HSP inhibits actin polymerization (Miron et al., 1991). Phosphorylation of the murine hsp25 abolishes its actin-depolymerizing activity (Benndorf et al., 1994). Hamster cells permanently transfected with human hsp27 mutants show an increased resistance to heat and/or to actin depolymerizing agents, but this effect is abolished upon replacement of the phosphorylated serines by glycines (Lavoie et al., 1995). In contrast, in transient transfections assays of murine fibroblasts, the phosphorylation does not contribute to the increased resistance to heat (Knauf et al., 1993). Overexpression of the murine hsp25 increases the cell survival after heat stress, but HSP mutant proteins in which the serines are replaced by alanines are as efficient as the wild-type protein.

The mammalian hsp27 kinase activating cascade, a stress-activated protein kinase cascade homologous to the yeast HOG osmosensing cascade

The small HSPs are phosphorylated by a specific kinase. This kinase was first described as a mitogen activated protein kinase (MAP kinase) activated protein kinase (MAPKAPK2) (Stokoe et al., 1992). Two proteins of 45-pKa/54-kDa showing heat-activated hsp27 kinase have been characterized (Huot et al., 1995). MAPKAPK2 can be activated *in vitro* by phosphorylation with MAP kinases of the ERK (extracellular regulated kinase) type. But, recent evidences have been brought that MAPKAPK2 can be activated by a distinct stress-activated MAP kinase defined as RK (Rouse et al., 1994) or p40 MAP kinase (Freshney et al., 1994). This new MAP kinase also termed p38 MAP kinase is highly homologous to the yeast *hog1* gene *(Saccharomyces cerevisiae)* product and complements *hog1* deficient yeast mutants (Han et al., 1994). In mammalian cells, the p38 MAP kinase is activated by a wide variety of stress including heat-shock, oxidative stress (Rouse et al., 1994), hyperosmolarity (Han et al., 1994), endotoxic lipopolysaccharide (LPS) (Han et al., 1994) and interleukin-1 (IL-1) (Freshney et al., 1994). In addition to mitogenic stimulation, all these treatments lead to the small HSP phosphorylation (reviewed in Arrigo and Landry, 1994). Most fascinating is the discovery of a family of bicyclic imidazole compounds which inhibit the p38 MAP kinase with an extremely high specificity (Lee et al., 1994). Stress-activation of MAP-KAPK2 and phosphorylation of hsp27 are suppressed in mammalian cells exposed to micromolar amounts of these compounds (Cuenda et al., 1995) which inhibit in turn, the production of IL-1 and tumor necrosis factor (TNF).

The involvement of the HOG kinase cascade in the regulation of stress gene expression is documented in yeast. A stress response element (STRE) having the consensus sequence CCCCT, distinct from the heat shock element (HSE), mediates the stress induction of transcription of a set of genes including the cytosolic catalase T *(cttl)* gene, the DNA damage responsive gene *(ddr2)* (Kobayashi and McEntee, 1993), the trehalose phosphate phosphatase *(tps2)* gene (Gounalaki and Thireos, 1994), the glycerol-3-phosphate dehydrogenase *(gpd1)* gene and is present in the promoter of other stress-induced genes such as polyubiquitin *(ubi4)*, *hsp12, hsp104* and a protein tyrosine phosphatase *(ptp2)*. In mammalian cells, the expression of a MAP kinase tyrosine phosphatase is also enhanced after stress and might contribute to a down-regulation of the MAP kinases (Keyse and Emslie, 1992). The HOG pathway controls the stress regulation of transcription via the STRE (Schüller et al., 1994). It is interesting to note that several of the genes controlled by the HOG pathway are involved in protection or in damage repair.

The c-jun kinase (JNK) MAP kinase cascade, a stress-activated cascade homologous to the HOG cascade

The ERK/MAP kinases were primitively thought to phosphorylate and activate the protooncogene c-jun which is part of the AP-1 transcriptional activator. However, it became recently apparent that c-jun was a target for a new class of stress-activated kinases presenting significant homologies with the ERK/MAP kinases, the JNKs (Dérijard et al., 1994). This family of closely related kinases is strongly induced by numerous stresses including heat-shock and are sometimes designated as stress-activated protein kinases (SAPK) (Kyriakis et al., 1994). The JNKs are homologous to the above-mentioned p38 MAP kinase and the HOG1 yeast gene product. In contrast to the ERK genes, but like the p38 MAP kinase gene, the mammalian JNK1 gene rescues a *hog1* deficient yeast mutant (Galcheva-Gargova et al., 1994).

As a possible consequence, heat and arsenite stress enhance the transcription of the *c-fos* proto-oncogene (Bukh et al., 1990; van Delft et al., 1993). This enhanced transcription is accompanied by a marked increase in c-fos mRNA stability (Andrews et al., 1987). The resulting c-fos mRNA accumulation could lead, in some cases, to a stimulated cell cycle progression after mild stress (van Wijk et al., 1993).

The extracellular regulated kinase (ERK) mitogen activated
protein kinases (MAPK) cascade

The MAP kinases of the ERK type were the first described (Cobb et al., 1994). They are activated by phosphorylation on tyrosine and threonine during heat stress (Chung et al., 1992; Dubois and Bensaude, 1993) and osmotic stress (Matsuda et al., 1995). In yeast, the ERK/MAP kinase homologous is coded by the *slt2/mpk1* gene (Kamada et al., 1995; Torres et al., 1991). The activity of Mpk1p is modulated by the growth temperature, it increases over a 170-fold range from 23° to 37°C (Kamada et al., 1995). This gene is essential for survival at high temperature (37°C) in normal medium. Cells disrupted for the *slt2/mpk1* gene lyse rapidly at 37°C with a phenotype recalling that of actin and polarized growth mutants (Mazzoni et al., 1993). The slt2/mpk1 cascade is distinct from the HOG cascade, but it is highly homologous. The temperature sensitivity of the *slt2/mpk1* gene can be rescued by growing the cells at high osmolarity (Kamada et al., 1995; Torres et al., 1991) however, activation of the HOG cascade by high osmolarity is not involved in this rescue.

Numerous putative targets of the ERK/MAP kinase have been described (L'Allemain, 1994) some of which but not all are involved in transcriptional regulation (Hill and Treisman, 1995). Three such targets have been examined in heat shocked cells: stathmin, the microtubule-associated protein tau and the largest subunit of RNA polymerase II.

Stathmin is an ubiquitous 19 kd protein phosphorylated in response to a wide range of stimuli (Sobel, 1992), its phosphorylation increases during heat stress on a serine phosphorylated by MAP kinases of the ERK type (Beretta et al., 1995).

The microtubule-associated protein tau isolated from brains of patients suffering from Alzheimer disease is phosphorylated on specific sites. Phosphorylation serves normally to regulate the binding of tau to microtubules. The same sites become phosphorylated in normal rat brain submitted to heat stress (Papasozomenos and Yuan, 1991) or in heat shocked PC12 cells (Johnson et al., 1993). These sites are phosphorylated *in vitro* by the ERK/MAP kinases (Drewes et al., 1991).

Phosphorylation state of RNA polymerase II largest subunit
in stressed cells

In unstressed cells, two forms of RNA polymerase II coexist in equivalent amounts, they differ in the phosphorylation of the C-terminal domain (CTD) of the largest subunit (Dahmus, 1994; Greenleaf, 1993). The underphosphorylated polymerase binds to the promoters and is phosphorylated upon entry into elongation of transcription (Dahmus, 1994). The polymerase pauses on heat-shock genes in its unphosphorylated form shortly after initiating transcription, but the CTD is phosphorylated upon entry into elongation of transcription (O'Brien et al., 1994; Weeks et al., 1993). In unstressed cells, equal amounts of phosphorylated and underphosphorylated subunits are found (Dubois et al., 1994a) resulting from the antagonistic action of the CTD kinases belonging to the general transcription factor TFIIH (Valay et al., 1995) and CTD phosphatases (Chambers et al., 1995). Upon mild heat-shocks, the polymerase has a tendency to accumulate in an underphosphorylated form (Venetianer et al., 1995). This dephosphorylation corresponds to the loss of TFIIH CTD kinase activity (Dubois et al., in preparation). Upon severe stress however, the hyperphosphorylated form accumulates (Dubois et al., 1994a; Venetianer et al., 1995). Heat-shock markedly enhances the activity of kinases phosphorylating the CTD (Dubois et al., 1994a). The major stress-inducible CTD-kinases are the stress-activated MAP/ERK kinases (Trigon and Morange, 1995; Venetianer et al., 1995). Indeed, the RNA polymerase II largest subunit is a good substrate for the MAP/ERK kinases (Dubois et al., 1994b). Hence, the stress-activated MAP kinases might rescue the inactivation of the TFIIH CTD kinase.

Phosphorylation of the ribosomal subunit S6

Phosphorylation of the S6 protein increases the translation efficiency of mRNAs coding for proteins of the translational apparatus such as the eukaryotic elongation factor, eEF-1α (Stewart and Thomas, 1994). Two S6 kinases, p70^{s6k} and the p85^{s6k}, are activated by mitogenic stimulation (Kozma and Thomas, 1994). The p90^{rsk} (or p85^{s6k}) is activated by the ERK/MAPK, it is thus also referred to as MAPK activated protein kinase 1 (MAPKAPK1). The p70^{s6k} is turned on by a transducing pathway which is specifically inhibited by the immunosuppressant rapamycin (Chung et al., 1992). In heat-shocked quiescent fibroblasts, the phosphorylation of S6 increases meanwhile both p70^{s6k} and p90^{rsk} S6 kinases are activated (Jurivich et al., 1991). Unexpectedly, heat shock induces rapid dephosphorylation of the ribosomal subunit protein S6 in growing cells (reviewed in Nover, 1991), but little is known about the S6 phosphatases.

Triggers of the MAP kinase cascades

The heat sensors which trigger the various MAP kinases cascades seem to be located in the plasma membrane. Inducers of the heat shock response also trigger an increase in the intracellular calcium concentration and stimulate enzymes such as phospholipase C and phospholipase A2 involved in triggering signal transduction cascades through the synthesis of second messengers (Calderwood et al., 1993; Kiang and McClain, 1993). The MAP/ERK kinases are activated by phosphorylation on tyrosine and threonine by MAP kinase kinases (MEK or MKK1/2) which can be activated by two distinct pathways, implicating either the Ras protooncogene or tyrosine kinases (Cobb et al., 1994). Interestingly, the activation of heat shock gene transcription is inhibited in cells in which the Ras pathway is constitutively active (Engelberg et al., 1994). Activation of the yeast *slt2/mpk1* cascade involves the PKC-1 gene upstream from the BCK1 and MKK1/MKK2 kinases (Kamada et al., 1995). Membrane stretch or increased membrane fluidity might be the primary event.

The JNK and the p38 MAP kinase are stress-activated by the MEK homologous MKK3 and MKK4 (or SEK1) (Dérijard et al., 1995; Lin et al., 1995; Sánchez et al., 1994). The SEK1 kinase is in turn activated through phosphorylation by MEKK-1 (Yan et al., 1994). The small GTP-binding proteins Rac1 and Cdc42 would be upstream from the JNK cascade (Coso et al., 1995). Interestingly, the JNK cascade is strongly activated by treatment with protein synthesis inhibitors (Cano and Mahadevan, 1995; Kyriakis et al., 1994) hence, its stress activation might be related to the stress inhibition of protein synthesis discussed below. The Hog1p protein is activated by phosphorylation on tyrosine and threonine by the products of the yeast genes *pbs2 (Saccharomyces cere-visiae)* and *wis1* (S. Pombe) which are protein kinase homologous to MEK, the mammalian MAP kinase kinases (reviewed in Levin and Errede, 1995). Two independent signals emanated from distinct cell-surface osmosensors activate the Pbs2p (Maeda et al., 1995). The *sln1* gene product is a histidine kinase upstream from the HOG cascade; it interacts with the *ssk1* gene product thus forming a two-component system involving the transfer of phosphate moieties from histidines to aspartate residues (Swanson et al., 1994). The sln1/ssk1 two-component sensor activates MAPKKKs encoded by genes *ssk2* and *ssk22*. Independently, Pbs2p is activated by binding to the Src homology 3 (SH3) domain of the osmosensor Sho1p. The activity of the *pbs2* and *wis1* gene products is counteracted by over-expression of the protein phosphatases 2C (PP2C) encoded by the *ptc1* gene (Shiozaki and Russel, 1995). Deletion of the *ptc1* gene markedly decreases the survival to a brief exposure to elevated temperature (Shiozaki et al., 1994).

Various kinases are known to associate to hsp90 in the inactive form, some of them dissociate from hsp90 upon heat-shock and become active (see below the case of HRI) (reviewed in Rutherford and Zuker, 1994). The catalytic domain of the raf proteins which are MAP kinase kinase kinases (MAPKKK or MEKK) form a heterocomplex with hsp90 (Stancato et al., 1994). The putative contribution to MAP kinase cascade activation by dissociation from hsp90 is further supported by the fact that the yeast *ptc1* gene is a multicopy suppressor of *swo1-26*, a temperature-sensitive muta-tion of a yeast hsp90 gene (Shiozaki et al., 1994). Overexpression of the

Ptc1p protein phosphatase might counteract the activation of an upstream kinase by thermal dissociation from hsp90.

Some of the MAP kinase cascades have first been presented as stress specific, however it turns out that such specificity depends on the cell system investigated rather than on the inducer.

Phosphorylation of the eukaryotic initiation factors

Protein synthesis is rapidly arrested upon submitting eukaryotic cells to heat-shock. During recovery from stress, the preferential translation of the HSP mRNAs is observed (McGarry and Lindquist, 1985). Changes in the phosphorylation state of various components of the translational apparatus might be involved in these alterations (Duncan and Hershey, 1989). For instance, heat-shock inactivates by dephosphorylation the cap-binding factor, eukaryotic initiation factor 4 (eIF-4E) (Lamphear and Panniers, 1991; Thach, 1992; Zapata et al., 1991). Inactivation of eIF-4E might contribute to the preferential translation of heat shock mRNAs by stressed cells (Joshi-Barve et al., 1992; Rhoads et al., 1993).

Lysates prepared from heat-shocked cells for *in vitro* translation are inactive. This inactivation may in part be due to the phosphorylation of the eukaryotic initiation factor 2 subunit α (eIF-2α) resulting from the stress-activation of the heme-regulated inhibitor (HRI) (De Benedetti and Baglioni, 1986; Farrell et al., 1977). The HRI is a protein kinase which phosphorylates eIF-2α (reviewed in Chen and London, 1995). Phosphorylation of this factor on serine 51 inactivates initiation of translation (Samuel, 1993). Although phosphorylation of eIF-2α is not the sole mechanism responsible for protein synthesis shut-off (Mariano and Siekierka, 1986), expression of a phosphorylation-resistant eIF-2α mutant (serine 51 is replaced by an alanine) partially protects cells from the inhibition of protein synthesis in response to heat-treatment (Murtha-Riel et al., 1993). Noteworthy is that eIF-2α itself is a heat-shock protein (Colbert et al., 1987).

In a reticulocyte lysate, the inactive HRI kinase is bound to hsp90 (Rose et al., 1987), hsp70 and hsp56 (Matts and Hurst, 1992). Addition of denatured proteins to the lysate activates the HRI kinase (Matts et al., 1993). Denatured proteins may bind to the hsp70 present in the lysate and deplete the lysate from "free" hsp70. The dissociation of hsp70 from HRI is strongly correlated with the activation of HRI and suggests that sequestration of hsp70 might be the signal that leads to activation of HRI in response to heat shock.

Since hsp70 chaperones the nascent polypeptide chains (Beckmann et al., 1990), the cellular economy might justify the arrest of protein synthesis when the pool of "free" hsp70 is depleted.

The AMP-activated kinase cascade

Severe heat shocks and other stress which result in ATP concentration drop and increase in 5′-AMP levels activate the mammalian AMP-activated protein kinase (AMPK) (Corton et al., 1994). AMP activates an AMPK kinase which phosphorylates AMPK and acts by direct allosteric activation of the AMPK. AMPK inactivates by phosphorylation HMG-CoA reductase (a key enzyme in sterol/isoprenoïd biosynthesis), acetyl-CoA carboxylase (involved in fatty acid biosynthesis) and glycogen synthase (require for glycogen synthesis). Thus, it may act as an emergency shut-off of energy consuming metabolic reactions (Corton et al., 1994). The catalytic subunit of AMPK is homologous to the yeast *snf1* gene product (Carling et al., 1994). The non-catalytic subunit of AMPK are homologous with proteins that interact with the snf1 protein (Stapleton et al., 1994). The *snf1* gene is essential to derepress the transcription of genes required for breakdown of sucrose upon glucose starvation. In particular, the *snf1* gene and its cofactor *snf4* are required for the transcriptional activation of the metallothionein gene *cup1* upon glucose starvation (Tamai et al., 1994). The homology suggests that despite the *snf1* yeast system does not seem to be activated by AMP, the AMPK cascade might also be involved in the control of gene transcription.

Conclusion

Thermally-induced conformational changes and phosphorylation regulate the activity of proteins controlling the cellular metabolism. Both mechanisms regulate the activity of the heat shock factors, which are key elements controlling the heat-shock response (see chapters by Morimoto, Voellmy). Thus, the HSPs and their constitutive cognates obviously play a central role both in preventing and repairing protein damage, but future studies might emphasize their involvement in the control of the kinase cascades (Rutherford and Zuker, 1994).

A major conclusion that might emerge from this chapter is that several signal transducing pathways triggered by heat stress are also triggered by other stress (Holbrook et al., this volume) and mitogenic stimulations. It is also apparent that a wide variety of factors, chaperones, protein kinases and other modifying enzymes are contributing to the cellular thermoresistance. But, non proteic factors such as small RNAs may also turn out to be important (Fung et al., 1995).

References

Andrews, G.K., Harding, M.A. Calvet, J.P. and Adamson E.D. (1987) The heat shock response in HeLa cells is accompanied by elevated expression of the *c-fos* proto-oncogene. *Mol. Cell. Biol.* 7:3452–3458.

Arnason, T. and Ellison M.J. (1994) Stress resistance in *Saccharomyces cerevisiae* is strongly correlated with assembly of a novel type of multiubiquitin chain. *Mol. Cell. Bio.* 14:7876–7883.

Arrigo, A.P. (1983) Acetylation and methylation patterns of core histones are modified after heat or arsenite treatment of *Drosophila* tissue culture cells. *Nucl. Acid. Res.* 11:1389–1404.

Arrigo, A.P. and Landry, J. (1994) Expression and function of the low-molecular-weight heat shock proteins. *In:* R. Morimoto, A. Tissières and C. Georgopoulos (eds): *The Biology of Heat Shock Proteins and Molecular Chaperones,* Cold Spring Harbor Laboratory Press, Cold Spring Harbor, NY, 335–373.

Beckmann, R.P., Mizzen, L.A. and Welch, W.J. (1990) Interaction of hsp70 with newly synthesized proteins: implications for protein folding and assembly. *Science* 248:850–854.

Beckmann, R.P., Lovett, M. and Welch W.J. (1992) Examining the function and regulation of hsp70 in cells subjected to metabolic stress. *J. Cell Biol.* 117:1137–1150.

Benndorf, R., Hayeß, K., Ryazantsev, S., Wieske, M., Behlke J. and Lutsch, G. (1994) Phosphorylation and supramolecular organization of murine small heat shock protein HSP25 abolish its actin polymerization-inhibiting activity. *J. Biol. Chem.* 269:20780–20784.

Bensaude, O., Pinto, M., Dubois, M.-F., Nguyen, V.T. and Morange, M. (1990) Protein denaturation during heat-shock and related stress. *In:* M.J. Schlesinger and G.Santoro (eds): *Heat Shock Proteins,* Springer Verlag, Berlin, pp 89–99.

Beretta, L., Dubois, M.-F., Sobel, A. and Bensaude, O. (1995) Stathmin, a major substrate for mitogen-activated protein kinase during heat shock and chemical stress. *Eur. J. Biochem.* 227:388–395.

Borelli, M.J., Lee, Y.J., Frey, H.E., Ofenstein, J.P. and Lepock, J.R. (1991) Cycloheximide increases the thermostability of proteins in chinese hamster ovary cells. *Biochem. Biophys. Res. Comm.* 177:575–581.

Brewster, J.L., Devaloir, T., Dwyer, N.D., Winter, E. and Gustin, M.C. (1993) An osmosensing signal transduction pathway in yeast. *Science* 259:1760–1763.

Bukh, A., Martinez-Valdez, H., Freedman, J., Freedman, M.H. and Cohen, A. (1990) The expression of c-fos, c-jun, and c-myc genes is regulated in human lymphoid cells. *J. Immunol.* 144:4835–4840.

Calderwood, S.K., Stevenson, M.A. and Price, B.D. (1993) Activation of phospholipase C by heat shock requires GTP analogs and is resistant to pertussis toxin. *J. Cell Physiol.* 156:153–159.

Cano, E. and Mahadevan, L.C. (1995) Parallel signal processing among mammalian MAPKs. *TIBS* 20:117–122.

Caplan, A.J., Tsai, J., Casey, P.J. and Douglas, M.G. (1992) Farnesylation of YDJ1 is required for function at elevated growth temperatures in *Saccharomyces cerevisiae. J. Biol. Chem.* 267:18890–18895.

Carling, D., Aguan, K., Woods, A., Verhoeven, A.J.M., Beri, R.K., Brennan, C.H., Sidebotton, C., Davison, M.D. and Scott, J. (1994) Mammalian AMP-activated protein kinase is homologous to yeast and plant protein kinases involved in the regulation of carbon metabolism. *J. Biol. Chem.* 269:11442–11448.

Chambers, R.S., Wang, B.Q., Burton, Z.F. and Dahmus, M.E. (1995) The activity of COOH-terminal domain phosphatase is regulated by a docking site on RNA polymerase II and by the general transcription factors IIF and IIB. *J. Biol. Chem.* 270:14962–14969.

Chen, J.-J. and London, I.M. (1995) Regulation of protein synthesis by heme-regulated eIF-2α kinase. *Trends in Biochem. Sci.* 20:105–108.

Cheng, K.H., Hui, S.W. and Lepock, J.R. (1987) Protection of the membrane calcium adenosine triphosphatase by cholesterol from thermal inactivation. *Cancer Res.* 47:1255–1262.

Cheng, T.-J. and Lai, Y.-K. (1994) Transient increase in vimentin phosphorylation and vimentin-hsc70 association in 9L rat brain tumor cells experiencing heat-shock. *J. Cell. Biochem.* 54:100–109.

Chung, J., Kuo, C.J., Crabtree, G.R. and Blenis, J. (1992) Rapamycin-FKBP specifically blocks growth-dependent activation and signaling. *Cell* 69:1227–1236.

Ciavarra, R.P., Duvall, W. and Castora, F. (1992) Induction of thermotolerance in T cells protects nuclear DNA topoisomerase I from heat stress. *Biochem. Biophys. Res. Comm.* 186:166–172.

Ciavarra, R.P., Goldman, C., Wen, K.-K., Tedeschi, B. and Castora, F.J. (1994) Heat stress induces hsc70/nuclear topoisomerase I complex formation *in vivo*: Evidence for hsc70-mediated, ATP-independent reactivation *in vitro*. *Proc. Natl. Acad. Sci. USA* 91:1751–1755.

Cobb, M.H., Hepler, J.E., Cheng, M. and Robbins, D. (1994) The mitogen-activated protein kinases, ERK1 and ERK2. *Cancer Biology* 5:261–268.

Colbert, R.A., Hucul, J.A., Scorsone, K.A. and Young, D.A. (1987) A subunit of eukaryotic translational initiation factor-2 is a heat-shock protein. *J. Biol. Chem.* 262:16763–16766.

Corton, J.M., Gillespie, J.G. and Hardie, D.G. (1994) Role of the AMP-activated protein kinase in the cellular stress response. *Curr. Biol.* 4:315–324.

Coso, O.A., Chiariello, M., Yu, J.-C., Teramoto, H., Crespo, P., Xu, N., Miki, T. and Gutkind, J.S. (1995) The small GTP-binding proteins Rac1 and Cdc42 regulate the activity of the JNK/SAPK signaling pathway. *Cell* 81:1137–1146.

Craig, E.A. and Gross, C.A. (1991) Is hsp70 the cellular thermometer? *TIBS* 16:135–140.

Csermely, P. and Kahn, C.R. (1991) The 90-kDa heat Shock Protein (hsp-90) possesses an ATP binding site and autophosphorylating activity. *J. Biol. Chem.* 266:4943–4950.

Cuenda, A., Rouse, J., Doza, Y.N., Meier, R., Cohen, P., Gallagher, T.F., Young, P.R. and Lee, J.C. (1995) SB 203580 is a specific inhibitor of a MAP kinase homologue which is stimulated by cellular stresses and interleukin-1. *FEBS Lett.* 364:229–233.

Dahmus, M.E. (1994) The role of multisite phosphorylation in the regulation of RNA polymerase II activity. *Prog. Nuc. Ac. Res. & Mol. Biol.* 48:143–179.

DeBenedetti, A. and Baglioni, C. (1986) Activation of Hemin-regulated initiation factor-2 kinase in heat-shocked HeLa cells. *J. Biol. Chem.* 261:338–342.

De Virgilio, C., Hottiger, T., Dominguez, J., Boller, T. and Wiemken, A. (1994) The role of trehalose synthesis for the acquisition of thermotolerance in yeast. I. Genetic evidence that trehalose is a thermoprotectant. *Eur. J. Biochem.* 219:179–186.

Dérijard, B., Hibi, M., Wu, I.-H., Barrett, T., Su, B., Deng, T., Karin, M. and Davis, R.J. (1994) JNK1: A protein kinase stimulated by UV light and Ha-Ras that binds and phosphorylates the c-jun activation domain. *Cell* 76:1025–1037.

Dérijard, B., Raingeaud, J., Barrett, T., Wu, I.-H., Han, J., Ulevitch, R.J. and Davis, R.J. (1995) Independent human MAP kinase signal transduction pathways defined by MEK and MKK isoforms. *Science* 267:682–685.

Desrosiers, R. and Tanguay, R.M. (1988) Methylation of *Drosophila* histones at proline, lysine, and arginine residues during heat shock. *J. Biol. Chem.* 263:4686–4692.

Dewey, W.C. (1989) The search for critical cellular targets damaged by heat. *Rad. Res.* 120:191–204.

Dey, N.B., Bounelis, P., Fritz, T.A., Bedwell, D.M. and Marchase, B. (1994) The glycosylation of phosphoglucomutase is modulated by carbon source and heat shock in *Saccharomyces cerevisiae*. *J. Biol. Chem.* 269:27143–27148.

Drewes, G., Lichtenberg-Kraag, B., Döring, F., Mandelkow, E.-M., Biernat, J., Goris, J., Dorée, M. and Mandelkow, E. (1992) Mitogen activated protein (MAP) kinase transforms tau protein into an Alzheimer-like state. *EMBO J.* 11:2131–2138.

Dubois, M.-F., Galabru, J., Lebon, P., Safer, B. and Hovanessian, A.G. (1989) Reduced activity of the interferon-induced double-stranded RNA-dependent protein kinase during a heat shock stress. *J. Biol. Chem.* 264:12165–12171.

Dubois, M.-F., Hovanessian, A.G. and Bensaude, O. (1991) Heat-shock-induced denaturation of proteins. Characterization of the insolubilization of the interferon-induced p68 kinase. *J. Biol. Chem.* 266:9707–9711.

Dubois, M.-F. and Bensaude, O. (1993) MAP kinase activation during heat shock in quiescent and exponentially growing mammalian cells. *FEBS Lett.* 324:191–195.

Dubois, M.-F., Bellier, S., Seo, S.-J. and Bensaude, O. (1994a) Phosphorylation of the RNA polymerase II largest subunit during heat-shock and inhibition of transcription in HeLa cells. *J. Cell. Physiol.* 158:417–426.

Dubois, M.-F., Nguyen, V.T., Dahmus, M.E., Pagès, G., Pouysségur, J. and Bensaude, O. (1994b) Enhanced phosphorylation of the C-terminal domain of RNA polymerase II upon serum stimulation of quiescent cell: possible involvement of MAP kinases. *EMBO J.* 13:4787–4797.

Duncan, R.F. and Hershey, J.W.B. (1989) Protein synthesis and protein phosphorylation during heat stress, recovery, and adaptation. *J. Cell. Biol.* 109:1467–1481.

Edington, B.V., Whelan, S.A. and Hightower, L.E. (1989) Inhibition of heat shock (stress) protein induction by deuterium oxide and glycerol: additional support for the abnormal protein hypothesis of induction. *J. Cell. Physiol.* 139:219–228.

Engelberg, D., Zandi, E., Parker, C.S. and Karin, M. (1994) The yeast and mammalian Ras pathways control transcription of heat shock genes independently of heat shock transcription factor. *Mol. Cell. Biol.* 14:4929–4937.

Farrell, P.J., Balkow, K., Hunt, T. and Jackson, R.J. (1977) Phosphorylation of initiation factor eIF-2 and the control of protein synthesis. *Cell.* 11:187–200.

Finley, D., Bartel, B. and Varshavsky, A. (1989) The tails of ubiquitin precursors are ribosomal proteins whose fusion to ubiquitin facilitates ribosome biogenesis. *Nature* 338:394–401.

Freshney, N.W., Rawlinson, L., Guesdon, F., Jones, E., Cowley, S., Hsuan, J. and Saklavatla, J. (1994) Interleukin-1 activates a novel protein kinase cascade than results in the phosphorylation of Hsp27. *Cell* 78:1039–1049.

Fung, P.A., Gaertig, J., Gorovsky, M.A. and Hallberg, R.L. (1995) Requirement of a small cytoplasmic RNA for the establishment of thermotolerance. *Science* 268:1036–1039.

Galcheva-Gargova, Z., Dérijard, B., Wu, I.-H. and Davis, R.J. (1994) An osmosensing signal transduction pathway in mammalian cells. *Science* 265:806–808.

Georgopoulos, C. and Welch, W.J. (1993) Role of the major heat shock proteins as molecular chaperones. *Annu. Rev. Cell Biol.* 9:601–634.

Gounalaki, N. and Thireos, G. (1994) Yap1, a yeast transcriptional activator that mediates multidrug resistance, regulates the metabolic stress response. *EMBO J.* 13:4036–4041.

Greenleaf, A. (1993) A positive addition to a negative tail's tale. *Proc. Natl. Acad. Sci. USA* 90:10896–10897.

Han, J., Lee, J.-D., Bibbs, L. and Ulevitch, R.J. (1994) A MAP kinase targeted by endotoxin and hyperosmolarity in mammalian cells. *Science* 265:808–811.

Hartl, F.-U., Hlodan, R. and Langer, T. (1994) Molecular chaperones in protein folding: the art of avoiding sticky situations. *TIBS* 19:20–25.

Haveman, J. and Hahn, G.M. (1981) The role of energy in hyperthermia-induced mammalian cell inactivation: a study of the effects of glucose starvation and an uncoupler of oxidative phosphorylation. *J. Cell. Physiol.* 107:237–241.

Hazel, J.R. (1995) Thermal adaptation in biological membranes: is homeoviscous adaptation the explanation? *Annu. Rev. Physiol.* 57:19–42.

Hengge-Aronis, R., Klein, W., Lange, R., Rimmele, M. and Boos, W. (1991) Trehalose synthesis genes are controlled by the putative sigma factor encoded by rpoS and are involved in stationary-phase thermotolerance in *Escherichia coli. J. Bact.* 173:7918–7924.

Henle, K.J., Monson, T.P. and Stone, A. (1990) Enhanced glycosyltransferase activity during thermotolerance development in mammalian cells. *J. Cell. Physiol.* 142:372–378.

Hill, C.S. and Treisman, R. (1995) Transcriptional regulation by extracellular signals: mechanisms and specificity. *Cell* 80:199–211.

Höll-Neugebauer, B., Rudolph, R., Schmidt, M. and Buchner, J. (1991) Reconstitution of a heat shock effect *in vitro*: Influence of GroE on the thermal aggregation of a-glucosidase from yeast. *Biochemistry* 30:11609–11614.

Hottiger, T., De Virgilio, C., Hall, M.N., Boller, T. and Wiemken, A. (1994) The role of trehalose synthesis for the acquisition of thermotolerance in yeast. II. Physiological concentrations of trehalose increase the thermal stability of proteins *in vitro. Eur. J. Biochem.* 219:187–193.

Huot, J., Lambert, H., Lavoie, J., Guimond, A., Houle, F. and Landry, J. (1995) Characterization of 45-kDa/54-kDa HSP27 kinase, a stress-sensitive kinase which may activate the phosphorylation-dependent protective function of mammalian 27-kDa heath-shock protein HSP27. *Eur. J. Biochem.* 227:416–427.

Jaenicke, R. (1991) Protein stability and molecular adaptation to extreme conditions. *Eur. J. Biochem.* 202:715–728.

Jakob, U. and Buchner, J. (1994) Assisting spontaneity: the role of Hsp90 and small Hsps as molecular chaperones. *Trends in Biochem. Sci.* 19:205–211.

Jennissen, H.P. (1995) Ubiquitin and the enigma of intracellular protein degradation. *Eur. J. Biochem.* 231:1–30.

Jentsch, S. (1992) Ubiquitin-dependent protein degradation: a cellular perspective. *Trends Cell. Biol.* 2:98–101.

Jethmalani, S.M., Henle, K.J. and Kaushal, G.P. (1994) Heat shock-induced prompt glycosylation. Identification of P-SG67 as calreticulin. *J. Biol. Chem.* 269:23603–23609.

Johnson, G., Refolo, L.M. and Wallace, W. (1993) Heat-shocked neuronal PC12 cells reveal Alzheimer's disease-associated alterations in amyloïd precursor protein and tau. *Ann. N.Y. Acad. Sci.* 695:194–197.

Joshi-Barve, S., De Benedetti, A. and Rhoads, R.E. (1992) Preferential translation of heat shock mRNAs in HeLa cells deficient in protein synthesis initiation factors eIF-4E and eIF-4γ. *J. Biol. Chem.* 267:21038–21043.

Jurivich, D.A., Chung, J. and Blenis, J. (1991) Heat shock induces two distinct S6 protein kinase activities in quiescent mammalian fibroblasts. *J. Cell. Physiol.* 148:252–259.

Kabakov, A.E. and Gabai, V.L. (1994) Heat-shock proteins maintain the viability of ATP-deprived cells: what is the mechanism? *Trends Cell. Biol.* 4:193–196.

Kabakov, A.E. and Gabai, V.L. (1995) Heat shock-induced accumulation of 70-kDa stress protein (hsp70) can protect ATP-depleted tumor cells from necrosis. *Exp. Cell Res.* 217: 15–21.

Kamada, Y., Sung Jung, U., Piotrowski, J. and Levin, D. (1995) The protein kinase C-activated MAP kinase pathway of *Saccharomyces cerevisiae* mediates a novel aspect of the heat shock response. *Genes & Development* 9:1559–1571.

Kampinga, H.H., Jorritsma, J.B.M. and Konings, A.W.T. (1985) Heat-induced alterations in DNA polymerase activity of HeLa cells and isolated nuclei. Relation to cell survival. *Int. J. Radiat. Biol.* 47:29–40.

Kampinga, H.H. (1993) Thermotolerance in mammalian cells. Protein denaturation and aggregation, and stress proteins. *J. Cell Science* 104:11–17.

Kato, K., Hasegawa, K., Goto, S. and Inaguma, Y. (1994) Dissociation as a result of phosphorylation of an aggregated form of the small stress protein, hsp27. *J. Biol. Chem.* 269: 11274–11278.

Keyse, S.M. and Emslie, E.A. (1992) *Nature* 359:644–646.

Khandjian, E.W. (1995) Heat treatment induces dephosphorylation of pRb and dissociation of T-antigen/pRb complex during transforming infection with SV40. *Oncogene* 10: 359–367.

Kiang, J.G. and McClain, D.E. (1993) Effect of heat shock, (Ca2$^+$)i, and cAMP on inositol tris-phosphate in human epidermoid A-431 cells. *Am. J. Phys.* 264:C1561–C1569.

Knauf, U., Jakob, U., Engel, K., Buchner, J. and Gaestel, M. (1993) Stress- and mitogen-induced phosphorylation of the small heat shock protein Hsp25 by MAPKAP kinase 2 is not essential for chaperone properties and cellular thermoresistane. *EMBO J.* 13:54–60.

Kobayashi, N. and McEntee, K. (1993) Identification of *cis* and *trans* components of a novel heat shock stress regulatory pathway in *Saccharomyces cerevisiae*. *Mol. Cell. Biol.* 13: 248–256.

Konstantinova, M.F., Bogomazova, A.N., Beliaeva, T.N., Bulychev, A.G. and Leont'eva, E.A. (1994) The protein spectrum and thermostability of the lysosomal hydrolases in heat-resistant sublines of chinese hamster cells. *Tsitologiia* 36:1113–1117.

Kozma, S.C. and Thomas, G. (1994) p70^{s6k}/p85^{s6k}: mechanism of activation and role in mitogenesis. *Semin. Cancer. Biol.* 5:255–260.

Kyriakis, J.M., Banerjee, P., Nikolakaki, E., Dai, T., Rubie, E.A., Ahmad, M.F., Avruch, J. and Woodgett, J.R. (1994) The stress-activated protein kinase subfamily of c-jun kinases. *Nature* 369:156–160.

L'Allemain, G. (1994) Deciphering the MAP kinase pathway. *Prog. in Growth Factor Res.* 5:291–334.

Ladino, C.A. and O'Connor, C.M. (1992) Methylation of atypical protein aspartyl residues during the stress response of HeLa cells. *J. Cell. Physiol.* 153:297–304.

Lamphear, B.J. and Panniers, R. (1991) Heat shock impairs the interaction of cap-binding protein complex with 5'mRNA cap. *J. Biol. Chem.* 266:2789–2794.

Laszlo, A. (1992a) The effects of hyperthermia on mammalian cell structure and function. *Cell Prolifer.* 25:59–87.

Laszlo, A. (1992b) The thermoresistant state: protection from initial damage of better repair? *Exp. Cell Res.* 202:519–531.

Laszlo, A., Wright, W. and Roti Roti, J.L. (1992) Initial characterization of heat-induced excess nuclear proteins in HeLa cells. *J. Cell. Physiol.* 151:519–532.

Lavoie, J., Lambert, H., Hickey, E., Weber, L.A. and Landry, J. (1995) Modulation of cellular thermoresistance and actin filament stability accompanies phosphorylation-induced changes in the oligomeric structure of heat shock protein 27. *Mol. Cell. Biol.* 15:505–516.

Lee, J.C., Laydon, J.T., et al. and Young, P.R. (1994) A protein kinase involved in the regulation of inflammatory cytokine biosynthesis. *Nature* 372:739–746.

Lee, Y.J., Zi-Zheng, H. and Corry, P.M. (1991) Effect of tunicamycin on glycosylation of a 50 kDa protein and thermotolerance development. *J. Cell. Physiol.* 149:202–207.

Legagneux, V., Morange, M. and Bensaude, O. (1991) Heat shock increases turnover of 90 kDa heat shock protein phosphate groups in HeLa cells. *FEBS Lett.* 291:359–362.

Lepock, J.R., Frey, H.E., Heynen, M.P., Nishio, J., Waters, B., Ritchie, K.P. and Kruuv, J. (1990) Increased thermostability of thermolerant CHL V769 cells as determined by differential scanning calorimetry. *J. Cell. Physiol.* 142:628–634.

Levin, D. E. and Errede, B. (1995) The proliferation of MAP kinase signaling pathways in yeast. *Curr. Opin. Cell Biol.* 7:197–202.

Li, C. and Clarke, S. (1992) A protein methyltransferase specific for altered aspartyl residues is important in *Escherichia coli* stationary-phase survival and heat-shock resistance. *Proc. Natl. Acad. Sci. USA* 89:9885–9889.

Li, G.C., Li, L., Liu, R.Y., Rehman, M. and Lee, W.M.F. (1992) Heat shock protein hsp70 protects cells from thermal stress even after deletion of its ATP-binding domain. *Proc. Natl. Acad. Sci. USA* 89:2036–2040.

Lin, A., Minden, A., Martinetto, H., Claret, F.-X., Lange-Carter, C., Mercurio, F., Johnson, G.L. and Karin, M. (1995) Identification of a dual specificity kinase that activates the jun kinases and p38-Mpk2. *Science* 268:286–290.

Maeda, T., Takekawa, M. and Saito, H. (1995) Activation of yeast PBS2 MAPKK by MAPKKKs or by binding of an SH3-containing osmosensor. *Science* 269:554–557.

Maresca, B. and Kobayashi, G. (1993) Changes in membrane fluidity modulate heat shock gene expression and produced attenuated strains in the dimorphic fungus *Histoplasma capsulatum. Arch. Med. Res.* 24:247–249.

Mariano, T. and Siekierka, J. (1986) Inhibition of HeLa cell protein synthesis under heat-shock conditions in the absence of initiation factor eIF-2α phosphorylation. *Biochem. Biophys. Res. Comm.* 138:519–525.

Martin, J., Horwich, A.L. and Hartl, F.U. (1992) Prevention of protein denaturation under heat stress by the chaperonin Hsp60. *Science* 258:995–998.

Matsuda, S., Kawasaki, H., Moriguchi, T., Gotoh, Y. and Nishida, E. (1995) Activation of protein kinase cascades by osmotic shock. *J. Biol. Chem.* 270:12781–12786.

Matts, R.L. and Hurst, R. (1992) The relationship between protein synthesis and heat shock proteins levels in rabbit reticulocyte lysates. *J. Biol. Chem.* 267:18168–18174.

Matts, R.L., Hurst, R. and Xu, Z. (1993) Denatured proteins inhibit translation in hemin-supplemented rabbit reticulocyte lysate by inducing the activation of the heme-regulated eIF-2α kinase. *Biochemistry* 32:7323–7328.

Mazzoni, C., Zarzov, P., Rambourg, A. and Mann, C. (1993) The SLT2 (MPK1) MAP kinase homolog is involved in polarized cell growth in *Saccaromyces cerevisiae. J. Cell. Biol.* 123:1821–1833.

McGarry, T.J. and Lindquist, S. (1985) The preferential translation of *Drosophila* hsp70 mRNA requires sequences in the untranslated leader. *Cell* 42:903–911.

Mehlen, P. and Arrigo, P. (1994) The serum-induced phosphorylation of mammalian hsp27 correlates with changes in its intracellular localization and levels of oligomerization. *Eur. J. Biochem.* 221:327–334.

Michels, A., Nguyen, V.T., Konings, A., Kampinga, H.H. and Bensaude, O. (1995) Thermo-stability of a nuclear targeted luciferase expressed in mammalian cells. Destabilizing influence of the intranuclear microenvironment. *Eur. J. Biochem.* 234:382–389.

Miron, T., Vancompernolle, K., Vandecherckhove, J., Wilchek, M. and Geiger, B. (1991) A 25-kD inhibitor of actin polymerization is a low molecular mass heat shock protein. *J. Cell Biol.* 114:255–261.

Mivechi, N.F. and Dewey, W.C. (1985) DNA polymerase α and β activities during the cell cycle and their role in heat radiosensitization in Chinese hamster ovary cells. *Radiat. Res.* 103:337–350.

Murtha-Riel, P., Davies, M.V., Scherer, B.J., Choi, S.-Y., Hershey, J.W.B. and Kaufman, R.J. (1993) Expression of a phosphorylation-resistant eukaryotic initiation factor 2α-subunit mitigates heat shock inhibition of protein synthesis. *J. Biol. Chem.* 268:12946–12951.

Nguyen, V.T. and Bensaude, O. (1994) Increased thermal aggregation of proteins in ATP-depleted mammalian cells. *Eur. J. Biochem.* 220:239–246.

Nguyen, V.T., Morange, M. and Bensaude, O. (1989) Protein denaturation during heat shock and related stress. *J. Biol. Chem.* 264:10487–10492.

Nover, L. (1991) *Heat Shock Response.* CRC Press, Boca Raton.

O'Brien, T., Hardin, S., Greenleaf, A. and Lis, J.T. (1994) Phosphorylation of RNA polymerase II C-terminal domain and transcriptional elongation. *Nature* 370:75–77.

Papasozomenos, S.C. and Yuan, S. (1991) Altered phosphorylation of t protein in heat-shocked rats and patients with Alzheimer disease. *Proc. Natl. Acad. Sci. USA* 88:4543–4547.

Parsell, D.A. and Lindquist, S. (1993) The function of heat shock proteins in stress tolerance: degradation and reactivation of damaged proteins. *Annu. Rev. Genet.* 27:437–496.

Parsell, D.A., Kowai, A.S., Singer, M.A. and Lindquist, S. (1994) Protein disaggregation mediated by heat-shock protein Hsp104. *Nature* 372:475–478.

Pinto, M., Morange, M. and Bensaude, O. (1991) Denaturation of proteins during heat shock. *In vivo* recovery of solubility and activity of reporter enzymes. *J. Biol. Chem.* 266:13941–13946.

Raman, B., Ramakrishna, T. and Mohan Rao, C. (1995) Temperature dependent chaperone-like activity of alpha-crystallin. *FEBS Lett.* 365:133–136.

Rhoads, R.E., Joshi-Barve, S. and Rinker-Schaeffer, C. (1993) Mechanism of action and regulation of protein synthesis initiation factor 4E: effects on mRNA discrimination, cellular growth rate, and oncogenesis. *Prog. Nucleic Acids Res.* 46:183–220.

Ritossa, F. (1962) A new puffing pattern induced by heat shock and DNP in *Drosophila. Experientia* 18:571–573.

Rose, D.W., Wettenhall, R.E.H., Kudlicki, W., Kramer, G. and Hardesty, B. (1987) The 90-kilodalton peptide of the heme-regulated eIF-2α kinase has sequence similarity with the 90-kilodalton heat shock protein. *Biochemistry* 26:6583–6587.

Rouse, J., Cohen, P., Trigon, S., Morange, M., Alonso-Llamazares, A., Zamanillo, D., Hunt, T. and Nebreda, A.R. (1994) A novel kinase cascade triggered by stress and heat shock that stimulates MAPKAP Kinase-2 and phosphorylation of the small heat shock proteins. *Cell* 78:1027–1037.

Rutherford, S.L. and Zuker, C.S. (1994) Protein folding and the regulation of signalling pathways. *Cell* 79:1129–1132.

Samuel, C.E. (1993) The eIF-2a protein kinases, regulators of translation in eukaryotes from yeasts to humans. *J. Biol. Chem.* 268:7603–7606.

Sànchez, I., Hughes, R.T., Mayer, B.J., Yee, K., Woodgett, J.R., Avruch, J., Kyriakis, J.M. and Zon, L.I. (1994) Role of SAPK/ERK kinase-1 in the stress-activated pathway regulating transcription factor c-jun. *Nature* 372:794–798.

Schröder, H., Langer, T., Hartl, F.-H. and Bukau, B. (1993) DnaK, DnaJ and GrpE form a cellular chaperone machinery capable of repairing heat-induced protein damage. *EMBO J.* 12:4137–4144.

Schüller, C., Brewster, J.L., Alexander, M.R., Gustin, M.C. and Ruis, H. (1994) The HOG pathway controls osmotic regulation of transcription via the stress response element (STRE) of the *Saccharomyces cerevisia* CTT1 gene. *EMBO J.* 13:4382–4389.

Schumacher, R.J., Hurst, R., Sullivan, W.P., McMahon, N.J., Toft, D.O. and Matts, R.L. (1994) ATP-dependent chaperoning activity of reticulocyte lysate. *J. Biol. Chem.* 269:9493–9499.

Sherman, M.Y. and Goldberg, A.L. (1993) Heat shock of *Escherichia coli* increases binding of dnaK (the hsp70 homolog) to polypeptides by promoting its phosphorylation. *Proc. Natl. Acad. Sci. USA* 90:8648–8652.

Sherman, M. and Goldberg, A.L. (1994) Heat shock-induced phosphorylation of GroEL alters its binding and dissociation from unfolded proteins. *J. Biol. Chem.* 269:31479–31483.

Shiozaki, K., Akhavan-Niaki, H., McGowan, C. and Russell, P. (1994) Protein phosphatase 2C, encoded by *ptc1⁺*, is important in the heat shock response of *Schizosaccharomyces pombe. Mol. Cell. Biol.* 14:3742–3751.

Shiozaki, K. and Russell, P. (1995) Counteractive roles of protein phosphatase 2C (PP2C) and a MAP kinase homolog in the osmoregulation of fission yeast. *EMBO J.* 14:492–502.

Siverio, J.M., González, C., Mendoza-Riquel, A., Pérez, M.D. and González, G. (1993) Reversible inactivation and binding to mitochondria of nitrate reductase by heat-shock in the yeast *Hansenula anomala. FEBS Lett.* 318:153–156.

Sobel, A. (1992) Stathmin: a relay phsophoprotein for multiple signal transduction? *Trends in Biochemical Sciences* 16:301–305.

Somero, G.N. (1995) Proteins and temperature. *Annu. Rev. Physiol.* 57:43–68.

Stancato, L.F., Chow, Y.-H., Owens-Grillo, J.K., Yem, A.W., Deibel, M.R., Jove, R. and Pratt, W.B. (1994) The native v-raf·hsp90·p50 heterocomplex contains a novel immunophilin of the FK506 binding class. *J. Biol. Chem.* 269:22157–22161.

Stapleton, D., Gao, G., Michell, B.J., Widmer, J., Mitchelhill, K., Teh, T., House, C.M., Witters, L.A. and Kemp, B.E. (1994) Mammalian 5′AMP-activated protein kinase non-catalytic subunits are homologs of proteins that interact with yeast Snf1 protein kinase. *J. Biol. Chem.* 269:29343–29346.

Stege, G.J.J., Li, L., Kampinga, H.H., Konings, A.W.T. and Li, G.C. (1994) Importance of the ATP-binding domain and nucleolar localization domain of HSP72 in the protection of nuclear proteins against heat-induced aggregation. *Exp. Cell Res.* 214:279–284.

Stewart, M.J. and Thomas, G. (1994) Mitogenesis and protein synthesis: A role for ribosomal protein S6 phosphorylation? *BioEssays* 16:809–815.

Stokoe, D., Engel, K., Campbell, D.G., Cohen, P. and Gaestel, M. (1992) Identification of MAPKAP kinase 2 as a major enzyme responsible for the phosphorylation of the small mammalian heat shock proteins. *FEBS Lett.* 313:307–313.

Swanson, R.V., Alex, L.A. and Simon, M.I. (1994) Histidine and aspartate phosphorylation: two-component systems and the limit of homology. *TIBS* 19:485–490.

Taguchi, H. and Yoshida, M. (1993) Chaperonin from *Thermus thermophilus* can protect several enzymes from irreversible heat denaturation by capturing denaturation intermediate. *J. Biol. Chem.* 268:5371–5375.

Tamai, K.T., Liu, X., Silar, P., Sosinowski, T. and Thiele, D.J. (1994) Heat shock transcription factor activates yeast metallothionein gene expression in response to heat and glucose starvation via distinct signalling pathways. *Mol. Cell. Biol.* 14:8155–8165.

Tanguay, R.M. and Desrosiers, R. (1990) Posttranslational methylation of histones and heat shock proteins in response to heat and chemical stresses. *In:* W.K. Paik and S. Kim (eds): *Protein Methylation.* CRC Press, Boca Raton, pp 139–153.

Thach, R.E. (1992) Cap recap: The involvement of eIF-4F in regulating gene expression. *Cell* 68:177–180.

Torres, L.H., Martin, M.I., Garcia-Saez, J., Arroyo, M., Molina, M., Sanchez, M. and Nombela, C. (1991) A protein kinase gene complements the lytic phenotype of *Saccharomyces cerevisiae lyt2* mutants. *Mol. Microbiol.* 5:2845–2854.

Trigon, S. and Morange, M. (1995) Different carboxyl-terminal domain kinase activities are induced by heat-shock and arsenite. *J. Biol. Chem.* 270:13091–13098.

Valay, J.-G., Simon, M., Dubois, M.-F., Bensaude, O., Facca, C. and Faye, G. (1995) The *KIN28* gene is required both for RNA polymerase II mediated transcription and phosphorylation of the Rpb1p CTD. *J. Mol. Biol.* 249:535–544.

van Delft, S., Coffer, P., Kruijer, W. and van Wijk, R. (1993) C-fos induction by stress can be mediated by the SRE. *Biochem. Biophys. Res. Comm.* 197:542–548.

van Dyk, T.K., Gatenby, A. and LaRossa, R.A. (1989) Demonstration by genetic suppression of interaction of GroE products with many proteins. *Nature* 342:451–453.

van Wijk, R., Welters, M., Souren, J.E.M., Ovelgonne, H. and Wiegant, F.A.C. (1993) Serum-stimulated cell cycle progression and stress protein synthesis in C3H10T1/2 fibroblasts treated with sodium arsenite. *J. Cell. Physiol.* 155:265–272.

Venetianer, A., Dubois, M.-F., Nguyen, V.T., Seo, S.J., Bellier, S. and Bensaude, O. (1995) Phosphorylation state of RNA Polymerase II C-Terminal Domain (CTD) in Heat-Shocked Cells. Possible involvement of the stress activated MAP kinases. *Eur. J. Biochem.* 233:83–92.

Wang, C., Gomer, R.H. and Lazarides, E. (1981) Heat shock proteins are methylated in avian and mammalian cells. *Proc. Natl. Acad. Sci. USA* 78:3531–3535.

Weeks, J.R., Hardin, S.E., Shen, J., Lee, J.M. and Greenleaf, A. (1993) Locus-specific variation in phosphorylation state of RNA polymerase II *in vivo*: correlations with gene activity and transcript processing. *Genes & Dev.* 7:2329–2344.

Welch, W.J. (1992) Mammalian stress response: cell physiology, structure/function of stress proteins, and implications for medicine and disease. *Physiol. Rev.* 72:1063–1081.

Widelitz, R.B., Magun, B.E. and Gerner, E.W. (1986) Effects of cycloheximide on thermotolerance expression, heat shock protein synthesis, and heat shock protein mRNA accumulation in rat fibroblast. *Mol. Cell. Biol.* 6:1088–1094.

Wittliff, J.L., Wenz, L.L., Dong, J., Nawaz, Z. and Butt, T.R. (1990) Expression and characterization of an active human estrogen receptor as a ubiquitin fusion protein from *Escherichia coli. J. Biol. Chem.* 265:22016–22022.

Yan, M., Dai, T., Deak, J.C., Kyriakis, J.M., Zon, L.I., Woodgett, J.R. and Templeton, D.J. (1994) Activation of stress-activated protein kinase by MEKK1 phosphorylation of its activator SEK1. *Nature* 372:798–800.

Zapata, J.M., Maroto, F.G. and Sierra, J.M. (1991) Inactivation of mRNA cap-binding protein complex in *Drosophila melanogaster* embryos under heat shock. *J. Biol. Chem.* 266: 16007–1614.

Ziemienowicz, A., Skowyra, D., Zeilstra-Ryalls, J., Fayet, O., Georgopoulos, C. and Zylicz, M. (1993) Both the *Escherichia coli* chaperone systems, GroEL/GroES and DnaK/DnaJ/GrpE, can reactivate heat treated RNA polymerase. *J. Biol. Chem.* 268:25425–25431.

Ziemienowicz, A., Zylicz, M., Floth, C. and Hübscher, U. (1995) Calf thymus Hsc70 protects and reactivates prokaryotic and eukaryotic enzymes. *J. Biol. Chem.* 270:15479–15484.

Stress-Inducible Cellular Responses
ed. by U. Feige, R.I. Morimoto, I. Yahara and B. Polla
© 1996 Birkhäuser Verlag Basel/Switzerland

SOS response as an adaptive response to DNA damage in prokaryotes

H. Shinagawa

Department of Molecular Microbiology, Research Institute for Microbial Diseases, Osaka University, Suita, Osaka 565, Japan

Summary. *Escherichia coli* possesses an elaborate adaptive mechanism called the "SOS response" to cope with various types of DNA damage. More than 20 SOS genes, most of which are known to be involved in the functions that promote the survival of DNA-damaged cells, are induced by treatments that damage DNA or inhibit DNA synthesis. All the SOS genes share similar sequences in the regulatory regions called the "SOS box", to which LexA repressor binds to repress the transcription in the absence of DNA damage. The SOS signal appears to be the single-stranded DNA produced in vicinity of DNA damage, to which RecA protein binds to be activated as a coprotease. The activated RecA promotes autocleavage of LexA protein by allosteric interaction, which activates the latent serine protease activity of LexA. The induced products of the SOS genes repair DNA lesions by various mechanisms, including recombination, excision repair and error-prone repair, and as the consequence, the SOS signal in the cell decreases and the repression of the SOS genes is restored.

Inducible responses to DNA damage in *Escherichia coli*

Every living organism has evolved elaborate mechanisms to cope with various types of DNA damage from outside and inside the cell. They are best studied in *Escherichia coli* regarding range and depth. Studies in the last four decades revealed that a variety of physiological responses is induced by treatments that damage DNA or arrest DNA synthesis (Walker, 1984; Walker et al., 1995). It has long been known that lytic growth of prophages represented by λ phage is induced by UV radiation and various chemicals that damage DNA. Weigle (1953) demonstrated that UV-irradiated λ phage formed more plaques when plated on *E. coli* cells that were previously UV-irradiated than on unirradiated cells. He also demonstrated that UV-irradiated phages give more mutants when plated on irradiated cells than on unirradiated cells. These results suggest that the abilities to repair DNA damage and to cause mutations in the phage are induced in host cell by UV-irradiation. Subsequent works demonstrated that *de novo* protein synthesis is required for the increase in repair and mutagenic activities, and they not only can target phages but also its host chromosome DNA (Witkin, 1976). Pardee and colleagues (Inouye and Pardee, 1970; Gudas and Pardee, 1975) demonstrated that synthesis of several proteins, including the well-known protein X (RecA), is induced by treatments that damage DNA, and the induction is dependent on the functions of the *recA* and *lexA* genes. Based upon these observations and findings, Radman (1973) proposed that *E. coli* possesses a DNA repair system which is repressed under normal conditions and induced by various DNA lesions, and coined the term SOS (distress signal) response for such repair system.

Regulatory mechanisms of the SOS regulon in *E. coli*

Prophage induction and repressor inactivation

Insight into the molecular mechanism of SOS regulation came from the studies on induction mechanism of λ prophage. Prophage stage of λ is maintained by the active cI repressor which represses the transcription at early stage by binding to the operators. It was demonstrated by repressor-operator filter binding assay that repressor activity disappears in lysogenic cells after inducing treatments such as UV-irradiation and mitomycin C, and this inactivation depends on the functional *recA* and *lexA* alleles (Shinagawa and Itoh, 1973). Subsequently, Roberts and Roberts (1975) showed that this inactivation is brought about by proteolytic cleavage of the repressor. Since the repressor cleavage was demonstrated *in vitro* in a manner that required RecA protein, ATP and single-stranded DNA, it was believed then that RecA is a DNA- and ATP-activated protease (Roberts et al., 1978; Craig and Roberts, 1981).

LexA repressor and SOS box

Although it had been known for a long time that the induction of the SOS responses depends on the functions of *recA* and *lexA*, the elucidation of their real roles in the regulation had been hampered by the mysteriously complex phenotypes of their mutants (Walker, 1984). Isolation and characterization of three kinds of *lexA* mutants, dominant induction negative (Ind⁻) (Mount et al., 1972), temperature sensitive (Ts) (Mount et al., 1973) and defective (Def) (Pacelli et al., 1979), led to the conclusion that LexA protein is the repressor of the SOS genes. Elucidation of the prophage induction mechanism and cloning of the *recA* and *lexA* genes facilitated the progress in the studies on the regulatory mechanism of the SOS system. *In vitro* cleavage of LexA repressor required RecA protein, ATP and single-stranded DNA as with the cleavage of λ repressor (Little et al., 1980). Based upon these studies, the following model for SOS regulation was proposed (Little and Mount, 1982): LexA is the repressor for all the SOS genes and represses the transcription of SOS genes under normal conditions. Upon DNA-damage, cellular level of SOS signal increases and activates RecA as a protease that cleaves LexA. This model differs from the current model for the SOS regulation (Fig. 1) in that activated RecA is not

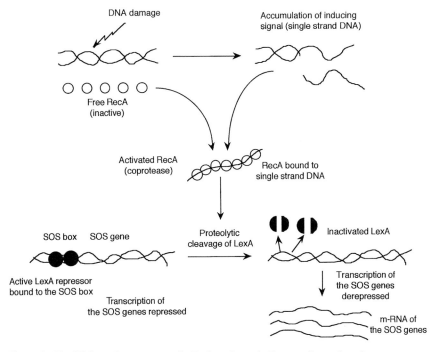

Figure 1. The SOS regulatory system in *Escherichia coli*. See text for explanation.

a protease, but it is considered as a "coprotease" that activates latent protease activity of LexA (see below).

Analysis of the regulatory regions of the SOS genes revealed a consensus sequence for LexA binding sites, which is called SOS box. It consists of an inverted repeat sequence of 20 nucleotides, taCTGtatatatataCTGta, where the sequences shown in capital letters are more highly conserved than those in small letters. Many of these sequences have been shown to be LexA binding sites, by footprinting and/or gel retardation experiments. In many cases, it was shown that the closer the sequence to the consensus, the higher the affinity of SOS box for the LexA repressor. Some SOS boxes such as those in *lexA*, *uvrA*, *uvrB* and *uvrD* bind LexA more weakly than those in *sfiA* and *umuDC*. The former genes are more easily turned on by lower doses of damaging treatments than the latter. This kind of differential expression allows *E. coli* to mobilize subsets of the SOS genes according to the physiological needs (Little, 1983). LexA also regulates the *lexA* and *recA* genes. This autoregulation enables the cell to maintain LexA concentration at a constant level in the cell under normal conditions and to achieve rapid recovery of the repression of the SOS genes after the DNA damage is repaired. The LexA control of the *recA* gene enables accelerated synthesis of RecA protein in the DNA damaged cells; DNA damage activates RecA in the form that promotes autocleavage of LexA, and this situation enhances synthesis of RecA further.

Nature of SOS signal

In vitro LexA cleavage requires RecA protein, ATP (or a nonhydrolyzable analog like ATP-γS) and single-stranded DNA. However, various treatments that damage DNA or inhibit DNA replication induce SOS genes *in vivo* (Walker, 1984). A large body of evidence suggests that mere presence of lesions in DNA is not sufficient to produce SOS-inducing signal in the cell. DNA replication of the damaged template is required to generate the SOS-inducing signal (Sassanfar and Roberts, 1990). Since DNA replication on damaged templates leads to the production of single-stranded DNA gaps (Rupp and Howard-Flanders, 1968), it was proposed that single-stranded DNA, to which RecA protein binds to be activated, is the cellular SOS-inducing signal (Sassanfar and Roberts, 1990). More direct evidence that single-stranded DNA is the SOS signal molecule *in vivo* was provided by the demonstration that the SOS response is induced in *E. coli* by infection with mutant filamentous phage that are defective in mediation of complementary-strand synthesis (Higashitani et al., 1992).

Mechanism of LexA repressor cleavage promoted by RecA protein

It had been believed that RecA protein was a protease which was activated by single-stranded DNA and ATP. Little (1984) accidentally discovered that LexA could be cleaved at high pH (~10) in the absence of RecA. λ repressor was also shown to be self-cleaved under alkaline conditions. The cleavage sites of these proteins were identical with those promoted by RecA at neutral pH. The cleavage sites of LexA and lambdoid phage repressors were Ala-Gly or Cys-Gly bond, giving the cleaved products of approximately equal sizes. Slilaty and Little (1987) proposed that these repressors were latent serine proteases having serine and lysine as active residues, which were shown to be essential for the activity by site-directed mutagenesis.

RecA does not provide the catalytic center for the cleavage, and thus it may not be a protease in a classical sense. However, it functions as a protein catalyst (enzyme) promoting the cleavage reaction at neutral conditions by allosteric interaction with the repressors. The term coprotease has been proposed for such RecA activity (Little, 1991).

Functions of the SOS genes in processing of DNA damage

Over the last two decades, more than 20 SOS genes have been identified. All of them appear to be involved in the functions that process DNA damage, tolerate the damage or promote survival of DNA damaged cells.

Nucleotide excision repair and recombination repair are two major mechanisms that repair bulky DNA lesions. To the cells that are defective in both repair functions such as *recA uvrA* double mutant, UV dose that produces a single pyrimidine dimer in the chromosome is fatal (Howard-Flanders and Boyce, 1966). *E. coli* developed a mechanism to transiently inhibit cell division when DNA is damaged to earn time for repairing DNA and to avoid producing annucleate cells. In these damage tolerance processes, many SOS genes participate together with non SOS genes (see Table 1).

Homologous recombination and recombination repair

Homologous DNA recombination plays two major biological roles; in mitosis, it functions in repairing DNA damage and, in meiosis, it enhances genetic diversity by reshuffling chromosome segments. The first recombination deficient mutant, *recA*, was highly sensitive to various treatments that damage DNA (Clark and Margulies, 1965), suggesting that recombinations play important roles in repairing DNA damage. This property was sometimes used to isolate recombination deficient mutants. Double-strand breaks can only be repaired by recombination. Daughter strand gaps, which are produced in the viscinity of pyrimidine dimers on the UV-irradiated chromosome ater replication reinitiation (Rupp and Howard-Flanders, 1968) are repaired by homologous recombination (Rupp et al., 1971). The SOS genes known to be involved in homologous recombination are *recA*, *recN*, *ruvA* and *ruvB* (Table 1). Since the putative SOS box of *recQ* previously suggested was not bound by LexA repressor (Lewis et al., 1994), it should no longer be considered as a SOS gene.

A current model of homologous recombination (Kowalczykowski et al., 1994) is shown in Figure 2. DNA molecules with 3′ single-strand regions, which are produced by either combination of helicase-exonuclease activities of RecB,C,D, or the combination of RecQ and/or HelD (helicases) with RecJ and/or RecN (nucleases), appear to be a recombinogenic DNA structure to which RecA binds and initiates homologous pairing and strand exchange with the help of RecF, RecO and RecR, and single-strand DNA binding protein (SSB). Extension of strand exchange reaction to the duplex region leads to the formation of a recombination intermediate called Holliday structure. This structure was proposed to explain gene conversion in fungi by R. Holliday (1964) as a central intermediate of recombination in which two homologous duplexes are linked by a single-stranded crossover.

Until a few years ago, RecA was solely considered to promote branch migration by catalyzing strand exchange reaction. However, RuvA and RuvB proteins, whose synthesis is regulated by the SOS system, promote branch migration more efficiently than RecA, and may play a major role in branch migration that produces heteroduplex DNA (Iwasaki et al., 1992; Tsaneva et al., 1992). RuvA is a Holliday junction-specific DNA binding

Table 1. Functions of the SOS genes and the related genes

Gene	Map location (min)	Regulation by LexA	Function
recA	58	Yes	Repressor cleavage, recombination, SOS mutagenesis
lexA	92	Yes	SOS repressor
ruvA, B	41	Yes	Recombination (strand-exchange at Holliday junctions)
ruvC	41	No	Recombination (Holliday junction resolvase)
recN	58	Yes	Recombination
recQ	85	No	Recombination (DNA helicase)
uvrA	92	Yes	Excision repair
uvrB	18	Yes	Excision repair
uvrC	42	No?	Excision repair
uvrD	86	Yes	Excision repair, recombination (DNA helicase II)
polB	1	Yes	DNA polymerase II
umuD, C	26	Yes	SOS mutagenesis
sulA (*sfiA*)	22	Yes	Inhibition of septum formation
ssb	92	Yes	DNA replication, recombination (single-strand DNA-binding protein)

protein which forms a functional complex with RuvB and targets it to the junction. RuvB possesses a DNA helicase activity and is a motor molecule that catalyzes strand exchange reaction. In the final step of homologous recombination, Holliday junction is resolved by endonuclease to give mature recombinant molecules. The product of *ruvC* gene which is located upstream of the *ruvAB* operon, was recently identified to be Holliday junction-specific endonuclease (Dunderdale et al., 1991; Iwasaki et al., 1991). *ruvC* belongs to a separate operon from the *ruvAB* operon and it is not regulated by the SOS system.

The *ruv* mutants were only mildly defective in recombination. However, when combined with other mutations such as *recG*, *recBCsbcA*, *recBCsbcBC*, the *ruv* derivatives were highly defective in recombination and were extremely sensitive to DNA damage (Otsuji et al., 1974; Lloyd et al., 1984; Lloyd and Buckman, 1991). These studies indicate that there are at least two pathways to process Holliday recombination intermediate, one promoted by Ruv proteins and the other by RecG protein. RecG protein exhibited very similar enzymatic activity to that of the RuvAB complex; it is a Holliday junction-specific helicase that promotes branch migration (Lloyd and Sharples, 1993). No counterpart of RuvC endonuclease which functions in conjunction with RecG has been found. Because under certain conditions,

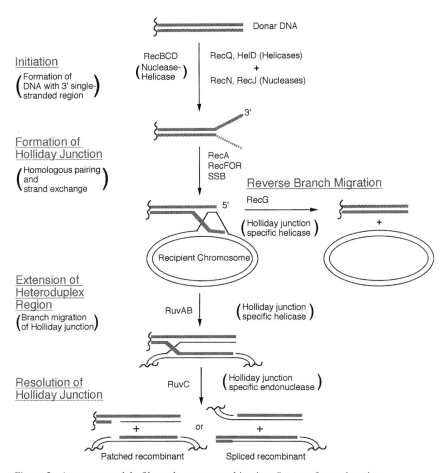

Figure 2. A current model of homologous recombination. See text for explanation.

RecG promoted branch migration in the opposite direction to that promoted by RecA, Lloyd and colleagues (Whithy et al., 1993) proposed that RecG dissociates the Holliday intermediates by promoting reverse branch migration into the prerecombinant molecules. This mechanism does not require the participation of a junction-specific endonuclease for the resolution.

Nucleotide excision repair

Several SOS genes such as *uvrA*, *uvrB* and *uvrD* genes are known to be involved in another important DNA repair mechanism, nucleotide excision repair. In addition to these genes, the products of non-SOS genes, *uvrC polA* and *ligA* (DNA ligase) are known to be essential for the efficient excision repair (Sancar and Sancar, 1988). *uvr* mutants (*uvrA*, *uvrB* and *uvrC*)

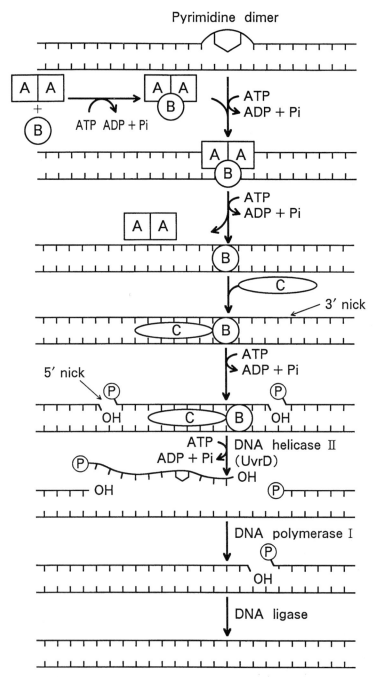

Figure 3. Molecular mechanism of nucleotide excision repair in *Escherichia coli*. Symbols: A A , dimer of UvrA protein; (B), UvrB protein; C , UvrC protein. See text for explanation.

are sensitive to UV-irradiation, various chemicals such as 4-nitroquinoline-1-oxide, N-acetoxy-2-acetylaminofluorene that produce bulky base adducts, and chemicals such as mitomycin C and psoralen that make cross-links in DNA. Major lethal UV-photoproducts are pyrimidine dimers and 6–4 pyrimidine photo adduct.

All of these genes have been cloned, and the products have been purified. The *in vitro* excision repair system has been reconstructed and the roles of individual proteins in the repair reactions have been assigned (Lin and Sancar, 1992) (Fig. 3). UvrA protein forms a complex with UvrB protein, which has not DNA-binding activity, and the complex binds to DNA lesions after sliding on DNA to search for distorted helices. UvrA is released after UvrB forms a complex with DNA. Formation of the UvrA-UvrB complex and the release of UvrA from the UvrB-DNA complex involves ATP hydrolysis. UvrC is incorporated into the UvrB-DNA complex, and incisions are incorporated into four nucleotides to the 3′ side of the lesions by UvrB and eight nucleotides to the 5′ side by UvrC. Release of approximately 12-mers of oligonucleotides is facilitated by the DNA helicase activity of UvrD protein. The resulting single strand gap is refilled by repair DNA synthesis catalyzed by DNA polymerase I, the *polA* gene product, and the remaining nick is sealed by DNA ligase.

Similar to the nucleotide excision repair in eukaryotes, *E. coli* also preferentially repair DNA lesions in the strands actively transcribed (Selby and Sancar, 1993). The protein required for this activity (TRCF: transcription-repair-coupling factor) was isolated from extracts of *E. coli* and it was shown to be encoded by a gene previously identified as *mfd* (mutation frequency decline). Experiments with TRCF, Uvr proteins and RNA polymerase are consistent with the hypothesis that TRCF removes RNA polymerase arrested at the DNA lesions and recruits the UvrAB complex to the lesions. This "stalled RNA polymerase" mechanism facilitates the preferential repair of the DNA strands actively transcribed and unequally reduces the mutation frequency in relation to the transcribed strands.

SOS mutagenesis

As mentioned earlier, UV rays and most chemically induced mutagenesis is an active process involving proteins whose synthesis is regulated by the SOS system. It was discovered that *recA* mutants are defective in UV-induced mutagenesis (Miura and Tomizawa, 1968). Subsequent search for the mutants defective in this process identified mutations in *recA, lexA* and *umu* genes; the former were thought to play regulatory roles (Kato and Shinoura, 1977). *recA* plays more than a regulatory role, since the *recA* activity is still required for the mutagenesis in a SOS constitutive *lexA* (Def) strain (Ennis et al., 1985). The *umu* locus was cloned and found to contain two genes, *umuD* and *umuC* that constitute SOS regulated operon

(Elledge and Walker, 1983; Shinagawa et al., 1983). Mutations induced by chemical reagents such as ethyl methanesulfonate and nitrosoguanidine that produce mispairing lesions are largely independent of the SOS functions (Kato et al., 1982). The current preferred view is that the SOS mutations are caused by translesion DNA synthesis in which DNA polymerase inserts wrong nucleotides leading to a misinstructional or non-instructional DNA synthesis (Echols and Goodman, 1991).

The second role of RecA protein in the SOS mutagenesis other than its role in inducing the expression of the *umu* operon and *recA* itself is to promote proteolytic activation of UmuD protein by its coprotease activity. Unlike the cleavage of LexA and phage repressors, cleavage removing the N-terminal 24 amino acids of UmuD activates it into a mutagenesis competent form (Burchhardt et al., 1988; Nohmi et al., 1988; Shinagawa et al., 1988). The direct involvement of RecA in the translesion DNA synthesis was suggested as the third role of RecA in mutagenesis, since the SOS mutation was not induced in the absence of the RecA function even under the conditions that allow the expression of the SOS genes and provide the processed UmuD. Among the three DNA polymerases in *E. coli*, only DNA polymerase III is required for translesion error-prone DNA replication. It has been proposed that RecA recognizes and binds to DNA lesions, recuirts UmuDC protein complex to the lesion and facillitates the translesion DNA synthesis by DNA polymerase III (Echols and Goodman, 1991). The exact roles of RecA, UmuD and UmuC in the mutagenesis remain largely to be elucidated. There appears to be two pathways for mutagenesis induced by x-rays and UV, one dependent on the UmuDC functions and the other by the functions of RuvAB (Sargentini and Smith, 1989).

Inducible stable DNA replication

Initiation of a new cycle of DNA replication requires protein synthesis. In other word, in the absence of protein synthesis, DNA replication can be completed, however, a new cycle of replication can not be initiated. Many cycles of replication are known to be initiated in the absence of protein synthesis under two different conditions (Asai and Kogoma, 1994), one in the strain lacking RNase H1 (constitutive stable DNA replication; cSDR) and the other in SOS induced cells (inducible stable DNA replication; iSDR). Both SDR is initiated from the sites different from the normal replication initiation site (oriC), and they do not require the initiator protein DnaA.

iSDR requires the function of RecA and RecBCD proteins suggesting that the requirement of recombination functions. Asai and Kogoma (1994) proposed that iSDR requires the formation of D-loop, which facilitates the opening of DNA duplex, and thus promotes the entry of primosome composed of PriA, PriB, DnaB, DnaC, and other proteins. Although PriA and

PriB proteins are required for the replication initiation of φx174 phage, they are dispensable for the cell viability. It has been a mystery that *E. coli* encodes proteins essential for the parasite but not necessary for itself. The finding that PriA is essential for both types of SDR should lead to the elucidation of important roles of the primosomal proteins in the replication initiation bypass mechanisms (Masai et al., 1994). Although iSDR is more resistant to DNA damage and mutagenic, it does not require the *umu* function (Witkin and Kogoma, 1984). iSDR is a stress-inducible function which could play important roles in survival and in increase of genetic diversity under the stress conditions.

SOS-induced inhibition of cell division

Progessions of cell division and DNA replication are well coordinated so that cell mass to DNA ratio is kept relatively constant. After exposure to treatments that damage DNA or stop DNA synthesis, *E. coli* cell delays the cell division by transiently inhibiting septum formation, leading to the formation of filamentous cells. A *recA* mutant (*recA441*), in which the coprotease activity of RecA is activated at 42°C, forms long filament at 42°C and loses viability. The *lon* mutants form long filaments without DNA-damaging treatments, and *lon* mutation aggravates filament formation and temperature sensitivity of the *recA441* mutant. Two mutants, *sfiA* (*sulA*) and *sifB* (*sulB*), were isolated that suppress the filament formation of the *rec A441 lon* double mutant and restore the viability at 42°C (George et al., 1975). It is now known that the SOS regulated *sfiA* gene encodes a protein that appears to directly interact with and inhibit the activity of FtsZ, a septum ring-forming protein (Lutkenhaus, 1990). The *sfiB* mutation is a *ftsZ* mutation that is refractory to the inhibition by SfiA protein. The *lon* gene is a stress inducible gene which is positively regulated by RpoH σ factor, and encodes a protease that proteolytically inactivates SfiA protein (Mizusawa and Gottesman, 1983). This reversible inhibition of cell division provides a mechanism to prevent the formation of anucleated cells by inhibiting septum formation while the cells repair DNA damage. Therefore, the regulation of septum formation is maintained by two kinds of stress response regulation, the SOS and heat shock systems.

Conclusion

Other than the SOS genes mentioned above, many SOS genes whose functions are not known have been identified as the damage inducible (din) genes using an *in vivo* gene fusion method (Kenyon and Walker, 1980), or by computer search for the SOS boxes in DNA data bases (Lewis et al., 1994). Extrapolation from the known SOS functions suggests that they may play roles in repairing DNA damage or in the functions related to the damage tolerance. Several genes which were once considered as the SOS genes were reexamined, and some of them like *phr*, *uvrC*, and *recQ*, are no longer considered as the SOS genes.

Since the *recA* and *lexA* genes have been cloned in many species of prokaryotes including gram-positive bacteria, the SOS system appears to be prevalent among prokaryotes. Although expression of the *E. coli recA* gene is inducible by DNA damage in most gram negative bacteria (de Henestrosa et al., 1991), the SOS boxes of gram-positive *Bacillus subtilis* (GAAG-N4-GttC) are substantially different from those of *E. coli* (Yasbin et al., 1992). Furthermore, the SOS genes of *B. subtilis* are also induced when the cells develop a competent state, which is another stressed condition in which cells could incorporate exogenic DNA and be genetically transformed.

Each of the SOS genes are induced differently. The genes which contain SOS boxes that are closest to the consensus sequence have the highest affinity for LexA repressor and they are induced only after the cells undergo severe DNA damage and cellular levels of LexA repressor decrease substantially (Little, 1983). The *sifA* gene which inhibits septum formation and the *umu* genes which are involved in error-prone DNA replication have the SOS boxes with highest affinities for LexA and their functions may be required for the situations of severe DNA damage. The heat shock regulon is also induced by UV light but only at high doses. Under severe conditions, the functions of the two stress responsible regulons may be mobilized.

Products of many SOS genes are involved in more than one function. RecA protein plays central roles in the regulation of the SOS system at transcriptional level by inactivating LexA repressor and at posttranslational level by proteolytically activating the mutator protein UmuD. RecA also plays direct roles in recombination, error-prone DNA replication, and stable DNA replication. The activation of RecA, which is brought about by binding to single strand DNA, is required for these reactions. RuvAB proteins are also required for recombination and mutagenesis. UvrD protein is involved in excision, recombination, and mismatch repairs. There are many SOS genes about whose functions we have no knowledge, and they undoubtedly should play important roles in DNA repair and cellular adaptation to the damage. Various aspects of DNA transactions such as DNA repair, recombination and replication are inseparable processes, and they should be studied as such.

DNA damage inducible systems are being rapidly unveiled in yeasts and animals (Holbrook et al., this volume). Although the regulatory mechanisms of these are quite different from that of *E. coli*, they achieve the same goal, adaptation to DNA damage. RecA and its eukaryotic homolog Rad51 appear to play similar roles in recombination (Shinohara et al., 1992). The two systems may share more common features than would be concluded from their superficial differences. It is important to examine to what degrees the paradigms established in *E. coli* are conserved in eukaryotes, and what new devices eukaryotes have developed during the evolution.

Acknowledgements
I thank Dr. H. Iwasaki for help in drawing the figures and Mr. T. Nishino for correcting the English. The work in our laboratory has been supported by Grants-in-Aids for Scientific Research from the Ministry of Education, Science and Culture of Japan.

References

Asai, T. and Kogoma, T. (1994) D-loops and R-loops: alternative mechanisms for the initiation of chromosome replication in *Escherichia coli. J. Bacteriol.* 176:1807–1812.

Brent, R. and Ptashne, M. (1980) The *lexA* gene product represses its own promoter. *Proc. Natl. Acad. Sci. USA* 77:1932–1936.

Burckhardt, S.E., Woodgate, R., Scheuermann, R.H. and Echols, H. (1988) UmuD mutagenesis protein of *Escherichia coli*: overproduction, purification, and cleavage by RecA. *Proc. Natl. Acad. Sci. USA* 85:1811–1815.

Craig, N.L. and Roberts, J.E. (1981) Function of nucleotide triphosphate and polynucleotide in *Escherichia coli recA* protein-directed cleavage of phage λ repressor. *J. Biol. Chem.* 256:8039–8044.

Clark, A.J. and Margulies, A.D. (1965) Isolation and characterization of recombination-deficient mutants of *Escherichia coli* K-12. *Proc. Natl. Acad. Sci. USA* 53:451–459.

de Henestrosa, A.F., Calero, S. and Barbe, J. (1991) Expression of the *recA* gene of *Escherichia coli* in several species of gram-negative bacteria. *Mol. Gen. Genet.* 226:503–506.

Dunderdale, H.J., Benson, F.E., Parson, C.A., Sharples, G.J., Lloyd, R.G. and West, S.C. (1991) Formation and resolution of recombination intermediates by *E. coli* RecA and RuvC proteins. *Nature* 354:506–510.

Echols, H. and Goodman, M.F. (1991) Fidelity mechanisms in DNA replication. *Annu. Rev. Biochem.* 60:477–511.

Elledge, S.J. and Walker, G.C. (1983) Proteins required for ultraviolet light and chemical mutagenesis: identification of the products of the *umuC* locus of *Escherichia coli. J. Mol. Biol.* 164:175–192.

Ennis, D.G., Fisher, B., Edmiston, S. and Mount, D.W. (1985) Dual role for *Escherichia coli* RecA protein in SOS mutagenesis. *Proc. Natl. Acad. Sci. USA* 82:3325–3329.

Friedberg, E.C., Walker, G.C. and Siede, W. (1995) *DNA Repair and Mutagenesis,* ASM Press, Washington D.C.

George, J., Casstellazzi, M. and Buttin, G. (1975) Prophage induction and cell division in *E. coli* III. Mutations *sfiA* and *sfiB* restore division in *tif* and *lon* strains and permit the expression of mutator properties of *tif. Mol. Gen. Genet.* 140:309–332.

Gudas, L.J. and Pardee, A.B. (1975) Model for regulation of *Escherichia coli* DNA repair functions. *Proc. Natl. Acad. Sci. USA* 72:23330–2334.

Higashitani, N., Higashitani, A., Roth, A. and Horiuchi, K. (1992) SOS induction in *Escherichia coli* by infection with mutant filamentous phage that are defective in initiation of complementary-strand DNA synthesis. *J. Bacteriol.* 174:1612–1618.

Holliday, R. (1964) A mechanism for gene conversion in fungi. *Genet. Res.* 5:282–304.

Horii, T., Ogawa, T. and Ogawa, H. (1980) Organization of the *recA* gene of *Escherichia coli. Proc. Natl. Acad. Sci. USA* 77:313–317.

Howard-Flanders, P. and Boyce, R.P. (1966) DNA repair and genetic recombination: studies on mutants of *Escherichia coli* defective in these processes. *Radiat. Res.* 6:156–184.

Inouye, M. and Pardee, A.B. (1970) Changes of membrane proteins and their relation to DNA synthesis and cell division of *Escherichia coli. J. Biol. Chem.* 245:5813–5819.

Iwasaki, H., Takahagi, M., Shiba, T., Nakata, A. and Shinagawa, H. (1991) *Escherichia coli* RuvC protein is an endonuclease that resolves the Holliday structure. *EMBO J.* 10:4381–4389.

Iwasaki, H., Takahagi, M., Nakata, A. and Shinagawa, H. (1992) *Escherichia coli* RuvA and RuvB proteins specifically interact with Holliday junctions and promote branch migration. *Genes Dev.* 6:2214–2220.

Kato, T. and Shinoura, Y. (1977) Isolation and characterization of mutants of *Escherichia coli* deficient in induction of mutations by ultraviolet light. *Mol. Gen. Gent.* 156:121–131.

Kato, T., Ise, T. and Shinagawa, H. (1982) Mutational specificity of the *umuC* mediated mutagenesis in *Escherichia coli*. *Biochemie* 64:731–733.

Kenyon, C.J. and Walker, G.C. (1980) DNA-damaging agents stimulate gene expression at specific loci in *Escherichia coli*. *Proc. Natl. Acad. Sci. USA* 77:2819–2823.

Kowalczykowski, S.C., Dixon, D.A., Eggleston, A.K., Lauder, S.D. and Rehrauer, W.M. (1994) Biochemistry of homologous recombination in *Escherichia coli*. *Microbiol. Rev.* 58: 401–465.

Lewis, L.K., Harlow, G.R., Gregg-Jolly, L.A. and Mount, D.W. (1994) Identification of high affinity binding sites for LexA which define new DNA damage-inducible genes in *Escherichia coli*. *J. Mol. Biol.* 241:507–523.

Lin, J.J. and Sancar, A. (1992) (A)BC excinuclease: the *Escherichia coli* nucleotide excision repair enzyme. *Mol. Microbiol.* 6:2219–2224.

Little, J. (1991) Mechanism of specific LexA cleavage: autodigestion and the role of RecA coprotease. *Biochimie* 73:411–422.

Little, J.W., Edmiston, S.H., Pacelli, Z. and Mount, D.W. (1980) Cleavage of the *Escherichia coli lexA* protein by the *recA* protease. *Proc. Natl. Acad. Sci. USA* 77:3225–3229.

Little, J.W. and Mount, D.W. (1982) The SOS regulatory system of *Escherichia coli*. *Cell* 29:11–22.

Little, J.W. (1983) The SOS regulatory system: control of its state by the level of *recA* protease. *J. Mol. Biol.* 167:791–808.

Little, J.W. (1984) Autodigestion of *lexA* and phage λ repressors. *Proc. Natl. Acad. Sci. USA* 91:1375–1379.

Lloyd, R.G., Benson F.E. and Shurvinton, C.E. (1984) Effect of *ruv* mutations on DNA repair in *Escherichia coli*. *Mol. Gen. Genet.* 194:303–309.

Lloyd, R.G. and Buckman, C. (1991) Genetic analysis of the *recG* locus of *Escherichia coli* K-12 and of its role in recombination and DNA repair. *J. Bacteriol.* 173:1004–1011.

Lloyd, R.G. and Sharples, G.J. (1993) Dissociation of synthetic Holliday junctions by *E. coli* RecG protein. *EMBO J.* 12:17–22.

Lutkenhause, J. (1990) Regulation of cell division in *E. coli*. *Trends Genet.* 6:22–25.

Masai, H., Asai, T., Kubota, Y., Arai, K. and Kogoma, T. (1994) *Escherichia coli* PriA protein is essential for inducible and constitutive stable DNA replication. *EMBO J.* 13:5338–5345.

Miura, A. and Tomizawa, J. (1968) Studies on radiation-sensitive mutants of *E. coli*. III. Participation of the Rec system in induction of mutation by ultraviolet irradiation. *Mol. Gen. Genet.* 103:1–10.

Mizusawa, S. and Gottesman, S. (1983) Protein degradation in *Escherichia coli*: the *lon* gene controls stability of SulA protein. *Proc. Natl. Acad. Sci. USA* 80:358–362.

Mount, D.W., Low, K.B. and Edmiston, S. (1972) Dominant mutation *(lex)* in *Escherichia coli* K-12 which affect radiation sensitivity and frequency of ultraviolet light-induced mutations. *J. Bacteriol.* 112:886–893.

Mount, D.W., Walker, A.C. and Kosel, C. (1973) Suppression of *lexA* mutations affecting deoxyribonucleic acid repair in *Escherichia coli* K-12 by closely linked thermosensitive muations. *J. Bacteriol.* 116:950–956.

Nohmi, T., Battista, J.R., Dodson, L.A. and Walker, G.C. (1988) RecA-mediated cleavage activates UmuD for mutagenesis: mechanistic relationship between transcriptional derepression and posttranslational activation. *Proc. Natl. Acad. Sci. USA* 85:1816–1820.

Otsuji, N., Iyehara, H. and Hideshima, Y. (1974) Isolation and characterization of an *Escherichia coli ruv* mutant which forms non-septate filaments after low doses of ultraviolet light irradiation. *J. Bacteriol.* 117:337–344.

Pacelli, L.A., Edmiston, S.H. and Mount, D.W. (1979) Isolation and characterization of amber mutations in the *lexA* gene of *Escherichia coli* K-12. *J. Bacteriol.* 137:568–573.

Radman, M. (1974) Phenomenology of an inducible mutagenic DNA repair pathway in *Escherichia coli*: SOS repair hypothesis. *In*: L. Prakash, F. Sherman, M. Miller, C. Lawrence, and H.W. Tabor (eds): *Molecular and Environmental Aspects of Mutagenesis*. Charles, C. Thomas, Springfield, IL., pp 128–142.

Robert, J.W., Roberts, C.W. (1975) Proteolytic cleaveage of bacteriophage λ repressor in induction. *Proc. Natl. Acad. Sci. USA* 72:147–151.

Robert, J.W., Roberts, C.W. and Craig, N.L. (1978) *Escherichia coli recA* gene product inactivates phage λ repressor. *Proc. Natl. Acad. Sci. USA* 755:4714–4718.

Rupp, W.D. and Howard-Flanders, P. (1968) Discontinuities in the DNA synthesized in an excision-defective strain of *Escherichia coli* following ultraviolet irradiation. *J. Mol. Biol.* 31:291–304.

Rupp, W.D., Wilde, C.E.I., Reno, D.L. and Howard-Flanders, P. (1971) Exchanges between DNA strands in ultraviolet-irradiated *Escherichia coli. J. Mol. Biol.* 61:25–44.

Sancar, A. and Sancar, G.B. (1988) DNA repair enzymes. *Annu. Rev. Biochem.* 57:29–67

Sargenini, N.J. and Smith, K.C. (1989) Role of *ruvAB* genes in UV- and γ-radiation and chemical mutagenesis in *Escherichia coli. Mutat. Res.* 215:115–129.

Sassanfar, M. and Roberts, J.W. (1990) Nature of the SOS inducing signal in *Escherichia coli:* the involvement of DNA replication. *J. Mol. Biol.* 212:79–96.

Selby, C.P. and Sancar, A. (1993) Molecular mechanism of transcription-repair coupling factor. *Science* 260:53–58.

Shinagawa, H. and Ito, T. (1973) Inactivation of DNA -binding activity of repressor in extracts of λ-lysogen treated with mitomycin C. *Mol. Gen. Genet.* 126:103–110.

Shinagawa, H., Kato, T., Ise, T., Makino, K. and Nakata, A. (1983) Cloning and characterization of the *umu* operon responsible for inducible mutagenesis in *Escherichia coli. Gene* 23:167–174.

Shinagawa, H., Iwasaki, H., Kato, T. and Nakata, A. (1988) RecA protein-dependent cleavage of UmuD protein and SOS mutagenesis. *Proc. Natl. Acad. Sci. USA* 85:1806–1810.

Shinohara, A., Ogawa, H. and Ogawa, T. (1992) Rad51 protein involved in repair and recombination in *S. cerevisiae* is a RecA-like protein. *Cell* 69:457–470.

Slilaty, S.N. and Little, J.W. (1987) Lysine-156 and serine-119 are required for LexA repressor cleavage: a possible mechanism. *Proc. Natl. Acad. Sci. USA* 84:3987–3991.

Tsneva, I.R., Müller, B. and West, S.C. (1992) ATP-dependent branch migration of Holliday junctions promoted by the RuvA and RuvB proteins of *Escherichia coli. Cell* 69:1171–1180.

Walker, G.C. (1984) Mutagenesis and inducible responses to deoxyribonucleic acid damage in *Escherichia coli. Microbiol. Rev.* 48:60–93.

Weigle, J.J. (1953) Induction of mutation in a bacterial virus. *Proc. Natl. Acad. Sci. USA* 39:628–636.

Whitby, M.C., Ryder, L. and Lloyd, R.G. (1993) Reverse branch migration of Holliday junctions by RecG protein: a new mechanism for resolution of Holliday junctions of intermediates in recombination. *Cell* 75:341–350.

Witkin, E.M. (1976) Ultraviolet mutagenesis and inducible DNA repair in *Escherichia coli. Bacteriol. Rev.* 40:869–907.

Witkin, E.M. and Kogoma, T. (1984) Involvement of the activated form of RecA protein in SOS mutagenesis and stable DNA replication in *Escherichia coli. Proc. Natl. Acad. Sci. USA* 81:7539–7543.

Yabsin, R.E., Cheo, D.L. and Bayles, K.W. (1992) Inducible DNA repair and differentiation in *Bacillus subtilis:* interactions between global regulons. *Mol. Microbiol.* 6:1263–1270.

Part III
Cellular responses to specific stresses

Introduction

B. S. Polla

Laboratoire de Physiologie Respiratoire, Université Paris 5, UFR Cochin Port-Royal, 24, rue du Faubourg Saint-Jacques, F-75014 Paris, France

Specific stresses induce specific as well as overlapping stress responses. Oxidants, UV radiation, metals and lymphotoxin share the fact that they play dual roles in biological systems, on the one hand serving as key cellular activators or chemical cofactors, raising essential second messengers, and on the other hand leading potentially to DNA damage and cell death by apoptosis or necrosis. Low-dose oxidants activate cell proliferation and contribute to antiinfectious defense Low-dose metals and low-dose tumor necrosis factor are essential in the regulation of proteins involved in antioxidant defenses such as SOD and metallothionins, respectively, while their toxic effects at high concentrations will overcome the induced protective mechanisms. Genotoxic agents in general evoke a series of events leading to the activation of a wide group of genes with various functions among which many are also rapidly induced by mitogenic stimulation. UV radiation and mitogens for example induce both shared and specific kinase phosphorylation cascades and similarly enhance both shared and specific gene expression. To prevent DNA damage, the cells have evolved a number of protective mechanisms including DNA repair, cell cycle arrest and apoptosis and the activation of p53 may play a pivotal role in the type of responses observed following DNA damage.

Oxidants induce HSP and SOD in both prokaryotes and eukaryotes, while the respective transcription factors are, in contrast, species-specific. Furthermore, SOS responses, which confer marked protection against a second exposure to genotoxic stress in bacteria, have not been found in mammalian cells, which might suggest a premium placed on survival in the former, genome stability being more critical in complex organisms such as mammals.

Each of the chapters of this part illustrates some aspects of the stress- and species-specificity of the cellular responses to a given stress as well as the overlapping of these responses.

Stress-Inducible Cellular Responses
ed. by U. Feige, R.I. Morimoto, I. Yahara and B. Polla
© 1996 Birkhäuser Verlag Basel/Switzerland

Transcriptional regulators of oxidative stress-inducible genes in prokaryotes and eukaryotes

G. Storz[1] and B. S. Polla[2]

[1] *Cell Biology and Metabolism Branch, National Institute of Child Health and Human Development, Bethesda, MD 20892, USA*
[2] *Laboratoire de Physiologie Respiratoire, UFR Cochin Port-Royal, Université Paris V, 24, rue du Faubourg Saint-Jacques, F-75014 Paris, France*

Summary. It appears that redox regulation is an important mechanism for the control of transcription factor activation. The role of oxidation-reduction is probably determined in part by the structure of the transcription factors. For example, the presence of cysteine residues within the DNA binding sites may sensitize a transcription factor to ROS. The ROS-mediated regulation of transcription factors is specific, some ROS are more efficient than other ROS in activating defined regulators. While the protective antioxidant responses induced by ROS in prokaryotes and eukaryotes are rather conserved (for example, SOD, HSP…), the regulators for these genes do not appear to be conserved. Further studies designed to fully characterize these regulators and understand the subtle mechanisms involved in redox gene regulation are ongoing, and should provide the theoretical basis for clinical approaches using antioxidant therapies in human diseases in which oxidative stress is implicated.

Introduction

Reactive oxygen species (ROS) are generated from molecular oxygen and include the free radicals superoxide ($\cdot O_2^-$), hydroxyl ($\cdot OH$) and nitric oxide ($NO\cdot$), as well as non-radical intermediates such as hydrogen peroxide (H_2O_2) and singlet oxygen ($^1O_2$). During normal cellular respiration, ROS are constantly produced at low rate in both eukaryotes and prokaryotes. At these low concentrations, ROS can act as second messengers, stimulate cell proliferation, and act as mediators for cell activation (Fridovich, 1978). However, during phagocytosis, infection or inflammation, or upon exposure to environmental stresses such as ionizing or non-ionizing radiations or metals, ROS can accumulate to toxic levels. At these elevated concentrations, ROS lead to oxidative stress, and may damage almost all cellular components (Fridovich, 1978; Halliwell and Gutteridge, 1990).

All organisms have mechanisms to detoxify the oxidants or to repair the damage caused by ROS, including superoxide dismutases, catalases, peroxidases, glutathione, thioredoxin, and heat shock proteins, which are quite conserved from prokaryotes to eukaryotes. The expression of the genes coding these proteins (oxidative stress genes) is induced by changes in the concentrations of ROS, suggesting that cells have developed mechanisms to sense the ROS. However, while there is an extensive similarity in the oxidative stress genes in prokaryotes and in eukaryotes, the regulators are quite different.

In this chapter, we review the properties of transcriptional regulators that are important for the induction of antioxidant defense genes in bacteria, yeast and mammalian cells. We also discuss the roles of these responses in disease and in the interactions between prokaryotes and eukaryotes.

Transcriptional regulators for oxidative stress inducible genes in prokaryotes

The adaptive responses to oxidative stress have been best characterized in *Escherichia coli* and *Salmonella typhimurium* (recently reviewed in Kullik and Storz, 1994; Hidalgo and Demple, 1995). The regulation of these responses occurs mainly on the level of transcription and several key transcription factors have been characterized (see Fig. 1).

OxyR

The OxyR protein is a transcriptional activator of genes induced by H_2O_2 including *katG* (hydroperoxidase I), *ahpCF* (an alkyl hydroperoxide

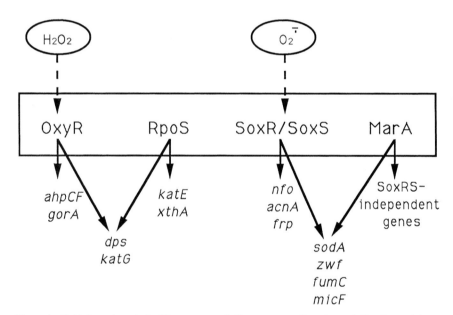

Figure 1. Oxidative stress-inducible genes and the corresponding transcriptional regulators in prokaryotes. The transcriptional regulators are indicated in the box. Genes induced by oxidative stress include *ahpCF* (an alkyl hydroperoxide reductase), *gorA* (glutathione reductase), *dps* (a non-specific DNA binding protein), *katG* (hydroperoxidase I), *kathE* (hydroperoxidase II), *xthA* (exonuclease III), *nfo* (endonuclease IV), *acnA* (aconitase), *frp* (ferredoxin reductase), *sodA* (MnSOD), *zwf* (glucose-6-phosphate dehydrogenase), *fumC* (fumarase C) and *micF* (an antisense RNA).

reductase), *dps* (a non-specific DNA binding protein), and *gorA* (gluta-thione reductase), all of which encode proteins likely to protect the cell against H_2O_2 (Kullik and Storz, 1994), OxyR also acts as a repressor of its own expression and the expression of the Mu phage *mom* gene. The OxyR protein is 34 kDa and shares homology to the LysR family of bacterial regulators.

The OxyR protein has been purified and characterized *in vitro*. Transcription assays with purified components showed OxyR exists in an oxidized and a reduced form, but only oxidized OxyR is able to activate transcription (Storz et al., 1990). Therefore, OxyR appears to be both the sensor and the transducer of the H_2O_2-stress signal and undergoes a conformational change upon oxidation which allows it to activate transcription. Oxidized OxyR binds to the promoters of its target genes, but the regions bound by OxyR are unusually long, covering 45 bp. Recent studies have shown that a consensus motif of four ATAGxt elements can be defined for oxidized OxyR, and the tetrameric protein contacts these elements by binding in four adjacent major grooves (Toledano et al, 1994). Reduced OxyR only binds to a subset of the sites bound by oxidized OxyR and contacts two pairs of adjacent major grooves separated by one helical turn. The oxidized OxyR protein has a cooperative effect on RNA polymerase binding and most likely activates transcription by recruiting RNA polymerase to the promoters (Tao et al., 1993).

The redox-reactive center of OxyR is postulated to be a cysteine residue at position 199, possibly oxidized to a sulfenic acid, since site-directed mutagenesis showed that Cys199 was critical for activity and several constitutive mutants isolated by random mutagenesis map to amino acids adjacent to Cys199 (Kullik et al., 1995a, b). However, biochemical experiments are needed to test for the presence of a sulfenic acid and to determine the state of OxyR *in vivo*. It is interesting to note that while OxyR is strongly activated by H_2O_2 in exponentially growing cells, the protein does not appear to be activated during stationary phase (Altuvia et al., 1994; Gonzalez-Flecha and Demple, 1995).

SoxR/SoxS

The two regulators required for the adaptive response to $\cdot O_2^-$ in *E. coli* are SoxR and SoxS. SoxR/SoxS-regulated genes include *sodA* (manganese superoxide dismutase, MnSOD), *nfo* (DNA repair endonuclease IV), *zwf* (glucose-6-phosphate dehydrogenase), *fumC* (fumarase C), *acnA* (aconitase), *fpr* (ferredoxin reductase), or *micF* (an antisense RNA regulator). SoxR codes for a 17 kDa protein with homology to the mercury-dependent MerR regulator, and the 13 kDa SoxS protein shares homology with the AraC family of transcriptional regulators.

In vivo studies have shown that the activation of target genes by SoxRS occurs in two steps. First, the SoxR protein, directly or indirectly activated upon exposure to $\cdot O_2^-$, induces expression of the SoxS protein. Then, SoxS binds to the promoters of the different target genes and induces their transcription. Therefore, SoxR is the redox-sensitive regulator and SoxS amplifies the signal in this cascade. Two forms of the SoxR protein have recently been purified, an apoprotein and an ironsulfur cluster containing form (Fe-SoxR) (Hildalgo and Demple, 1994; Wu et al., 1995). Both forms of SoxR can bind to the *soxS* promoter, but only the Fe-SoxR can activate transcription *in vitro*. The Fe-SoxR protein most likely acts to increase *soxS* transcription by facilitating promoter opening, most likely by distorting the *soxS* promoter (Hidalgo et al., 1995). The SoxS protein has also recently been purified and was shown to bind to the *micF, zwf, nfo,* and *sodA* promoters in regions found to be required for activation by SoxS (Fawcett and Wolf, 1994a, b; Li and Demple, 1994). SoxS activates transcription by facilitating promoter binding by RNA polymerase (Li and Demple, 1994).

The SoxR protein has been shown to contain four cysteine residues near the carboxy terminus and two {2Fe-2S} clusters per protein dimer (Hidalgo et al., 1995; Wu et al., 1995). Several models could explain how SoxR is modified upon exposure to $\cdot O_2^-$. Possibly, the SoxR protein is normally in a reduced {2Fe-2S}$^+$ state and is activated by oxidation to a {2Fe-2S}$^{2+}$ state through direct exposure to $\cdot O_2^-$ or through changes in the levels of NADPH or reduced flavodoxins or ferredoxins (Liochev et al., 1994; Wu et al., 1995). Alternatively, SoxR may exist as an apoprotein and the {2Fe-2S}$^+$ clusters are assembled upon exposure to oxidants (Hidalgo et al., 1995). All models for the activation of SoxR, must take into account the observation that the SoxRS target genes are not only induced by $\cdot O_2^-$ generating compounds but also by the free radical NO$\cdot$ generated by activated macrophages via the inducible NO$\cdot$ synthase (Nunoshiba et al., 1995).

MarA, RpoS, ArcA, Fnr, Fur

Another transcriptional activator important for the *E. coli* defense against $\cdot O_2^-$ generating compounds is MarA (multiple antibiotic resistance) (Hidalgo and Demple, 1995). MarA shows significant homology to the transcriptional regulator SoxS, and increased levels of the MarA protein appear to induce the expression of *sodA, zwf, fumC,* and *micF* as well as some SoxRS-independent genes. The expression of MarA is induced by a variety of compounds including antibiotics and redox-cycling drugs, but the mechanism of induction is just beginning to be studied.

The *rpoS* (initially denoted *katF*) gene encodes a sigma factor (σ^S; σ^{38}) subunit of RNA polymerase and is important for the expression of a large group of genes that are induced when cells enter starvation or stationary phase. RpoS has been shown to regulate the expression of *katE* (hydroperoxidase II), *xthA* (exonuclease III), *dps* (non-specific) DNA binding protein) and *katG* (hydroperoxidase I) (Kullik and Storz, 1994). The expression of *dps* and *katG* is also regulated by OxyR, suggesting that some defense genes are induced by H_2O_2 during exponential growth and are expressed constitutively when the cells enter stationary phase.

ArcA (aerobic respiration control), Fnr (fumarate nitrate reductase), and Fur (ferric uptake regulation) also impinge on the oxidative stress response in *E. coli* and have been shown to regulate *sodA* (MnSOD) expression in conjunction with SoxS and MarA (Compan and Touati, 1993). ArcA shows homology to the regulator component of the two-component family of regulators and is post-translationally activated by phosphorylation *via* ArcB. The ArcB sensor in turn may be stimulated by an electron carrier in the transport chain. Fnr carries an iron cofactor and is a redox-sensitive transcriptional regulator of anaerobically-induced genes. The Fur protein also binds iron and acts to repress genes involved in iron assimilation in the presence of high iron.

Transcriptional regulators for oxidative stress inducible genes in yeast

Due to the well developed genetic tools in yeast, *Saccharomyces cerevisiae* is receiving increasing attention as a model for studying the oxidative stress response in eukaryotes (reviewed by Gralla and Kosman, 1992; Jamieson, 1995), and the following transcriptional regulators of antioxidant genes have been identified.

YAP1, YAP2

The YAP1 (yeast AP-1) protein of *S. cerevisiae* is homologous to the AP-1 family of eukaryotic transcription factors (described below), and was identified by its ability to specifically bind to AP-1 binding sites as well as by its ability to confer resistance to a variety of compounds upon overexpression (Jamieson, 1995). The *yap1* mutant cells are more sensitive to H_2O_2 and to cadmium, show a decreased adaptive response to H_2O_2 and have decreased levels of glucose-6-phosphate dehydrogenase, SOD, and glutathione reductase activity. It is not known whether the genes encoding these enzymes are direct targets of YAP1, however, YAP1 has been shown to bind and activate the promoters of *TRX2* (one of the two thioredoxin genes in *S. cervisiae*), *GSH1* (γ-glutamylcysteine synthetase), and YCF1 (an ABC-type transporter protein) (Kuge and Jones, 1994; Wemmie et al., 1994; Wu and Moye-Rowley, 1994). The YAP1-dependent transcriptional activation of *TRX2* is stimulated by hydroperoxides and thiol-oxidants and is achieved by stimulating preexisting YAP1 protein rather than by increasing expression of YAP1 (Kuge and Jones, 1994). The mechanism of YAP1 activation in response to oxidants remains unclear though genetic studies suggest that activation may involve the RAS2-protein kinase A pathway (Gounalaki and Thireos, 1994).

Recently, another AP-1 binding protein has been identified and denoted both CAD1 and YAP2 (Bossier et al., 1993; Wu et al., 1993; Hirata et al., 1994). As for YAP1, the overexpression of CAD1/YAP2 leads to resistance to a number of stresses and *cad1/yap2* mutant strains are somewhat sensitive to H_2O_2 (Stephen et al., 1995). The targets for YAP2 have not yet been identified, however future studies to elucidate the roles of YAP1, YAP2 and other AP-1 homologs in the yeast stress response should be extremely interesting.

MAC1, ACE1

Another regulator of oxidative stress-inducible genes in yeast, fortuitously identified by sequencing, is MAC1 (Jungmann et al., 1993). Strains carrying disruptions of the *MAC1* gene are hypersensitive to heat, cadmium, zinc, lead and H_2O_2 and show decreases H_2O_2-dependent induction of the *CTT1* (cytosolic catalase T) gene. MAC1 also regulates the expression of *FRE1* (encoding ferric reductase which is involved in iron reduction and uptake). The MAC1 protein carries a "copper fist" domain known to be required for copper and DNA binding, but has not yet been shown to bind to promoter sequences.

The ACE1 activator was identified as a regulator of the yeast *CUP1* (metallothionein) gene but was also found to mediate the copper induction of *SOD1* (CuZnSOD) (Gralla and Kosman, 1992). The ACE protein also carries a "copper fist" DNA binding domain and binds to a region of the *SOD1* promoter which is essential for the copper-induced transcription.

HAP1

Both yeast catalases and superoxide dismutases appear to be induced by oxygen, and the two heme activation protein complexes, HAP1 (= CYP1) and HAP2/3/4 have been implicated in the regulation of *CTT1* and *SOD2* (MnSOD) expression (Gralla and Kosman, 1992). The HAP1 activator has a zinc finger DNA binding motif and binds upstream of the *CTT1* gene. The HAP2/3/4 protein complex responds to heme and carbon-source availability, and a HAP2/3/4-specific upstream activation site was identified upstream of the *SOD2* (MnSOD) gene.

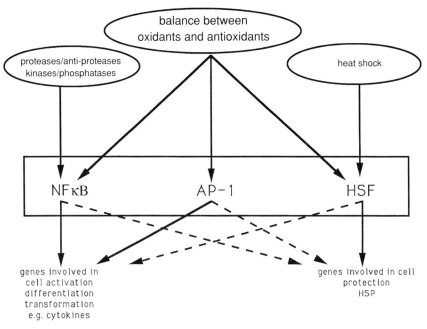

Figure 2. Transcriptional regulators for oxidative and other stress-inducible genes in mammalian cells. The transcriptional regulators are indicated in the box. NFκB and AP-1 predominantly induce genes involved in cell differentiation, activation and transformation (plain arrows), while HSF induces genes involved in cell protection, i.e., HSP genes (plain arrow). HSF may also contribute to the activation of genes involved in cell differentiation (broken arrow) and NFκB and AP-1 to the activation of genes involved in protection (broken arrows).

Transcriptional regulators for oxidative stress inducible genes in mammalian cells

The mechanisms and factors regulating oxidant and antioxidant responses in mammalian cells are still poorly understood. Here, we discuss the roles of three ubiquitous transcriptional regulators in the cellular responses to oxidative stress, i.e., NFκB, AP-1, and the heat shock (HS) transcription factor, HSF (see Fig. 2). There are a number of similarities, but also differences in the activation mechanisms described for these three factors, which illustrate the complexity of the response to oxidative stress in mammalian cells.

The transcription of the heme oxygenase gene is also regulated by redox pathways, but will not be discussed here, since it is the subject of the chapter by Tyrrell (this volume). Posttranscriptional regulation of oxidative stress responses is also beyond the scope of this chapter, but it should at least be mentioned that the activities of several iron regulatory proteins are posttranscriptionally regulated by oxidative stress and that a number of these proteins play a role in cellular protection against oxidative stress (Pantopoulos and Hentze, 1995; Vile and Tyrrell, 1995; see also Rouault and Klausner, this volume).

NFκB

The nuclear transcription factor NFκB was initially described as an activity that specifically bound DNA fragments containing the decameric DNA sequence motif 5'-GGGACTTTCC-3' (reviewed in Grimm and Baeuerle, 1993, and in Verma et al., 1995). This motif was first identified as a B-cell specific element, but it soon became evident that it was also functional in many other cells. NFκB is now recognized as a ubiquitous factor whose activation is independent of protein synthesis.

The number of genes which are activated by NFκB in many different cell types (T- and B-lymphocytes, monocytes-macrophages) is impressive, and include a number of genes which code for products, such as tumor necrosis factor α (TNFα) and other cytokines, that are more relevant to the immune responses and to cell proliferation rather than to cellular protection against oxidants. However, although it is usually considered as a proinflammatory cytokine, TNFα may also play a major role in protective antioxidant mechanisms by inducing the mitochondrial MnSOD (discussed in detail by Wong and Kaspar, this volume). Other genes induced by NFκB are Interleukines 1, 6, and 8, c-*fos*, leukocyte growth factors, the inducible NO· synthase and most likely other enzymes and factors involved in the control of the oxidoreductive state of the activated cell (Cogswell et al., 1994).

NFκB consists of two polypeptides with apparent molecular sizes of 50 and 65 kDa (refered to as p50 and p65). In unstimulated cells, these two polypeptides are associated with an inhibitory factor, termed inhibitor of NFκB, IκB (a number of IκBs have been identified, including IκBα and IκBβ). The transcriptional activation by NFκB requires the dissociation of the inhibitory protein IκB from the NFκB p50/p65 heterodimer in the cytoplasm and the subsequent translocation of active NFκB into the nucleus (Baeuerle, 1991). NFκB and other related factors of the v- and c-Rel oncoprotein family share a characteristic sequence motif with a cysteine and three arginine residues in the DNA binding region.

A great variety of agents can activate NFκB *in vivo* including phorbol esters, antigens, superantigens and antibodies, pro-inflammatory cytokines, viral (HIV) and bacterial proteins, lipopolysaccharides, ionizing and non-ionizing radiations (Staal et al., 1990; Schreck et al., 1991; Schreck et al., 1992a; Trede et al., 1993). Interestingly, all these factors share the potential to induce the production of low, non toxic, cell-activating concentrations of ROS. The effects of the ROS are specific since NFκB is activated by H_2O_2 but not by $\cdot O_2^-$, and cell lines overexpressing catalase, but not SOD are deficient in activating NFκB (Schmidt et al., 1995). The ROS may modify NFκB activity at two different levels. *In vivo* ROS activate NFκB by stimulating release from IκB, while *in vitro* ROS have also been shown to inhibit NFκB binding to DNA (Toledano and Leonard, 1991). Gluta-thione (GSSG) was the first physiologically relevant modulator of the redox-activation of NFκB to be identified, but other endogenous or ex-ogenous antioxidants, including thioredoxin and N-acetylcysteine (NAC), may also decrease NFκB activation, perhaps in conjunction with GSSG (Hayashi et al., 1993).

Although redox regulation appears central in the control of NFκB acti-vation, other factors, such as kinases and proteases also activate the tran-scription factor, possibly in cooperation with oxidants (Costello et al., 1995). As a consequence, phosphatases and antiproteases, together with antioxidants, are also involved in limiting NFκB activation (Costello et al., 1995).

AP-1

The Fos and Jun proteins are members of a family of nuclear transcription factors that are involved in the regulation of growth, differentiation, and cellular responses to stress. They form homodimeric (Jun/Jun) or hetero-dimeric (Jun/Fos) complexes that interact with a DNA sequence known as the activator protein-1 (AP-1) binding site or TPA-responsive element (5'ATGAGTCA-3') (Frame et al., 1991). As for NFκB, AP-1 activates, in a redox-controlled fashion, genes important in signal transduction rather than in antioxidant defenses (Angel and Karin, 1991).

The Jun and Fos subunits dimerize *via* a leucine zipper motif. The DNA binding regions of Fos, Jun and related transcription factors also contain a characteristic sequence with a cysteine and as many as seven arginine resi-dues in close proximity. Redox modifications of this conserved cysteine residue, located within the DNA-binding domain of both Fos and Jun, control, at least in part, the DNA binding of AP-1 (Abate et al., 1990; Schenk et al., 1994).

The critical cysteine of AP-1 has considerably lower redox potential that that of NFκB, and GSSG inhibits the activation of NFκB more effectively than the activation of AP-1 (Galter et al., 1994). Other antioxidants also have distinct effects on NFκB and AP-1 activity (Meyer et al., 1993; Galter et al., 1994; Schenk et al., 1994). For example, NAC abolishes NFκB and HSF (see below) activation while acting as positive modulator of AP-1. Thioredoxin, a molecular chaperone which might catalyze the isomeriza-tion, the oxidation or the reduction of protein thiol/disulfides, inhibits NFκB activation (Holmgren 1985; Fernando et al., 1992; Schenk et al., 1994). In contrast, thioredoxin strongly induces the activity of AP-1, indi-cating that both oxidants and antioxidants can promote AP-1 activation (Schenk et al., 1994). The mechanisms for AP-1 regulation by the oxidants and antioxidants appear to be distinct however, since the activation by thioredoxin but not the activation by NAC, requires *de novo* synthesis of the AP-1 subunits (Meyer et al., 1993). Interestingly, AP-1 activity was also found to be regulated by redox factor-1 (Ref-1), a nuclear factor corresponding to an apurinic/apyridimic endonuclease (Xanthoudakis and Curran, 1992; Xanthoudakis et al., 1992).

HSF

Human cells respond to heat stress and other cellular injuries by inducing the binding of a preexisting transcriptional activator(s), the heat shock factor(s), HSF, and subsequently synthesizing a set of protective proteins, the HS proteins (HSP) (reviewed in Wu, 1995; see also Morimoto et al., this volume). HSP function, at least in part, to maintain cellular homeostasis by acting as molecular chaperones. They participate in the folding and assembly of nascent and unfolded polypeptides and facilitate protein transport and degradation (Liu et al., this volume).

The induction of HSP has been observed as an adaptive response to oxidative stress in both prokaryotes and eukaryotes. Therefore, while the genes regulated by NFκB and AP-1 appear to be primarily involved in cell differentiation, the gene products induced after HSF activation play an established role in survival against oxidative stress (reviewed in Polla et al., 1995). For example, HSP can protect cells under conditions of ischemia-reperfusion or TNFα toxicity. A major subcellular target for HSP-mediated protection against ROS appears to be the mitochondria, which is the site of ROS production and MnSOD induction upon cell exposure to TNFα (Wong and Kaspar, this volume). While HSP most likely act to prevent the accumulation of oxidized and damaged proteins, it should be emphasised that the HSF-mediated stress responses induced by HS and by ROS are probably distinct. ROS lead to HSF activation in both *Drosophila* and human cells, but oxidative stress induces significantly less HSP expression than HS (Becker et al., 1990; Bruce et al., 1993, Polla et al., 1987; Jacquier-Sarlin and Polla, 1996).

Transcriptional induction of HS genes requires the activation of the HSF which binds to the consensus HS element (HSE) (5′-n-GAA-nn-TTC-n-3′). In unstressed cells, HSF exists in a non-DNA binding cytoplasmic form and activation leads to oligomerization and nuclear translocation of HSF (Sarge et al., 1991). To date, three HSF (HSF-1, HSF-2, HSF-4) have been characterized in mammalian cells and a fourth (HSF-3) has been identified in avian cells (Rabindran et al., 1991; Schuetz et al., 1991; Sistonen et al., 1994). Interestingly, only some HSF contain cysteine residues within their DNA-binding domains (for example, human HSF-2 and yeast HSF do not). The selective activation of one or the other HSF might thus depend upon its molecular structure, human HSF-1 being regulated by oxidative stress and

HSF-2, which has threonine or tyrosine residues in place of the cysteines, being regulated by phosphorylation.

As for NFκB, the activation of HSF-1 by oxidants is selective for the type of oxidant, H_2O_2 being more effective than $\cdot O_2^-$ in activating HSF-1 (Jacquier-Sarlin et al., 1995). Similar to what has been observed for NFκB, ROS may also act to modulate HSF activity at two different levels. Under some conditions, ROS may activate HSF and lead to nuclear translocation and DNA binding while under other conditions ROS could lead to oxidative alteration of the factor with subsequent inhibition of DNA binding (Jacquier-Sarlin and Polla, 1996). This inhibition of binding is likely to be similar to the inhibition seen for NFκB and AP-1 since all three of the mammalian transcription factors discussed here have cysteine residues within their DNA binding domains. Consistent with the presence of a critical cysteine, a number of redox and thiol-reactive agents such as thioredoxin have been found to affect HSF-HSE complex formation (Holmgren 1985; Fernando et al., 1992; Huang et al., 1994). One attractive hypothesis is that thioredoxin, due to its catalytic mechanism and ubiquitous distribution, exerts a protective role during oxidative stress and restores the activity of oxidatively damaged HSF and other redox-regulated transcription factors (Hayashi et al., 1993; Kuge and Jones, 1992; Mathews et al., 1992, Okamoto et al., 1992).

Regulation of oxidative stress inducible genes in response to disease

NFκB regulation is probably central to a number of diseases in humans, in particular, inflammation, HIV infection, and bacterial infection. Other diseases or physiological conditions, for which an imbalance between oxidants and antioxidants has been proposed, may be added to this list, including cancer and aging (Cross et al., 1987; Ames and Shinegawa, 1992).

Inflammation

There are many different initial triggers for inflammation, including physical and chemical agents, bacterial and viral infections, and exposure to antigens, superantigens or allergens. All of these have the potential to generate ROS and to activate NFκB. Thus, it has been suggested that this transcription factor is central to the initiation of an acute inflammatory reaction (Dall'Ava and Polla, submitted). Indeed, secondary to NFκB activation, a series of genes coding for proinflammatory cytokines are induced (Beg et al., 1993; Cogswell et al., 1994). The products of these genes in turn activate NFκB (see above), thus leading to a proinflammatory amplification loop, and possibly to chronic inflammation. One relevant clinical example is arterosclerosis, a chronic inflammatory state of arterial endothelium during the development of which free radical reactions are involved at multiple steps (McCord, 1985; Polla et al., 1993). Therapies aiming at downregulating the activity of NFκB are thus contemplated as an approach to the control of inflammation during both its early and late phases.

HIV infection

Therapeutical approaches which include the use of antioxidants, are also being considered in HIV infection (Dröge et al., 1992). Indeed, NFκB activates not only endogenous genes involved in the inflammatory and immune responses, but also HIV (Staal et al., 1990, Israël et al., 1992; Ryser et al., 1994; Aillet et al., 1994). The cellular redox status influences HIV activity in both human T cells and monocytic cell lines, and agents that interfere with thiol-disulfide interchange, including inhibitors of, and antibodies against, protein disulfide isomerase, modulate the outcome of infection by HIV, at least *in vitro* (Israël et al., 1992; Ryser et al., 1994). Protein disulfide isomerase-mediated thiol-disulfide interchange also interferes with the interaction of HIV and the target cells' receptor (Ryser et al., 1994). On the other hand, exposure of infected cells to HS, or to other inducers of stress responses, activate latent viruses such as cytomegalovirus and HIV, and it has been suggested that this activation occurs *via* an HES-like sequence within the long terminal repeat of these viruses (Geelen et al., 1987; Geelen et al., 1988). Whether or not this stress-induced viral activation results from concomitant stress-induced alterations in cellular redox state remains to be determined. Whether or not anti-oxidants, such as NAC or vitamin E, administered *in vivo*, might modulate HIV infection or its consequences, also remains to be established.

Bacterial infection

Another situation of potential clinical relevance where regulation of oxidative stress inducible genes modulates cellular responses occurs during bacterial infection. When infected, human phagocytes activate the complex respiratory burst enzyme NADHP oxidase, leading to massive production of $\cdot O_2^-$ (generated from molecular oxygen at the expense of NADPH) by the host cell. The ROS produced during phagocytosis of, and infection by, microorganisms, are usually considered as aiming to kill the responsible microorganism. However, it appears that infection-associated ROS also activate a number of redox-regulated transcription factors and play in role in the inducible protective responses (HSP, SOD, etc.) in both prokaryotes and eukaryotes (Kantengwa et al., 1993; Kantengwa et al., 1995; Polla et al., 1995). The respective induction of host and pathogen HSP appears tightly regulated by the type and the subcellular location of ROS production, $\cdot O_2^-$ being insufficient in itself for host HSF activation, while it does increase pathogen HSP, and $\cdot OH$ being most efficient in host HSP induction. Respective host and pathogen inducible protective responses such as HSP might determine the outcome of infections and further studies on the regulation of redox inducible genes in prokaryotes and eukaryotes during infection should be of great interest.

References

Abate, C., Patel, L., Rauscher, F.J., III and Curran, T. (1990) Redox regulation of Fos and Jun DNA-binding activity *in vitro. Science* 249:1157–1161.

Aillet, F., Gougerot-Pocidalo, M.-A., Virelizier, J.-L., Israël, N. (1994) Appraisal of potential therapeutic index of antioxidants on the basis of their *in vitro* effects on HIV replication in monocytes and interleukin-2 induced lymphocyte proliferation. *Aids Research and Human Retroviruses* 10:405–411.

Altuvia, S., Almiron, M., Huisman, G., Kolter, R. and Storz G. (1994) The dps promoter is activated by OxyR during growth and by IHF and ss in stationary phase. *Mol. Microbiol.* 13:265–272.

Ames, B.N. and Shigenaga, M.K. (1992) Oxidants are a major contributor to aging. *In:* C. Franceschi, G. Crepaldi, V.J. Cristofalo and J. Vijg (eds): *Aging and Cellular Defense Mechanisms.* Ann. New York Acad. Sciences, pp 85–96.

Angel, P. and Karin, M. (1991) The role of jun, fos and AP-1 complex in cell-proliferation and transformation. *Biochim. Biophys. Acta* 1072:129–157.

Baeuerle, P.A. (1991) The inducible transcription activator NF-κB: regulation by distinct protein subunits. *Biochim. Biophys. Acta* 1072:63–80.

Becker, J., Metzger, V., Courgeon, A.-M., Best-Belpomme, M. (1990) Hydrogen peroxide activates immediate binding of a *Drosophila* factor to DNA heat-shock regulatory element *in vivo* and *in vitro. J. Biochem.* 189:553–558.

Beg, A.A., Finco, T.S., Nantermet, P.V. and Baldwin, Jr., A.S. (1993) Tumor necrosis factor and interleukin-1 lead to phosphorylation and loss of IκBα: a mechanism for NFκB activation. *Mol. Cell Biol.* 13:3301–3310.

Bossier, P., Fernandes, L., Rocha, D. and Rodrigues-Pousada, C. (1993) Overexpression of YAP2, coding for a new YAP protein, and YAP1 in *Saccharomyces cerevisiae* alleviates growth inhibition caused by 1,10-Phenanthroline, *J. Biol. Chem.* 268:23640–23645.

Bruce, J.L., Price, B.D.,, Coleman, C.N. and Calderwood, S.K. (1993) Oxidative injury rapidly activates the heat shock transcription factor but fails to increase levels of heat shock proteins. *Cancer Res.* 53:12–15.

Cogswell, J.P., Godlevski, M.M., Wisley, G.B., Clay, W.C., Leesnitzer L.M., Ways, J.P., and Gray J.G. (1994) NFκB regulates IL-1β transcription through a consensus NFκB site and a non-consensus CRE-like site. *J. Immunol.* 153:712–723.

Compan, I., Touati, D. (1993) Interaction of six global transcription regulators in expression of manganese superoxide dismutase in *Escherichia coli. J. Bacteriol.* 175:1687–1696.

Costello, R., Lecine, P., Kahn-Perlès, B., Algarté M., Lipcey C., Olive D. and Imbert J. Activation du système de facteurs de transcription Rel/NFκB. *Médecine/Sciences* 7:957–976.

Cross, C.E., Halliwell, B., Borish, E.T., Pryor, W.A., Ames B.N., Saul, R.L., McCord J.M. and Harman, D. (1987) Oxygen radicals and human disease. *Ann. Int. Med.* 107:526–545.

Dröge, W., Eck, H.P. and Mihm, S. (1992) HIV-induced cystein deficiency and T-cell dysfunction – a rationale for treatment with *N*-acetylcysteine. *Immunol. Today* 13:211–214.

Fawcett, W.P. and Wolf, Jr., R.E. (1994a) Purification of a MalE-SoxS fusion protein and identification of the control sites of *Escherichia coli* superoxide-inducible genes. *Mol. Microbiol.* 14:669–679.

Fawcett, W.P. and Wolf, Jr., R.E. (1994b) Genetic definition of the *Escherichia coli* zwf "Soxbox", the DNA binding site for SoxS-mediated induction of glucose 6-phosphate dehydrogenase in response to superoxide. *J. Bacteriol.* 177:1742–1750.

Fernando, M.R., Nanri, H., Yoshitake, S., Nagata-Kuno, K. and Minakami, S. (1992) Thioredoxin regenerates proteins inactivated by oxidative stress in endothelial cells. *Eur. J. Biochem.* 209:917–922.

Frame, M.C., Wilkie, N.M., Darling, A.J., Chudleigh, A., Pintzas, A., Lang, J.C. and Gillespie, D.A.F. (1991) Regulation of AP-1/DNA complex formation in vitro. *Oncogene* 6:205–209.

Fridovich, I. (1978) The biology of oxygen radicals. *Science* 201:875–879.

Galter, G., Mihm, S. and Dröge, W. (1994) Distinct effects of glutathione disulphide on the nuclear transcription factors κB and the activator protein-1. *Eur. J. Biochem.* 221:639–648.

Geelen, J.L.M.C., Boom, R., Klaver, G.P.M., Minnaar, R.P., Feltkamp, M.C.W., Van Milligen, F.J., Sol, C.J.A. and Van Der Noordaa, J. (1987) Transcriptional activation of the major immediate early transcription unit of human cytomegalovirus by heat shock, arsenite and protein synthesis inhibitors. *J. Gen. Virol.* 68:2925–2931.

Geelen, J.L.M.C., Minnaar, R.P., Boom, R., Van Der Noordaa, J. and Goudsmit, J. (1988) Heat shock induction of the human immunodeficiency virus long terminal repeat. *J. Gen. Virol.* 69:2913–2917.

Gonzalez-Flecha, B. and Demple, B. (1995) Metabolic sources of hydrogen peroxide aerobically growing *Escherichia coli*. *J. Biol. Chem.* 270:13681–13687.

Gounalaki, N. and Thireos, G. (1994) Yap1p, a yeast transcriptional activator that mediates multidrug resistance, regulates metabolic stress response. *EMBO J.* 13:4036–4041.

Gralla, E.B. and Kosman, D.J. (1992) Molecular genetics of superoxide dismutases in yeasts and related fungi. *Adv. Genet.* 30:251–319.

Grimm, S. and Baeuerle, P.A. (1993) The inducible transcription factor NFκB: structure-function relationship of its protein subunits. *Biochem. J.* 290:297–308.

Halliwell, B. and Gutteridge, J.M.C. (1990) Role of free radicals and catalytic metal ions in human disease: An overview. *In:* L. Packer and A.N. Glazer (eds): *Methods in Enzymology.* Academic Press, New York, Vol. 86, pp 1–85.

Hayashi, T., Ueno, Y. and Okamoto, T. (1993) Oxidative regulation of nuclear factor κB. Involvement of a cellular reducing catalyst thioredoxin *J. Biol. Chem.* 268:11380–11388.

Hidalgo, E. and Demple, B. (1994) An iron-sulfur center essential for transcriptional activation by the redox-sensing SoxR protein. *EMBO J.* 13:138–146.

Hidalgo, E. and Demple, B. (1995). Adaptive responses to oxidative stress: the soxRS and oxyR regulon. *In:* E.C.C. Lin (ed): *Regulation of Gene Expression in Echerichia coli.* R.G. Landes Co., Austin TX; *in press.*

Hidalgo, E., Bollinger, Jr., J., Walsh T.M. and Demple, B. (1995) Binuclear [2Fe-2S] clusters in the *Escherichia coli* SoxR protein and role of the metal centers in transcription. *J. Biol. Chem.* 270:20908–20914.

Hirata, D., Yano, K. and Miyakawa, T. (1994) Stress-induced transcriptional activation medicated by YAP1 and YAP2 genes that encode the Jun family of transcriptional activators in *Saccharomyces cerevisiae.* *Mol. Gen. Genet.* 242:250–256.

Holmgren, A. (1985) Thioredoxin. *Annu. Rev. Biochem.* 54:237–271.

Huang, I.E., Zhang, H., Bae, S.W., Liu, A.Y.-C. (1994) Thiol reducing reagents inhibit the heat shock response. Involvement of a redox mechanism in the heat shock signal transduction pathway. *J. Biol. Chem.* 48:80718–80725

Israël, N., Gougerot-Pocidalo, M.-A., Aillet, F., Virelizier, J.-L. (1992) Redox status of cells influences constitutive or induced NF-κB translocation and HIV long terminal repeat activity in human T- and monocytic cell lines. *J. Immunol.* 149:3386–3393.

Jacquier-Sarlin, M.J., Jornot L. and Polla, B.S. (1995) Differential expression and regulation of hsp70 and hsp90 by phorbol ester and heat shock. *J. Bio. Chem.* 270:14094–14099.

Jacquier-Sarlin, M.J. and Polla, B.S. (1996) Dual regulation of heat-shock transcription factor (HSF) activation and DNA-bindung activity by H_2O_2: role of thioredoxin. *Biochem. J.* 318:187–193.

Jamieson, D.J. (1995) The effect of oxidative stress on *Saccharomyces cerevisiae. Redox Report* 1:89–95.

Jungmann, J., Reins, H.A., Lee, J., Romeo, A., Hasset, R., Kosman, D. and Jentsch, S. (1993) MAC1, a nuclear regulatory protein related to Cu-dependent transcription factors is involved in Cu/Fe utilisation and stress resistance in yeast. *EMBO J.* 12:5051–5056.

Kantengwa, S. and Polla, B.S. (1993) Phagocytosis of *Staphylococcus aureus* induces a selective stress response in human monocytes-macrophages (Mφ):modulation by Mφ differentiation and by iron. *Inf. Immun.* 61:1281–1287.

Kantengwa, S., Müller, I., Louis, J. and Polla, B.S. (1995) Infection of human and murine macrophages with *Leishmania major* is associated with early parasite heat shock protein synthesis but fails to induce a host cell stress response. *Immunol. Cell Biol.* 73:73–80.

Kuge, S. and Jones, N. (1994) YAP1 dependent activation of TRX2 is essential for the response of *Saccharomyces cerevisiae* to oxidative stress by hydroperoxides. *EMBO J.* 13:655–664.

Kullik, I. and Storz, G. (1994) Transcriptional regulators of the oxidative stress response in prokaryotes and eukaryotes. *Redox Report* 1:23–29.

Kullik, I., Stevens, J., Toledano, M.B. and Storz, G. (1995a) Mutational analysis of the redox-sensitive transcriptional regulator OxyR: Regions important for DNA binding and multimerization. *J. Bacteriol.* 177:1285–1291.

Kullik, I., Toledano, M.B., Tartaglia, L.A. and Storz, G. (1995b) Mutational analysis of the redox-sensitive transcriptional regulator OxyR: Regions important for oxidation and transcriptional activation. *J. Bacteriol.* 177:1275–1284.

Li, Z. and Demple, B. (1994) SoxS, an activator of superoxide stress genes in *Escherichia coli:* Purification and interaction with DNA. *J. Biol. Chem.* 269:18371–18377.

Liochev, S.I., Hausladen, A., Beyer, Jr., W.F. and Frodovich, I. (1994) NADPH: ferredoxin oxidoreductase acts as a paraquat diaphorase and is a member of the soxRS regulon. *Proc. Natl. Acad. Sci. USA* 91:1328–1331.

Matthews, J.R., Wakasugi, N., Virelizier, J.-L., Yodoi, J. and Hay, R.T. (1992) Thioredoxin regulates the DNA binding activity of NF-κB by reduction of a disulphide bond involving cysteine 62. *Nucleic. Acids Res.* 20:2005–2015.

McCord, J.M. (1985) Oxygen-derived free raduculs in postischemic tissue injury. *New Engl. J. Med.* 312:159–163.

Meyer, M., Schreck, R. and Baeuerle, P.A. (1993) H_2O_2 and antioxidants have opposite effects on activation of NF-κB and AP-1 in intact cells: AP-1 as secondary antioxidant-responsive factor. *EMBO J.* 12:2005–2015.

Nunoshiba, T., DeRojas-Walker, Y., Tannenbaum, S.R. and Demple, B. (1995) Roles of nitric oxide in inducible resistance of *Escherichia coli* to activated murine macrophages. *Infect. Immun.* 63:794–798.

Okamoto, T., Ogiwara, H., Hayashi, T., Mitsui, A., Kawabe, T. and Yodoi, J. (1992) Human thioredoxin/adult T cell leukemia-derived factor activates the enhancer binding protein of human immunodeficiency virus type 1 by thiol redox control mechanism. *Int. Immunol.* 4:811–819.

Pantopoulos, K. and Hentze, M.W. (1995) Rapid responses to oxidative stress mediated by iron regulatory protein. *EMBO J.* 14:2917–2924.

Polla, B.S., Healy A.M., Wojno, W.C. and Krane, S.M. (1987) Hormone 1α,25-dihydroxy-vitamin D3 modulates heat shock response in monocytes. *Am. J. Physiol.* 252 (Cell Physiol. 21):C640–C649.

Polla, B.S., Mili, N. and Kantengwa S. (1991) Heat shock and oxidative injury in human cells. *In:* B. Maresca and S. Lindquist (eds): *Heat Shock,* Springer Verlag, Berlin, pp 279–290.

Polla, B.S., Bornman L. and Perin M. (1993). Heat shock proteins, oxidative stress and atherosclerosis. *In:* M.M. Galteau, J. Henny and G. Siest (eds): *Biologie Prospective.* John Libbey Eurotext, Paris, pp 223–231.

Polla, B.S., Mariéthoz, E., Hubert, D. and Barazzone, C. (1995) Heat shock proteins in host-pathogen interactions: implications for cystic fibrosis. *Trends in Microbiology* 3:392–396.

Rabindran, S.K., Giorgi, G., Clos, J. and Wu, C. (1991) Molecular cloning and expression of human heat shock factor, HSF1. *Proc. Natl. Acad. Sci. USA* 88:6906–6910.

Ryser, J.-P.H., Levy, E.M., Mandel, R. and DiSciullo, G.J. (1994) Inhibition of human immunodeficiency virus infection by agents that interfere with thiol-disulfide interchange upon virus-receptor interaction. *Proc. Natl. Acad. Sci. USA* 91:4559–4563.

Sarge, K.D., Zimarino, V., Holm, K., Wu, C. and Morimoto, R.I. (1991) Activation of heat shock gene transcription by heat shock factor 1 involves oligomerization, acquisition of DNA-binding activity, and nuclear localization and can occur in the absence of stress. *Genes Dev.* 5:1902–1911.

Schenk, H., Klein, M., Erdbrügger, W., Dröge, W. and Schulze-Osthoff, K. (1994) Distinct effects of thioredoxin and antioxidants on the activation of transcription factors NF-κB and AP-1. *Proc. Natl. Acad. Sci. USA* 91:1672–1676.

Schmidt, K.N., Amstad, P., Cerutti, P. and Baeuerle, P.A. (1995) The roles of hydrogen peroxide and superoxide as messengers in the activation of transcription factor NF-κB. *Chem. & Biol.* 2:13–22.

Schreck, R., Rieber, P. and Baeuerle, P.A. (1991) Reactive oxygen intermediates as apparently widely used messengers in the activation of the NF-κB transcription factor and HIV-1. *EMBO J.* 10, 2247–2258.

Schreck, R., Meier, B., Männel, D., Dröge, W. and Baeuerle, P.A. 81992) Dithiocarbamates as potent inhibitors of nuclear factor κB activation in intact cells. *J. Exp. Med.* 175:1181–1194.

Schreck, R., Albermann, K. and Baeuerle, P.A. (1992 a) Nuclear factor κB: an oxidative stress responsive transcription factor of eukaryotic cells. *Free Rad. Res. Commun.* 17:221–237.

Schuetz, T.J., Gallo, G.J., Sheldon, L., Tempst, P. and Kingston, R.E. (1991) Isolation of a cDNA for HSF2: evidence for two heat shock factor genes in humans. *Proc. Natl. Acad. Sci. USA* 88:6911–6915.

Sistonen, L., Sarge K.D. and Morimoto R.I. (1994) Human heat shock factors 1 and 2 are differentially activated and can synergistically induce hsp70 gene transcription. *Mol. Cell. Biol.* 14:2087–2099.

Staal, F.J.T., Roederer, M., Herzenberg, L.A. and Herzenberg, L.A. (1990) Intracellular thiols regulate activation of nuclear factor κB and transcription of human immunodeficiency virus. *Proc. Natl. Acad. Sci. USA* 87:9943–9947.

Stephen, D.W.S., Rivers, S.L. and Jamieson, D.J. (1995) The role of YAP1 and YAP2 genes in the regulation of the adaptive oxidative stress responses of *Saccharomyces cerevisiae*. *Mol. Microbiol.* 16:415–423.

Storz, G., Tartaglia, L.A. and Ames, B.N. (1990) Transcriptional regulator of oxidative stress-inducible genes: direct activation by oxidation. *Science* 248:189–194.

Tao, K., Fujita, N. and Ishihama, A. (1993) Involvement of the RNA polymerase a subunit C-terminal region in co-operative interaction and transcriptional activation with OxyR protein. *Mol. Microbiol.* 7:859–864.

Toledano, M.B., and Leonard, W.J. (1991) Modulation of transcription factor NF-κB binding activity by oxidation-reduction *in vitro*. *Proc. Natl. Acad. Sci. USA* 88:4328–4332.

Toledano, M.B., Kullik, I., Trinh, F., Baird, P.T., Schneider, T.D. and Storz G. (1994) Redox dependent shift of OxyR-DNA contacts along an extended DNA binding site: A mechanism for differential promoter selection. *Cell* 78:897–909.

Trede, N.S., Castigli, E., Geha R.S. and Chatila, T. (1992) Microbial superantigens induce NFκB in the human monocytic cell line THP-1. *J. Immunol.* 150:5604–5613.

Verma I.M., Stevenson J.K., Schwarz E.M., Van Antwerp, D. and Miyamoto S. (1995) Rel/NF-κB/IκB family: intimate tales of association and dissociation. *Genes & Dev.* 9: 2723–2735.

Vile, G.F. and Tyrrell, R.M. (1995) Oxidative stress resulting from ultraviolet A irradiation of human skin fibroblasts leads to a heme oxygenase-dependent increase in ferritin. *J. Biol. Chem.* 268:14678–14681.

Wemmie, J.A., Szczypka, M.S., Thiele, D.J. and Moye-Rowley, W.S. (1994) Cadmium tolerance mediated by the yeast AP-1 protein requires the presence of an ATP-binding cassette transporter-encoding gene, YCF1. *J. Biol. Chem.* 269:32592–42597.

Wu, A.L. and Moye-Rowley, W.S. (1994) GSH1, which encodes g-glutamylcysteine synthetase, is a target gene for YAP-1 transcriptional regulation. *Mol. Cell. Biol.* 14:5832–5839.

Wu, C. (1995) Heat shock transcription factors. Structure and regulation. *Annu. Rev. Cell Dev. Biol.* 11:441–469.

Wu, J., Dunham, W.R. and Weiss, B. (1995) Overproduction and physical characterisation of SoxR a (2Fe-2S) protein that governs an oxidative response regulon in *Escherichia coli*. *J. Biol. Chem.* 270:10323–10327.

Xanthoudakis, S. and Curran, T. (1992) Identification and characterization of Ref-1, a nuclear protein that facilitates AP1 DNA binding activity. *EMBO J.* 11:653–665.

Xanthoudakis, S., Miao, G., Wang, F., Pan, Y.C.E. and Curran, T. (1992) Redox activation of Fos-jun DNA binding activity is mediated by a DNA repair enzyme. *EMBO J.* 11: 3323–3335.

Stress-Inducible Cellular Responses
ed. by U. Feige, R.I. Morimoto, I. Yahara and B. Polla
© 1996 Birkhäuser Verlag Basel/Switzerland

UV activation of mammalian stress proteins

R. M. Tyrrell

Swiss Institute for Experimental Cancer Research, CH-1006 Epalinges, Switzerland

Summary. Ultraviolet radiation may be divided into the non-solar UVC region, the solar UVB (290–320 nm) region which is strongly absorbed by nucleic acids, and the solar UVA (320–380 nm) region which is less strongly absorbed by nucleic acids and proteins but causes a variety of oxidative events. As a consequence of these different properties, UVC/UVB radiations induce an array of stress proteins quite distinct from those induced by UVA radiations. Although many studies with UVC and UVB radiations involve lethal doses, it is clear that these radiations have the property of mimicking growth factor responses and stimulate various signal transduction pathways that lead to gene activation including transcriptional activation of the jun and fos proto-oncogenes. Furthermore, UVB irradiation of skin, at physiologically relevant doses can increase the levels of various stress proteins including ornithine decarboxylase, various cytokines, the p53 tumor suppressor protein and to a limited extent, nuclear oncogene products. Non-cytoxic exposures of UVA radiation can lead to the up-regulation of several genes including collagenase, heme oxygenase 1, a specific protein phosphatase (CL100) and phospholipases. At least for heme oxygenase 1, there is evidence that the alteration may be involved in a pathway of defense against oxidative stress. However, much information is lacking in the quest to build up a complete picture of the physiological and pathological significance of the many UV inducible stress responses reported.

Introduction

UV radiation can cause both death and mutation of cells as well as more subtle effects such as changes in cell cycle progression and alterations in membrane permeability. In addition there are numerous examples in which UV radiation has been shown to activate or modify gene expression and thereby lead to synthesis of stress proteins. Before describing examples of such processes in more detail, it is crucial to underline several photobiological concepts which are essential to understanding both the interaction of UV with cells and the potential biological relevance of the phenomena observed.

Firstly, the interaction of UV with biological material changes as a function of wavelength. Many studies are performed with low-pressure mercury lamp sources with a primary emission at 254 nm, a UVC wavelength which is absorbed more strongly by biomolecules (including DNA and proteins) than the longer UVB (290–320 nm) or UVA (320–380 nm) wavelengths present in natural sunlight incident on the earth's surface. While such studies provide valuable information, it is important to remember that the spectrum of damage of all kinds changes with wavelength so that extrapolation from such studies must be done with considerable care. In general, there is a decreasing contribution of effects due to

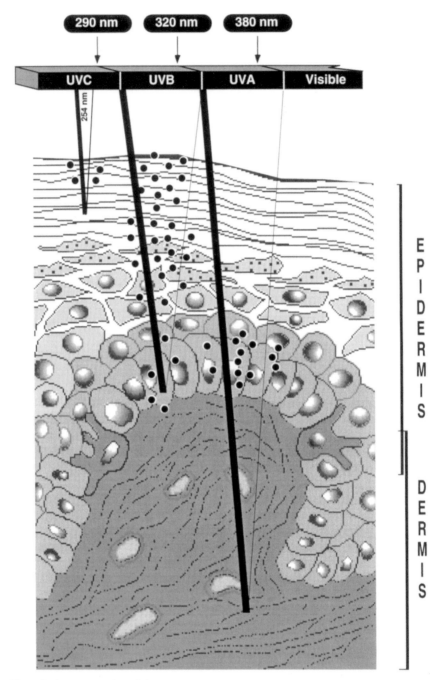

Figure 1. A representation of the UV spectrum and the increasing penetration through skin as a function of wavelength. UV activation (stabilization) of the p53 protein (●) shows the tissue distribution expected from the wavelength region employed except in the UVA region where modulation is restricted to the basal layer of the epidermis (Campbell et al., 1993).

direct absorption by target molecules with increase in wavelength and an increasing component of effects due to the generation of reactive species such as hydroxyl radical and singlet oxygen (for review, see Tyrrell, 1991). A second factor to consider is that cells and tissue transmit an increasing fraction of UV as the wavelength increases so that longer wavelengths penetrate much more deeply and can cause effects at targets different from those of the shorter wavelengths (see Fig. 1). This is crucial when extrapolating or comparing experiments done with single cells to tissue models. Not only can the longer wavelengths penetrate deep into skin, but a small amount will be absorbed by blood components. Finally, UV irradiation leads to a dose-dependent accumulation of damage to biomolecules within individual cells. Above a certain threshold dose, largely determined by cellular repair and antioxidant defense capacity, cells can no longer divide and will undergo either necrotic or programmed (apoptotic) cell death. The observation of activation of genes above this threshold (or sub-lethal) dose range may have limited physiological relevance and could even be a secondary consequence of the death process. The UVC dose to inactivate 90% of a population of cultured human skin fibroblasts lies in the lower end of the $5-20$ Jm^{-2} range and most mammalian cell types have a similar sensitivity. Many of the experiments on UVC activation of gene expression have therefore been done with mostly dead cells.

Since UV interacts quite differently with cells according to the wavelength range employed, the "stress" responses elicited by UVA, UVB and UVC radiations will be considered separately for the purpose of this overview and, where pertinent, comments will be made on their possible functional significance.

UVC induction of stress proteins

UVC is strongly absorbed by DNA and therefore leads to a large variety of DNA damage, a major component of which involves dimerization of adjacent pyrimidines in the same DNA strand. In several bacterial species this leads to an "SOS" response which involves the rapid induction of a type of DNA repair that is error-prone and therefore leads to an enhancement in spontaneous and induced mutation frequencies (see Walker, 1987). Similar, although much less dramatic, phenomena have been reported in eucaryotes, but the underlying mechanism has not been established. UVC radiation does lead to increased expression of many genes including those for collagenase (Wlaschek et al., 1993, 1994), plasminogen activator (Miskin and Ben-Ishai, 1982; Rotem et al., 1987), certain cytokines including IL1, IL-3, IL6, IL8 and IL10 (Schwarz and Luger, 1989; Ullrich, 1995), nuclear oncogenes such as c-*jun* and c-*fos* (Herrlich et al., 1994) and the p53 tumour suppressor gene (Maltzman and Czyzyk, 1984).

Many of the genes now characterized as UVC-inducible were recognized by screening cDNA libraries constructed to enrich for clones containing UVC-inducible genes. Among these are the diverse families of growth arrest and DNA damage inducible (GADD) genes (see following chapter by Holbrook et al.). Another set of genes that resulted from screening for UVC-inducible genes in human keratinocytes was originally designated as the small proline rich (spr) family (Kartasova and van de Putte, 1988). The spr's proved to be good markers of keratinocyte differentiation (Kartasova et al., 1988) and the N- and C-terminal regions of the proteins were later shown to be homologous to the corresponding domains of loricrin and involucrin which are cornified envelope precursor proteins (Backendorf and Hohl, 1992). The spr genes have now been divided into three spr families of which the spr2 and spr3 series are UVC-inducible (Gibbs, 1992). Little is known about their regulation, except that the promoter of the spr2 gene contains an AP-1 element (Gibbs et al., 1990). Spr2 messenger RNA levels are much lower in keratinocytes derived from chronically sun-exposed skin than cells cultured from non-exposed sites, an observation that has been taken to indicate that chronic exposure to solar UV radiation leads to a less differentiated state (Garmyn et al., 1992).

Although the entire signaling pathway for UV activation of gene expression has not been resolved, the literature on the subject is expanding and models have naturally been developed within the general context of our current understanding of signal transduction. The expanding network of kinase/phosphatase pathways will be considered in detail in the chapter that follows so that only essentials relating to UVC will be referred to here. Damage to DNA was originally considered to be the primary candidate for the initiating signal, mainly because UVC led to activation of certain genes faster in repair deficient human cell lines (Miskin and Ben-Ishai, 1982; Stein et al., 1989). However the picture is now less clear. Certain early steps in UV activation of gene expression can occur in enucleated cells (Devary et al., 1993; Vile et al., 1995) and rapidly activated membrane-associated kinases have been implicated in several of the pathways (Devary et al., 1992; Radler-Pohl 1993). Recently, evidence has been outlined for a UVC-induced ligand-independent dimerization of the epidermal growth factor receptor with associated phosphorylation of critical tyrosine residues as a crucial early step in activation (Sachsenmaier et al., 1994). Furthermore, using a dominant negative epidermal growth factor receptor mutant and an inhibitor that blocks growth factor binding, the authors demonstrated a relationship between receptor activation, activation of MAP kinases and the stimulation of responsive promoters (c-*jun* and c-*fos*). However, the UV doses employed are in a range where significant protein and lipid damage is induced and it is possible that the receptor dimerization merely reflects a photochemical event that will mimic the consequences of ligand-mediated growth factor activation at high doses, but which would occur only rarely at sub-lethal UV doses. Nevertheless, it is clear that at sufficient doses,

UVC can activate a pathway resembling that activated by growth factors and which involves the small Ha-*ras* protein as well as the *raf, src* and eventually MAP (ERK1 and ERK2) kinases of which the latter can phosphorylate the terniary complex factor, Elk-1 (Janknecht et al., 1993). The serum response factor/Elk 1 transcription factor complex is responsible for activation of the c-*fos* promoter (Hipskind et al., 1991; Büscher et al., 1988).

Fos and *jun* heterodimers were the prototype form of the AP-1 transcription factor complex which binds to specific sequence (cis-acting) elements in many genes. However, c-*jun* is transcriptionally activated to much higher levels by UV than c-*fos* (see Devary et al., 1991 for review) and there is reasonable evidence that the signaling pathway that leads to increased expression of these two oncogene products is quite different. Among the family of stress activated protein kinases (SAPK's, see Hill and Treisman, 1995, for review) activated by UVC are the proline-directed *jun* kinases (JNK's, Derijard et al., 1994) which lead to increased phosphorylation of serines 73 and 63 in the transactivation domain of c-*jun* and thereby activate the transcription-promoting activity of the protein. The *jun* promoter itself contains two *cis*-acting elements which bind to *jun*/ATF-2 complexes, both components of which require phosphorylation by SAPK's for maximum transactivation activity. The kinases that activate ATF-2 overlap in activity with but are not identical to the SAPK's which activate *jun* (Van Dam et al., in press). It is not clear whether the activation of SAPK's involves a membrane component, but at least the activation of JNK's by UVC can occur either dependently on or independently of *ras* (Minden et al., 1994).

A second family of transcription factors that can be activated by high doses of UVC is the nuclear factor kappa B (NFkB) series. UV radiation is among a large number of factors that can release this factor from its cytoplasmic anchor protein (IkB) and thereby allow it to rapidly migrate to the nucleus. Upstream events appear to involve both *src* tyrosine kinases and the Ha-*ras* protein (see Hill and Treisman, 1995, for review). The phosphorylation of IkB prior to its proteolytic degradation (Palombella et al., 1994) appears to be a crucial step (Naumann and Scheidereit, 1994). NFkB responsive DNA sequence elements occur in the promoters of many genes including several involved in the inflammatory response. The sequence also occurs in the U3 region of the long terminal repeat sequence of the promoter of the human immunodeficiency virus. However, recent data has indicated that UVC activation of transcription on the HIV promoter occurs independently of NFkB and arises as a result of activation of a pre-bound transcription factor complex.

Both the activation of *fos* and *jun* transcription by UVC and the consequent accumulation of the AP-1 factor and the activation of NFkB potentially lead to the up-regulation of many genes whose promoters contain the appropriate elements. Levels of the p53 protein, the product of a tumor suppressor gene which can act as a transcription factor, also increase after UVC treatment of cells (Maltzman and Czyzyk, 1984, Fig. 1) as a result of protein stabilization (Kastan et al., 1991). Again, this sharply increases the number of genes that would be expected to be activated by UV since a rapidly increasing inventory of genes contains promoters that respond to p53. Important recent examples of p53 activated genes are the human bax gene (Myashita and Reed, 1995) which is known to be involved in acceleration of apoptosis and the p21 (waf) gene which is an inhibitor of a cyclin dependent kinase.

As described in the first section of this book, many "stress" proteins have functions in unstressed cells. When proteins are induced in response to a low level of stress, it is appropriate to seek a function of such proteins in cellular defense against such stress. However, at the high levels of UVC exposure that are usually required to observe activation of genes and associated phenomena, it is often questionable whether the protein products are useful to the cell or whether they simply serve as markers of the destructive effect of the particular stress applied to the cell. UVC activation of multiple kinases may simply be a manifestation of aberrant signaling that develops as cells are damaged beyond repair and the up-regulation of dependent genes may therefore be an accident. The definition of the role of most UVC-activated genes in protection of eucaryotic cells is therefore inevitably elusive. For the most part, there is no equivalent to the UV-induced SOS pathway observed in bacteria (see above), although UV-enhancement of p53 levels may have several protective functions (see later). There is also recent evidence that the c-*fos* protein may protect against the genotoxic consequences of UVC radiation (Haas and Kaina, 1995) in contrast to a previous study consistent with a mutagenic role of c-*fos* (van den Berg et al., 1991).

UVA induction of stress proteins

Ultraviolet A radiation interacts with cells in an entirely different manner from the shorter wavelengths. DNA and proteins do not absorb this radiation strongly, but a variety of small molecules in cells does participate in photochemical reactions to generate free radicals and other active species such as singlet oxygen (for review, see Tyrrell, 1991) that can react with any susceptible target molecule within range. Thus, it is not surprising that radiation at the longer wavelengths activates a different spectrum of genes from UVC radiation. Since these are quite diverse and no unifying pathway of induction has been identified, several examples of UVA-inducible gene expression will be described separately.

Heme oxygenase 1

UVA radiation leads to a very strong increase in the rate of transcription and the eventual level of expression of the heme oxygenase 1 (HO-1) gene in cells from all mammalian species so far tested including human skin fibroblasts (Keyse and Tyrrell, 1989; Keyse et al., 1991; Applegate at al.; 1991) in which the phenomena was originally observed. Both basal level expression and UVA activation of this gene are low in epidermal keratino-cytes which express high levels of the constitutive HO-2 isozyme (Apple-gate et al., 1995). The enhanced expression of HO-1 almost certainly results from the oxidizing properties of UVA radiation since expression occurs at much higher levels after depletion of cellular glutathione (Lautier et al., 1992). Furthermore, the stress response is also induced strongly by oxidants including hydrogen peroxide.

At least for UVA induction of the gene, the effector species appears to be singlet oxygen (Basu-Modak and Tyrrell, 1993). It is not a heat shock pro-tein in humans and it is only slightly induced by UVB and UVC radiations. Efforts to understand the functional role of this stress response have focus-ed on a possible contribution of the increase in HO-1 enzymatic activity to the antioxidant status of the cell. Although the end products of heme cata-bolism, biliverdin and bilirubin, have antioxidant properties, they are un-likely to be generated in sufficient quantities to block oxidative processes in cells. On the other hand, increases in HO-1 enzymatic activity will lead to increased heme breakdown and the consequent increase in free intra-cellular iron appears to be responsible for the enhanced levels of newly synthesized ferritin that occur in cells one day after UVA treatment (Vile and Tyrrell, 1993). This increased iron scavenging capacity will lower oxidative reactions in the cell and has been shown to increase cellular resistance to oxidative membrane damage (Vile et al.,1994). Nevertheless, it is likely that this ubiquitous stress response has further functional signi-ficance that has not yet been revealed.

The protein phosphatase, CL100

A second gene that is strongly up-regulated by UVA radiation and other oxidants is a dual specificity (thr, tyr) protein phosphatase (CL100, Keyse and Emslie, 1992). Although the gene was originally isolated from a cDNA library from human fibroblasts treated with hydrogen peroxide, it is also activated by growth factors and heat shock. Interestingly, the only known target proteins for the phosphatase activity are the 42 and 44 kDa (ERK2 and ERK1) MAP kinases (although other stress activated protein kinases are potential substrates). The clear implication from this observation is that this enzyme (together with a growing family of protein phosphatases, see Keyse, 1995) may be involved in down-regulation of pathways activated by stress and mitogens. The phosphatase is only activated at very high doses of UVC radiation (Keyse, 1995) so that, as for UVC activation of the ras/ERK pathway, this is unlikely to be of physiological significance. In contrast, the response to UVA occurs at doses in the sub-lethal range and it is therefore likely that this enzyme plays a role in the response of cells to the oxidizing component of solar UV.

Matrix metalloproteinases

The gene for collagenase was among the first to be shown to be up-regulated in human fibroblast cell populations as a late response to UVC radiation. Collagenase activity is also induced by UVA radiation (Scharfetter et al., 1991; Petersen et al, 1992; Hermann et al., 1993). Interestingly, there is evidence for cytokine involvement in the delayed induction of collagenase by both wavelength ranges. High doses of UVC radiation lead to the eventual secretion of basic fibroblast growth factor and interleukin 1 from cells and these appear to constitute an autocrine loop partially responsible for up-regulation of the collagenase gene (Krämer et al., 1993). IL-6 synthesis and release can be stimulated by both IL-1 and by UVA radiation and it has been argued that IL-6 is at least partly responsible for the activation of collagenase following UVA irradiation of human fibroblast cultures (Wlascheck et al., 1993). UVA penetrates deep into the skin and activation of collagenase by repeated exposures is likely to be involved in collagen breakdown which is an underlying cause of actinic elastosis, a prominent feature of photoaging of the skin.

Phospholipases

In addition to the CL100 phosphatase, several of the proteins whose activity is increased following UVA irradiation may play a role in signal transduction pathways. Among these is phospholipase A2 (PLA_2) (de Leo et al., 1985) which acts on lipid membranes to release arachidonic acid which is then catabolized by cyclooxygenases and lipoxygenases to generate inflammatory mediators that include leukotrienes and prostaglandins. There is indirect evidence (based on arachidonate release) that UVA radiation can activate PLA_2 in human epidermal keratinocytes (Hanson and DeLeo, 1990) which is consistent with later evidence obtained using mouse macrophages that shows that reactive oxygen species can be involved in activation of the gene (Goldman et al., 1992). Activation appears to involve phosphorylation by a MAP kinase (Lin et al., 1993) (although UVA irradiation of cells can also slightly increase PLA_2 messenger RNA levels). Phospholipase activation may be involved in defense pathways at two levels. Firstly, the enzyme appears to play a critical role in rapid repair of membrane damage, and secondly, it is clearly involved in the acute inflammatory response in skin. The activation of this enzyme by UVA radiation at exposure levels expected from natural sunlight appears to provide a further example of an adaptive defense pathway.

Transcription factor activation

In contrast with the literature concerning UVC radiation, there is only limited evidence to show that UVA radiation can activate common transcription factors. The proximal promoter of the rodent and human HO-1 genes contains an element that binds upstream stimulatory factor (USF). Although UVA radiation was shown to change the pattern of binding of USF to its DNA binding site (Nascimento et al., 1993) it now appears that this may have been due to an artifactual proteolytic truncation during the extraction procedure as a result of UVA-induced release of proteases (Waltner and Tyrrell, unpublished data). Regions 4 kb and 10 kb upstream of the start site of the HO-1 gene have been shown to contain sequence elements (including AP-1 sites) that are important in activation of the gene by cadmium chloride (Alam et al., 1995) but the role of these regions in UVA activation has not yet been clarified.

UVA radiation at sublethal doses does activate binding of NFkB to its well-characterized DNA recognition sequence (Vile et al., 1995) by a pathway inhibitable by membrane antioxidants. However, there is no evidence as yet that this activation has *in vivo* significance in up-regulation of genes (see also Smudzka and Beer, 1990) UVC and UVB radiation do activate NFkB-dependent promoter activity when applied at doses high enough to cause membrane damage (see also Storz and Polla, this volume).

It should be mentioned that there is an expanding literature on the UV activation of immunomodulatory molecules (see Ullrich, 1995, and later) and that both ICAM-1 (Krutmann and Trefzer, 1992) and IL-10 (Grewe et al., 1995) have been shown to be induced by UVA radiation.

UVB induction of stress proteins

The short UVB wavelengths in sunlight are not only absorbed by DNA and cause DNA damage, but also penetrate sufficiently into skin to cause severe damage in the epidermis and even the dermis. In unprotected human skin, these are therefore the most important wavelengths for causing severe acute effects such as sunburn and chronic effects that include basal and squamous cell carcinomas. In terms of gene activation, UVB would be expected to have UVC-like properties because of its strong absorption by critical macromolecules and UVA-like properties because of its oxidative component. Investigations to date show that, in general, UVB activates a similar set of genes to UVC. The genes that are most strongly up-regulated by UVA radiation (HO-1 and the CL-100 phosphatase) are only slightly activated by short UV wavelengths and high doses are required that are outside of the physiologically relevant range. Because of its penetrating properties and its value as a model for solar irradiation, UVB from broad spectrum lamps such as solar simulators is most commonly used in animal studies. The possible biological relevance of gene activation studies originally done with cultured cells using UVC radiation is often deduced from subsequent studies using UVB irradiation of rodent or even human skin.

The up-regulation of the c-*fos* nuclear oncogene that occurs in cultured cells after UVC irradiation (Büscher et al., 1988) also occurs after UVB irradiation (Stein et al., 1989; Ghosh et al., 1993; Garmyn et al., 1991) of cultured fibroblasts and epidermal cells. RNA messenger levels have also been measured after broad spectrum UVB irradiation of human (Roddey et al., 1994) and rodent (Brunet and Giacomoni, 1990; Gillardon et al., 1994) skin at erythemogenic doses. However, the induction in mouse skin is at most two-fold and in the human study was only seen in half (three out of six) of the cases studied. The levels of *jun* mRNA are enhanced up to four-fold after UVB treatment of rodent skin (Gillardon et al., 1994). It is clearly important to further investigate the possible biological relevance of the UVB enhancement of *fos* and *jun* protein levels *in vivo*. Based on recent experimental evidence, it has been proposed that by overcoming the UV-induced block to replication, *fos* may actually play a decisive role in cellular defense following UV radiation (Haas and Kaina, 1995). Levels of the c-*myc* protein, another nuclear oncogene and component of a transcription factor complex, are also increased in the human epidermis following two times the minimum erythemal dose of UVB (Takahashi et al., 1993), although a decrease has been observed in an independent study (Garmyn et al., 1991).

Ornithine decarboxylase (ODC) is another protein which can lead to stimulation of both DNA synthesis and cell proliferation, in this case by virtue of its involvement in the synthesis of polyamines. It is therefore noteworthy that this protein is induced much more strongly than the c-*fos* protein by UVB irradiation of skin (Lowe et al., 1978; Hillebrand et al., 1990). No evidence has so far been presented for a protective role of induction of this enzyme which, on the contrary, is often used as a marker of skin damage and tumor promoting ability. Indeed, a new class of anti-cancer drugs is currently being developed based on their ability to inhibit ODC gene transcription (Gerhäuser et al., 1995). Although there is some evidence both for and against a role of an oxidative component in UVB activation of ODC (Black and Mathews-Roth, 1991; Kono et al., 1992), UV activation of expression appears to be limited to the short wavelength regions and, as far as it has been sought, there is no evidence of induction by oxidizing UVA radiations (Young et al., 1986).

Activation of ODC expression in mouse keratinocytes by UVB occurs at least partly as a result of an increase in messenger accumulation. Recent information on the genomic sequence of the gene has led to the recognition of potential myc/max binding elements in the promoter and first intron of the gene, and it appears that myc protein (but probably not max) is involved in transactivation of the gene (Pena et al., 1993; Bello-Fernandez et al., 1993).

The product of the tumor suppressor gene, p53, also increases in skin after UVB irradiation (Campbell et al., 1993, Fig. 1), presumably as a result of protein stabilization (Maltzmann and Czyzyk, 1984). Enhanced levels of the p53 protein may be protective in at least two ways. Firstly, in contrast to the stimulating effects of fos and ODC, p53 appears to be involved in a pathway which temporily blocks DNA synthesis and mitosis by a G1 arrest and allows time for repair of DNA damage (see Canman and Kastan, 1995, for a recent review). However, the p53 protein can also induce apoptosis and the p53 apoptotic pathway appears to be involved in suppression of tumor growth and progression *in vivo* (Symonds et al., 1994). Finally, the protein is a transcription factor with many dependent genes (see above), several of which are involved in either cell cycle arrest or apoptosis. Interestingly, UVA irradiation also leads to an increase in the level of p53 protein in the basal layer of the epidermis, but nothing is known about the mechanism by which this occurs.

Keratinocytes synthesize many cytokines at relatively low levels. UVB irradiation of both cultured keratinocytes and the epidermis *in vivo* leads to the up-regulation of expression of a wide-range of cytokines (for reviews, see Schwarz and Luger, 1989; Ullrich, 1995) including the interleukins IL-1 α and β, IL-3, IL-6, IL-8, IL-10 tumor necrosis factor (TNF) and granulocyte/macrophage colony stimulating factor (GM-CSF). In several cases, the increased synthesis and secretion of these immunomodulators has been shown to involve up-regulation of transcription. IL-1 itself may up-regulate several of the other cytokines including GM-CSF. Many of these cytokines play a key role in inflammation and therefore cellular defense. They may also act locally by infiltration into the dermis to trigger other gene activation pathways (in fibroblasts for example) but there is relatively little information available in this area. Certainly the same pathways of gene activation observed in fibroblasts that involve autocrine stimulation by IL-1 (Krämer et al., 1993) could be affected by the high levels of IL-1 secreted by epidermal keratinocytes in response to UVB radiation.

Intracellular adhesion molecule 1 (ICAM-1) is an immunomodulating molecule which mediates adhesion between epidermal keratinocytes and inflammatory leukocytes and is critical in the development of the inflammatory response. Both UVB and UVA radiation can up-regulate expression of ICAM-1 in human keratinocytes (Krutmann and Trefzer, 1992). However, unlike UVA radiation, UVB radiation shows a biphasic response and initially suppresses cytokine-induced ICAM1 mRNA accumulation over the first 24 h followed by a later stimulation. Indeed, the existence of a balance between gene activation and UVB induced suppression of gene activation has also been demonstrated for c-*fos* (Ghosh et al., 1993) and ODC (Zheng et al., 1993) and there is evidence that it depends on the induction of DNA damage (Krutmann et al., 1995). It is evident that the delicate balance between suppression and activation of gene expression needs to be taken into account in all models of gene activation by UVB.

Studies in cultured keratinocytes have indicated additional pathways by which UVB can generate signaling intermediates which could lead to gene activation. It has been known for some time that UVB radiation can prevent epidermal growth factor (EGF) binding to its membrane receptor. There is now evidence that UVB irradiation of keratinocytes leads to ligand-independent phosphorylation and activation of the EGF receptor (Miller et al., 1994). The same treatment leads to a sharp increase in prostaglandin E_2, an important inflammatory mediator. Epidermal phospholipase activity is essential to break down membrane phospholipids to arachidonate and thus provide the substrate for prostaglandin synthesis via cyclooxygenase. Since this activity is also increased by similar UVB treatment (De Leo et al., 1985; Kang-Rotondo et al., 1993) and there is evidence that EGF receptor kinase activation is a critical step in the kinase cascade that leads to MAP kinase activation and phosphorylation of PLA_2 (Goldberg et al., 1990; Hack et al., 1991; Lin et al., 1993), it appears that an important pathway by which UVB can generate proliferative and inflammatory responses is by mimicking a growth factor response. As noted earlier, high doses of UVC can also generate such responses but it is not known whether UVA can activate the EGF receptor. Furthermore, an independent study has provided evidence that increased PGE_2 synthesis following UVB irradiation of transformed human epidermal cells is, at least in part, mediated by tumor necrosis factor (TNF-α) and IL-1 acting at specific cell surface receptors (Grewe et al., 1993).

In conclusion, UV radiation can induce a variety of stress proteins which differ sharply between the short UVC wavelengths and the long oxidizing UVA wavelengths present in natural sunlight. Although activation of genes by UVB radiation more closely resembles the pattern of activation by UVC, many of the studies with the non-solar wavelength are irrelevant to the physiological or even pathological response of tissues to solar UVB irradiation because of the high doses used. Nevertheless, UVB radiation does induce the expression of several genes in skin at dose levels relevant to natural solar exposure. The stress proteins that do appear are often involved in growth stimulation, possibly because an important property of UVB is to mimic growth factor responses. An understanding of the involvement of UVB modulated genes in apoptosis will also help to build up a picture of how UVB can act as a complete carcinogen. Arresting or stimulating growth at appropriate times may also be protective in the short term. In addition to these effects, UVB can stimulate the inflammatory response by up-regulation of cytokines and other inflammatory mediators. However the precise balance between the positive and negative effects of the induction of stress proteins by UVB radiation in cells and tissue has yet to be elucidated. UVA radiation may also have positive short-term effects related to stress protein induction such as the boosting of protective mechanisms. Nevertheless, chronic exposure to both the UVA and UVB components of sunlight is a central factor not only in photoaging of the skin but in the development of both non-melanoma and melanoma-type skin cancers in man.

Acknowledgements
The work of the author is supported by the Association for International Cancer Research (UK), the League Against Cancer of Central Switzerland, La Ligue Neuchâteloise contre le Cancer and the Swiss National Science Foundation (31-37148.93, 31-040720.94).

References

Alam, J., Camhi, S. and Choi, A.M.K. (1995) Identification of a second region upstream of the mouse heme oxygenase-1 gene that functions as a basal level and inducer-dependent transcription enhancer. *J. Biol., Chem.* 270:11977–11984.

Applegate, L.A., Luscher, P. and Tyrrell R.M. (1991) Induction of heme oxygenase: A general response to oxidant stress in cultured mammalian cells. *Cancer Res.* 51:974–978.

Applegate, L.A., Noel, A., Vile, G.F., Frenk, E. and Tyrrell, R.M. (1995) Two genes contribute to different extents to the heme oxygenase enzyme activity measured in cultured human skin fibroblasts and keratinocytes: implications for protection against oxidant stress. *Photochem. Photobiol.* 61:285–291.

Backendorf, C. and Hohl, D. (1992) A common origin for cornified envelope proteins? *Nature Genetics* 2:91.

Basu-Modak, S. and Tyrrell, R.M. (1993) Singlet oxygen. A primary effector in the UVA/near visible light induction of the human heme oxygenase gene. *Cancer Res.* 53:4505–4510.

Bello-Fernandez, C., Packham, G. and Cleveland, J.L. (1993) The ornithine decarboxylase gene is a transcriptional target of c-Myc. *Proc. Natl. Acad. Sci. USA* 90:7804–7808.

Black, H.S. and Mathews-Roth, M.M. (1991) Protective role of butylated hydroxytoluene and certain carotenoids in photocarcinogenesis. *Photochem. Photobiol.* 53:707–716.

Brunet, S. and Giacomoni, P.U. (1990) Specific mRNAs accumulate in long-wavelength UV-irradiated mouse epidermis. *J. Photochem. Photobiol. B. Biol.* 6:431–441.

Büscher, M., Rahmsdorf, H.J., Litfin, M., Karin, M. and Herrlich, P. (1988) Activation of the c-fos gene by UV and phorbol ester: different signal transduction pathways converge to the same enhancer element. *Oncogene* 3:301–311.

Campbell, C., Quinn, A.G., Angus, B., Farr, P.M. and Rees, J.L. (1993) Wavelength specific patterns of p53 induction in human skin following exposure to UV radiation. *Cancer Res.* 53:2697–2699.

Canman, C.E. and Kastan, M.B. (1995) Induction of apoptosis by tumor suppressor genes and oncogenes. *Seminars in Cancer Biology* 6:17–25.

DeLeo, V.A., Hanson, D., Weinstein, I.B. and Harber, L.C. (1985) Ultraviolet radiation stimulates the release of arachidonic acid from mammalian cells in culture. *Photochem. Photobiol.* 41:51–56.

Derijard B., Hibi, M., Wu, I.H., Barrett, T., Su, B., Deng, T., Karin, M. and Davis, R.J. (1994) JNK1: a protein kinase stimulated by UV light and Ha-Ras that binds an phosphorylates the c-Jun activation domain. *Cell* 76:1025–1037.

Devary, Y., Gottlieb, R.A., Lau, L.F. and Karin, M. (1991) Rapid and preferential activation of the c-jun gene during the mammalian UV response. *Mol. Cell. Biol.* 11:2804–2811.

Devary, Y, Gottlieb, R.A., Smeal, T. and Karin, M. (1992) The mammalian ultraviolet response is triggered by activation of Src tyrosine kinases. *Cell* 71:1081–1091.

Devary, Y., Rosette, C., DiDonato, J.A. and Karin, M. (1993) NF-κB activation by ultraviolet light not dependent on a nuclear signal. *Science* 261:1442–1445.

Garmyn, M., Yaar, M., Holbrook, N., and Gilchrest, B.A. (1991) Immediate and delayed molecular response of human keratinocytes to solar-simulated irradiation. *J. Lab. Invest.* 65:471–478.

Garmyn, M., Yaar, M. Boileau, N., Backendorf, C. and Gilchrest, B. (1992) Effect of aging and habitual sun exposure on the genetic response of cultured human keratinocytes to solar-stimulated irradiation. *J. Invest. Dermatol.* 99:743–748.

Gerhäuser, C., Mar, W., Lee, S.K., Suh, N., Luo Y., Kosmeder, J., Luyeng, L., Fong, H.H.S., King-Horn, A., Moriarty, R.M., Mehta, R.G., Constantinou, A., Moon, R.C. and Pezzuto, J.M. (1995) Rotenoids mediate potent cancer chemopreventive activity through transcriptional regulation of ornithine decarboxylase. *Nature Med.* 1:260–266.

Ghosh, R., Amstad, P. and Cerutti, P. (1993) UVB-induced DNA breaks interfere with tran-scriptional induction of c-fos. *Mol. Cell. Biol.* 13:6992–6999.

Gibbs, S., Lohman, F., Teubel, W., van de Putte, P. and Backendorf, C. (1990) Characteriza-tion of the human spr2 promoter: induction after UV irradiation or TPA treatment and regulation during differentiation of cultured primary keratinocytes. *Nucleic Acids Res.* 18:4401–4407.

Gibbs, S. (1992) Interference of light with gene regulation in epidermal keratinocytes; PhD thesis, University of Leiden.

Gillardon, F., Eschenfelder, C., Uhlmann, E., Hartschuh, W. and Zimmermann, M. (1994) Differential regulation of c-fos, fosB, c-jun, junB, bcl-2 and bax expression in rat skin follow-ing single or chronic ultraviolet irradiation and *in vivo* modulation by antisense oligo-deoxynucleotide superfusion. *Oncogene* 9:3219–3225.

Goldman, R., Ferber, E. and Zort, U. (1992) Reactive oxygen species are involved in the activa-tion of cellular phospholipase AZ. *FEBS Lett.* 309:190–192.

Grewe, M., Treftzer, U., Ballhorn, A., Gyufko, K., Henninger, H. and Krutmann, J. (1993) Analysis of the mechanisms of ultraviolet (UV) B radiation-induced prostaglandin E2 syn-thesis by human epidermoid caricnoma cells. *J. Invest. Dermatol.* 101:528–531.

Grewe, M., Gyufki, K. and Krutmann, J. (1995) Interleukin-10 production by cultured human keratinocytes: regulation by ultraviolet B and ultraviolet A1 radiation. *J. Invest. Dermatol.* 104:3–6.

Haas, S. and Kaina, B. (1995) c-Fos is involved in the cellular defence against the genotoxic effect of UV radiation. *Carcinogenesis* 16:985–991.

Hack, N., Margolis, B.L., Ullrich, A., Schlessinger, J. and Skorecki, K.L. (1991) Distinct struc-tural specificities for functional coupling of the epidermal growth factor receptor to calcium-signalling versus phospholipase A2 responses. *Biochemical J.* 275:563–567.

Hanson, D. and DeLeo, V. (1990) Long-wave ultraviolet light induces phospholipase activation in cultured human epidermal keratinocytes. *J. Invest. Dermatol.* 95:158–163.

Hermann, G., Wlaschek, M., Lange, T.S., Prenzel, K., Goerz, G., Scharffetter-Kochanek, K. (1993) UVA irradiation stimulates the synthesis of various matrix-metalloproteinases (MMPS) in cultured human fibroblasts. *Exp. Dermatol.* 2:92–97.

Herrlich, P., Sachsenmaier, C., Radler-Pohl, A., Gebel, S., Blattner, C. and Rahmsdorf, H.J. (1994) The mammalian UV response: mechanism of DNA damaged induced gene expres-sion. *Advances in Enzymes Regulation* 34:381–395.

Hillebrand, G.G., Winslow, M.S., Benzinger, M.J., Heitmeyer, D.A. and Bisset, D.L. (1990) Acute and chronic ultraviolet radiation induction of epidermal ornithine decarboxylase activity in hairless mice. *Cancer Res.* 50:1580–1584.

Hipskind, R.A., Rao, V.N., Mueller, C.G., Reddy, E.S. and Nordheim, A. (1991) Ets-related protein Elk-1 is homologus to the c-fos regulatory factor p62TCF. *Nature* 354:531–534.

Janknecht, R., Ernst, W.H., Pingoud, V. and Nordheim, A. (1993) Activation of ternary complex factor Elk-1 BY map kinases. *EMBO J.* 12:5097–5104.

Kartasova, T. and van de Putte, P. (1988) Isolation, characterization, and UV-stimulated expres-sion of two families of genes encoding polypeptides of related structure in human epidermal keratinocytes. *Mol. Cell. Biol.* 8:2195–2203.

Kartasova, T., Goos, van Muijen, G.N.P., van Pelt-Heerschap, H. and van de Putte, P. (1988) Novel protein in human epidermal keratinocytes: regulation of expression during differ-entiation. *Mol. Cell. Biol.* 8:2204–2210.

Kastan, M.B., Onyekwere, O., Sidransky, D., Vogelstein, B. and Craig, R.W. (1991) Parti-cipation of p53 protein in the cellular response to DNA damage. *Cancer Res.* 51:6304–6311.

Keyse, S.M. and Tyrrell, R.M. (1989) Heme oxygenase is the major 32-kDa stress protein induced in human skin fibroblasts by UVA radiation, hydrogen peroxide and sodium arsenite. *Proc. Natl. Acad. Sci. USA* 86:99–103.

Keyse, S.M. and Emslie, E.A. (1992) Oxidative stress and heat shock induce a human gene encoding a protein-tyrosine phosphatase. *Nature* 359:644–647.

Keyse, S.M. (1995) An emerging family of dual specificity MAP kinase phosphatases. *Bio-chim. Biophys. Acta* 1265:152–160.

Kono, T., Taniguchi, S., Mizuno, N., Fukuda, M., Maekawa, N., Hisa, T., Ishii, M., Otani S. and Hamada, T. (1992) Effects of butylated hydroxyanisole on ornithine decarboxylase activity induced by ultraviolet-B and PUVA in mouse skin. *J. Dermatol.* 19:389–392.

Krämer, M., Sachsenmaier, C., Herrlich, P. and Rahmsdorf, H.J. (1993) UV-irradiation-induced interleukin-1 and basic fibroblast growth factor synthesis and release mediate part of the UV response. *J. Biol. Chem.* 268:6734–6741.

Krutmann, J. and Trefzer, U. (1992) Modulation of the expression of intercellular adhesion molecule-1 (UCAM-1) in human keratinocytes by ultraviolet (UV) radiation. *Springer Semin. Immunopathol.* 13:333–344.

Krutmann, J. Bohnert E. and Jung, E.G. (1994) Evidence that DNA damage is a mediate in ultraviolet B radiation-induced inhibition of human gene expression: ultraviolet B radiation effects on intercellular adhesion molecule-1 (ICAM-1) expression. *J. Invest. Dermatol.* 102; 428–432.

Lautier, D., Lüscher, P. and Tyrrell, R.M. (1992) Endogenous glutathione levels modulate both constitutive and UVA radiation/oxidant-inducible expression of the human heme oxygenase gene. *Carcinogenesis* 13:227–232.

Lin, L.-L., Wartmann, M., Lin, A.Y., Knopf, J.L., Seth, A. and Davis, R.J. (1993) cPLA2 is phosphorylated and activated by MAP kinase. *Cell* 72:269–278.

Lowe, M., Verma, A.K. and Boutwell, R.K. (1978) Ultraviolet light induces epidermal ornithine decarboxylase activity. *J. Invest. Dermatol.* 71:417–418.

Maltzman, W. and Czyzyk, L. (1984) UV irradiation stimulates levels of p53 cellular tumor antigen in nontransformed mouse cells. *Mol. Cell. Biol.* 4:1689–1694.

Miller, C.C., Hale, P. and Pentland, A.P. (1994) Ultraviolet B injury increases prostaglandin synthesis through a tyrosine kinase-dependent pathway. Evidence for UVB-induced epidermal growth factor receptor activation. *J. Biol. Chem.* 269:3529–3533.

Minden, A., Lin, A., Smeal, T., Derijard, B., Cobb, M., Davis, R. and Karin, M. (1994) c-jun N-terminal phosphorylation correlates with activation of the JNK subgroup but not the ERK subgroup of mitogen-activated protein kinases. *Mol. Cell. Biol.* 14:6683–6688.

Miskin, R. and Ben-Ishai, R. (1982) Induction of plasminogen activator by UV light in normal and *Xeroderma pigmentosum* fibroblasts. *Proc. Natl. Acad. Sci. USA* 78:6236–6240.

Myashita, T. and Reed, J.C. (1995) Tumor suppressor p53 is a direct transcriptional activator of the human box gene. *Cell* 80:243–299.

Nascimento, A.L.T.O., Luscher, P. and Tyrrell, R.M. (1993) Ultraviolet A (320–380 nm) radiation causes the binding of a specific protein complex to a short region of the promoter of the human heme oxygenase 1 gene. *Nucleic Acids Res.* 21:1103–1109.

Naumann, M. and Scheidereit, C. (1994) Activation of NF-kB *in vivo* is regulated by multiple phosphorylations. *EMBO J.* 13:4597–4607.

Palombella, V.J., Rando, O.J., Goldberg, A.L. and Maniatis, T. (1994) The ubiquitin-proteasome pathway is required for processing the NF-kB1 precursor protein and the activation of NF-kB. *Cell* 78:773–785.

Pena, A., Reddy, C.D., Wu, S., Hickcok, N.J., Reddy, E.P., Yumet, G., Soprano, D.R. and Soprano, K.J. (1993) Regulation of human ornithine decarboxylase expression by the c-Myc.Max protein complex. *J. Biol. Chem.* 258:27277–27285.

Petersen, M.J., Hansen, C. and Craig, S. (1992) Ultraviolet A irradiation stimulates collagenase production in cultured human fibroblasts. *J. Invest. Dermatol.* 99:440–444.

Radler-Pohl, A., Sachsenmaier, C., Gebel, S., Auer, H.P., Bruder, J.T., Rapp, U., Angel, P., Rahmsdorf, H.J. and Herrlich, P. (1993) UV-induced activation of AP-1 involves obligatory extranuclear steps including Raf-1 kinase. *EMBO J.* 12:1005–1012.

Roddey, P.K., Garmyn, M., Park, H.-Y., Bhawan, J. and Gilchrest, B.A. (1994) Ultraviolet irradiation induces c-fos but not c-Ha-ras proto-oncogene expression in human epidermins. *J. Invest. Dermatol.* 102:296–299.

Ronai, Z.A., Okin, E. and Weinstein, I.B. (1988) Ultraviolet light induces the expression of oncogenes in rat fibroblast and human keratinocyte cells. *Oncogene* 2:201–204.

Rosen, C.F., Gajic, D. and Drucker, D.J. (1990) Ultraviolet radiation induction of ornithine decarboxylase in rat keratinocytes. *Cancer Res.* 50:2631–2635.

Rosen, C.F., Gajic, D., Jia, Q. and Drucker, D.J. (1990) Ultraviolet B radiation induction of ornithine decarboxylase gene expression in mouse epidermis. *Biochem. J.* 270:565–568.

Rosenstein, B.S. and Mitchell, D.L. (1987) Action spectra for the induction of pyrimidine (6-4)pyrimidone photoproducts and cyclobutane pyrimidine dimers in normal human skin fibroblasts. *Photochem. Photobiol.* 45:775–780.

Roshchupkin, D.J., Pelenitsyn, A.B., Potapenko, A.Y., Talitsky, V.V. and Vladimirov, Y.A. (1975) Study of the effects of ultraviolet light on biomembranes – IV. The effect of oxygen on UV-induced hemolysis and lipid photoperoxidation in rat erythrocytes and liposomes. *Photochem. Photobiol.* 21:63–69.

Rotem, N., Axelrod, J.H. and Miskin, R. (1987). Induction of urokinase-type plasminogen activator by UV light in human fetal fibroblasts is mediated through a UV-induced secreted protein. *Mol. Cell. Biol.* 7:622–631.

Sachsenmaier, C., Radler-Pohl, A., Zinck, R., Nordheim, A., Herrlich, P. and Rahmsdorf, H.J. (1994) Involvement of growth factor receptors in the mammalian UVC response. *Cell* 78:963–972.

Sarna, J. (1992) Properties and function of ocular melanin – a photobiophysical view. *J. Photochem. Photobiol. B. Biol.* 12:215–228.

Sato M., Ishizawa, S., Yoshida, T. and Shibahara, S. (1990) Interaction of upstream stimulatory factor with the human heme oxygenase promoter. *Eur. J. Biochem.* 188:231–237.

Schaeffer, L., Roy, R., Humbert, S., Moncollin, V., Vermeulen, W., Hoeijmakers, J.H.J., Chambon, P. and Egly, J.M. (1993) DNA repair helicase: a component of BTF2 (TFIIH) basic transcription factor. *Science* 260:58–63.

Scharffetter, K., Wlaschek, M., Hogg, A., Bolsen, K., Schothorst, A., Goerz, G., Krieg, T. and Plewig, G. (1991) UVA irradiation induces collagenase in human dermal fibroblasts *in vitro* and *in vivo. Arch. Dermatol. Res.* 283:506–511.

Scharffetter-Kochanek, K., Wlaschek, M., Briviba, K. and Sies, H. (1993) Singlet oxygen induces collagenase expression in human skin fibroblasts. *FEBS Lett.* 331:304–306.

Schwarz, T. and Luger, T.A. (1989) New trends in photobiology (invited review). Effect of UV irradiation on epidermal cell cytokine production. *J. Photochem. Photobiol. B. Biol.* 4:1–13.

Smudzka, B.Z. and Beer, J.Z. (1990) Activation of human immunodeficiency virus by ultraviolet radiation. *Photochem. Photobiol.* 52:1153–1162.

Stein, B., Rahmsdorf, H.J., Steffen, A., Litfin, M. and Herrlich, P. (1989) UV-induced DNA damage is an intermediate step in UV-induced expression of human immunodeficiency virus type 1, collagenase, c-fos, and metallothionein. *Mol. Cell. Biol.* 9:5169–5181.

Symonds, H., Krall, L., Remington, L., Saenz-Roble, S.M., Lowe, S., Jacks, T. and van Dyke, T. (1994) p53-dependent apoptosis suppresses tumour growth and progression *in vivo. Cell* 78:703–711.

Takahashi, S., Pearse, A.D. and Marks, R. (1993) The acute effects of ultraviolet-B radiation on c-myc and c-Ha ras expression in normal human epidermis. *J. Dermatol. Science* 6:165–171.

Tyrrell, R.M. (1991) UVA (320–380 nm) radiation as an oxidative stress. *In:* H. Sies (ed.) *Oxidative Stress: Oxidants and Antioxidants*, Academic Press, London, pp 57–83.

Ullrich, S.E. (1995) The role of epidermal cytokines in the generation of cutaneous immune reactions and ultraviolet radiation-induced immune suppression. *Photochem. Photobiol.* 62:389–401.

Van Dam H., Willhelm, D., Herr, I., Steffen, A., Herrlich, P. and Angel, P. (1995) ATF-2 is preferentially activated by stress-activated protein kinases to mediate c-jun induction in response to genotoxic agents. *EMBO J.* 14:1798–1811.

Van den Berg, S., Kaina, B., Rahmsdorf, H.J., Ponta, H. and Herrlich, P. (1991) Involvement of fos in spontaneous and ultraviolet light-induced genetic changes. *Mol. Carcinogenesis* 4:460–466.

Vile, G.F. and Tyrrell, R.M. (1993) Oxidative stress resulting from ultraviolet A irradiation of human skin fibroblasts leads to a heme oxygenase-dependent increase in ferritin. *J. Biol. Chem.* 268:14678–14681.

Vile, G.F., Basu-Modak, S., Waltner, C. and Tyrrell, R.M. (1994) Haeme oxygenase 1 mediates an adaptive response to oxidative stress in human skin fibroblasts. *Proc. Natl. Acad. Sci. USA* 91:2607–2610.

Vile, G.F., Tanew-Iliitschew, A. and Tyrrell, R.M. (1995) Activation of NF-kB in human skin fibroblasts by the oxidative stress generated by UVA radiation. *Photochem. Photobiol.* 62:463–468.

Walker, G.C. (1987) The SOS response of *Escherichia coli. In:* F.C. Niedhardt, J.L. Ingraham, K.B. Low, B. Magasanik, M. Shaechter and H.E. Umbarger (eds): *Escherichia coli* and *Salmonella typhimurium. Cellular and Molecular Biology.* American Society of Microbiology, Washington DC, pp 1346–1357.

Wlaschek, M., Bolsen, K., Herrmann, G., Schwarz, A., Wilmroth, F., Herrlich, P.C., Goerz, G. and Scharffetter-Kochanek, K. (1993) UVA-induced autocrine stimulation of fibroblast-derived collagenase by IL-6: a possible mechanism in dermal photodamage. *J. Invest. Dermatol.* 101:164–168.

Wlaschek, M., Heinen, G., Poswig, A., Schwarz, A., Krieg, T. and Scharffetter-Kochanek, K. (1994) UVA-induced autocrine stimulation of fibroblast-derived collagenase/MMP-1 by interrelated loops of interleukin-1 and interleukin-6. *Photochem. Photobiol.* 59:550–556.

Young, A.R., Connor, M.T. and Lowe, N.J. (1986) UV wavelength dependence for the induction of ornithine decarboxylase activity in hairless mouse epidermis. *Carcinogenesis* 7: 601–604.

Zheng, Z.S., Chen, R.Z. and Prystowsky, J.H. (1993) UVB radiation induces phosphorylation of the epidermal growth factor receptor, decreases EGF binding and blocks EGF induction of ornithine decarboxylase gene expression in SV40-transformed human keratinocytes. *Exp. Dermatol.* 2:257–265.

Stress-Inducible Cellular Responses
ed. by U. Feige, R.I. Morimoto, I. Yahara and B. Polla
© 1996 Birkhäuser Verlag Basel/Switzerland

Signaling events controlling the molecular response to genotoxic stress

N. J. Holbrook[1], Y. Liu[1] and A. J. Fornace, Jr.[2]

[1]*Gene Expression and Aging Section, National Institute on Aging, Baltimore, MD 21244, USA*
[2]*Laboratory of Molecular Pharmacology, National Cancer Institute, Bethesda, MD 20892, USA*

Summary. Recently, much progress has been made in defining the signal transduction pathways mediating the cellular response to genotoxic stress. Multiple pathways involving several distinct MAP kinases (ERK, JNK/SAPK, and p38/HOG1) as well as the tumor suppressor protein p53 contribute to the response; the various pathways being differentially activated by particular genotoxic agents. Although both DNA damage and extranuclear events are important in initiating the response, recent evidence suggests the response is controlled primarily through events occurring at the plasma membrane, overlapping significantly with those important in initiating mitogenic responses. Attenuation of the responses appears to be largely controlled through feedback mechanisms involving gene products produced during the activation process.

Exposure of cells to genotoxic agents evokes a series of events leading to the activation of a wide group of genes with diverse functions (Holbrook and Fornace, 1991). Among these are a number of transcription factors as well as other gene products that are also rapidly and highly induced in response to mitogenic stimulation, illustrating the commonalities of the pathways involved in mediating the cellular response to stress and proliferative signals.

Much of what we know concerning the cellular events controlling the genotoxic stress response has come from studies utilizing UVC irradiation. However, given the large number of agents that damage DNA, the diversity of these agents with respect to their structures and mechanisms of action, and their ability to inflict damage on other cellular components (aside from DNA), it is not surprising that the genotoxic response is complex, involving multiple signaling pathways running in parallel to produce the final outcome. This review will briefly summarize the current information regarding the molecular response to genotoxic stress addressing four issues: (1) the signaling pathways employed, (2) the damage sensing and signal initiating events, (3) the attenuation of the response, and (4) the consequences of the response, emphasizing its relationship to cell cycle control and influence on cell survival and/or cell death.

Signal transduction pathways mediating the response

Role of mitogen-activated protein kinases

As depicted in Figure 1, at least four pathways appear to be involved in mediating the cellular response to genotoxins. Three of these rely on the activation of mitogen-activated protein (MAP) kinases which include extra-cellular signal-regulated kinases (ERK), stress activated protein kinases (SAPK)/c-Jun N-terminal kinases (JNK), and p38/RK/CSBP kinases (Cano and Mahadevan, 1995). These kinases play a key role in the activation of transcription factors and other regulatory proteins involved in activating gene expression. A common feature of MAP kinases is their activation through phosphorylation of threonine and tyrosine residues in a homologous kinase subdomain VIII (Cano and Mahadevan, 1995). They are distinguished by (1) the tripeptide motifs specifying their phosphorylation (TEY, TPY, and TGY; for ERK, JNK and p38, respectively), (2) the phosphorylation cascades leading to their activation, and (3) the proteins which they in turn phosphorylate.

Extracellular signal-regulated protein kinases (ERK)
It has long been appreciated that there is considerable overlap in the cellular responses to mitogenic (i.e. growth factor) stimulation and UVC (Blattner et al., 1994; Herrlich et al., 1994). This overlap can be attributed, in large part, to the activation of either or both of two ERK isoforms, ERK1 and ERK2, in both responses. Events leading to the activation of ERK1 and ERK2 have been most extensively studied with regard to mitogenic responses and more detailed reviews are available on this topic (Davis, 1993; Avruch et al., 1994). The first step involves the activation of growth factor receptor tyrosine kinases resulting in their autophosphorylation. This triggers a cascade of phosphorylation events which involves sequential activation of Ras, Raf, MAP kinase kinase (MEK) and finally ERKs. ERKs phosphorylate a variety of regulatory proteins leading to their activation (Davis, 1993). Among the best characterized of these is the Elk1/TCF transcription factor complex which forms a ternary complex with the serum response factor p67[SRF] (Gille et al., 1992; Hill et al., 1993). Phosphorylation enhances complex formation and its binding to the serum response element located in the promoters of stress response genes such as c-*fos*. Other proteins regulated by ERK include p90[RSK], activating transcription factor-2 (ATF-2; a cAMP response element-binding protein), NF-IL6 (nuclear factor for the activation of Interleukin 6) and c-Myc (Davis, 1993).

This same phosphorylation cascade is involved in activating genes in response to genotoxic stresses including UVC irradiation (Sachsenmaier et al., 1994), arsenite (Liu et al., 1996), and hydrogen peroxide (Guyton et al., 1996). Although variable from one cell type to another, the magnitude of ERK activation occurring in response to stress is generally less than that seen with mitogen treatment.

Stress-activated protein/c-Jun N-terminal kinases (SAPK/JNK)
The SAPK/JNK family consists of at least six isoforms, with JNK1 and JNK2, being the most extensively studied (Derijard et al., 1994; Sluss et al., 1994; Kyriakis et al., 1994). As implied by their name, these kinases are

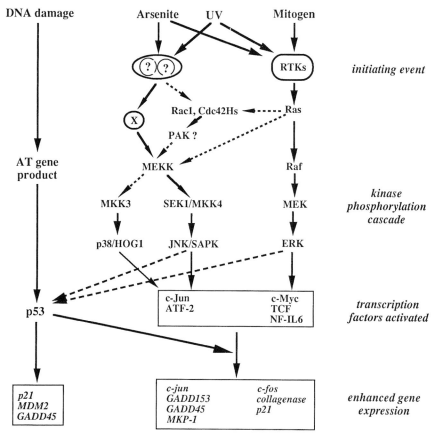

Figure 1. Signaling pathways mediating cellular response to genotoxic stress. Continuous lines indicate established pathways; broken lines represent presumed but less well established events. Transcription factors and activated genes listed are a partial (representative) list of those affected by the pathways. The abbreviations used are UV, short wavelength ultraviolet light; RTK, receptor tyrosine kinase; ERK, extracellular signal-regulated kinase; SAPK, stress-activated protein kinases (identical to JNK); JNK, c-Jun N-terminal kinase (identical to SAPK); HOG1, high osmolarity gene 1 product of yeast; p38, mammalian HOG1 homologue; MEK, ERK kinase, MKK3, mitogen-activated protein kinase kinase 3; MKK4, mitogen activated protein kinase kinase 4 (identical to SEK1); SEK1, SAPK/ERK kinase 1 (identical to MKK4); MEKK, MEK kinase; Rac1, Cdc42Hs, subfamily members of Rho family of GTP-binding proteins; PAK, p21(GTP-binding protein)-activated kinase; ATF-2, activating transcription factor-2; NF-IL6, nuclear factor for interleukin 6; TCF, ternary complex factor (which in association with serum response factor binds to serum response element); AT gene product, product of gene absent or mutated in ataxia telangectasia; *MDM2*, murine double mutant 2 gene; p21, 21 kDa cyclin dependent kinase inhibitory protein; *GADD45* and *GADD153*, growth arrest and DNA damage inducible genes 45 and 153; *MKP-1*, MAP kinase phosphatase 1 gene.

highly activated in response to stresses including UVC irradiation, heat shock, the DNA alkylating agent methyl methanesulfonate (MMS), inflammatory cytokines and inhibitors of protein synthesis. In contrast to ERKs, JNKs show little activation in response to most growth factors (Derijard et al., 1994; Kyriakis et al., 1994). The immediate upstream activator of JNK is SEK1/MKK4, a kinase with significant homology to MEK (involved in the ERK activation) (Sanchez et al., 1994; Derijard et al., 1995; Lin et al., 1995). SEK1/MKK4 is phosphorylated by MAP kinase kinase kinase (MEKK) (Lange-Carter et al., 1993; Minden et al., 1994; Yan et al., 1994). Despite its name, MEKK does not appear to be involved in the phosphorylation of MEK under physiologic conditions. Rather, as depicted in Figure 1, MEKK appears to occupy a position in the JNK pathway analogous to Raf in the ERK pathway. The role of Ras in activating JNK is not entirely clear, but current evidence favors the view that both Ras-dependent (as for UVC and growth factors) and Ras-independent (as for arsenite and TNFα) pathways exist (Lange-Carter and Johnson, 1994; Minden et al., 1994). Very recently, Rac1 and Cdc42Hs (small GTP-binding proteins within the Rho family of Ras-related proteins), have been implicated as intermediates in JNK activation in response to growth factors and inflammatory cytokines (Minden et al., 1995; Coso et al., 1995). Rac1 and Cdc42Hs are responsible for activation of a kinase p65[PAK] (PAK) in mammalian cells (Manser et al., 1994). Since PAK is homologous to the yeast protein kinases Ste20 and Cla4, responsible for phosphorylating yeast MEKK homologues Ste11 and Ssk2/22, respectively (Herskowitz, 1995), it has been proposed that PAK, or a related kinase, may link Rac1 and Cdc42Hs to MEKK (see Fig. 1). However, a role for these intermediates in JNK activation by genotoxic stress has not been demonstrated.

JNK was identified based on its ability to phosphorylate the c-Jun oncoprotein leading to its enhanced transcriptional activity (Hibi et al., 1993). It has also recently been shown to be capable of phosphorylating the transcription factor ATF-2 (Abdel-Hafiz et al., 1992; Gupta et al., 1995; van Dam et al., 1995).

p38/RK Family
This third group of MAP kinases consists of the endotoxin and osmotic stress-activated protein kinase p38 (Han et al., 1994), MAP kinase activated protein kinase-2 (MAPKAPK-2) reactivating kinase (RK) (Rose et al., 1994), and two kinases involved in the regulation of inflammatory cytokine biosynthesis (CSBP1 and CSBP2) (Lee et al., 1994). In general, they appear to be activated by the same conditions resulting in the activation of SAPK/ JNKs, but little is known concerning the upstream events controlling their phosphorylation. A dual specificity kinase designated RKK has been purified from arsenite-treated PC12 cells and shown to be capable of phosphorylating RK (Rose et al., 1994). In addition, a dual specificity kinase, MKK3, capable of activating p38 *in vitro* has been cloned (Derijard et al.,

1995). Whether RKK and MKK3 are identical or related proteins is unclear. MKK3 does share significant homology with MKK4/SEK1 (the upstream activator of JNK). *In vitro* studies indicate that MKK4/SEK1 can also activate p38, but MKK3 cannot activate JNK (Derijard et al., 1995; Lin et al., 1995). Given the similarity in circumstances leading to the activation of the p38 and JNK, and the homologies between MKK3 and MKK4/SEK, it will be of great interest to determine whether MEKK (the upstream activator of MKK4/SEK) can likewise phosphorylate MKK3.

The physiologic role of p38/RK is also not well understood. RK can activate MAPKAPK-2, a kinase responsible for the phosphorylation of small heat shock protein HSP25/27 (Rose et al., 1994). Hence, p38 has been thought to be important for extranuclear events occurring in response to stress. However, recent studies indicate that p38 also resides in the nucleus of stressed cells, raising the possibility that it might function in the phosphorylation of transcription factors. Indeed, p38 can phosphorylate ATF-2, although less efficiently than either ERK or JNK (Raingeaud et al., 1995; Liu and Holbrook, unpublished findings).

Role of p53 tumor suppressor protein

The p53 tumor suppressor protein appears to serve numerous functions during the cellular response to genotoxic stress, some of which involve its ability to act as a transcription factor to activate gene expression (Sang et al., 1995). Although this pathway initially appeared distinct from those involving the MAP kinases described above there is increasing evidence for an overlap in the responses.

The role for p53 in the induction of gene expression is best established for ionizing radiation. The current view is that DNA strand breaks in some way signal the activation of the p53 protein increasing both its stability and site-specific DNA binding activity (Nelson and Kastan, 1994; Jayaraman et al., 1995). p53 then acts as a transcription factor to induce downstream effector genes. These include GADD45, p21/WAF1/CIP1, MDM2, and BAX (Kastan, 1992; El-Deiry et al., 1993, Wu et al., 1993; Miyashita and Reed, 1995). Induction of these genes by ionizing radiation is absent (or greatly reduced) in cells lacking functional p53. In addition, the promoter of the human p53 gene itself contains a p53 responsive element, suggesting the possibility for autoregulation (Deffie et al., 1993). The mechanisms mediating p53 activation are unclear but appear to involve gene products that are defective in the cancer-prone and radiosensitive disorder ataxia telangectasia (AT), as cells from AT patients respond poorly to ionizing radiation (Kastan et al., 1992). The gene defective in AT has recently been identified (Savitsky et al., 1995). There is little doubt that it will prove to be an important link between kinase signaling cascades and the p53 response.

p53 is also induced in response to other forms of genotoxic stress (e.g. UVC irradiation, MMS treatment), and although not essential (as it is for the response to ionizing radiation) appears to contribute to the activation of at least some genes induced by these treatments (Zhan et al., 1993; Zhan et al., 1995).

The p53 protein has been shown to be phosphorylated at multiple sites *in vivo* (Meek, 1994). Various kinases including double-stranded DNA-activated protein kinase (DNA-PK), cyclin A- and cyclin B-associated kinase Cdc2, casein kinase I, casein kinase II, ERK, and JNK1 have been implicated in p53 phosphorylation (Anderson, 1993; Lees-Miller et al., 1992; Bischoff et al., 1990; Milne et al., 1992; Milne et al., 1994; Milne et al., 1995; Herrmann et al., 1991). The physiological significance of these phosphorylations remains unclear, but some of these kinases could participate in regulating p53 activity during genotoxic stress. In particular, phosphorylation of p53 by either ERK or JNK, would provide a direct link between the MAP kinase and p53-dependent gene activation pathways shown in Figure 1. Interestingly, as also indicated in Figure 1, p53 can contribute to the expression of genes whose transcriptional activation does not involve a p53-binding site, as loss of p53 activity results in reduced transcriptional activation of both *GADD45* and *GADD153* in response to DNA damaging treatments (Zhan et al., unpublished findings). Since neither of these genes contains a p53 binding site in the promoter regions analyzed for transcriptional activity (the p53 binding site in the GADD45 gene lies in the third intron), p53 must interact with other components of the transcriptional machinery to produce these effects.

Sensing of damage and initiation of the response

Role of DNA damage

A number of studies over the years have supported the contention that DNA damage is an important signal for gene induction by genotoxic agents. At least four independent laboratories have demonstrated a modification in the dose-response relations for various genes activated in response to UVC irradiation in DNA excision repair-deficient cells (Stein et al., 1989; Gibbs et al., 1990; Valerie and Rosenberg, 1990; Luethy and Holbrook, 1991). That is, substantially smaller doses of UVC radiation are required to produce an induction in uvr-xeroderma pigmetosum cells equivalent to that observed in normal cells (presumably because damaged DNA accumulates to a higher level in the repair deficient cells). More recently, it was shown that direct transfection of UVC-irradiated DNA into unirradiated cells harboring a stably integrated HIV-CAT reporter construct resulted in enhanced CAT activity (Yarosh et al., 1993). Although small, this effect could be abrogated by treatment of the UVC-treated plasmid with T4 endonuclease V. Taken together, these results indicate that damage to DNA itself does contribute to gene activation in response to UVC. Importantly, however, there is no evidence that DNA damage leads to activation of MAP kinases seen with various genotoxic compounds. In fact it appears unlikely that DNA damage contributes significantly to gene induction attributed to MAP kinases, because while significant DNA damage and gene induction can occur with UVC doses as low as 10 J/m^2, little MAP kinase activation occurs with doses below 20 J/m^2. In these instances the DNA damage must be acting through an alternative mechanism to enhance transcription.

There is good evidence that direct damage to DNA, in the form of strand breaks, leads to activation of p53 (Nelson and Kastan, 1994; Jayaraman et al., 1995). p53 protein levels have been shown to increase temporally in cells treated with a variety of agents which produce strand breaks, and electroporation of nucleases into cells leads to a rapid increase in p53 protein levels (Nelson and Kastan, 1994). Furthermore, radiosensitizers enhance both the activation of p53, and the expression of GADD45 and WAF1/CIP1, genes whose induction following ionizing radiation is dependent on p53 (Zhan et al., 1995). However, as already noted, the effect of DNA damage is unlikely to be direct, but rather mediated via the AT gene product.

Extranuclear signals controlling the response

The involvement of extranuclear events in triggering gene activation in response to high doses of UVC irradiation has been suggested by a variety of experiments over recent years (Devary et al., 1992; Herrlich et al. 1994). That nuclear events are not required for transcription factor activation in response to UVC was directly demonstrated by the activation of the transcription factor NF-κB in enucleated HeLa cells (Devary et al., 1993).

Role of growth factor receptors

Recent studies focusing on the epidermal growth factor receptor (EGFR) have provided strong evidence that the response to UVC is mediated at least in part through growth factor receptors (Sachsenmaier et al., 1994). UVC treatment leads to the rapid phosphorylation of EGFR. Pre-treatment of cells with the growth factor receptor poison suramin (Stein et al., 1993), prevents this phosphorylation and blocks both the activation of ERK and expression of the ERK-dependent gene c-*fos* in response to UVC. In addition, both prestimulation of cells with EGF, as well as overexpression of a dominant negative mutant of the EGFR, results in a partial reduction in the UVC response. The importance of growth factor receptors in mediating the UVC response is not limited to those for EGF, but can be extended to at least fibroblast growth factor and IL-1α, because prestimulation of cells with these agents also reduces the response to UVC (Sachsenmaier et al., 1994).

Activation of JNK following UVC treatment is also prevented by suramin (Liu et al., 1996). The level of MAP kinase activation in a given cell type is, therefore, likely to be dependent on the specificity and density of affected growth factor receptors present on the cell. The degree to which such findings can be extended to other genotoxic agents is not clear. For example, we have found that activation of both ERK and JNK by hydrogen peroxide and MMS is likewise inhibited in the presence of suramin. However, with sodium arsenite, ERK activation is inhibited while JNK activation is unaffected by suramin (Liu et al., 1996).

Role for oxidative stress in initiating the response
Many genotoxic agents appear to trigger their effects through an oxidative stress mechanism. The free radical scavenger N-acetyl-L-cysteine (NAC), which elevates cellular glutathione levels, can block the activation of transcription factors (e.g., NFκB, AP-1) as well as gene induction in response to genotoxic agents including UVC, hydrogen peroxide, MMS and arsenite (Schreck et al., 1991; Devary et al., 1992; Devary et al., 1993; Liu et al., 1996). In addition, activation of either or both ERK and JNK in response to several genotoxic agents (UVC, sodium arsenite and hydrogen peroxide) can be prevented by prior treatment of the cells with NAC (Guyton et al., 1996; Liu et al., 1996). Finally, we have found that treatment of cells with buthionine sulfoximine, which depletes intracellular glutathione levels (resulting in reduced free radical scavenging ability), potentiates the response to genotoxic agents.

Attenuation of the response

The role of MAP kinase phosphatases

As MAP kinases play a central role in mediating the response to genotoxic stress, their inactivation is key to the attenuation of the response. Inactivation appears to be accomplished largely through the activities of a growing family of threonine/tyrosine dual specificity phosphatases that include MKP-1 (3CH134, CL100), MKP-2, PAC-1 and B23) (Ishibashi et al., 1994; Keyse and Emslie, 1992; Missra-Press et al., 1995; Sun et al., 1993; Ward et al., 1994) (Fig. 2). Most members of the family were cloned based either on their induction in response to growth factors or in response to stress, and thus are themselves products of the MAP kinase signaling pathways.

    Evidence that the MKP-1 and PAC-1 phosphatases can dephosphorylate and thereby inactivate ERK both *in vivo* and *in vitro* has been provided by several different laboratories (Ward et al., 1994; Sun et al., 1993; Sun et al., 1994). Further, forced expression of exogenous MKP-1 prevents activation of ERK2 by various stimuli including mitogens, tetradecanoyl phorbol acetate (TPA), and DNA damaging agents (UVC and NMS) as well as by

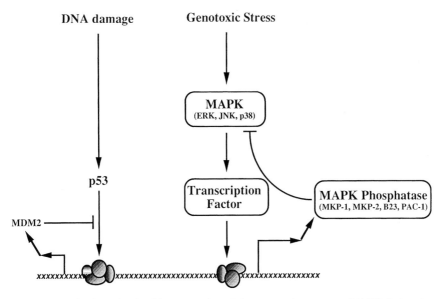

Figure 2. Mechanisms involved in attenuation of the genotoxic response. MAPK designates the mitogen activated protein kinase.

v-Ras and activated Raf, and blocks induction of ERK-dependent reporter gene expression by the same treatments (Liu et al., 1995; Sun et al., 1994). That MKP-1 is important in the normal attenuation of ERK activity is further supported by the kinetics of endogenous MKP-1 expression in response to both mitogenic stimulation and treatment with genotoxic agents; increased MKP-1 expression is correlated with a decline in ERK activity (Sun et al., 1993; Liu et al., 1995). In addition, treatment of cells with cycloheximide to prevent MKP-1 expression results in prolonged activation of ERK in mitogen stimulated cells (Sun et al., 1993).

Recent studies have provided evidence that MKP-1 also plays a similar role in down-regulating JNK and p38 activities in response to genotoxic stress (Liu et al., 1995; Raingeaud et al., 1995). Whether other members of the MKP family show similar activities has not been determined. It is possible that other phosphatases also contribute to inactivation of MAP kinases as protein phosphatase-2A as well as several other protein phosphatases have been shown to be capable of inactivating ERKs *in vitro* (Rose et al., 1994).

MDM2 inactivation of p53

MDM2 is a novel p53-regulated gene that appears to function in a "negative feedback loop" to regulate p53 activity (Fig. 2). The MDM2 gene product binds p53 to conceal the activation domain and inhibit its transcriptional activity (Momand et al., 1992; Oliner et al., 1993). Thus, p53-dependent accumulation of MDM2 protein limits the duration of p53 activity. The function of this "feedback loop" might be to limit the duration of p53-mediated G_1 arrest (discussed below). One might predict then that constitutive over-expression of MDM2 would permanently inactivate p53 and lead to the same phenotype as seen in p53 mutations and deletions. In fact, amplification of the *MDM2* locus has been found in several types of sarcomas (Oliner et al., 1992). Increased levels of MDM2 in these tumors and in *MDM2*-transfected cells leads to attenuation of the G_1 checkpoint (Finlay, 1993; Chen et al., 1994).

Consequences of the acute cellular response to stress

Role in cell cycle checkpoints and growth arrest

Cellular stress, particularly that involving damage to the cell's replicative machinery, elicits multiple cell cycle delays in virtually all living organisms (discussed in Hartwell and Kastan, 1994 and references therein). The G_1 and G_2 cell cycle delays or checkpoints in particular are considered important after DNA damage since they allow the cell time to recover and/or repair damaged DNA prior to replication and mitosis (for further discussion of DNA repair processes the reader is referred to the next chapter in this volume by ap Rhys and Bohr). It is not surprising then, that a variety of mammalian genes induced by genotoxic stress have been found to be associated with growth-inhibitory processes (Fornace, 1992). While a complete discussion of these genes and their functions is beyond the scope of this review, a common feature in mammalian cell checkpoints appears to be p53 and signaling pathways involving p53 (Kastan et al., 1991; Kuerbitz et al., 1992). Growth suppression by p53 requires sequence-specific transcriptional activation (Pietenpol et al., 1994). Among the most prominent of the p53-regulated proteins involved in growth suppression is the cyclin dependent kinase (cdk) inhibitor p21/WAF1/CIP1 which, through its ability to inhibit various cdks, prevents progression to S phase (Harper et al., 1993; El-Deiry et al., 1993; Xiong and Beach, 1993). GADD45, another p53-regulated gene is similarly associated with growth arrest following DNA damage (Zhan et al., 1995). Importantly, like numerous other stress-inducible genes, both p21 and GADD45 are also regulated in a p53-independent fashion through mechanisms involving MAP kinase signaling pathways (unpublished results).

Role in apoptosis

Programmed cell death is an important response to genotoxic stress in mammalian cells and is mediated by both transcription-dependent and transcription-independent mechanisms. A variety of well conserved genes have important roles in apoptosis and several are regulated by stress. For example, BCL2, which promotes cell survival and contributes to tumorigenesis, and its alter ego BAX, which promotes apoptosis (Oltvai et al., 1993), both show altered expression after exposure to DNA damaging agents; BCL2 expression is reduced, while BAX expression is increased (Zhan et al., 1994). Interestingly, induction of BAX is p53-dependent and the BAX promoter contains functional p53-binding sites that confer p53 responsiveness (Selvakumaran et al., 1994; Zhan et al., 1994; Miyashita and Reed, 1995). In addition to its well-defined role as a transcription factor, evidence exists that p53 may also contribute to certain apoptotic responses independent of new protein synthesis (Caelles et al., 1994). Finally, p53-independent apoptosis also occurs in response to genotoxic stress, and it is likely that some of the damage-inducible gene products contribute to this effect.

Roles in determining cytotoxicity

The search for SOS responses in mammalian cells has been a controversial topic for many years. In bacteria, DNA damage triggers the SOS and similar responses that confer marked protection against a second exposure to genotoxic stress (discussed in Fornace et al., 1992). While some evidence exists for such responses in mammalian cells (Protic et al., 1988), the magnitude is substantially less than in simple one-cell organisms like bacteria where a premium is probably placed on survival rather than other factors such as genome stability which is critical for more complex organisms. In the case of the mammalian Ras signaling pathway, it appears to be conserved appreciably even in yeast and contributes to cellular resistance to UV radiation (Engelberg et al., 1994). Another cytoplasmic/membrane signaling event triggered by DNA damage involves induction of TNFα which when secreted, increases the radiation sensitivity of neighboring cells (Hallahan et al., 1989). Here, the response actually works against cell survival, but perhaps lessens the likelihood of damaged cells progressing to malignancy. In the case of p53, the most prominent effect in certain cell types, such as myeloid, lymphoid, and thymic, is to increase cytotoxicity by triggering apoptosis. However, recent evidence indicates that functional p53 can confer a small survival advantage to other cell types after exposure to UV radiation and cis-Pt diammine dichloride, whose damage is repaired by a UV-type mechanism (Smith et al., 1995; Fan et al., 1995).

Therapeutic implications

Elucidation of the components of mammalian genotoxic stress pathways should provide more specific targets that can be exploited therapeutically.

In the case of p53, substantial effort is now underway to develop pharmacological agents that can force mutant p53 into a wt functional configuration; presumably this could trigger p53 apoptosis. The cell cycle defects found in p53 deficient cells make appealing targets in cancer therapy. Elucidation of the intracellular kinase cascades will also likely provide additional targets that can be exploited therapeutically. One such approach could involve the use of TNFα, and perhaps other cytokines, as a tumor radiosensitizer (Hallahan et al., 1989). Considering that genotoxic stress has also been implicated in oxidant injury, atherosclerosis, neurodegenerative processes, and aging, elucidation of the response mechanisms may well have wide-ranging therapeutic implications in the future.

References

Abdel-Hafiz, H.A.M., Heasley, L.E., Kyriakis, J.M., Avruch, J., Kroll, D.J., Johnson, G.L. and Hoeffler, J.P. (1992) Activating transcription factor-2 DNA-binding activity is stimulated by phosphorylation catalyzed by p42 and p54 microtubule-associated protein kinases. *Mol. Endocrinol.* 16:2079–2089.

Anderson, C.W. (1993) DNA damage and the DNA-activated protein kinase. *TIBS* 18:433–437.

Avruch, J., Zhang, X.F. and Kyriakis, J.M. (1994) Raf meets Ras: completing the framework of a signal transduction pathway. *Trends in Biochem. Sci.* 19:279–283.

Bischoff, J.R., Friedman, P.N., Marshak, D.R., Prives, C. and Beach, D. (1990) Human p53 is phosphorylated by p60-cdc2. *Proc. Natl. Acad. Sci. USA* 87:4766–4770.

Blattner, C., Knebel, A., Radler-Pohl, A., Sachsenmaier, C., Herrlich, P. and Rahms-dorf, H.J. (1994) DNA damaging agents and growth factors induced changes in the program of expressed gene products through common routes. *Environ. Mol. Mutagen* 24:3–10.

Caelles, C., Heimberg, A. and Karin, M. (1994) p53-dependent apoptosis in the absence of transcriptional activation of p53-target genes. *Nature* 370:220–223.

Cano, E., Hazzalin, C.A. and Mahadevan, L.C. (1994) Anisomycin-activated protein kinases p45 and p55 but not mitogen-activated protein kinases ERK-1 and -2 are implicated in the induction of c-fos and c-jun. *Mol. Cell. Biol.* 14:7252–7362.

Cano, E. and Mahadevan, L.C. (1995) Parallel signal processing among mammalian MAPKs. *Trends in Biochem. Sci.* 20:117–122.

Chen, C.Y., Oliner, J.D., Zhan, Q., Fornace, Jr., A.J., Vogelstein, B. and Kastan, M.B. (1994) Interactions between p53 and MDM2 in a mammalian cell cycle checkpoint pathway. *Proc. Natl. Acad. Sci. USA* 91:2684–2688.

Coso, O.A., Chiariello, M., Yu, J.C., Teramoto, H., Crespo, P., Xu, N., Miki, T. and Gutkind, J.S. (1995) The small GTP-binding proteins Rac1 and Cdc42 regulate the activity of the JNK/SAPK Signaling Pathway. *Cell* 81:1137–1146.

Davis, R.J. (1993) The mitogen-activated protein kinase signal transduction pathway. *J. Biol. Chem.* 268:1553–14556.

Deffie, A., Wu, H., Reinke, V. and Lozano, G. (1993) The tumour suppressor p53 regulates its own transcription. *Mol. Cell. Biol.* 13:3415–3423.

Dérijard, B., Hibi, M., Wu, I.-H., Barrett, T., Su, B., Deng, T., Karin, M. and Davis, R.J. (1994) JNK1: a protein kinase stimulated by UV light and Ha-Ras that binds and phosphorylates the c-Jun activation domain. *Cell* 76:1025–1037.

Dérijard, B., Raingeaud, J., Barrett, T., Wu, I.-H., Han, J., Ulevitch, R.J. and Davis, R.J. (1995) Independent human MAP kinase signal transduction pathways defined by MEK and MKK isoforms. *Science* 267:682–685.

Devary, Y., Gottlieb, R.A., Smeal, T. and Karin, M. (1992) The mammalian ultraviolet response is triggered by activation of Src tyrosine kinases. *Cell* 71:1081–1091.

Devary, Y., Rosette, C., DiDonato, J.A. and Karin, M. (1993) NF-κB activation by ultraviolet light not dependent on a nuclear signal. *Science* 261:1442–1445.

El-Deiry, W.S., Tokino, T., Velculescu, V.E., Levy, D.B., Parson, R., Trent, J.M., Lin, D., Mercer, W.E., Kinzler, K.W. and Vogelstein, B. (1993) WAF1, a potential mediator of p53 tumor suppression. *Cell* 75:817–825.

Engelberg, D., Klein, C., Martinetto, H., Struhl, K. and Karin, M. (1994) The UV response involving the Ras signaling pathway and AP-1 transcription factors is conserved between yeast and mammals. *Cell* 77:381–390.

Fan, S., Smith, M.L., Rivet, II, D.J., Duba, D., Zhan, Q., Kohn, K.W., Fornace, Jr., A.J. and O'Connor, P.M. (1995) Disruption of p53 function sensitizes breast cancer MCF-7 cells to cisplatin and pentoxifylline. *Cancer Res.* 55:1649–1654.

Finlay, C.A. (1993) The MDM-2 oncogene can overcome wild-type p53 suppression of transformed cell growth. *Mol. Cell. Biol.* 13:301–306.

Fornace, J., A.J. (1992) Mammalian genes induced by radiation; activation of genes associated with growth control. *Annual. Rev. Genetics* 26:507–526.

Gibbs, S., Lohman, F., Teabel, W., vande Putte, P. and Backendort, C. (1990) Characterization of the human spr2 promoter: induction after UV irradiation or TPA treatment and regulation during differentiation of cultured keratinocytes. *Nucleic Acids Res.* 18:4401–4407.

Gille, H., Sharrocks, A.D. and Shaw, P.E. (1992) Phosphorylation of transcription factor p62[TCF] by MAP kinase stimulates ternary complex formation at c-fos promoter. *Nature* 358:414–417.

Gupta, S., Campbell, D., Dérijard, B. and Davis, R.J. (1995) Transcription factor ATF2 regulation by the JNK signal transduction pathway. *Science* 267:389–393.

Guyton, K.Z., Liu, Y., Gorospe, M., Xu, Q. and Holbrook, N.J. (1996) Activation of mitogen-activated protein kinase by H_2O_2. *J. Biol. Chem.* 271:4138–4142.

Hallahan, D.E., Spriggs, D.R., Beckett, M.A., Kufe, D.W. and Weichselbaum, R.R. (1989) Increased tumor necrosis factor alpha mRNA after cellular exposure to ionizing radiation. *Proc. Natl. Acad. Sci. USA* 86:10104–10107.

Han, J., Lee, J.-D., Bibbs, L. and Ulevitch, R.J. (1994) A MAP kinase targeted by endotoxin and hyperosmolarity in mammalian cells. *Science* 265:808–811.

Harper, J.W., Adami, G.R., Wei, N., Keyomarsi, K. and Elledge, S.J. (1993) The p21 Cdk-interacting protein Cip1 is a potent inhibitor of G1 cyclin-dependent kinases. *Cell* 75:805–816.

Hartwell, L.H. and Kastan, M.B. (1994) Cell cycle control and cancer. *Science* 266:1821–1828.

Herrlich, P., Sachsenmaier C., Radler-Pohl, A., Gebel, S., Blattner, C. and Rahmsdorf, H.J. (1994) The mammalian UV response: mechanism of DNA damage induced gene expression. *Adv. Enzyme Regul.* 34:381–395.

Herrmann, C.P.E., Kraiss, S. and Montenarh, M. (1991) Protein kinase activity associated with immunopurified p53 protein. *Oncogene* 6:877–884.

Herskowitz, I. (1995) MAP kinase pathways in yeast: for mating and more. *Cell* 80:187–197.

Hibi, M., Lin, A., Smeal, T., Minden, A. and Karin, M. (1993) Identification of an oncoprotein- and UV-responsive protein kinase that bind and potentiate the c-Jun activation domain. *Genes Dev.* 7:2135–2148.

Hill, C.S., Marais, R., John, S., Wynne, J., Dalton, S. and Treisman, R. (1993) Functional analysis of a growth factor-responsive transcription factor complex. *Cell* 73:395–406.

Holbrook, N.J. and Fornace, A.J., Jr. (1991) Response to adversity: molecular control of gene activation following genotoxic stress. *New Biologist* 3:825–833.

Ishibashi, T., Bottaro, D.P., Michieli, P., Kelley, C.A. and Aaronson, S.A. (1994) A novel dual specificity phosphate induced by serum stimulation and heat shock. *J. Biol. Chem.* 47:29897–29902.

Jayaraman, L. and Prives, C. (1995) Activation of p53 sequence-specific DNA binding by short single strand of DNA requires the p53 C-terminus. *Cell* 81:10231–11029.

Kastan, M.B., Onyekwere, O., Sidransky, D., Vogelstein, B. and Craig, R.W. (1991) Participation of p53 protein in the cellular response to DNA damage. *Cancer Res.* 51:6304–6311.

Kastan, M.B., Zhan, Q., El-Deiry, W.S., Carrier, F., Jacks, T., Wlash, W.V., Plunkett, B.S., Vogelstein, B. and Fornace, A.J., Jr. (1992) A mammalian cell cycle checkpoint pathway utilizing p53 and GADD45 is defective in Ataxia-Telangiectasia. *Cell* 71:587–597.

Keyse, S.M. and Emslie, E.A. (1992) Oxidative stress and heat shock induce a human gene encoding a protein tyrosine phosphatase. *Nature* 359:644–647.

Kyriakis, J.M. Banerjee, P., Nikolakaki, E., Dai, T., Rubie, E.A., Ahmad, M.F., Avruch, J. and Woodgett, J.R. (1994) The stress-activated protein kinase subfamily of c-Jun kinases. *Nature* 369:156–160.

Kuerbitz, S.J., Plunkett, B.S., Walsh, W.V. and Kastan, M.B. (1992) Wild type p53 is a cell cycle checkpoint determinant following irradiation. *Proc. Natl. Acad. Sci. USA* 89:7491–7495.

Lange-Carter, C.A., Pleiman, C.M., Gardner, A.M., Blumer, K.J. and Johnson, G.L. (1993) A divergence in the MAP kinase regulatory network defined by MEK kinase and Raf. *Science* 260:315–319.

Lange-Carter, C.A. and Johnson, G.L. (1994) Ras-dependent growth factor regulation of MEK kinase in PC12 cells. *Science* 265:1458–1461.

Lee, J.C., Laydon, J.T., McDonnell, P.C., Gallagher, T.F., Kumar, S., Green, D., McNulty, D., Blumenthal, M.J., Heys, J.R., Landvatter, S.W., Strickler, J.E., McLaughlin, M.M., Siemens, I.R., Fisher, R.M., Livi, G.P., White, J.R., Adams J.L. and Young, P.R. (1994) A protein kinase involved in the regulation of inflammatory cytokine biosynthesis. *Nature* 372:739–800.

Lees-Miller, S.P., Sakaguchi, K., Ullrich, S.J., Appella, E. and Anderson, C.W. (1992) Human DNA-activated protein kinase phosphorylates serines 15 and 37 in the amino-terminal trans-activation domain of human p53. *Mol. Cell. Biol.* 12:5041–5049.

Lin, A., Minden, A., Martinetto, H., Claret, F.X., Lange-Carter, C., Mercurio, F., Johnson, G.L. and Karin, M. (1995) Identification of a dual specificity kinase that activates the Jun kinases and p38-Mpk2. *Science* 268:286–290.

Liu, Y., Gorospe, M., Yang, C. and Holbrook, N.J. (1995) Role of mitogen-activated protein kinase phosphatase during the cellular response to genotoxic stress. Inhibition of c-Jun N-terminal kinase activity and Ap-1-dependent gene activation. *J. Biol. Chem.* 270:8377–8380.

Liu, Y., Guyton, K.Z., Gorospe, M., Xu, Q., Lee, J.C. and Holbook, N.J. (1996) Differential activation of ERK, JNK/SAPK and p38/CSBP/RK MAP kinase family members during the cellular response to arsenite. *Free Radicals Bio. Med.;* in press.

Luethy, J.D. and Holbrook, N.J. (1992) Activation of the *gadd153* promoter by genotoxic agents; a rapid and specific response to DNA damage. *Cancer Res.* 52:5–10.

Manser, E., Leung, T., Salihuddin, H., Zhao, Z. and Lim, L. (1994) A brain serine/threonine protein kinase activated by Cdc42 and Rac1. *Nature* 367:40–46.

Meek, D.W. (1994) Post-translational modification of p53. *Semin. Cancer Biol.* 5:203–210.

Milne, D.M., Palmer, R.H., Campbell, D.G. and Meek, D.W. (1992) Phosphorylation of the p53 tumour suppressor protein at three N-terminal residues by a novel casein kinase I-like enzyme. *Oncogene* 7:1361–1369.

Milne, D.M., Campbell, D.G., Caudwell, F.B. and Meek, D.W. (1994) Phosphorylation of the tumour suppressor protein p53 by mitogen activated protein (MAP) kinases. *J. Biol. Chem.* 269:9253–9260.

Milne, D., Campbell, L.E., Campbell, D.G. and Meek, D.W. (1995) p53 is phosphorylated *in vitro* and *in vivo* by an ultraviolet radiation-induced protein kinase characteristic of the c-Jun Kinase, JNK1. *J. Biol. Chem.* 270:5511–5518.

Minden, A., Lin, A., McMahon, M., Lange-Carter, C., Dérijard, B., Davis, R.J., Johnson, G.L.and Karin, M. (1994) Differential activation of ERK and JNK mitogen-activated protein kinases by Raf-1 and MEKK. *Science* 266:1719–1723.

Minden, A., Lin, A., Claret, F.X., Abo, A. and Karin, M. (1995) Selective activation of the JNK signaling cascade and c-Jun transcriptional activity by the small GTPases Rac and Cdc42Hs. *Cell* 81:1147–1157.

Misra-Press, A., Rim, C.S., Yao, H., Roberson, M.S. and Stork, P.J.S. (1995) A novel mitogen-activated protein kinase phosphatase. *J. Biol. Chem.* 270:14587–24596.

Miyashita, T. and Reed, J.C. (1995) Tumor suppressor p53 is a direct transcriptional activator of the human bax gene. *Cell* 80:293–299.

Momand, J., Zambetti, G.P., Olson, D.C., George, D. and Levine, A.J. (1992) The MDM-2 onco-gene product forms a complex with the p53 protein and inhibits p53 mediated transactiva-tion. *Cell* 69:1237–1245.

Nelson, W.G. and Kastan, M.B. (1994) DNA strand breaks: the DNA template alterations that trigger p53-dependent DNA damage response pathways. *Mol. Cell. Biol.* 14:1815–1823.

Oliner, J.D., Kinzler, K.W., Meltzer, P.S., George, D.L. and Vogelstein, B. (1992) Amplification of a gene encoding a p53-associated protein in human sarcomas. *Nature* 358:80–83.

Oliner, J.D., Pietenpol, J.A., Thiagalingam, S., Gyuris, J., Kinzler, K.W. and Vogelstein, B. (1993) Oncoprotein MDM2 conceals the activation domain of tumor suppressor p53. *Nature* 362:857–860.

Oltvai, Z.N., Milliman, C.L. and Korsmeyer, S.J. (1993) Bcl-2 heterodimerizes *in vivo* with a conserved homolog, Bax, that accelerates programmed cell death. *Cell* 74:609–619.

Pietenpol, J.A., Tokino, T., Thiagalingam, S., el-Deiry, W.S., Kinzler, K.W. and Vogelstein, B. (1994) Sequence-specific transcriptional activation is essential for growth suppression by p53. *Proc. Natl. Acad. Sci. USA* 91:1998–2002.

Protic, M., Roilides, E., Levine, A.S. and Dixon, K. (1988) Enhancement of DNA repair capacity of mammalian cells by carcinogen treatment. *Somat. Cell. Mol. Genet.* 14:351–357.

Raingeaud, J., Gupta, S., Rogers, J.S., Dickens, M., Han, J., Ulevitch, R.J. and Davis, R.J. (1995) Pro-inflammatory cytokines and environmental stress cause p38 mitogen-acitvated protein kinase activation by dual phosphorylation on tyrosine and threonine. *J. Biol. Chem.* 270:7420–7426

Rouse, J., Cohen, P., Trigon, S., Morange, M., Alonso-Liamazares, A., Zamanillo, D., Hunt, T. and Nebrada, A.R. (1994) A novel kinase cascade triggered by stress and heat shock that stimulates MAPKAP kinase-2 and phosphorylation of the small heat shock proteins. *Cell* 78:1027–1037.

Sachsenmaier, C., Radler-Pohl, A., Zinck, R., Nordheim, A., Herrlich, P. and Rahmsdorf, H.J. (1994) Involvement of growth factor receptors in the mammalian UVC response. *Cell* 78:963–972.

Sanchez, I., Hughes, R.T., Mayer, B.J., Yee, K., Woodgett, J.R., Avruch, J., Kyriakis, J.M. and Zon, L.I. (1994) Role of SAPK/ERK kinase-1 in the stress-activated pathway regulating transcription factor c-Jun. *Nature* 372:794–798.

Sang, N., Baldi, A. and Giordano, A. (1995) The roles of tumor suppressors pRB and p53 in cell proliferation and cancer. *Mol. Cell. Differen.* 3:1–29.

Savitsky, K., Bar-Shira, A., Gilad, S., Rotman, G., Ziv, Y., Vanagaite, M., Pecker, I., Frydman, M., Harnik, R., Patanjali, S.R., Simmons, A., Clines, G.A., Sartiel, A., Gatti, R.A., Chessa, L., Sanal, O., Lavin, M.F., Jaspers, N.G.J., Malcom, A., Taylor, R., Arlett, C.F., Miki, T., Weissman, S.M., Lovett, M., Collins, F.S. and Shiloh, Y. (1995) A single ataxia telangiectasia gene with a product similar to pI-3 kinase. *Science* 268:1749–1753.

Schieven, G.L. and Ledbetter, J.A. (1994) Activation of tyrosine kinase signal pathways by radiation and oxidative stress. *Trends in Endocrinol. Metab.* 5:383–388.

Schreck, R., Rieber, P. and Baeuerle, P.A. (1991) Reactive oxygen intermediates as apparently widely used messengers in the activation of the NF-κB transcription factor and HIV-1. *EMBO J.* 10:2247–2258.

Selvakumaran, M., Lin, H.-K., Miyashita, T., Wang, H.G., Krajewski, S., Reed, J.C., Hoffman, B. and Liebermann, D. (1994) Immediate early up-regulation of bax expression by p53 but not TGFβ1: a paradigm for distinct apoptotic pathways. *Oncogene* 9:1791–1798.

Sluss, H.K., Barrett, T., Dérijard, B. and Davis, R.J. (1994) Signal transduction by tumor necrosis factor mediated by JNK protein kinases. *Mol. Cell. Biol.* 14:8376–8384.

Smith, M.L., Chen, I.T., Zhan, Q., O'Connor, P.M. and Fornace, Jr., A.J. (1995) Involvement of the p53 tumor suppressor in repair of UV-type DNA damage. *Oncogene* 10:1053–1059.

Stein, B., Rahmsdorf, H.J., Steffen, A., Litfin, M. and Herrlich, P. (1989) UV-induced DNA damage is an intermediate step in UV-induced expression of human immunodeficiency virus Type 1, collagenase, c-fos, and metallothionein. *Mol. Cell. Biol.* 9:5169–5181.

Stein, C.A. (1993) Suramin: a novel antineoplastic agent with multiple potential mechanisms of action. *Cancer Res.* 2239–2248.

Sun, H., Charles, C.H., Lau, L.F. and Tonks, N.K. (1993) MKP-1 (3CH134), an immediate early gene product, is a dual specificity phosphatase that dephosphorylates MAP kinase *in vivo*. *Cell* 75:487–493.

Sun, H., Tonks, N.K. and Bar-Sagi, D. (1994) Inhibition of Ras-induced DNA synthesis by expression of the phosphatase MKP-1. *Science* 266:285–288.

Valerie K. and Rosenberg, M. (1990) Chromatin structure implicated in activation of HIV-1 gene expression by ultraviolet light. *New Biol.* 2:712–718.

Van Dam, H., Wilhelm, D., Herr, I.N., Steffen, A., Herrlich, P. and Angel, P. (1995) ATF-2 is preferentially activated by stress-activated protein kinases to mediate c-jun induction in response to genotoxic agents. *EMBO J.* 14:1798–1811.

Ward, Y., Gupta, S., Jensen, P., Wartman, M., Davis, R.J. and Kelly, K. (1994) Control of MAP kinase activation by the mitogen-induced threonine/tyrosine phosphatase PAC1. *Nature* 367:651–654.

Wu, X., Bayle, J.H., Olson, D. and Levine, A.J. (1993) The p53-mdm-2 autoregulatory feedback loop. *Genes Dev.* 7:1126–1132.

Yan, M., Dai, T., Deak, J.C., Kyriakis, J.M., Zon, L.I., Woodgett, J.R. and Templeton, D.J. (1994) Activation of stress-activated protein kinase by MEKK1 phosphorylation of its activator SEK1. *Nature (London)* 372:798–800.

Yarosh, D.B., Alas, L., Kibitel, J., O'Connor, A., Carrier, F. and Fornace, Jr., A.J. (1993) Cyclobutane pyrimidine dimers in UV-DNA induce release of soluble mediators that activate the human immunodeficiency virus promoter. *J. Invest. Dermatol.* 100:190–194.

Zhan, Q., Carrier, F. and Fornace, Jr., A.J. (1993) Induction of cellular p53 activity by DNA damaging agents and growth arrest. *Mol. Cell. Biol.* 13:4242–4250.

Zhan, Q., Fan, S., Bae, I., Guillouf, C., Liebermann, D.A., O'Connor, P.M. and Fornace, Jr., A.J. (1994) Induction of BAX by genotoxic stress in human cells correlates with normal p53 status and apoptosis. *Oncogene* 9:3743–3751.

Zhan, Q., El-Deiry, W., Bae, I., Alamo, Jr., I., Kastan, M.B., Vogelstein, B. and Fornace, Jr. A.J. (1995) Similarity of the DNA-damage responsiveness and growth suppressive properties of WAF1 to GADD45. *Int. J. Oncology* 6:937–946.

Stress-Inducible Cellular Responses
ed. by U. Feige, R.I. Morimoto, I. Yahara and B. Polla
© 1996 Birkhäuser Verlag Basel/Switzerland

Mammalian DNA repair responses and genomic instability

C. M. J. ap Rhys and V. A. Bohr

Laboratory of Molecular Genetics, The National Institute on Aging, National Institutes of Health, NIH, 4940 Eastern Avenue, Baltimore, MD 21224, USA

Summary. A cell responds to damage to its DNA in one of three ways: by tolerating the damage, by repairing the damage or by undergoing apoptosis. The latter two responses represent defenses against genomic instability and tumorigenesis resulting from unrepaired damage. There are multiple DNA repair pathways to cope with a variety of damage reflecting the importance of DNA repair in maintaining both cell viability and genomic stability. These include base excision repair, mismatch repair, double-strand break repair and nucleotide excision repair. Several signal transduction pathways are activated by DNA damage resulting in cell-cycle arrest. Cell-cycle arrest increases the time available for DNA repair before DNA replication and mutation fixation. Recently, there has been tremendous progress in our understanding of the molecular components repair processes and to examine recently observed interactions between DNA repair, signal transduction pathways and other cellular processes such as cell-cycle control, transcription, replication and recombination.

Introduction

Cellular stress results from a variety of insults to the cell from both exogenous and endogenous sources. Environmental stresses such as exposure to UV light, carcinogens, chemotherapeutic agents, heat and physical trauma can elicit a variety of cellular responses. Many of these traumas as well as metabolic by-products cause DNA damage, mutagenesis and even tumorigenesis.

DNA damage

A cell may encounter various forms of DNA damage throughout its lifespan. The ultraviolet rays in sunlight cause two major photolesions, the cyclobutane pyrimidine dimer and the pyrimidine-pyrimidone (6–4) photoproduct. Ionizing radiation can cause single- and double-strand DNA breaks. Reactive oxygen species cause the most common forms of DNA damage. More than 100 different types of oxidative lesions have been identified (Dizdaroglu, 1986). These can be generated by radiation, drugs, chemicals, or cellular metabolism, e.g., mitochondrial respiration. Additionally, a variety of monofunctional and bifunctional alkylating agents

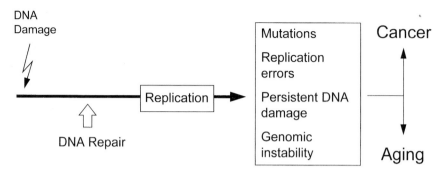

Figure 1. Schematic representation of stresses incurred by a cell and its possible responses.

can react with DNA, methylating nucleotide residues or forming inter- and intra-strand crosslinks.

There are many ways that a cell responds to DNA damage. Severe damage can activate the cell's apoptosis pathway leading to cell death. The cell may develop tolerance mechanisms which lead to mutations in the genome or the cell may repair its DNA. If the DNA damage persists through replication, the result is genomic instability. This can take the form of simple point mutations, deletions, insertions, duplications, or inversions. Serious defects in DNA repair can result in chromosomal rearrangements and even chromosomal loss. These processes are schematized in Figure 1.

DNA repair

DNA repair occurs by several mechanisms. One is by direct reversal of DNA damage and another is by excision of the damage. An example of DNA repair by a direct reversal mechanism is demonstrated by the procaryotic enzyme, photolyase. This enzyme specifically recognizes pyrimidine dimers and splits a T-T dimer after activation with light between 300–500 nm in wavelength. A human equivalent has not been identified. Another example is the transfer and acceptance of a methyl group from O^6-methylguanine by the enzyme O^6-methylguanine transferase.

Base excision repair

Excision repair is responsible for the majority of DNA repair in the cell. Some types of damaged bases are simply removed in a process termed *base excision repair.* This is a multistep process initiated by the action of a class of enzymes referred to as DNA glycosylases. Different DNA glycosylases recognize different forms of damaged bases and hydrolyze the N-glycosyl bond linking the base to the sugar-phosphate backbone. Their action usually results in an apurinic or apyrimidinic site which is incised at the 5′ side by an apurinic/apyrimidinic (AP) endonuclease. The 5′-terminal sugar phosphate residue is finally removed by a DNA deoxyribosephosphodiesterase. The missing nucleotide is replaced by DNA polymerase and DNA ligase (Dianov and Lindahl, 1994; Friedberg et al., 1995).

Mismatch repair

Mispairing of bases results from the insertion of an incorrect base during DNA replication. These errors are corrected by a form of excision repair termed *mismatch repair*. In *E. coli*, mismatch repair strand-specificity is provided by methylation of adenine residues in GATC sequences in the parental strand. A mismatched base in newly synthesized DNA is bound by the MutS protein. This initiates the assembly of a repair complex containing the MutS, MutL and MutH proteins. The MutH protein then incises at a GATC sequence in the unmethylated strand. Next, a MutS, MutL, and MutU dependent excision step removes a section of DNA containing the GATC site and the mismatch. The resulting single-stranded gap is filled in by DNA polymerase III (Loeb, 1994; Modrich, 1994). The *E. coli* mismatch repair system has also been shown to be responsible for maintaining the stability of tracts of repeated sequences such as $(GT)_n$ which are prone to instability due to DNA polymerase slippage during replication (Levinson and Gutman, 1987).

Homologous proteins have been identified in humans. The hMLH1, hPMS1 and hPMS2 proteins are homologs of the bacterial MutL protein and the hMSH2 protein is a homolog of the bacterial MutS protein (Fishel et al., 1993; Papadopoulos et al., 1994; Bronner et al., 1994). In human cells, mutations in hMSH2, hMLH1, hPMS2, and hPMS1 have been found to be associated with a common type of malignancy, hereditary nonpolyposis colorectal cancer (HNPCC). This disease is characterized by microsatellite instability of the genome. As in *E. coli* mismatch repair mutants, regions of the genome containing mono-, di-, and trinucleotide repeats change length at a relatively high frequency (Loeb, 1994; Modrich, 1994). Defects in hMSH2 activity prevent correction of G-T mispairs, single displaced bases or displaced loops of two to four bases. Recently, GTBP/p160 has been identified as a protein that forms a heterodimer with hMSH2 (Drummond et al., 1995; Palombo et al., 1995). Defects in GTBP/p160 activity prevent correction of G-T mismatches and single displaced bases, but allow loops larger than a single base pair to be corrected efficiently (Papadopoulos et al., 1995).

Nucleotide excision repair

A broader spectrum of damage is repaired by another multistep pathway termed *nucleotide excision repair* (NER). This process is responsible for the removal of the majority of bulky lesions occurring in DNA. These include DNA damages induced by UV, cis-platinum, carcinogens and 4-Nitroquinoline-1-oxide (4NQO). This process is characterized by the recognition of the damage and the excision of an oligonucleotide fragment containing the damaged DNA. In *E. coli* cells, an incision occurs at the $3^{rd}-5^{th}$ phosphodiester bond on the 3' side of the lesion and the 8th phosphodiester bond on the 5' side, resulting in the release of a $12-13$ base oligonucleotide (Sancar, 1995; Yeung et al., 1986). In human cells, an incision occurs at the $3^{rd}-5^{th}$ phosphodiester bond ion the 3' side of the lesion and the 24th on the 5' side resulting in the release of a 27–29 base oligonucleotide (Huang et al., 1992).

Many insights into the importance and the mechanism of human NER have been gained from the study of cells from patients with the hereditary disorder, xeroderma pigmentosum (XP). These individuals are abnormally sensitive to sun exposure presenting with skin melanomas at an early age. They are also prone to other malignancies caused by known carcinogens. It has been established that the defects in XP patients are a result of a defective NER pathway. Seven XP complementation groups (A-G) and their respective protein defects have been identified. Additionally, UV-sensitive rodent cell lines derived by mutagenesis of Chinese hamster ovary (CHO) cell lines have provided 11 NER complementation groups. The human genes that correct the defect in these cell lines were termed excision repair cross complementing rodent repair deficiency genes (ERCC), but are now named after their XP complementation group (Reardon et al., 1993; Hoeijmakers, 1993; Weeda et al., 1993; Sancar and Tang, 1993; Barnes et al., 1993; Hanawalt, 1994; Bootsma and Hoeijmakers, 1993).

There has been tremendous progress in the field of DNA repair over the last few years. A major advance is the characterization of the individual molecular steps involved in NER. A model of human NER is shown in Figure 2. Recent work in the field of NER implicates at least 17 proteins in the process of damage excision (Sancar, 1995). The XPA protein is involved in the initial damage recognition step and binds to the lesion. It then forms a protein complex with the heterodimer, XPF-ERCC1 and replication protein A (RPA). Another protein complex, TFIIH, is recruited to the lesion site in association with the XPC protein and the XPG protein. The XPG protein incises 3' of the lesion and the XPF protein incises 5' of the lesion. The resulting single-strand gap is filled in by DNA polymerase δ or DNA polymerase ε and sealed by DNA ligase. There appears to be a requirement for proliferating cell nuclear antigen (PCNA) in NER possibly in stimulating polymerase activity or increasing the catalytic turnover of the enzymes involved in excision.

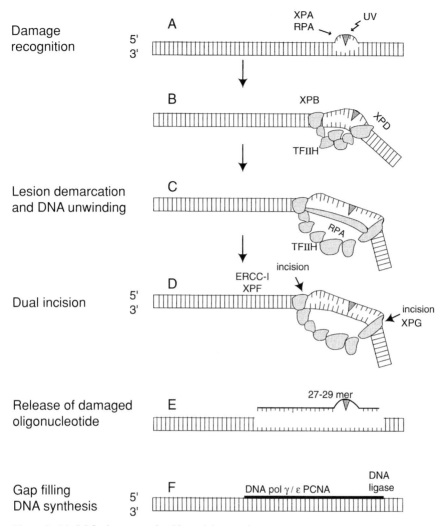

Figure 2. Model for human nucleotide excision repair.

Preferential repair of genes and the connection to transcription

The efficiency of DNA repair by NER varies within the cell's genome. Measurements of the average DNA repair in the total genome generally reflect repair of the noncoding regions which constitute 99% of the genome. Protein coding regions are preferentially repaired when compared to the bulk of the genome. This preferential repair phenotype has been demonstrated in many biological systems including hamster and human cells. In human cells, preferential repair is characterized by faster repair in transcriptionally active genes than in transcriptionally inactive genes or in the general bulk of the genome. In hamster cells and in human cells from individuals with the disease, XP complementation group C, there is repair in active parts of the genome but not in the general bulk of the genome (Evans et al., 1993a). Since hamster cells are as resistant to UV irradiation as human cells, active gene repair must be the most biologically important component of the repair process. It is now evident that there are different biochemical pathways for repair of active and inactive genomic regions. Gene specific repair enables cells to rapidly reestablish mRNA synthesis of essential genes after DNA damage. Importantly, it provides a mechanism for the rapid repair of genes that when mutated could lead to cellular transformation. There appears to be a hierarchy of gene specific repair. Certain genes are repaired more efficiently than others and this repair efficiency can change with the gene's transcriptional activity and other parameters (Okumoto and Bohr, 1987; Leadon and Snowdon, 1988). The repair efficiency can also vary within a gene. Within the hamster dihydrofolate reductase (DHFR) gene, there is more efficient repair at the 5′ end of the gene than at the 3′ end (Bohr et al. 1986) and within the p532 gene there is considerable heterogeneity of repair throughout the gene (Tornaletti and Pfeifer, 1994). Repair differences may also be influenced by chromatin structure, by the level of CpG methylation (Ho et al. 1989) or by proximity to nuclear matrix attachment sites (Mullenders et al. 1984).

The repair of transcriptionally active genes containing UV or other bulky lesions has been shown to be biased toward the transcribed strand (template strand). This specific type of nucleotide excision repair is termed transcription coupled repair. Fibroblasts from patients with the hereditary DNA repair disorder, Cockayne's syndrome (CS), have been shown to be defective in transcription coupled repair but not in bulk repair of the genome. These patients suffer from a variety of conditions including mental retardation, cachetic dwarfism and premature aging (Nance and Berry, 1992). Interestingly, CS patients do not appear more prone to cancer (Lehmann, 1987). Two complementation groups have been identified, CSA and CSB (Lehmann, 1982; Tanaka et al., 1981). *In vitro* studies have shown that the CSA protein and the CSB protein form a heterodimer. CSA binds the p44 subunit of TFIIH and CSB binds XPG (Henning et al., 1995). In a model designed by Aziz Sancar, RNA polymerase II (pol. II) stalled at a lesion in its template strand is recognized by CSA/CSB, which causes the polymerase to back off the lesion. CSA/CSB then recruits TFIIH and other repair proteins to the damage site (Sancar, 1995). It is believed that TFIIH is required for local unwinding of the DNA and for assembling repair proteins at the damage site.

The following protein subunits of TFIIH have been identified: XPB/ ERCC3/p89, XPD/ERCC2/p87, p62, p44, p34, Cdk7/p41, and CycH/ p38/MO15 (Fischer et al., 1992; Humbert et al., 1994; Roy et al., 1994;

Schaeffer et al., 1993; Schaeffer et al., 1994; Shiekhattar et al., 1995). Cells expressing mutant XPB protein or mutant XPD protein are not only defective in NER but also in mRNA regulation (Schaeffer et al., 1993; Schaeffer et al., 1994). This defect in mRNA metabolism is a reflection of an additional requirement for the helicase activity of TFIIH in the transition of pol II from the transcriptional initiation state to the elongation state commonly referred to as promoter clearance (Goodrich and Tjian, 1994). The Cdk7 and CycH subunits constitute a TFIIH associated kinase which phosphorylates the carboxy terminal domain (CTD) of the largest subunit of RNA pol II (Roy et al., 1994; Makela et al., 1995; Shiekhattar et al., 1995; Serizawa et al., 1995). Since phosphorylation of the CTD occurs during the initiation step prior to elongation, this phosphorylation is believed to be mechanistically important for promoter clearance (Laybourn and Dahmus, 1990; Lu et al., 1991). Additionally, this associated kinase activity has been shown to phosphorylate and activate the cyclin-dependent kinases (CDK). Cdc2 and Cdk2. It has been speculated that TFIIH, by phosphorylating CDKs may play a role in the modulation of transcription and DNA repair during the cell cycle (Shiekhattar et al., 1995). Despite these apparent disparate roles, TFIIH provides a functional linkage connecting transcription, DNA repair and cell cycle control.

Double-strand break repair

Ionizing radiation causes a variety of DNA lesions including single- and double-strand breaks. Most likely, it is the extent of double-strand breakage of DNA that is responsible for the lethality associated with ionizing radiation. Double-strand breaks also arise after excision of UV photoproducts when two T-T dimers are closely spaced but on opposing strands. A fraction of this damage is rapidly repaired by direct rejoining and ligation. The mechanism of slow repair is not clear but it may involve some sort of recombinational processing (Weibezahn et al., 1985).

A direct connection between double-strand break repair and a DNA-dependent Ser/Thr protein kinase (DNA-PK) is demonstrated by the Chinese hamster ovary (CHO) cell lines, xrs-6, XR-1 and V3. These cells are extremely radiosensitive, defective in double-strand break repair and deficient in DNA-dependent protein kinase (DNA-PK) activity. The DNA-PK complex comprises three proteins represented in equimolar amounts (Anderson, 1993). The autoantigens, Ku70 and Ku80, form a stable heterodimer, Ku, which binds DNA and targets p350, the catalytic subunit to DNA (Dvir et al., 1993; Gottlieb and Jackson, 1993; Griffith et al., 1992; Finnie et al., 1995). Each of these mutant cell lines is defective in one of the three proteins comprising the DNA-PK complex. The cell lines, xrs-6 and XR-1 are defective in DNA binding activity, Ku80 and Ku70 respectively (Taccioli et al., 1994b). V3 and a mouse cell line, SCID, appear to be defective in the catalytic subunit, p350 (Peterson et al., 1995). Interestingly, each of these cell lines also displays an immunological deficiency (Taccioli et al., 1994a).

During the process of V(D)J recombination, double-strand breaks are generated by cleavage at recombination signal sequences by the RAG1 and RAG2 proteins (Roth et al., 1993; Schlissel et al., 1993; van Gent et al., 1995). These sequences consist of a conserved heptamer sequence separated from a conserved nonamer sequence by a 12 or 23 bas pair spacer region. The cleavage occurs adjacent to the heptamer sequence generating a signal end and a coding end. Ligation of two signal ends results in a precise signal joint containing two heptamer sequences arranged in a head-to-head fashion. Ligation of two coding ends results in a coding joint which often contains a deletion or an addition of a few nucleotides (Tonegawa et al., 1983). Each of the double-strand break repair deficient cell lines described is significantly impaired in this process of V(D)J rejoining of double strand breaks. The xrs-6 and XR-1 cell lines are deficient in both signal and coding joint formation (Taccioli et al., 1994b). The V3 cell line is deficient in coding joint formation and is phenotypically similar to the SCID mouse cell line (Blunt et al., 1995).

The characteristics of these cell lines suggest that the catalytic subunit of DNA-PK is important in coding joint formation while Ku70/Ku80 is also important for signal sequence joint formation (Blunt et al., 1995). In keeping with the high affinity of Ku for double-strand ends, mutations in Ku80 that disrupt DNA binding may result in lack of free end protection resulting in degradation by endogenous exonucleases. This may also partially account for the defect in double-strand break repair. However, it appears that the catalytic activity of DNA-PK is very important in double-strand break repair since loss of this activity either by loss of Ku70/Ku80 binding or by loss of p350 correlates with a severe impairment of double-strand break repair. DNA-PK has been shown to phosphorylate a number of nuclear proteins (Anderson, 1993) and it is likely that this function plays an important role in double-strand break repair. It remains to be determined which are its target proteins in this process. The above is an example of an emerging understanding of the relationship between DNA repair and other cellular processes.

Responses to DNA damage involving p53

Although there is no evidence for a mammalian SOS response analogous to that seen in procaryotes, there does appear to be several cellular pathways that are activated in response to DNA damage. Many of these function to arrest the cell at certain cell cycle checkpoints to allow DNA repair before the cell undergoes mitosis or replication. Others lead to apoptosis preventing tumor formation by cell death. The activation of p53

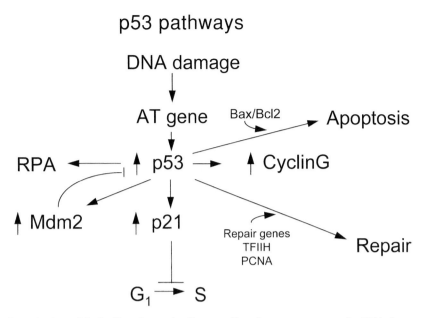

Figure 3. A model of p53 pathways leading to cell cycle arrest or apoptosis. DNA damage increases the stability of the p53 protein either directly or through an AT signal transduction pathway. Increased levels of p53 activate the expression of p21 and BAX. The p21 protein arrests the cell cycle permitting DNA repair before DNA replication. The Mdm2 protein acts directly with p53 abolishing its transactivating function. Prolonged expression of p53 results in an excess of the Bax protein relative to the Bcl2 protein and leads to apoptosis.

following DNA damage plays a pivotal role in either of these cellular responses as depicted in Figure 3. The p53 gene has been found to be missing or defective in over half of all human tumors reflecting its vital function in cellular regulation (Hollstein et al., 1991). After cells are exposed to ionizing radiation or UV light, the levels of p53 protein increase (Maltzman and Dzyzk, 1984). After ionizing irradiation, the presence of DNA strand breaks seems to signal a p53 response, since single and double strand break inducing agents including the restriction enzyme, *PvuII*, elicit a strong p53 response (Lu and Lane, 1993; Nelson and Kastan, 1994). The p53 response after UV irradiation may depend on the persistence of lesions in actively transcribed genes. Comparison of the minimum UV response dose (MRD) for the induction of p53 in several XP repair deficient cells with that of normal cells showed that there was a marked reduction in a cell's MRD that correlated with a defect in transcription coupled repair but not with bulk genomic repair (Yamaizumi and Sugano, 1994). This suggests that UV induction of p53 is signaled by the actual damage present in actively transcribed genes rather than by the repair process generating single strand gaps after excision.

The p53 response after DNA damage does not appear to be due to gene induction. The p53 gene is actively transcribed but mRNA levels remain constant after treatment with ionizing radiation (Kastan et al., 1991). The p53 gene itself is rapidly repaired in human cells in a gene-specific fashion. In XPC cells the transcribed strand bof the gene is repaired very efficiently but not the nontranscribed strand (Evans et al., 1993b). In studies of p53 mutations in patients with this disorder, all mutations were found in the nontranscribed strand (Dumaz et al., 1993) reflecting the cell's capacity to maintain the correct coding for an essential protein and the importance of DNA repair in the generation of the mutational footprint.

It is believed that the increase in p53 levels is a result of posttranslational modifications resulting in stabilization of the protein possibly through its phosphorylation. In fact, p53 is phosphorylated at multiple sites *in vivo* and potentially by many different protein kinases, suggesting the convergence of several pathways at the point of increasing p53 levels.

Part of the function of the p53 protein in DNA-damage induced G1 arrest and apoptosis is to act as a transcriptional activator. The p53 protein binds to specific DNA sequences within several genes inducing their mRNA expression, e.g., p21/WAF1/CIP1, MDM2, GADD45, BAX (Kastan et al., 1992; El-Deiry et al., 1993; Juven et al., 1993). Two of these, p21 and MDM2, encode proteins that are intimately involved in the cell-cycle arrest pathway as illustrated in Figure 2. The p21 protein has been shown to complex with cyclin-Cdk and PCNA causing cell cycle arrest but not inhibition of PCNA dependent NER (Waga et al., 1994; Chen et al., 1995; Luo et al., 1995; Li et al., 1994). The Mdm2 protein acts to end this arrest by binding directly to p53 in a manner that prevents p53 further activating the p21 gene (Chen et al., 1994). Another gene, BAX, encodes a protein which complexes with Bcl2 and increases the cell's susceptibility to DNA damage induced apoptosis (Selvakumaran, 1994). The regulation of p53 stability and in turn the p53 regulation of gene activity thus modulates the balance between cell cycle arrest and apoptosis.

In view of the importance of p53, one would predict that an impairment of a DNA damage signal transduction pathway would adversely affect a cell. For instance, a delay in stabilization of the p53 protein after DNA damage would result in a failure to arrest the cell cycle before the damage has been corrected. Evidence for more than one such pathway is observed from patients suffering from ataxia telangiectasia (AT). These individuals are profoundly sensitive to ionizing radiation and present with malignancies at an early age (Gotoff et al., 1967; Morgan et al., 1968). Cells from these patients are very sensitive to ionizing radiation and radiomimetic agents. They show a pronounced delay in p53 stabilization after ionizing irradiation but not after UV irradiation (Cohen et al., 1975; Kojis et al., 1991; Pandita and Hittleman, 1992; Kastan et al., 1992; Khanna and Lavin, 1993). The gene, ATM, responsible for these defects has been identified. It contains a domain that shows strong similarity to the lipid kinase gene of phosphatidylinositol-3' kinase (PI-3 kinase), a mediator of cellular responses to various mitogenic growth factors (Savitsky et al., 1995). The ATM gene may be part of a DNA damage signal transduction pathway that responds to ionizing irradiation.

In addition to its role in cell cycle arrest, p53 may also play a direct role in the DNA repair process. There is evidence that p53 binds directly to two subunits of TFIIH, the XPB and XPD proteins, to the CSB protein and to RPA (Wang et al., 1995a, Wang et al., 1995b; Dutta et al., 1993). Furthermore, cells from patients with Li-Fraumeni Syndrome with a germline defect in one p53 allele show decreased repair of UV damage within several genes at different loci (Wang et al., 1995a). Cell lines lacking a wild-type p53 allele show rapid repair of the transcribed strand but not of the nontranscribed strand, indicating a defect in overall NER (Ford et al., 1995). This repair defect could be observed within 2 h after UV irradiation during which time the majority of cells would not have reached the G1/S checkpoint, suggesting that it is not a result of DNA synthesis and replication interfering with repair.

DNA-PK

Recent evidence suggests DNA-PK participates in DNA excision repair. We have recently observed a gene-specific repair defect in cell lines derived from a SCID mouse (Beecham et al., unpublished data). This may be expected since the repair process itself generates single base gaps by base excision and 29 base gaps by nucleotide excision repair. Ku has been shown to bind single-to double-strand transitions including minicircles with nicks or larger single strand gaps (Morozov et al., 1994; Blier et al., 1993). Furthermore, the human replication protein (RPA) is phosphorylated by DNA-PK and the phosphorylation occurs during single-to double-strand transitions such as occur during replication or repair (Brush et al., 1994). Since RPA is involved in the strand displacement and DNA synthesis steps of NER as well as in the initial damage recognition in cooperation with XPA (Matsuda et al., 1995), phosphorylation by DNA-PK may occur during the excision step. This phosphorylation may increase the efficiency of RPA in the final steps of NER and account for the NER defects in the SCID mouse cell lines.

DNA-PK can also bind to minicircles containing a bubble created by a 30 base pair mismatched region or to intact dumbbell-like structures made up of a central duplex region terminated with single stranded loops (Paillard and Strauss, 1991; Falzon et al, 1993). These properties and its nuclear localization suggest DNA-PK may be involved in transcription, replication and/or recombination as well as in DNA repair. Since there is an intimate relationship between DNA repair and each of these activities, affecting one of these processes would indirectly affect DNA repair. DNA-PK is a prime

candidate for modulating transcription. DNA-PK has been shown to phosphorylate a variety of transcription factors including Sp1, fos, jun and p53
(Jackson et al., 1990; Bannister et al., 1993; Lees-Miller et al., 1990).
Interestingly, DNA-PK has been shown to phosphorylate RNA pol II
(Dvir et al., 1992) within its carboxy terminal domain (CTD). This occurs
when RNA pol II is in close association with a DNA template containing
bound DNA-PK (Peterson et al., 1992). Since CTD phosphorylation
inhibits entry of pol II into the initiation complex stimulating abortive
initiation, one would expect that DNA-PK bound to a gene containing a
single-stranded region would halt further transcriptional initiation of that
gene (Chestnut et al., 1992).

Conclusions

Many mechanisms deal with DNA damage and the resulting genomic
instability. These include multiple repair pathways such as base excision
repair, nucleotide excision repair, mismatch repair and double-strand break
repair. Additionally, there are multiple signal transduction pathways acting
to prevent fixation of a mutation, some of which are depicted in Figure 4.

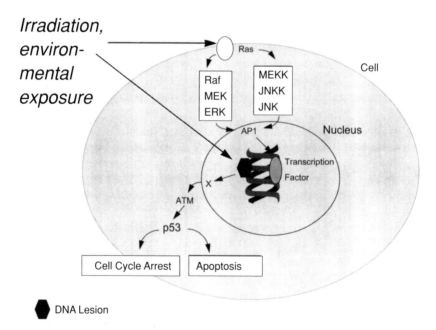

Figure 4. A model illustrating a cell's signal transduction pathways that are activated as a result
of DNA damage or as a modification of a protein component in a signal transduction cascade.
The parallel protein kinase cascades leading to activation of ERK and JNK result in transcriptional activation of AP1 responsive genes. Damaged DNA increases the stability of the p53
protein leading to cell cycle arrest or apoptosis.

There are pathways that lead to cell cycle arrest at G1 or G2 allowing time for DNA repair before the cell undergoes replication or mitosis. If the damage to DNA exceeds the repair capacity of the cell, other pathways are evoked that induce apoptosis. Some of these pathways transduce extra-cellular signals into intracellular responses as described in the previous chapter entitled, "Signalling events controlling the molecular response to genotoxic stress". For example, the parallel mitogen-activated protein kinase (MAPK) cascades activated in response to UV exposure induce the transcriptional activation of AP-1 and ATF responsive genes. Other pathways activated by DNA damage induce p53 expression resulting in cell cycle arrest or apoptosis. It is apparent that signal transduction pathways and DNA repair processes act together to maintain genomic fidelity. Elucidating these interactions provides a major challenge for future work.

References

Anderson, C.W. (1993) DNA damage and the DNA-activated protein kinase. *TIBS* 18:433–437.

Bannister, A.J., Gottlieb, T.M., Kouzarides, T. and Jackson, S.P. (1993) c-jun is phosphorylated by the DNA-dependent protein kinase *in vitro*; definition of the minimal kinase recognition motif. *Nucleic Acids Res.* 21:1289–1295.

Barnes, D.E., Lindahl, T. and Sedgwick, B. (1993) DNA repair. *Curr. Opin. Cell. Biol.* 5: 424–433.

Blier, P.R., Griffith, A.J., Craft, J. and Hardin, J.A. (1993) Binding of Ku protein to DNA. Measurement of affinity for ends and demonstration of binding to nicks. *J. Biol. Chem.* 268: 7594–7601.

Blunt, T., Finnie, N.J., Taccioli, G.E., Smith, G.C.M., Demengeot, J., Gottlieb, T.M., Mizuta, R., Varghese, A.J., Alt, F.W., Jeggo, P.A. and Jackson, S.P. (1995) Defective DNA-dependent protein kinase activity is linked to V(D)J recombination and DNA repair defects associated with the murine *scid* mutation. *Cell* 80:813–823.

Bohr, V.A., Okumoto, D.S., Ho, L. and Hanawalt, P.C. (1986) Characterization of a DNA repair domain containing the dihydrofolate reductase gene in chinese hamster ovary cells. *J. Biol. Chem.* 261:16666–16672.

Bootsma, D. and Hoeijmakers, J.H. (1993) DNA repair. Engagement with transcription [news; comment]. *Nature* 363:114–115.

Bronner, C., Baker, S., Morrison, P., Warrren, G., Smith, L., Lescoe, M., Kane, M., Earabine, C., Lipford, J., Lindblom, A., Tannergard, P., Bollag, J., Godwin, A., Ward, D., Nordenskjold, M., Fishel R., Kolodner, R. and Liskay, R. (1994) Mutation in the DNA mismatch repair gene homologue *hMLH1* is associated with hereditary nonpolyposis colon cancer. *Nature* 368: 258–261.

Brush, G., Anderson, C. and Kelly, T. (1994) The DNA-activated protein kinase is required for the phosphorylation of replication protein A during simian virus 40 DNA replication. *Proc. Natl. Acad. Sci. USA* 91:12520–12524.

Chen, C., Oliner, J., Zhan, Q., Fornace, A., Vogelstein, B. and Kastan, M. (1994) Interactions between p53 and MDM2 in a mammalian cell cycle checkpoint pathway. *Proc. Natl. Acad. Sci. USA* 91:2684–2688.

Chen, K., Jackson, P., Kirschner, M. and Dutta, A. (1995) Separate domains of p21 involved in the inhibition of Cdk kinase and PCNA. *Nature* 374:386–388.

Chestnut, J.D., Stephens, J.H. and Dahmus, M.E. (1992) The interaction of RNA polymerase II with the Adenovirus-2 major late promoter is precluded by phosphorylation of the C-terminal domain of subunit IIa. *J. Biol. Chem.* 267:10500–10506.

Cohen, M., Shaham, M., Dagan, J., Shumeli, E. and Kohn, G. (1975) Cytogenetic investigations in families with ataxia telangiectasia. *Cytogent. Cell Genet.* 15:338–356.

Dianov, G. and Lindahl, T. (1994) Reconstitution of base-excision repair pathway for DNA containing uracil residues. *Current Biol.* 4:1069–1107.

Dizdaroglu, M. (1986) Characterization of free radical induced DNA damage to DNA by the combined use of enzymatic hydrolysis and gas chromatography-mass spectrometry. *J. Chromatography* 367:357–366.

Drummond, J., Li, G., Longley, M. and Modrich P. (1995) Isolation of an hMSH2-p160 heterodimer that restores DNA mismatch repair to tumor cells. *Science* 268:1909–1912.

Dumaz, N., Drougard, C., Sarasin, A. and Daya-Grosjean, L. (1993) Specific UV-induced mutation spectrum in the p53 gene of skin tumors from DNA-repair-deficient xeroderma pigmentosum patients. *Proc. Natl. Acad. Sci. USA* 90:10529–10533.

Dutta, A., Ruppert, J., Aster, J. and Winchester, E. (1993) Inhibition of DNA replication factor RPA by p53. *Nature* 365:79–82.

Dvir, A., Peterson, S., Knuth, M., Lu, H. and Dynan, W. (1992) Ku autoantigen is the regulatory component of a template-associated protein kinase that phosphorylates RNA polymerase II. *Proc. Natl. Acad. Sci. USA* 89:11920–11924.

Dvir, A., Stein, L., Calore, B. and Dynan, W. (1993) Purification and characterization of a template associated protein kinase that phosphorylates RNA polymerase II. *J. Biol. Chem.* 268:10440–10447.

El-Deiry, W.S., Tokino, T., Velculescu, V.E., Levy, D.B., Parsons, R., Trent, J.M., Lin, D., Mercer, W.E., Kinzler, K.W. and Vogelstein, B. (1993) WAF1 a potential mediator of p53 tumor suppression. *Cell* 75:817–825.

Evans, M.K., Robbins, J.H., Ganges, M.B., Tarone, R.E., Nairn, R.S. and Bohr, V.A. (1993a) Gene-specific DNA repair in xeroderma pigmentosum complementation groups A, C, D, and F. Relation to cellular survival and clinical features. *J. Biol. Chem.* 268:4839–4847.

Evans, M.K., Taffe, B.G., Harris, C.C. and Bohr, V.A. (1993b) DNA strand bias in the repair of the p53 gene in normal human and xeroderma pigmentosum group C fibroblasts. *Cancer Res.* 53:5377–5382.

Falzon, M., Fewell, J.W. and Kuff, E.L. (1993) EBP-80, a transcription factor closely resembling the human autoantigen Ku, recognizes single- do double-strand transitions in DNA. *J. Biol. Chem.* 268:10546–10552.

Finnie, N., Gottlieb, T., Blunt, T., Jeggo, P. and Jackson, S. (1995) DNA-dependent protein kinase activity is absent in xrs-6 cells: Implications for site-specific recombination and DNA double-strand break repair. *Proc. Natl. Acad. Sci. USA* 92:320–324.

Fischer, L., Gerard, M., Chalut, C., Lutz, Y., Humbert, S., Kanno, M., Chambon, P. and Egly, J.L. (1992) Cloning of the 62-kilodalton component of basic transcription factor BTF2. *Science* 257:1392–1395.

Fishel, R., Lescoe, M., Rao, M., Copeland, N., Jenkins, N., Garber, J., Kane, M. and Kolodner, R. (1993). The human mutator gene homolog *MSH2* and its association with hereditary non-polyposis cancer. *Cell* 75:1027–1038.

Ford, J.M. and Hanawalt, P.C. (1995) Li-Fraumeni syndrome fibroblasts homozygous for p53 mutations are deficient in global DNA repair but exhibit normal transcription-coupled repair and enhanced UV-resistance. *Proc. Natl. Acad. Sci. USA* 92:8876–8880.

Friedberg, E.C., Walker, G.C. and Siede, W. (1995) Base excision repair. *In:* E.C. Friedberg, G.C. Walker and W. Siede (eds): *DNA Repair and Mutagenesis.* ASM Washington, DC, pp 135–190.

Goodrich, J.A. and Tjian, R. (1994) Transcription factors IIE and IIH and ATP hydrolysis direct promoter clearance by RNA polymerase II. *Cell* 77:145–156.

Gotoff, S., Amirmokri, E. and Leibner, E. (1967) Ataxia telangiectasia untoward response to X-irradiation and tuberous sclerosis. *Am. J. Dis. Child.* 114:617–625.

Gottlieb, T. and Jackson, S. (1993) The DNA-dependent protein kinase: requirement for DNA ends and association with Ku antigen. *Cell* 72:131–142.

Griffith, A., Craft, J., Evans, J., Mimori, T. and Hardin, J. (1992) Nucleotide sequence and genomic structure analysis of the p70 subunit of the human Ku autoantigen: evidence for a family of genes encoding Ku(p70)-related polypeptides. *Mol. Biol. Rep.* 16:91–97.

Hanawalt, P.C. (1994) Evolution of concepts in DNA repair. *Environ, Mol. Mutagen.* 23 Suppl 24:78–85.

Henning, K., Li, L., Legerski, R., Iyer, N., McDaniel, L., Schultz, R., Stefanini, M., Lehmann, A., Mayne, L. and Friedberg, E. (1995) The Cockayne syndrome complementation group A gene encodes a WD-repeat protein which interacts with CSB protein and a subunit of the RNA pol II transcription factor IIH. *Cell* 82:555–564.

Ho, L., Bohr, V.A. and Hanawalt, P.C. (1989) Demethylation enhances removal of UV damage from the overall genome and from specific DNA sequences in CHO cells. *Mol. Cell. Biol.* 9:1594–1603.

Hoeijmakers, J.H. (1993) Nucleotide excision repair. II: From yeast to mammals. *Trends Genet.* 9:211–217.

Hollstein, M., Sindransky, D., Vogelstein, B. and Harris, C. (1991) p53 Mutations in human cancers. *Science* 253:49–53.

Huang, J.C., Svoboda, D.L., Reardon, J.T. and Sancar, A. (1992) Human nucleotide excision nuclease removes thymine dimers by hydrolyzing the 22nd phosphodiester bond 5′ and the 6th phosphodiester bond 3′ to the photodimer. *Proc. Natl. Acad. Sci. USA* 89: 3664–3668.

Humbert, S., Van Vuuren, H., Lutz, Y., Hoeimakers, J.H.J., Egly, J.M. and Moncollin, V. (1994) Characterization of p44 and p34 subunits of the BTF2/TFIIH transcription repair factor. *EMBO J.* 13:2393–2398.

Jackson, S.P., MacDonald, J.J., Lees-Miller, S. and Tjian, R. (1990) GC box binding induces phosphorylation of Sp1 by a DNA-dependent protein kinase. *Cell* 63:155–165.

Juven, T., Barak, Y., Zauberman, A., George, D. and Oren, M. (1993) Wild type p53 can mediate sequence-specific transactivation of an internal promoter within the MDM gene. *Oncogene* 8:3411–3416.

Kastan, M., Onkywere, O., Sidransky, D., Vogelstein, B. and Craig, R. (1991) Participation of p53 in the cellular response to DNA damage. *Cancer Res.* 51:6304–6311.

Kastan, M.B., Zhen, Q., El-Deiry, W.S., Carrier, F., Jacks, T., Walsh, W.V., Plunkett, B.S., Vogelstein B. and Fornace Jr., A.J. (1992) A mammalian cell cycle ceckpoint pathway utilizing p53 and GADD45 is defective in ataxia telangiectasia. *Cell* 7:587–596.

Khanna, K.K. and Lavin, M.F. (1993) Ionizing radiation and UV induction of p53 protein by different pathways in ataxia telangiectasia cells. *Oncogene* 8:3307–3312.

Kojis, T., Gatti, R. and Sparks, R. (1991) The cytogenetics of ataxia telangiectasia. *Cancer Genet. Cytogenet.* 56:143–156.

Laybourn, P.J. and Dahmus, M.E. (1990) Phosphorylation of RNA polymerase IIA occurs subsequent to interaction with the promoter and before the initiation of transcription. *J. Biol. Chem.* 265:13165–13173.

Leadon, S.A. and Snowdon, M.M. (1988) Differential repair of DNA damage in the human metallothionein gene family. *Mol. Cell. Biol.* 8:5331–5338.

Lees-Miller, S., Chen, Y. and Anderson, C. (1990) Human cells contain a DNA-activated protein kinase that phosphorylates simian virus 40 T antigen, mouse p53, and the human Ku autoantigen. *Mol. Cell. Biol.* 10:6472–6481.

Lehmann, A. (1982) Three complementation groups Cokayne syndrome. *Mutat. Res.* 106: 347–356.

Lehmann, A.R. (1987) Cockayne's syndrome and trichothiodystrophy: defective repair without cancer. *Cancer Rev.* 7:82–103.

Levinson, G. and Gutman, G.A. (1987) High frequencies of short frameshifts in poly-CA/TG tandem repeats borne by bacteriophage M13 in *Escherichia coli* K-12. *Nucleic Acids Res.* 15:5323–5338.

Li, R., Waga, S., Hannon, G., Beach, D. and Stillman, B. (1994) Differential effects by the p21 CDK inhibitor on PCNA-dependent DNA replication and repair. *Nature* 371:534–537.

Loeb, L.A. (1994) Microsatellite instability: marker of a mutator phenotype in cancer. *Cancer Res.* 54:5059–5063.

Lu, H., Flores, O., Weinmann, R. and Reinberg, D. (1991) The nonphosphorylated form of RNA polymerase II preferentially associates with the preinitiation complex. *Proc. Natl. Acad. Sci. USA* 88:10004–10008.

Lu, X. and Lane, D.P. (1993) Differential induction of transcriptionally active p53 following UV or ionizing radiation: defects in chromosome instability syndromes? *Cell* 75:765–778.

Luo, Y., Hurwitz, J. and Massague, J. (1995) Cell-cycle inhibition by independent CDK and PCNA binding domains in p21. *Nature* 375:159–161.

Makela, T.P., Parvin, J.D., Kim, J., Huber, L.J., Sharp, P.A. and Weinberg, R.A. (1995) A kinasedeficient transcription factor TFIIH is functional in basal and activated transcription. *Proc. Natl. Acad. Sci. USA* 92:5174–5178.

Maltzman, W. and Czyzk, L. (1984) UV irradiation stimulates levels of p53 cellular tumor antigen in nontransformed mouse cells. *Mol. Cell. Biol.* 4:1689–1694.

Matsuda, T., Masafumi, S., Kuraoka, I., Kobayashi, T., Nakatsu, Y., Nagai, A., Enjoji, T., Masutani, C., Sugasawa, K., Hanaoka, F., Yasui, A. and Tanaka, K. (1995) DNA repair protein XPA binds replication protein A (RPA). *J. Biol. Chem.* 270:4152–4157.

Modrich, P. (1994) Mismatch repair, genetic stability, and cancer. *Science* 266:1959–1960.

Morgan, J., Holcomb, T. and Morrissey, R. (1968) Radiation reactions in ataxia telangiectasia. *Am. J. Dis. Child.* 116:557–558.

Morozov, V., Falzon, M., Anderson, C. and Kuff, E. (1994) DNA-dependent protein kinase is activated by nicks and larger single-stranded gaps. *J. Biol. Chem.* 269:16684–16688.

Mullenders, L.H.F., van Kesteren, A.C., Bussman, C.J.M., van Zeeland, A.A. and Natarajan, A.T. (1984) Preferential repair of nuclear matrix associated DNA in Xeroderma pigmentosum group C. *Mutat. Res.* 141:75–82.

Nance, M.A. and Berry, S.A. (1992) Cockayne syndrome: review of 140 cases. *Am. J. Med. Genet.* 42:68–84.

Nelson, W. and Kastan, M. (1994) DNA strand breaks: the DNA template alterations that trigger p53-dependent DNA damage response pathways. *Mol. Cell. Biol.* 14:1815–1823.

Okumoto, D.S. and Bohr, V.A. (1987) DNA repair in the metallothionein gene in CHO cells increases with transcriptional activation. *Nuclei Acids Res.* 15:10021–10030.

Paillard, S. and Strauss, F. (1991) Analysis of the mechanisms of interaction of simian Ku protein with DNA. *Nucleic Acids Res.* 19:5619–5624.

Palombo, F., Gallinari, P., Iaccarino, I., Lettieri, T., Hughes, M., D'Arrigo, A., Truong, O., Hsuan, J. and Jiricny, J. (1995) GTBP, a 160-kilodalton protein essential for mismatch-binding activity in human cells. *Science* 268:1912–1914.

Pandita, T. and Hittleman, W. (1992) Initial chromosome damage but not DNA damage is greater in ataxia telangiectasia cells. *Radiat. Res.* 130:94–103.

Papadopoulos, N., Nicolaides, N., Wei, Y., Ruben, S., Carter, K., Rosen, C., Haseltine, W., Fleischmann, R., Fraser, C., Adams, M., Venter, J., Hamilton, S., Petersen, G., Watson, P., Lynch, H., Peltomaki, P., Mecklin, J., De la Chapelle, A., Kinzler, K. and Vogelstein, B. (1994) Mutation of a *mutL* homolog in hereditary colon cancer. *Science* 263:1625–1629.

Papadopoulos, N., Nicolaides, N., Liu, B., Parsons, R., Lengauer, C., Palombo, F., D'Arrigo, A., Markowitz, S., Wilson, J., Kinzler, K., Jiricny, J. and Vogelstein, B. (1995) Mutations of GTBP in genetically unstable cells. *Science* 268:1915–1917.

Peterson, S.R., Dvir, A., Anderson, C.W. and Dynan, W.S. (1992) DNA binding provides a signal for phosphorylation of the RNA polymerase II heptapeptide repeats. *Genes & Dev.* 6:426–438.

Peterson, S.R., Kurimasa, A., Oshimura, M., Dynan, W.S., Bradbury, E.M., and Chen, D.J. (1995) Loss of the catalytic subunit of the DNA-dependent protein kinase in DNA double-strand break repair mutant mammalian cells. *Proc. Natl. Acad. Sci. USA* 92:3171–3174.

Reardon, J.T., Nichols, A.F., Keeney, S., Smith, C.A., Taylor, J.S. and Linn, S. (1993) Comparative analysis of binding of human damaged DNA-binding protein (XPE) and *Escherichia coli* damage recognition protein (UvrA) to the major ultraviolet photoproducts: T[c,s]T, T[t,s]T, T[6–4]T, and T[Dewar]T. *J. Biol. Chem.* 268:21301–21308.

Roth, D.B., Zhu, C. and Gellert, M. (1993) Characterization of broken DNA molecules associated with V(D)J recombination. *Proc. Natl. Acad. Sci. USA* 90:10788–10792.

Roy, R., Adamczewski, J.P., Seroz, T., Vermeulen, W., Tassan,, J.-P., Schaeffer, L., Nigg, E.A., Hoeijmakers, J.H.J. and Egly, J.-M. (1994) The MO15 Cell cycle kinase is associated with the TFIIH transcription-DNA repair factor. *Cell* 79:1093–1101.

Sancar, A. (1995) Excision repair in mammalian cells. *J. Biol. Chem.* 270:15915–15918.

Sancar, A. and Tang, M.S. (1993) Nucleotide excision repair. *Photochem. Photobiol.* 57:905–921.

Savitsky, K., Bar-Shira, A., Gilad, S., Rotman, G., Ziv, Y., Vanagaite, L., Tagle, D., Smith, S., Uziel, T., Sfez, S., Ashkenazi, M., Pecker, I., Frydman, M., Harnik, R., Patanjali, S., Simmons, A., Clines, G., Sartiel, A., Gatti, R., Chessa, L., Sanal, O., Lavin, M., Jaspers, N., Taylor, A., Arlett, C., Miki, T., Weissman, S., Lovett, M., Collins, F. and Shiloh, Y. (1995) A single ataxia telangiectasia gene with a product similar to PI-3 kinase. *Science* 268:1749–1753.

Schaeffer, L., Roy, R., Humbert, S., Moncollin, V., Vermeulen, W., Hoeijmakers, J.H., Chambon, P. and Egly, J.M. (1993) DNA repair helicase: a component of BTF2 (TFIIH) basic transcription factor. *Science* 260:58–63.

Schaeffer, L., Moncollin, V., Roy, R., Staub, A., Mezzina, M., Sarasin, A., Weeda, G., Hoeij-makers, J.H.J. and Egly, J.M. (1994) The ERCC2/DNA repair protein is associated with the class II BTF2/TFIIH transcription factor. *EMBO J.* 13:2388–2392.

Schlissel, M., Constantinescu, A., Morrow, T., Baxter, M. and Peng, A. (1993) Double-strand signal sequence breaks in V(D)J recombination are blunt, 5′ phosphorylated, RAG-dependent, and cell-cycle-regulated. *Genes & Dev.* 7:2520–2532.

Selvakumaran, M. (1994) Immediate early up-regulation of *bax* expression by p53 but not TGF beta 1: a paradigm for distinct apoptotic pathways. *Oncogene* 9:1791–1798.

Serizawa, H., Makela, T.P., Conaway, J.W., Conaway, R.C., Weinberg, R.A. and Young, R.A. (1995) Association of Cdk-activating kinase subunits with transcription factor TFIIH. *Nature* 374:280–282.

Shiekhattar, R., Mermelstein, F., Fisher, R.P., Drapkin, R., Dynlacht, B., Wessling, H.C., Morgan, D.O. and Reinberg, D. (1995) Cdk-activating kinase complex is a component of human transcription factor TFIIH. *Nature* 374:283–287.

Taccioli, G., Cheng, H., Varghese, A., Whitmore, G. and Alt, F. (1994 a). A DNA repair defect in Chinese hamster ovary cells affects V(D)J recombination similarly to the murine *scid* mutation. *J. Biol. Chem.* 269:7439–7442.

Taccioli, G., Gottlieb, T., Blunt, T., Priestley, A., Demengeot, J., Mizuta, R., Lehmann, A., Alt, F., Jackson, S. and Jeggo, P. (1994 b) Ku80: product of the XRCC5 gene and its role in DNA repair and V(D)J recombination. *Science* 265:1442–1445.

Tanaka, K., Kawai, K., Kumahara, Y., Ikenaga, M. and Okada, Y. (1981) Genetic complementation groups in Cockayne syndrome. *Somat. Cell Mol. Genet.* 7:445–455.

Tonegawa, S. (1983) Somatic generation of antibody diversity. *Nature* 302:575–581.

Tornaletti, S. and Pfeifer, G. (1994) Slow repair of pyrimidine dimers at *p53* mutation hotspots in skin cancer. *Science* 263:1436–1438.

Waga, S., Hannon, G.J., Beach, D. and Stillman, B. (1994) The p21 inhibitor of cyclin-dependent kinases controls DNA replication by interaction with PCNA. *Nature* 369:574–578.

Wang, X.W., Yeh, H., Schaeffer, L., Roy, R., Moncollin, V., Egly, J.-M., Wang, Z., Friedberg, E.C., Evans, M.K., Bohr, V.A., Weeda, G., Hoeijmakers, J.H.J., Forrester, K. and Harris, C.C. (1995 a) p53 modulation of TFIIH-associated nucleotide excision repair activity. *Nature* 10:188–195.

Wang, Z., Buratowski, S., Svejstrup, J.Q., Feaver, W.J., Wu, X., Kornberg, R.D., Donahue, T.F. and Friedberg, E.C. (1995 b) The yeast *TFB1* and *SSL1* genes, which encode subunits of Transcription Factor IIH, are required for nucleotide excision repair and RNA polymerase II transcription. *Mol. Cell. Biol.* 15:2288–2293.

Weeda, G., Hoeijmakers, J.H. and Bootsma, D. (1993) Genes controlling nucleotide excision repair in eukaryotic cells. *Bioessays* 15:249–258.

Weibezahn, K.F., Lohrer H. and Herrlich, P. (1985) Double-strand break repair and G2 block in Chinese hamster ovary cells and their radiosensitive mutants. *Mutat. Res.* 145:177–183.

Yamaizumi, M. and Sugano, T. (1994) UV-induced nuclear accumulation of p53 is evoked through DNA damage of actively transcribed genes independent of the cell cycle. *Oncogene* 9:2775–27784.

Yeung, A.T., Mattes, W.B., Oh, E.Y., Yoakum, G.H. and Grossman, L. (1986) The purification of the *Escherichia coli* UvrABC incision system. *Nucleic Acids Res.* 14:8535–8556.

Stress-Inducible Cellular Responses
ed. by U. Feige, R.I. Morimoto, I. Yahara and B. Polla
© 1996 Birkhäuser Verlag Basel/Switzerland

Toxic metal-responsive gene transcription

Z. Zhu and D.J. Thiele

*Department of Biological Chemistry, University of Michigan Medical School,
Ann Arbor, MI 48109-0606, USA*

Summary. Metals play a dual role in biological systems, serving as essential co-factors for a wide range of biochemical reactions yet these same metals may be extremely toxic to cells. To cope with the stress of increases in environmental metal concentrations, eukaryotic cells have developed sophisticated toxic metal sensing proteins which respond to elevations in metal concentrations. This signal is transmitted to stimulate the cellular transcriptional machinery to activate expression of metal detoxification and homeostasis genes. This review summarizes our current understanding of the biochemical and genetic mechanisms which underlie cellular responses to toxic metals via metalloregulatory transcription factors.

Introduction

Inorganic properties of metal ions

Metal ions are versatile biological constituents (Williams, 1978; Phipps, 1976; Lippard and Berg, 1994). Highly mobile and weakly bound ions such as Na^+ and K^+ serve as charge carriers, while Ca^{2+} and Mg^{2+}, of moderate affinity, as structure formers. In contrast, static metal ions such as Fe^{2+} and Cu^{2+}, both of which bind with high affinity to biological ligands, often function as redox catalysts. The functions of metal ions in biological systems are governed by their chemical properties such as ion sizes, electron affinities and geometric demands, which also allow biochemical differentiation between similar metal ions (Cotton and Wilkinson, 1980).

Metals in biology and as toxins

The functions of metal ions in biological systems can be generalized as a means of activating bio-molecules through coordination for further biological reactions. The activation of biomolecules by metal ion co-ordination can mainly be categorized into two classes: structural or catalytic. Metal ion coordination is a well established biochemical mechanism for creating a structural scaffold for proper protein conformation. With Zn^{2+}, for example, among different types of Zn-finger transcription factor proteins, Zn^{2+} itself does not directly interact with DNA, rather through properly positioned amino acid residues triggered by Zn^{2+} coordination of these proteins (Harrison, 1991). Coordination of metal ions also generates more reactive electrophiles or nucleophiles, therefore fulfilling catalytic functions of enzymes such as ribozymes, a class of RNA metalloenzymes catalyzing the hydrolysis of phosphate diester bonds in RNA (Pyle, 1993). As a cofactor, metal ions such as Fe^{2+}/Fe^{3+} in iron-sulfur cluster proteins and Cu^+/Cu^{2+} in blue copper proteins are often activated through protein binding by altering redox potentials, therefore, facilitating electron-transfer events (Phipps, 1976). The intrinsic complexity of proteins together with the rich and versatile chemistry of metal ions account for molecular recognition with a high degree of selectivity.

As manifested in the periodic table, many metal ions resemble one another in chemistry and physical chemistry, such as ionic radii and stereochemical demands, due to similarity in electronic configuration (Cotton and Wilkinson, 1980). However, the similarity is often deleterious in biological systems. To biological ligands, metal ions of high affinity such as Cd^{2+} and Pb^{2+}, can compete or displace weakly bound essential metal ions such as Ca^{2+} and Mg^{2+}, thereby inhibiting functions of metallo-enzymes. For example, Cd^{2+} has been shown to be able to disrupt a trans-membrane flux of Ca^{2+} and inhibit postsynaptic Ca^{2+} channels (Giles et al. 1993). Low concentrations of Pb^{2+} can substitute for Ca^{2+} in second messenger metabolism, activate protein kinase C (PKC) and cause neuro-toxicity (Goldstein, 1993). Replacement of Zn^{2+} in the human estrogen receptor DNA-binding domain by copper can abolish the DNA-binding ability and severely inhibit the function of this important signal transduc-tion molecule (Hutchens et al., 1992; Predki and Sarkar, 1993). Redox active metal ions, such Fe^{2+} and Cu^{2+}, can cause toxicities through generat-ing damaging free radicals from metabolites such as H_2O_2 or O_2 via Fenton reactions (Cheeseman and Slater, 1993). Therefore, even metal ions which are essential to biological systems can be toxic at abnormally elevated con-centrations. Indeed, cells must be able to distinguish essential metal ions from toxic metal ions and to differentiate between normal and toxic levels for an essential metal ion. In this review, we discuss the molecular me-chanisms by which yeast and mammalian cells sense and transcriptionally respond to toxic metals by activating the transcription of metallothionein genes.

Yeast: A model for copper-responsive gene transcription

Yeast cells are excellent models for understanding the precise biochemical details underlying biological processes due to their genetic manipulability, ease of growth and amenability to a wide range of molecular approaches. Importantly, yeast cells carry out similar biochemical reactions to those required in humans such as DNA replication and repair, protein secretion, cell cycle regulation and other key processes.

Yeast metallothioneins: Structure and function

Since copper is essential yet toxic at an elevated concentration, an important question is how cells maintain a delicate balance between normal and toxic intracellular copper levels. In the baker's yeast, *S. cerevisiae*, copper detoxification is in large part carried out by the class of low molecular weight cysteine-rich metal binding proteins known as metallothioneins (MT), encoded by the *CUP1* gene (Hamer, 1986; Brenes-Pomales et al., 1965; Butt and Ecker, 1987). CUP1 shares very limited sequence homology to known mammalian MTs, however, the metal-binding Cys-X-Cys and Cys-X-X-Cys motifs are retained. Surprisingly, NMR studies of silver-substituted CUP1 protein show that only 10 of the 12 cysteines are involved in coordination of 7 Ag^+ in a mixture of tridental and bidental coordination geometry (Narula et al., 1993). The role of the other two cysteine thiolates in MT function is currently unknown. *CUP1* gene deletion experiments in *S. cerevisiae* have demonstrated that MT is not required for growth under standard laboratory conditions, however, *CUP1* is critical when yeast cells are exposed to high concentration of exogenous copper (Hamer et al., 1985). These observations show that yeast MT is an essential component in the copper detoxification pathway. Unlike mammalian MTs (see below), transcription of *CUP1* is only induced in response to the metal Cu^+ and Ag^+ (Karin et al., 1984; Butt et al., 1984).

Metallothioneins have also been shown to be able to scavenge free radicals efficiently *in vitro* (Thornalley and Vasak, 1985). Tamai et al. have demonstrated that CUP1 can functionally substitute Cu,ZnSOD, an enzyme which catalyzes the conversion of O_2^- to H_2O_2 and O_2, and that *CUP1* is transcriptionally activated by oxidative stress (Tamai et al., 1993). Recently, a second MT-like protein encoded by the *CRS5* gene in *S. cerevisiae* has been identified (Culotta et al., 1994). Deletion analysis shows that *CRS5* confers a slight increase in copper resistance in *cup1* deletion strains at very low copper concentrations (Culotta et al., 1994). These results demonstrate that *CUP1* is dominant in copper detoxification in *S. cerevisiae*.

In contrast to baker's yeast and resembling higher eukaryotes, the opportunistic pathogenic yeast *C. glabrata* harbors a large *MT* gene family containing a single *MT-I* gene, a tandemly amplified *MT-IIa* gene, an unlinked *MT-IIb* gene and perhaps additional genes encoding MT isoforms (Mehra et al., 1989 and 1990). *MT-I* and *MT-II* encode 62 and 57 amino acid polypeptides, containing 18 and 16 cysteines, respectively, which are arranged in Cys-X-Cys and Cys-X-X-Cys motifs characteristic of all known metallothioneins. Titration analysis of apo-MT-I and MT-II has established that *in vitro* MT-I binds 12 Cu^+ ions and MT-II binds 10 Cu^+ ions (Mehra et al., 1989). Both MT-I and MT-II have been shown to play crucial roles in the *C. glabrata* defense against copper toxicity and their transcription is induced only by the metals Cu⁺ or Ag⁺ (Mehra et al., 1992). However, not only *MT-I* and *MT-II* but also *MT-IIa* and *MT-IIb* function differentially in copper detoxification. Upon addition of copper, *MT-I* shows an extremely rapid kinetics of mRNA accumulation, while high level of transcription activation of *MT-II* reaches maximum levels only after a 1-h exposure to copper (Zhou and Thiele, 1993). Genetic analysis has demonstrated that the disruption of the single copy *MT-IIb* gene causes much more severe defects in copper tolerance than disruption of amplified *MT-IIa* locus (Mehra et al., 1992).

Yeast copper-metalloregulatory transcription factors

Two lines of experimentation strongly suggested that the *S. cerevisiae* *CUP1* gene is transcriptionally activated by one or more positive-acting Cu-responsive, regulatory factors. First, the analysis of *CUP1* promoter deletion mutants demonstrated the presence of a large Cu-responsive promoter region (Upstream Activation Sequence for *CUP1*, UAScUP1) between – 105 and approximately – 200 nucleotides upstream from the start site for transcription which is necessary for Cu-inducible transcription (Thiele and Hamer, 1986). Since deletions within this region progressively diminished the response to Cu, the presence of multiple *cis*-acting Cu-responsive promoter elements in UAScUP1 was postulated. Consistent with these elements responding to a positive-acting factor, fusion of synthetic oligonucleotides within this region to a *CYC1* (iso-1 cytochrome c) minimal promoter-*galK* (*E. coli* galactokinase) gene rendered this promoter Cu-inducible.

The second observation indicating the presence of a positive transcription factor stemmed from studies of *CUP1* gene function (Hamer et al., 1985). The introduction of a *CUP1* promoter-*galK* fusion gene into *CUP1* wild-type and isogenic *cup1* deletion strains revealed that *cup1* deletion strains express high constitutive levels of galK mRNA that are only modestly induced by Cu addition. In the *CUP1* wild-type strain, however, galK mRNA levels are highly induced after Cu administration as expected. Therefore, *CUP1* gene transcription is subject to negative autoregulation. Both the high constitutive levels observed in the *cup1* deletion strain and the typical Cu inducible transcription observed in the wild-type strain required the presence of UAScUP1, suggesting that both modes of activation required Cu-responsive factors. Further studies demonstrated that the negative autoregulation of the *CUP1* gene directly correlated with the ability of the *CUP1*-encoded MT to bind Cu and protect cells from Cu toxicity (Wright et al., 1988). Taken together, the promoter deletion-fusion studies and the observation of *CUP1* gene negative autoregulation suggest that basal levels of *CUP1*-encoded MT sequesters Cu from a positive trans-acting Cu-responsive transcription factor.

Early genetic studies resulted in the identification and isolation of a locus, designated *ACE1* (or *CUP2*), which plays a pivotal role in copper-responsive transcription of *CUP1* in *S. cerevisiae* (Thiele, 1988; Welch et al., 1989). Interestingly, ACE1 is also a cysteine-rich, Cu(I)-binding protein much like the MTs. ACE1 functions both as a copper sensor and as a sequence-specific DNA binding transcription factor. The primary structure of ACE1 can be divided into two domains: an amino-terminal DNA-binding domain and an acidic carboxyl-terminal transcription activation domain (Szczypka and Thiele, 1989; Furst et al., 1988). The intracellular copper sensing function of ACE1 is constituted by Cu(I) coordination through cysteinyl thiolates interdigitated within the ACE1 Cu^+ activated DNA binding domain. The coordination of Cu^+ by ACE1 is transduced into a biological signal by altering the ACE1 DNA binding domain conformation which triggers sequence-specific binding of monomeric ACE1 to four independent Metal Responsive Elements (MREs) in the *CUP1* promoter. Consistent with the binding of a monomer, the yeast MT MRE sequence do not exhibit two-fold symmetry, nor do they resemble the MREs of mammals (see below). Once bound, ACE1 activates *CUP1* transcription rapidly and potently. Therefore, the Cu^+ sensing domain provides a direct link between sensing and transcriptional activation and constitutes a simple and economical cellular response to sudden changes in environmental Cu^+ levels. The selectivity of this biological switch is governed by the precise copper coordination chemistry, which accounts for the specificity of copper-responsive gene transcription of *CUP1* in *S. cerevisiae*. By considering the differences in coordination chemistries between Cu^+ or Ag^+ and other heavy metal ions such Cd^{2+} or Zn^{2+}, and the requirement of sequence-specific interaction between DNA and the ACE1 protein, it makes sense why the biosynthesis of CUP1 is only induced by Cu^+ or Ag^+. Furthermore, the less potent induction of *CUP1* transcription activation by Ag^+ can be explained by the ionic size difference between Cu^+ and Ag^+. Even though, Cu^+ and Ag^+ can assume identical connectivities to the cysteinyl thiolates of ACE1 protein, the larger ionic radii of Ag^+, especially when bound to ligands, can certainly cause spatial changes in respect to the distances between the residues interacting with DNA of the *CUP1* promoter, compared to Cu^+ which has a smaller ionic radius (Cotton and Wilkinson, 1980). It is also possible that Cu^{2+} is transported more efficiently into yeast cells than is Ag^+, resulting in Ag^+ being a less potent activator of *CUP1* transcription *in vivo*.

The sequence specific interaction between ACE1 and the *CUP1* promoter has been established by detailed biochemical experiments including *in vitro* and *in vivo* DNase I footprinting, *in vitro* hydroxyl radical footprinting and mobility shift assays (Evans et al., 1990; Huibregtse et al., 1989; Furst and Hamer, 1989; Buchman et al., 1990). Recent experiments utilizing ACE1 protein partially modified at cysteine residues by carboxy methylation and methylation interference footprinting analysis, suggest that ACE1 comprises two distinct domains which interact with the *CUP1* promoter through a minor groove binding located between two major groove bindings (Dobi et al., 1995).

Despite functioning as a Cu-sensing transcription activating molecule, ACE1 is constitutively expressed both at transcription and translation levels. This is thought to poise this molecule for rapid transcriptional responses to sudden elevations in the levels of toxic copper ions (Szczypka and Thiele, 1989). By contrast, the Cu metalloregulatory transcription factor gene *AMT1* of *C. glabrata* is positively auto-regulated by its own product, AMT1 protein, in response to copper (Zhou and Thiele, 1993). AMT1 shows a high degree of sequence homology to ACE1 with 39% identity and 61% of similarity (Zhou and Thiele, 1991). Compared to ACE1, the positioning of the 11 cysteine residues located within amino-terminal half of AMT1 are conserved. AMT1 activates not only the *AMT1* gene but also the entire family of metallothionein genes in *C. glabrata* (Zhou et al., 1992). A single MRE located in the *AMT1* promoter is required for both the binding of AMT1 *in vitro* and Cu and AMT1-dependent transcriptional activation *in vivo* (Zhou and Thiele, 1993). This homologous *cis*-acting regulatory element shares the same consensus of a core sequence 5′-GCTG-3′, which is preceded by an AT-rich region, with the MREs found in the MT gene promoters of both *S. cerevisiae* and *C. glabrata*. The same consensus MREs between AMT1 and ACE1 binding sequences reflects the structural similarities between AMT1 and ACE1 proteins. Similar to ACE1, the localization of the Cu^+- and DNA-binding functions of AMT1 within the same amino-terminal region (1–115) establishes that Cu^+-binding confers proper conformation and therefore directs the sequence specific DNA-AMT1 interactions. The inducible transcription activation of *AMT1, MT-I* and *MT-II* by Cu(I) and iso-electronic Ag(I), but not by Cd(II) or Zn(II), further establish the structural requirement for Cu(I) in the DNA-binding and transcriptional regulation. Recently, Thorvaldsen et al. have shown that AMT1 protein expressed in and purified from bacteria has a stoichiometry of Cu_4Zn: AMT1. However, the Zn^{2+}-coordination does not affect the DNA-binding of Cu_4: AMT1 (Thorvaldsen et al., 1994). Whether or not this Zn^{2+}-coordination occurs *in vivo* and the corresponding biological relevance have yet to be determined. The functional importance of *AMT1* auto-activation in copper detoxification has been demonstrated by the findings that a mutant incapable of autoactivation fails to grow even at low copper concentration of 25 μM and the copper-sensitive *amt1-1* mutant strain transformed with a plasmid containing a wild-type *AMT1* gene which contains a mutation in the *AMT1* binding site on its own promoter, showed dramatic decreases in copper tolerance (Zhou and Thiele, 1993). The observed copper hypersensitivity of *amt1* mutant in part is due to a marked loss of *MT-II* gene expression while *MT-I* mRNA levels were only modestly reduced (Zhou and Thiele, 1993). A schematic representation of *MT* gene transcription via Cu and AMT1 in *C. glabrata* is shown in Figure 1.

Metallothionein Gene Transcription

Candida glabrata

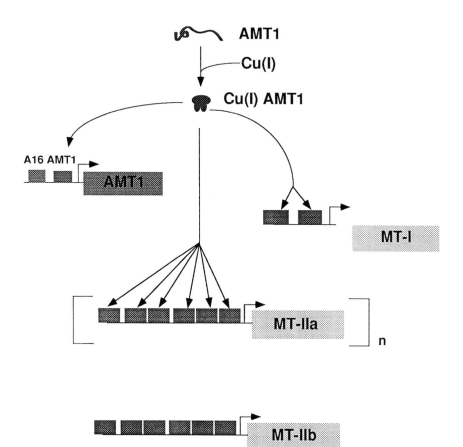

Figure 1. A schematic representation of *MT* gene transcription activation via Cu⁺ and AMT1 in *C. glabrata*. The dark boxes are Metal Responsive Elements (MRE). As a copper sensing molecule, AMT1 binds Cu⁺ through its cysteinyl thiolates, which alters its DNA-binding domain conformation and triggers the sequence-specific interactions of AMT1 with the MREs. Binding therefore positively autoregulates the *AMT1* gene and transcriptionally activates the metallothionein gene family: *MT-I, MT-IIa* and *MT-IIb*, in response to elevated copper concentrations. The metallothionein proteins in turn chelate free intracellular copper preventing cytotoxicity.

Mammalian metal-responsive gene transcription

In mammals the effects of toxic metals such as Cd, Hg, Cu and others are well documented and exposure to toxic metals is known to lead to renal failure, mental retardation and other illnesses. Furthermore, metals form the basis for several pharmacological agents used to treat cancers and for biomedical imaging analysis (Abrams and Murrer, 1993). Therefore, it is crucial to understand how mammals sense the presence of elevated concentrations of toxic metals and modulate target gene transcription to elicit the appropriate protective or homeostatic response. Indeed, metals may be sensed through proteins designed by nature for this process such as ACE1 or AMT1 described above, or through the inactivation or alteration of proteins which, under normal unstressed physiological conditions, do not ordinarily interact with toxic metals. Below we summarize very recent work in mammalian systems on metal-responsive MT gene transcription.

Similar to the studies described for yeast, both transfection studies in cell culture and either overexpression or gene deletion studies in transgenic mice have clearly demonstrated that the expression of mammalian MT genes is critical for resistance to high levels of toxic metals such as Cd (Masters et al., 1994; Michalska and Choo, 1993). Indeed, since mammalian MT genes are highly transcriptionally induced by metals such as Cd, Zn, Cu and others, these genes represent the best understood examples of toxic metal-responsive gene expression in mammals. Nonetheless, although metal-responsive MT gene transcription in mammals has been under intensive study for many years, the precise mechanisms by which mammalian cells sense and respond to the large array of chemically distinct metal ions to activate gene expression remain to be elucidated. This work has, however, provided a fundamental framework for ultimately understanding these biochemical details.

All mammalian MT gene promoters harbor multiple copies of a small *cis*-acting promoter Metal Responsive Element (MRE) which contains as a core sequence the consensus 5′-TGCPuCXC-3′ typically flanked by a non-conserved GC-rich element (Westin and Schaffner, 1988). The MREs of higher eukaryotes bear no sequence similarity to those described above for yeast (Thiele, 1992). Mammalian MREs activate metal-responsive transcription weakly if at all when in single copy, however, these elements confer potent metal-responsive transcription of heterologous promoters when present in two or more copies upstream of the transcriptional initiation site. The mouse MT-I promoter harbors six MREs (MRE a through f) within approximately 200 bp upstream of the transcriptional start site, which differ in their respective transcriptional potency (Stuart et al., 1985). The mMT-I MREd has been shown by promoter fusion studies to be the strongest MRE with respect to metal-responsive transcriptional activation (Stuart et al., 1985). Point mutagenesis of the mouse MT-I promoter MREd demonstrated two important features of this MRE. First, the residues in the core sequence, but not the GC-rich flanking element are critical for metal responsive transcription. Secondly, a single MRE is capable of responding to a range of metals such as Cd, Zn and Cu (Culotta and Hamer, 1989). The fact that MRE sequences are conserved in humans, mice, sea urchins, fish, flies and other higher eukaryotes strongly suggests that these properties of MREs, as well as MRE DNA binding proteins, may be conserved in their mode of action in metal-responsive transcription.

In both mouse and humans, four genes and at least 14 MT isoform genes, respectively, are known to be tightly linked (Searle et al., 1984). Although MT-I or MT-II promoters driving heterologous genes in either transfection assays or in transgenic mice do give rise to metal responsiveness, this regulation does not reflect the tissue-specific expression levels of the endogenous MT genes, nor does it typically correlate with trans-gene copy number. Palmiter and co-workers reported that both the mouse MT-I and MT-II promoters harbor crucial distal regulatory elements, which when adjacent to the metal-responsive promoter regions improved both the expression levels, regulation of expression and appropriate tissue-specific expression in transgenic mice (Palmiter et al., 1993). The presence of such distal regulatory elements is similar to their occurence as critical regulatory regions in the developmentally regulated expression of fetal and adult globin genes (Forrester et al., 1986). Encompassed within these flanking sequenced are DNase I hypersensitive sites which, although necessary for this control of MT expression, are not sufficient. Therefore, the mouse MT genes harbor important distal regulatory elements which modulate the efficacy with which metals mediate transcription via MREs. Alternatively these elements may alter chromatin structure to allow access to metalloregulatory factors or other features of the DNA structure which confer proper regulation independent of the site at which the MT promoter is integrated in genomic DNA.

Although we currently know that in yeast cells the toxic metal sensor and MRE binding protein are identical, the precise mechanisms by which metalloregulation occurs via mammalian MREs is complex and poorly understood. Since a single MRE sequence mediates transcriptional responses to a wide variety of metals, a number of possibilities exists for the function of mammalian *trans*-acting factor action through these MREs. Elegant *in vivo* transcriptional competition experiments with the mouse MT-I promoter gave the first indication for the existence, in mouse cells, of a titratable positively acting factor or factors required for MRE-dependent metal-responsive gene transcription (Seguin et al., 1984). Furthermore, that such factors are induced to bind to the five MREs in the mouse MT-I promoter, in response to administration of Cd or Zn at concentrations which induce gene transcription, was demonstrated by *in vivo* footprinting (Muller et al., 1988). Taken together, these data strongly indicated the presence of one or more metal-responsive MRE binding activities in mammalian cells as candidates for mammalian metalloregulatory transcription factors. Indeed, early experiments demonstrated that metal-responsive transcription in mammalian cells occurs in the presence of protein synthesis inhibitors, indicating that the transcription factors involved in this process are pre-existing (Karin et al., 1980). A series of biochemical and genetic hunts ensued which resulted in the identification of several potential candidates for mammalian metalloregulatory transcription factors. Here we will focus on one of these factors which, on the basis of genetic and biochemical experiments, is currently the most thoroughly understood in terms of its structure and mechanism of action. However, it is highly likely that several metal responsive factors may activate mammalian MT genes.

Schaffner and colleagues initially detected a Zn inducible MREd DNA binding activity, termed MTF-1, in HeLa cell extracts which bound only transcriptionally competent forms of the MRE but not to inactive mutant derivatives (Westin and Schaffner, 1988). Subsequently, a synthetic oligonucleotide was utilized to screen a mouse cDNA library, expressed in *E. coli,* to search for clones expressing a protein which binds tightly to the MRE. One such cDNA was isolated and the sequence of this cDNA demonstrated the presence of an open reading frame encoding a protein of 675 amino acids, denoted MTF-1, containing six TF-IIIA-like Zn fingers (Cys2-His2) (Radtke et al., 1993). Furthermore, consistent with MTF-1 functioning as a transcription factor, three potential transcription activation domains are located carboxyl-terminal to the Zn fingers and consist of an acidic region, a proline-rich region and a serine-threonine-rich region. When fused to the yeast GAL4 protein DNA binding domain the acidic region of MTF-1 was able to activate transcription *in vivo* (Xu et al., 1994). Indeed, consistent with MTF-1 playing an important role in metal responsive induction of MT gene transcription, Schaffner and colleagues demonstrated that mouse embryonic stem cells harboring homozygous disruptions of the MTF-1 locus are defective in activation of MT-I and MT-II gene transcription in response to a variety of metals including Zn, Cd, Cu and Pb. It is striking that even basal levels of MT-I or MT-II mRNA in the MTF-1 -/- ES cells are almost undetectable, suggesting that the MTF-1 mediated MT transcription may also sense endogenous levels of free metals as well. The introduction of the MTF-1 cDNA back into the -/- ES cells completely restored metal-responsive activation of MT gene transcription, clearly supporting a critical role for MTF-1 in the activation of MT gene transcription in response to metals (Heuchel et al., 1994). It will be very interesting to identify other target genes for the action of MTF-1, as well as to ascertain whether MTF-1, like the MT genes, is critical for toxic metal resistance in mice via gene knockout technology.

The binding of mouse MTF-1 to MREs *in vitro* has been demonstrated to be stimulated by Zn in a dose-dependent manner and inactivated by Zn chelation (Radtke et al., 1993). Therefore, like the yeast Cu metalloregulatory transcription factors ACE1 and AMT1, the metal sensing function may be an integral part of the transcriptional activator protein. The human MTF-1 protein, purified to homogeneity in an elegant study using a bio-tinylated MRE probe absorbed with streptavidin beads, also binds to synthetic MREs in a Zn-dependent manner *in vitro* and to MREs a and b in the human MT-IIa promoter (Otsuka et al., 1994). The corresponding human MTF-1 cDNA has also been obtained and encodes a protein which is similar in structure to that of the mouse, although the protein is 78 amino acids longer at the carboxyl-terminus (Brugnera et al., 1994). Interestingly, when the transfected human MTF-1 cDNA is expressed in human cells, the Zn-dependent transcriptional induction of MRE-driven reporter genes is much more responsive than that of the mouse MTF-1 (Brugnera et al., 1994). Perhaps regions of the human MTF-1 DNA binding domain or *trans*-activation domains, distinct from mouse, mediate this stronger Zn-responsiveness.

Although based on its primary structure it is unclear how MTF-1 may respond to a wide array of chemically distinct metals to activate MT gene transcription, Palmiter has suggested the presence of an MTF-1-associated protein termed MTF-1-inhibitor (MTI), which is released from MTF-1 by metals (Palmiter, 1994). Interestingly, in an effort to use somatic cell genetics to isolate candidate genes encoding MTI-like activities, two cDNAs encoding Zn transporters have been isolated, providing important tools for future studies aimed at understanding how the homeostasis of this metal occurs in mammalian cells (Palmiter and Findley, 1995).

Conclusion

In this review we have underscored the importance of metals in biology and the consequences of metal toxicity in all known organisms. Indeed, it is vital that toxic metals be appropriately sensed and that transcriptional responses ensue rapidly to allow organisms to effectively deal with these stressful condition. Yeast cells utilize transcription factors which also serve as metal sensors to ensure that the sensory machinery is rapidly coupled to transcription of metal detoxification genes. The selectivity of the sensor is highly dependent on the coordination chemistry of the toxic metals. Although in mammalian systems these mechanisms are just beginning to unfold, it is tempting to speculate that these mechanisms, although perhaps not strictly conserved between microorganisms and mammals, are likely to have conserved fundamental biochemical sensing processes for toxic metals. Therefore, work in model systems is likely to provide a solid mechanistic basis for understanding the processes. The mechanisms utilized by mammalian cells for sensing and responding to toxic metal stress will clearly be a major area for investigations over the next several years. Perhaps an important outcome of this work will be the linkage of metalloregulatory responses to human disease states which are either genetically inherited or induced by environmental factors. Furthermore, perhaps variability in the allele status of human genes encoding toxic metal sensors could predispose or protect individuals from toxic metals by altering the sensitivity of the sensors or the efficacy of transcriptional responses to toxic metal stress.

Acknowledgments
The authors are grateful to the United States Public Health Service for an Individual National Research Service Award to Z. Zhu, and National Institutes of Health grant NIH-RO1-GM41840 and to the Burroughs Wellcome Fund for a Burroughs Wellcome Toxicology Scholar Award to D.J. Thiele.

References

Abrams, M.J. and Murrer, B.A. (1993) Metal compounds in therapy and diagnosis. *Science* 261:725–730.

Brenes-Pomales, A., Lindegren, G. and Lindegren, C.C. (1955) Gene control of copper-sensitivity in *Saccharomyces*. *Nature* 136:841–842.

Brugnera, E., Georgiev, O., Radtke, F., Heuchel, R., Baker, E., Sutherland, G.R. and Schaffner, W. (1994) Cloning, chromosomal mapping and characterization of the human metal-regulatory transcription factor MTF-1. *Nucleic Acids Res.* 22:3167–3173.

Buchman, C., Skroch, P., Dixon, W., Tillius, T.D. and Karin, M. (1990) A single amino acid change in CUP2 alters its mode of DNA binding. *Mol. Cell. Biol.* 10:4778–4787.

Butt, T.R., Sternberg, E.J., Gorman, J.A., Clark, P., Hamer, D., Rosenberg, M. and Crooke, S.T. (1984) Copper metallothionein of yeast, structure of the gene, and regulation of expression. *Proc. Natl. Acad. Sci. USA* 81:3332–3336.

Butt, T.R. and Ecker, D.J. (1987) Yeast metallothionein and application in biotechnology. *Microbiol. Rev.* 51:351–364.

Cheeseman, K.H. and Slater, T.F. (1993) An introduction to free radical biochemistry. *British Medical Bulletin* 49:491–493.

Cotton, F.A. and Wilkinson, G. (1980) *Advanced Inorganic Chemistry, A Comprehensive Text.* John Wiley and Son, New York.

Culotta, V.C. and Hamer, D.H. (1989) Fine mapping of a mouse metallothionein gene metal response element. *Mol. Cell. Biol.* 9:1376–1380.

Culotta, V.C., Howard, W.R. and Liu, X.F. (1994) *CRS5* encodes a metallothionein-like protein in *Saccharomyces cerevisiae*. *J. Biol. Chem.* 269:25295–25302.

Dobi, A., Dameron, C., Hu, S., Hamer, D. and Winge, D.R. (1995) Distinct regions of Cu(I)·ACE1 contact two spatially resolved DNA major groove sites. *J. Biol. Chem.* 270:10171–10178.

Evans, C.F., Engelke, D.R. and Thiele, D.J. (1990) ACE1 transcription factor produced in *Escherichia coli* binds multiple regions within yeast metallothionein upstream activation sequences. *Mol. Cell. Biol.* 10:426–429.

Forrester, W.C., Thompson, C., Elder, J.T. and Groudine, M. (1986) A developmentally stable chromatin structure in the human beta-globin gene cluster. *Proc. Natl. Acad. Sci. USA* 83:1359–1363.

Furst, P., Hu, S., Hackett, R. and Hamer, D. (1988) Copper activates metallothionein gene transcription by altering the conformation of a specific DNA binding protein. *Cell* 55:705–717.

Furst, P. and Hamer, D. (1989) Cooperative activation of a eukaryotic transcription factor: interaction between Cu(I) and yeast ACE1 protein. *Proc. Natl. Acad. Sci. USA* 86:5267–5271.

Giles, W., Hume, J.R. and Shiata, E.F. (1993) Presynaptic and postsynaptic actions of cadmium in cardiac muscle. *Federation Proc.* 42(13):2994–2997.

Goldstein, G.W. (1993) Evidence that lead acts as a calcium substitute in second messenger metabolism. *Neurotoxicology* 14(2–3):97–101.

Hamer, D.H., Thiele, D.J. and Lemontt, J.E. (1985) Function and autoregulation of yeast coppertionein. *Science* 228:685–690.

Hamer, D.H. (1986) Metallothionein. *Annu. Rev. Biochem.* 55:913–951.

Harrison, S.C. (1991) A structural taxonomy of DNA-binding domains. *Nature* 353:715–719.

Heuchel, R., Radtke, F., Georgiev, O., Stark, G., Aguet, M. and Schaffner, W. (1994) The transcription factor MTF-1 is essential for basal and heavy metal-induced metallothionein gene expression. *EMBO J.* 13:2870–2875.

Huibregtse, J.M., Engelke, D.R. and Thiele, D.J. (1989) Copper-induced binding of cellular factors to yeast metallothionein upstream activation sequences. *Proc. Natl. Acad. Sci. USA* 86:65–69.

Hutchens, T.W., Allen, M.H., Li, C.M. and Yip, T.-T. (1992) Occupancy of a C_2-C_2 type "zinc-finger" protein domain by copper. *FEBS Lett.* 309(2):170–174.

Karin, M., Andersen, R.D., Slater, E., Smith, K. and Herschman, H.R. (1980) Metallothionein mRNA induction in HeLa cells in response to zinc or dexamethasone is a primary induction response. *Nature* 286:295–297.

Karin, M., Najarian, R., Haslinger, A., Valenzuela, P., Welch, J. and Fogel, S. (1984) Primary structure and transcription of an amplified genetic locus: the CUP1 locus of yeast. *Proc. Natl. Acad. Sci. USA* 81:337–341.

Lippard, S.J. and Berg, J.M. (1994) *Principles of Bio-inorganic Chemistry,* University Science Books, Mill Valley, CA, p 2.

Masters, B.A., Kelly, E.J., Quaife, C.J., Brinster, R.L. and Palmiter, R.D. (1994) Targeted disruption of metallothionein I and II genes increases sensitivity to cadmium. *Proc. Natl. Acad. Sci. USA* 91:584–588.

Mehra, R.K., Garey, J.R., Butt, T.R., Gray, W.R. and Winge, D.R. (1989) *Candida glabrada* metallothioneins. *J. Biol. Chem.* 264:19747–19753.

Mehra, R.K., Garey, J.R. and Winge, D.R. (1990) Selective and tandem amplification of a member of the metallothionein gene family in *Candida glabrata. J. Biol. Chem.* 265:6369–6375.

Mehra, R.K., Thorvaldsen, J.L., Macreadie, I.G. and Winge, D.R. (1992) Disruption analysis of metallothionein-encoding genes in *Candida glabrata. Gene* 114:75–80.

Michalska, A.E. and Choo, K.H.A. (1993) Targeting and germ-line transmission of a null mutation at the metallothionein I and II loci in mouse. *Proc. Natl. Acad. Sci. USA* 90:8088–8092.

Mueller, P.R., Saler, S.J. and Wold, B. (1988) Constitutive and metal-inducible protein: DNA interactions at the mouse metallothionein I promoter examined by *in vivo* and *in vitro* footprinting. *Genes & Dev.* 2:412–427.

Narula, S.S., Winge, D.R. and Armitage, I.M. (1993) Copper- and silver-substituted yeast metallothionein: Sequential ¹HNMR assignments reflecting conformational heterogeneity at the C terminus. *Biochemistry* 32:6773–6787.

Otsuka, F., Iwamatsu, A., Suzuki, K., Ohsawa, M., Hamer, D. and Koizumi, S. (1994) Purification and characterization of a protein that binds to metal responsive elements of the human metallothionein IIa gene. *J. Biol. Chem.* 269:23700–23707.

Palmiter, R.D., Sandgren, E.P., Koeller, D.M. and Brinster, R. (1993) Distal regulatory elements from the mouse metallothionein locus stimulate gene expression in transgenic mice. *Mol. Cell. Biol.* 13:5266–5275.

Palmiter, R.D. (1994) Regulation of metallothionein genes by heavy metals appears to be mediated by a zinc-sensitive inhibitor that interacts with a constitutively active transcription factor, MTF-1. *Proc. Natl. Acad. Sci. USA* 91:1219–1223.

Palmiter, R.D. and Findley, S. (1995) Cloning and functional characterization of a mammalian zinc transporter that confers resistance to zinc. *EMBO J.* 14:639–649.

Phipps, D.A. (1976) *Metals and Metabolism,* Oxford University Press, Ely House, London.

Predki, P.F. and Sarkar, B. (1992) Effect of replacement of "zinc-finger" zinc on estrogen receptor DNA interactions. *J. Biol. Chem.* 267(9):5842–5846.

Pyle, A.M. (1993) Ribozymes: A distinct class of metalloenzymes. *Science* 261:709–714.

Radtke, F., Heuchel, R., Georgiev, O., Hergersberg, M., Gariglio, M., Dembic, Z. and Schaffner, W. (1993) Cloned transcription factor MTF-1 activates the mouse metallothionein I promoter. *EMBO J.* 12:1355–1362.

Searle, P.F., Davison, B.L., Strart, G.W., Wilkie, T.M., Norstedt, G. and Palmiter, R.D. (1984) Regulation, linkage and sequence of mouse metallothionein I and II genes. *Mol. Cell. Biol.* 4:1221–1230.

Seguin, C., Felber, B.K., Carter, A.D. and Hamer, D.H. (1984) Competition for cellular factors that activate metallothionein gene transcription. *Nature* 312:781–785.

Stuart, G.W., Searle, P.F. and Palmiter, R.D. (1985) Identification of multiple metal regulatory elements in mouse metallothionein-I promoter by assaying synthetic sequences. *Nature* 317:828–831.

Szczypka, M.S. and Thiele, D.J. (1989) A cysteine-rich nuclear protein activates yeast metallothionein gene transcription. *Mol. Cell. Biol.* 9:421–429.

Tamai, K.T., Gralla, E.B., Ellerby, L.M., Valentine, J.S. and Thiele, D.J. (1993) Yeast and mammalian metallothioneins functionally substitute for yeast copper-zinc superoxide dismutase. *Proc. Natl. Acad. Sci. USA* 90:8013–8017.

Thiele, D.J. and Hamer, D.H. (1986) Tandemly duplicated upstream control sequences mediate copper-induced transcription of the *Saccharomyces cerevisiae* copper-metallothionein gene. *Mol. Cell. Biol.* 6:1158–1163.

Thiele, D.J. (1988) ACE1 regulates expression of the *Saccharomyces cerrevisiae* metallothionein gene. *Mol. Cell. Biol.* 8:2745–2752.

Thiele, D.J. (1992) Metal-regulated transcription in eukaryotes. *Nucleic Acids Res.* 20:1183–1191.

Thornalley, P.J. and Vasak, M. (1985) Possible role for metallothionein in protection against radiation-induced oxidative stress: Kinetics and mechanism of its reaction with superoxide and hydroxyl radicals. *Biochim. Biophys. Acta* 827:36–44.

Thorvaldsen, J.L., Sewell, A.K., Tanner, A.M., Peltier, J.M., Pickering, I.J., George, G.N. and Winge, D.R. (1994) Mixed Cu^+ and Zn^{2+} coordinating in the DNA-binding domain of the AMT1 transcription factor from *Candida glabrata. Biochemistry* 33:9566–9577.

Xu, L., Rungger, D., Georgiev, O., Seipel, K. and Schaffner, W. (1994) Different potential of cellular and viral activators of transcription revealed in oocytes and early embryos of *Xenopus laevis. Biol. Chem. Hoppe-Seyler* 375:105–112.

Welch, J., Fogel, S., Buchman, C. and Karin, M. (1989) The CUP2 gene product regulates the expression of the CUP1 gene, coding for yeast metallothionein. *EMBO J.* 8:255–260.

Westin, G. and Schaffner, W. (1988) A zinc-responsive factor interacts with a metal-regulated enhancer element (MRE) of the mouse metallothionein-I gene. *EMBO J.* 7:3763–3770.

Williams, R.J.P. (1978) *In*: R.J.P. Williams and J.R.F. Da Silva (eds) *New Trends in Bio-Inorganic Chemistry.* Academic Press Inc., New York, pp 1–10.

Wright, C.F., Hamer, D.H. and McKenney, K. (1988) Autoregulation of the yeast metallothionein gene depends on metal binding. *J. Biol. Chem.* 263:1570–1574.

Zhou, P. and Thiele, D.J. (1991) Isolation of a metal-activated transcription factor gene from *Candida glabrata* by complementation in *Saccharomyces cerevisiae. Proc. Natl. Acad. Sci. USA* 88:6112–6116.

Zhou, P., Szczypka, M.S., Sosinowski, T. and Thiele, D.J. (1992) Expression of a yeast metallothionein gene family is activated by a single metalloregulatory transcription factor. *Mol. Cell. Biol.* 12:3766–3775.

Zhou, P. and Thiele, D.J. (1993) Rapid transcriptional autoregulation of a yeast metalloregulatory transcription factor is essential for high-level copper detoxification. *Genes & Dev.* 7:1824–1835.

Stress-Inducible Cellular Responses
ed. by U. Feige, R.I. Morimoto, I. Yahara and B. Polla
© 1996 Birkhäuser Verlag Basel/Switzerland

Tumor necrosis factor and lymphotoxin: Protection against oxidative stress through induction of MnSOD

G. H. W. Wong[2], R. L. Kaspar[1] and G. Vehar[2]

[1] *Department of Chemistry and Biochemistry, Brigham Young University, Provo, UT 84602, USA*
[2] *Genentech, Inc., 460 Point San Bruno Blvd., South San Francisco, CA 94080, USA*

Summary. Tumor necrosis factor (TNF) and lymphotoxin (LT) are related cytokines produced in response to infection or oxidative insults such as radiation. These cytokines bind to the same receptors and have pleiotropic effects on a variety of cell types. TNF or LT pretreatment, which can induce the synthesis of "protective" proteins such as mitochondrial manganese superoxide dismutase (MnSOD), protects animals from lethal doses of radiation or the chemotherapeutic drug doxorubicin. In contrast, TNF or LT pretreatment of tumor cells, which do not express MnSOD, results in sensitization to these insults. Therefore, radio- or chemoprotection of normal cells may act partially through enhanced expression of MnSOD. On the other hand, tumor sensitization may result from activation of "killing" proteins such as interleukin-1β converting enzyme (ICE) or other ICE-like proteases, possibly through TNF/LT-induced oxygen free radicals. In addition to their originally described anti-tumor activity, these cytokines may have new therapeutic indications in protecting normal cells while sensitizing tumor cells to radiation or chemotherapeutic drugs.

Introduction

Production of oxygen free radicals is an unavoidable by-product of cellular metabolism in all respiring cells. One of the oxygen free radicals produced is superoxide whose toxicity can be mitigated by a family of super-oxide dismutases (SOD) (Weisinger and Fridovich, 1973). There are three distinct mammalian SOD genes including MnSOD, CuZn-SOD and a secreted extracellular (EC-SOD) form. MnSOD is found mainly in the mitochondria whereas CuZn-SOD is found predominantly in the cytoplasm. The expression of CuZn-SOD and EC-SOD is constitutive while MnSOD levels can be induced by cytokines such as tumor necrosis factor (TNF) or lymphotoxin (LT) (Wong and Goeddel, 1988a).

TNF and LT are cytokines that play key roles in protecting cells against oxidative stress (Wong et al., 1992b). Paradoxically, TNF has been shown to induce oxygen free radical formation (Wong et al., 1992b). TNF and LT are encoded by linked genes located on chromosome 6 and functionally exist as homo-trimers (Goeddel et al., 1986). The production of these cytokines is tightly regulated and is induced by pathogens or oxidative insults (Goeddel et al., 1986; Beutler, 1992). TNF can be produced by a number of cell types including activated macrophages, mast cells, and lymphocytes, while LT is produced only by activated lymphocytes (Beutler, 1992).

Despite the low homology between TNF and LT (approximately 30% at the amino acid level (Goeddel et al., 1986), they bind to the same receptors and elicit similar biological activities (Goeddel et al. 1986; Fiers, 1991). Most cells express both p55 and p75 receptors which can be activated by agonist antibodies to either receptor in the absence of ligand. Experiments using agonist antibodies against p55 indicate that this receptor is responsible for the majority of LT/TNF effects. These functions include triggering apoptosis (Wong and Goeddel, 1994a), mediating antiviral activity (Wong and Goeddel, 1986; Wong et al., 1992c), production of oxygen free radicals (Wong et al., 1992b), induction of MnSOD (Tartaglia et al.; 1991; Wong and Goeddel, 1994b) and intercellular adhesion molecule-1 (Trefzer et al., 1991; Mackay et al., 1993) and other cellular genes (Tartaglia et al., 1991). Activation of the p75 receptor results in proliferation of lymphocytes (Tartaglia et al., 1993). Recently, a new LT receptor has been identified (Ware et al., 1995). Interestingly, LT cannot bind to this new receptor directly, but requires interaction with another member of the LT family (LT-β). The function of this new receptor is not known (Browning et al., 1993).

TNF or LT pretreatment protects normal tissue against oxidative stress while sensitizing tumor cells

In addition to stimulating free radical production, TNF pretreatment can simultaneously protect animals and tissue culture cells from some insults that appear to be associated with the generation of free radicals. Thus, high TNF doses may induce toxicity, while lower TNF concentrations might induce "protective" proteins. TNF pretreatment has been shown to protect animals against radiation (Neta et al., 1988; Slordal et al., 1989), cytotoxic drugs (Slordal et al., 1990), ischemia reperfusion (Eddy et al., 1992), endotoxin shock (Sheppard et al., 1989), hyperoxia (Tsan et al., 1990) and cytosine arabinoside-induced alopecia (Jimenez et al., 1991).TNF has also been shown to protect animals against systemic lupus erythematosus (Jacob et al., 1990b), type I diabetes (Jacob et al., 1990a) and lethal doses of TNF (Patton et al., 1987). Injection of TNF antibodies increases the sensitivity of animals to radiation (Neta et al., 1991), cytotoxic drugs (Moreb et al., 1990), experimental peritonitis (Alexander et al., 1991), and stress hyperthermia (Long et al., 1990). Furthermore, pretreatment of established cell lines with TNF protects against radiation and heat (Wong et al., 1991; Wong et al., 1992d), H_2O_2, and oxygen free radical-generating agents such as paraquat (Warner et al., 1991), adriamycin and menadione (G. H. W. Wong, unpublished data). These results suggest that TNF may act directly on target cells to protect them from oxidative damage.

Like TNF, LT can also induce the synthesis of MnSOD (Wong and Goeddel, 1988a). Thus, LT pretreatment may also have protective effects simi-

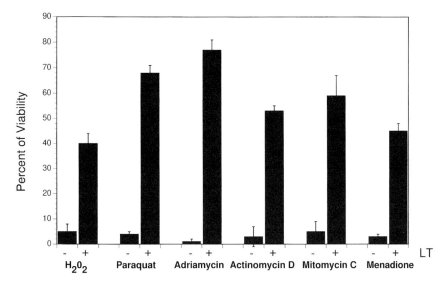

Figure 1. LT protects cells against killing by agents that generate oxygen free radicals. Human A549 lung carcinoma cells were treated with LT (1 μg/ml) for 24 h prior to various oxidative insults including hydrogen peroxide (0.3%), paraquat (10 μM), adriamycin (100 μg/ml), actinomycin D (100 μg/ml), mitomycin C (100 μg/ml), and menadione (10 μM). Cell viability was measured after 36 h by trypan blue staining.

lar to those seen with TNF (Fig. 1). In this regard, we have shown that LT pretreatment can protect cells *in vitro* (Wong et al., 1992a, Wong et al., 1992b; Wong, 1995) as well as animals against oxidative insults (Eddy et al., 1992; Wong et al., 1992a; Wong et al., 1992b; Nelson et al., 1995). LT pretreatment also protects mice against lethal doses of ionizing radiation (Wong et al., 1992a). The mechanism of this protection is not clear although our results indicate that LT pretreatment prevents radiation-induced damage to several organs including progenitor cells in the bone marrow and crypt cells in the gut, enhances the recovery of platelets, and protects mice against chemotherapy-induced hair loss (G.H.W. Wong et al., submitted).

TNF and LT do not induce MnSOD in HIV-infected (Wong et al., 1991) or some tumor (Wong and Goeddel, 1988a; Wong, 1995) cells. HIV-infected cells have low basal levels of MnSOD and are more sensitive to killing by radiation and heat than the uninfected parental cells (Wong et al., 1991). It will be interesting to see how HIV suppresses MnSOD expression and induction by TNF and LT. In this regard, McCord and colleagues (Flores et al., 1993) have shown that the HIV Tat protein can repress the expression of MnSOD. Also, we have shown that TNF pretreatment sensitizes HIV-infected cells to radiation or heat damage (Wong et al., 1991). If the model that MnSOD protects cells from radiation and/or chemotherapy is correct, then one would expect that TNF/LT pretreatment should have no radioprotective effect on tumor cells that do not express MnSOD. Surprisingly, rather than having no effect, pretreatment with TNF or LT actually enhanced the sensitivity of human breast, lung and colon carcinoma cell lines to killing by radiation or cytotoxic drugs such as doxorubicin and 5-fluorouracil *in vitro* and *in vivo* (Wong et al., submitted). Even though both TNF and LT share similar biological activities (Goeddel et al., 1986; Beutler, 1992), TNF appears to be much more potent than LT in releasing IL-1 from human endothelial cells (Locksley et al., 1987) and inducing the expression of adhesion molecules (Broudy et al., 1987; Desch et al., 1990). Furthermore, LT has been shown to be less toxic than TNF *in vivo* (Qin et al., 1995). These studies raise the hope that LT may have more therapeutic potential than TNF to protect normal cells from radiation or chemotherapy damage while sensitizing tumor cells to these treatments.

Role of MnSOD: Protection versus sensitization

The mechanisms by which TNF or LT pretreatment protect cells against the toxic effects of radiation or chemotherapy are not clear but may involve the induction of protective proteins. Inhibition of protein synthesis by cycloheximide increases the sensitivity of cells to TNF killing (Wallach, 1984), indicating that newly synthesized proteins such as MnSOD are necessary for protection (Wong et al., 1989). The expression of MnSOD can be up-regulated by several cytokines such as TNF (Wong and Goeddel, 1988a; Asoh et al., 1989; Kawaguchi et al., 1990; Shaffer et al., 1990; Valentine and Nick, 1990; Visner et al., 1990; Visner et al., 1992), LT (Wong and Goeddel 1988a; Wong, 1995), IL-1 (Masuda et al., 1988; Wong and Goeddel, 1988a; Visner et al., 1992), IL-6 (Valentine and Nick, 1991) and IFN-γ (Harris et al., 1991). The MnSOD induction by TNF/LT is rapid, long lived, and occurs in cultured cells and in the whole animal (Wong and Goeddel, 1988a). The regulation of MnSOD gene expression is complex and may occur at several levels including transcription (Wong et al., 1992d), splicing and export to the cytoplasm, stability of the transcripts (Melendez and Baglioni, 1993), and transport (Wispe et al., 1989) of the protein to the mitochondria (Fig. 2). Other cytokines such as stem cell factor, IL-2, IL-3, IFN-α, IFN-β, GM-CSF or TGF-β do not induce MnSOD (Wong and Goeddel, 1988b; G. H. W. Wong, unpublished data).

Engineered overexpression of mitochondrial MnSOD (but not cytosolic CuZn-SOD nor extracellular EC-SOD) protected cells against radiation damage (Wong, 1995). These data suggest that the mitochondria might be particularly sensitive to oxidative insult. We tested this hypothesis by constructing several forms of SODs. Deletion of the mitochondrial localization signal sequence of MnSOD should generate a cytoplasmic-localized form. In addition, the MnSOD mitochondrial localization sequence was inserted into CuZn-SOD to allow mitochondrial accumulation. Trans-

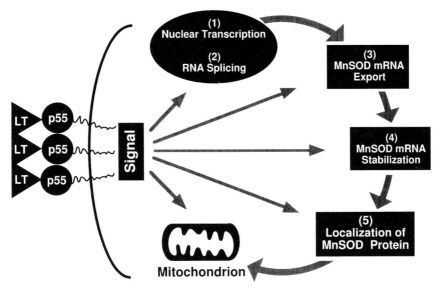

Figure 2. Regulation of MnSOD gene expression after binding of LT to three receptor molecules. The signal generated through receptor binding can potentially regulate MnSOD gene expression at the level of nuclear transcription, processing of the RNA transcript, export of the mRNA from the nucleus, stability of the mRNA transcripts, and MnSOD protein localization to the mitochondria.

fecting cells with MnSOD lacking its mitochondrial matrix signal did not provide protection against radiation (Wong, 1995). However, insertion of the MnSOD-derived mitochondrial targeting sequence into CuZn-SOD cDNA resulted in significant radioprotection (Wong, 1995). These data indicate that the mitochondrial localization of MnSOD is critical for its unique protective property. Experiments in which *Escherichia coli* iron superoxide dismutase was targeted to the yeast mitochondria have shown similar protective results (Balzan et al., 1995). These results suggest that mitochondria may be especially sensitive to oxidative damage (Ames et al., 1993). This is in contrast to the traditional view that the nuclear DNA is the target of lethal damage from radiation. Mitochondrial DNA may be particularly sensitive to this damage, possibly due to a lack of sufficient DNA repair mechanisms in the mitochondria (Ames, et al., 1993). Indeed, cells with low or undetectable mitochondrial MnSOD were found to be more sensitive to oxidative insult (van Loon et al., 1986; Wong et al., 1992a; Hirose et al., 1993).

The crucial role of MnSOD is illustrated by MnSOD knock-out-mice. These mice exhibit heart, liver, and muscle abnormalities and die within the first ten days after birth (Li et al., 1995). It will be interesting to test whether these mice are more sensitive to radiation damage and whether LT can still protect them. If LT can enhance the survival of these MnSOD-deficient mice, LT must induce other protective proteins in addition to MnSOD.

In addition to its known anti-oxidant function, MnSOD has also been implicated as a differentiation marker. For example, cellular transformation reduces or abolishes constitutive MnSOD levels whereas vitamin D3 treatment, which induces cellular differentiation, leads to higher levels of MnSOD expression (Iwamoto et al., 1990). We tested the tumorigenicity of several tumor cell lines including A549 (human lung carcinoma), 293 (human kidney embryonic carcinoma), and CHO (chinese hamster ovary) cells that had been engineered to overexpress either MnSOD, CuZn-SOD, or EC-SOD in nude mice. Our results suggest that cells that overexpress MnSOD do not form tumors in nude mice, whereas those cell lines that express other non-mitochondrial-localized forms of SOD are capable of tumor formation (G. H. W. Wong, unpublished data). These results suggest that MnSOD could be a tumor suppressor gene. Similar results using human melanoma cells have been reported to suppress tumor formation in nude mice (Church et al., 1993; Safford et al., 1994).

In contrast to its role as a protective protein, MnSOD may also play a role in increased toxicity of TNF (Melendez and Baglioni, 1992). For example, while moderate induction of MnSOD may give a protective effect, high MnSOD levels may be lethal. In our experience, cell clones that overexpress high levels of MnSOD (100 fold over wild-type) were unable to grow in culture (G. H. W. Wong, unpublished data), suggesting that expression of extremely high levels of MnSOD is lethal. It is possible that excess production of hydrogen peroxide by SOD (in cells lacking sufficient catalase or glutathione peroxidase) leads to production of more toxic hydroxyl radicals. Thus, induction of MnSOD can be beneficial (at moderate levels) on the one hand, and detrimental (at extremely high levels) on the other.

Possible protective and sensitization mechanisms of TNF or LT

TNF/LT-mediated protection of damage induced by oxidative stress functions at least in part through induction of MnSOD (Wong and Goeddel, 1988 a; Wong et al., 1992 b). In addition to MnSOD, TNF/LT may mediate radioprotection through induction of other protective proteins such as DNA repair enzymes, Bcl-2 (Tallex et al., 1995), p53 and p21 (P. Hainaut and G. H. W. Wong, unpublished data), A20 (Opipari, Jr. et al., 1992), metal-lothioneins (Wong and Goeddel, 1998 b; Sciavolino et al., 1992), heat shock proteins (Simon et al., 1995), ferritin (Torti et al., 1988), poly (ADP-ribose) polymerase (Lichtenstein et al., 1991) and protease inhibitors (Kumar and Baglioni, 1991) (Fig. 3). Preliminary data suggest that TNF/LT had no effect on DNA glycosylase mRNA or protein levels (G. H. W. Wong and J. Hau, unpublished data). Treatment with TNF or LT increases p53 or

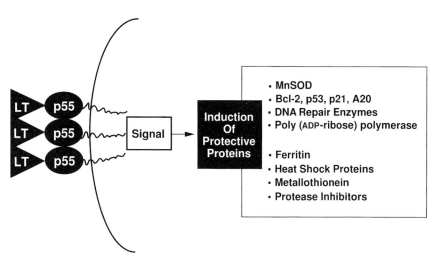

Figure 3. Possible protective mechanisms of LT. Trimers of LT complexed with three receptor molecules result in induction of protective proteins which may act to prevent DNA damage or protein degradation. Proteins with known function such as MnSOD, ferritin, and metallothionein appear to scavenge oxygen free radicals which are known to be harmful to DNA, RNA, lipids and proteins. Proteins with uncertain functions including poly (ADP-ribose) polymerase, Bcl-2, p53, p21, A20, and heat shock proteins are also candidates for protective proteins.

p21 protein in the absence of increased mRNA levels (P. Hainaut and G. H. W. Wong, unpublished data). Thus, these cytokines may also act at a post-transcriptional level.

The susceptibility of tumor cells to killing by TNF/LT, radiation, or other oxidative insults may be determined by the relative balance of the synthesis of protective proteins and activation of killing proteins. One such killing protein could be the interleukin-1β converting enzyme (ICE), homologous to the *C. elegans* cell-death protein *ced*-3 (Miura et al., 1993). Overexpression of ICE (Miura et al., 1993) and other ICE-like proteases (e.g. YAMA/Cpp32) (Tewari and Dixit, 1995; Tewari et al., 1995) have been shown to induce apoptosis in some cells. This "killing" activity can be inhibited by the cowpox virus *Crm*A gene product (Talley et al., 1995; Tewari and Dixit, 1995; Tewari et al., 1995). Interestingly, *Crm*A also blocks TNF/TRADD-induced apoptosis (Hsu et al., 1995; Tewari and Dixit, 1995; Tewari et al., 1995). These results suggest that ICE and/or YAMA or other proteases are candidate killing proteins or are involved in activating killing proteins.

We have shown that TNF-sensitive cells express high levels of ICE mRNA, whereas MnSOD mRNA levels are low (R. L. Kaspar and G. H. W. Wong, unpublished data). Conversely, TNF-resistant cells show an opposite relationship: low ICE mRNA levels with high levels of MnSOD mRNA. Un-like MnSOD, the levels of ICE mRNA are not regulated by either TNF or LT treatment. Thus an inverse relationship between ICE and

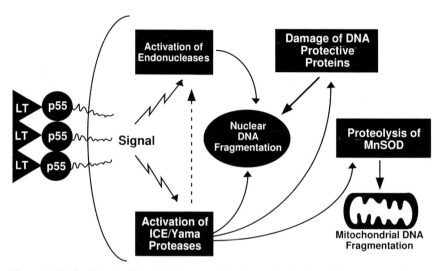

Figure 4. LT signaling leads to DNA fragmentation through activation of proteases and endo-nucleases. Activation of ICE or ICE-like proteases may lead to proteolytic degradation of DNA protective proteins and thereby increase the susceptibility of DNA to endonucleolytic cleavage. On the other hand, these proteases may degrade MnSOD leading to increased sensitivity of mito-chondrial DNA and proteins to oxidative insult.

MnSOD may determine the susceptibility of the cells to TNF/LT apoptosis and possibly TNF/LT tumor sensitization to oxidative stress.

An alternative hypothesis is that MnSOD exerts radioprotective effects by inactivating ICE or ICE-like proteases through inactivation of superoxides. Conversely, these proteases may degrade MnSOD and thereby sensitize tumor cells to oxidative insult (Fig. 4). Whether protective proteins such as MnSOD can abrogate the harmful effects of killing proteins such as ICE/YAMA is of great interest. On the other hand, TNF/LT-tumor sensitiza-tion may occur through an interaction of TRADD with the receptor (Hsu et al., 1995) and result in signal transduction which could activate ICF-like proteases and other potential killing proteins such as phospholipase A2 (Neale et al. 1988), lysosomal enzymes (Liddil et al., 1989), and endonuclea-ses. These cytosolic proteins may enter the nucleus by degrading the nuclear lamina or by passive diffusion through nuclear pores. This could result in triggering characteristic apoptotic events such as DNA fragmentation, presumably through proteolytic cleavage of DNA protective proteins, DNA repair enzymes or activation of endonucleases (Fig. 4). In addition, it is pos-sible that these proteases could mediate mitochondrial DNA fragmentation after degrading MnSOD (Fig. 4). These hypotheses are directly testable.

In summary, inducers of oxidative stress such as irradiation can induce the synthesis of TNF and LT which in turn can increase the levels of MnSOD and possibly p53 and other protective proteins in normal cells and thereby enhance resistance to oxidative stress. In contrast, TNF/LT neither increases MnSOD in some tumor cells nor protects them from oxidative insult. In fact, these cytokines can actually sensitize these tumor cells to killing by radiation or chemotherapy, possibly through activation of killing proteins such as ICE or ICE-like proteases. Thus, in principle, LT or TNF potentially could be used as adjuncts to standard cancer therapy (G. H. W. Wong et al., unpublished data).

Acknowledgements
We thank the manufacturing group at Genentech for providing pure recombinant human TNF and LT, Warren Young, Cathy Carlson and Sharon Fong for help with animal work, and Nora Kolthoff for art drawing. GHW thanks Dr. Jim Quinn and Alexander David Kamb for support.

References

Alexander, H.R., Doherty, G.M., Fraker, D.L., Block, M.I., Swendenborg, J.A. and Norton, J.A. (1991) Human recombinant interleukin-1 alpha: Protection against the lethality of endotoxin and experimental sepsis in mice. *J. Surg. Res.* 50:421–424.

Ames, B.N., Shigenaga, M.K. and Hagen, T.M. (1993) Oxidants, antioxidants, and the degenerative diseases of aging. *Proc. Natl. Acad. Sci. USA* 90:7915–7922.

Asoh, K.-I., Watanabe, Y., Mizuguchi, H., Mawatari, M., Ono, M., Kohno, K. and Kuwano, M. (1989) Induction of manganese superoxide dismutase by tumor necrosis factor in human breast cancer MCF-7 cell line and its TNF-resistant variant. *Biochem. Biophys. Res. Commun.* 162:794–801.

Balzan, R., Bannister, W.H., Hunter, G.J. and Bannister, J.V. (1995) *Escherichia coli* iron superoxide dismutase targeted to the mitochondria of yeast cells protects the cells against oxidative stress. *Proc. Natl. Acad. Sci. USA* 92:4219–4223.

Beutler, B. (ed.) (1992) *Tumor Necrosis Factors: The Molecules and Their Emerging Role in Medicine.* New York, Raven Press.

Broudy, V.C., Harlan, J.M. and Adamson, J.W. (1987) Disparate effects of tumor necrosis-factor-alpha/cachectin and tumor necrosis factor-beta/lymphotoxin on hematopoietic growth factor production and neutrophil adhesion molecule expression by cultured human endothelial cells. *J. Immunology* 138:298.

Browning, J.L., Ngam-ek, A., Lawton, P., DeMarinis, J., Tizard, R., Chow, E.P., Hession, C., O'Brine-Greco, B., Foley, S.F. and Ware, C.F. (1993) Lymphotoxin β, a novel member of the TNF family that forms a heteromeric complex with lymphotoxin on the cell surface. *Cell* 72:847–856.

Church, S.L., Grant, J.W., Ridnour, L.A., Oberley, L.W., Swanson, P.E., Meltzer, P.S. and Trent, J.M. (1993) Increased manganese superoxide dismutase expression suppresses the malignant phenotype of human melanoma cells. *Proc. Natl. Acad. Sci. USA* 90:3113–3117.

Desch, C.E., Dobrina, A., Aggarwal, B.B. and Halran, J.M. (1990) Tumor necrosis factor-alpha exhibits greater proinflammatory activity than lymphotoxin in vitro. *Blood* 75:2030–2034.

Eddy, L.J., Goeddel, D.V. and Wong, G.H. (1992) Tumor necrosis factor-alpha pretreatment is protective in a rat model of myocardial ischemia-reperfusion injury. *Biochem. Biophys. Res. Commun.* 184:1056–9.

Fiers, W. (1991) Tumor necrosis factor characterization at the molecular, cellular and *in vivo* level. *FEBS Lett.* 285:199–212.

Flores, S.C., Marecki, J.C., Harper, K.P., Bose, S.K., Nelson, S.K. and McCord, J.M. (1993) Tat protein of human immunodeficiency virus type 1 represses expression of manganese superoxide dismutase in HeLa cells. *Proc. Natl. Acad. Sci. USA* 90:7632–7636.

Goeddel, D.V., Aggarwal, B.B., Gray, P.W., Leung, D.W., Nedwin, G.E., Palladino, M.A., Patton, J.S., Pennica, D., Shepard, H.M., Sugarman, B.J. and Wong, G.H.W. (1986) Tumor necrosis factors: gene structure and biological activities. *Cold Spring Harb. Symp. Quant. Biol.* 1:597–609.

Harris, C.A., Derbin, K.S., Hunte-McDonough, B., Krauss, M.R., Chen, K.T., Smith, D.M. and Epstein, L.B. (1991) Manganese superoxide dismutase is induced by IFN-gamma in multiple cell types. Synergistic induction by IFN-gamma and tumor necrosis factor or IL-1. *J. Immunol.* 147:149–154.

Hirose, K., Longo, D.L., Oppenheim, J.J. and Matsushima, K. (1993) Overexpression of mito-chondrial manganese superoxide dismutase promotes the survival of tumor cells exposed to interleukin-1, tumor necrosis factor, selected anticancer drugs, and ionizing radiation. *FASEB J.* 7:361–368.

Hsu, H., Xiong, J. and Goeddel, D.V. (1995) The TNF receptor 1-associated protein TRADD signals cell death and NF-κB activation. *Cell* 81:495–504.

Iwamoto, S., Takeda, K., Kamijo, R. and Konno, K. (1990) Induction of resistance to TNF cyto-toxicity and mitochondrial superoxide dismutase on U-937 cells by 1,25-dihydroxyvitamin D3. *Biochem. Biophys. Res. Commun.* 170:73–79.

Jacob, C.O., Aiso, S., Michie, S.A., McDevitt, H.O. and Acha-Orbea, H. (1990a) Prevention of diabetes in nonobese diabetic mice by tumor necrosis factor (TNF): similarities between TNF-alpha and interleukin 1. *Proc. Natl. Acad. Sci.* 87:968–972.

Jacob, C.O., Fronek, Z., Lewis, G.D., Koo, M., Hansen, J.A. and McDevitt, H.O. (1990b) Heritable major histocompatibility complex class II-associated differences in production of tumor necrosis factor alpha: relevance to genetic predisposition to systemic lupus erythema-tosus. *Proc. Natl. Acad. Sci.* 87:1233–1237.

Jimenez, J.J., Wong, G.H.W. and Yunis, A.A. (1991) Interleukin 1 protects from cytosine arabinoside-induced alopecia in the rat model. *FASEB J.* 5:2456–2468.

Kawaguchi, T., Takeyasu, A., Matsunobu, K., Uda, T., Ishizawa, M., Suzuki, K., Nishiura, T., Ishikawa, M. and Taniguchi, N. (1990) Stimulation of Mn-superoxide dismutase expression by tumor necrosis factor-alpha: quantitative determination of Mn-SOD protein levels in TNF-resistant and sensitive cells by ELISA. *Biochem. Biophys. Res. Commun.* 171:1378–1386.

Kumar, S. and Baglioni, C. (1991) Protection from tumor necrosis factor-mediated cytolysis by overexpression of plasminogen activator inhibitor type-2. *J. Biol. Chem.* 266:20960–20964.

Li, Y., Huang, T.T., Carlson, E.J., Melov, S., Ursell, P.C., Olson, J.L., Noble, L.J., Yoshimura, M.P., Berger, C., Chan, P.H., Wallace, D.C. and Epstein, C.J. (1995) Dilated cardiomyopathy and neonatal lethality in mutant mice lacking manganese superoxide dismutase. *Nature Genetics* 11:376–381.

Lichtenstein, A., Gera, J.F., Andrews, J., Berenson, J. and Ware, C.F. (1991) Inhibitors of ADP-ribose polymerase decrease the resistance of HER2/neu-expressing cancer cells to the cytotoxic effects of tumor necrosis factor. *J. Immunol.* 146:2052–2058.

Liddil, J.D., Dorr, R.T. and Scuderi, P. (1989) Association of lysosomal activity with sensitivity and resistance to tumor necrosis factor in murine L929 cells. *Cancer Res.* 49:2722–2728.

Locksley, R.M., Heinzel, F.P., Shepard, H.M., Agosti, J., Eessalu, T.E., Aggarwal, B.B. and Harlan, J.M. (1987) Tumor necrosis factors alpha and beta differ in their capacities to gener-ate interleukin 1 release from human endothelial cells. *J. Immunology* 139:1891–1895.

Long, N.C., Kunkel, S.L., Vander, A.J. and Kluger, M.J. (1990) Antiserum against tumor necro-sis factor enhances lipopolysaccharide fever in rats. *Am. J. Physiol.* 258:R332–R337.

Mackay, F., Loetscher, H., Stueber, D., Gehr, G. and Lesslauer, W. (1993) Tumor necrosis fac-tor alpha (TNF-alpha)-induced cell adhesion to human endothelial cells is under dominant control of one TNF receptor type, TNF-R55. *J. Exp. Med.* 177:1277–1286.

Masuda, A., Longo, D.L., Kobayashi, Y., Appella, E., Oppenheim, J.J. and Matsushima, K. (1988) Induction of mitochondrial manganese superoxide dismutase by interleukin-1. *FASEB J.* 2:3087–3091.

Melendez, J.A. and Baglioni, C. (1992) Reduced expression of manganese superoxide dismutase in cells resistant to cytolysis by tumor necrosis factor. *Free Radic. Biol. Med.* 12:151–159.

Melendez, J.A. and Baglioni, C. (1993) Differential induction and decay of manganese super-oxide dismutase mRNAs. *Free Radic. Biol. Med.* 14:601–608.

Miura, M., Zhu, H., Rotello, R., Hartwieg, E.A. and Junying, Y. (1993) Induction of apoptosis in fibroblasts by IL-1β-converting enzyme, a mammalian homolog of the *C. elegans* cell death gene ced-3. *Cell* 75:653–660.

Moreb, J., Zucali, J.R. and Rueth, S. (1990) The effects of tumor necrosis factor-alpha on early human hematopoietic progenitor cells treated with 4-hydroperoxycyclophosphamide. *Blood* 76:681–689.

Neale, M.L., Fiera, R.A. and Matthews, N. (1988) Involvement of phospholipase A₂ activation in tumor cell killing by tumor necrosis factor. *Immunology* 64:81–85.

Nelson, S.K., Wong, G.H. and McCord, J.M. (1995) Leukemia inhibitory factor and tumor necrosis factor induce manganese superoxide dismutase and protect rabbit hearts from reperfusion injury. *J. Mol. Cell. Cardiol.* 27:223–229.

Neta, R., Oppenheim, J.J. and Douches, S.D. (1988) Interdependence of the radioprotective effects of human recombinant IL-1, TNF, G-CSF, and murine recombinant G-CSF, *J. Immunology* 140:108–111.

Neta, R., Oppenheim, J.J., Schreiber, R.D., Chizzonite, R., Ledney, G.D. and MacVittie, T.J. (1991) Role of cytokines (interleukin 1, tumor necrosis factor and transforming growth factor β) in natural and lipopolysaccharide-enhanced radioresistance. *J. Exp. Med.* 173: 1177–1182.

Opipari Jr., A.W., Hu, H.M., Yabkowitz, R. and Dixit, V.M. (1992) The A20 zinc finger protects cells from tumor necrosis factor cytotoxicity. *J. Biol. Chem.* 267: 12424–12427.

Patton, J.S., Peters, P.M., McCabe, J., Crase, D., Hansen, S., Chen, A.B. and Liggitt, D. (1987) Development of partial tolerance to the gastrointestinal effects of high doses of recombinant tumor necrosis factor-alpha in rodents. *J. Clin. Invest.* 80:1587–1596.

Qin, Z., van Tits, L.J., Buurman, W.A. and Blankenstein, T. (1995) Human lymphotoxin has a least equal antitumor activity in comparison to human tumor necrosis factor but is less toxic in mice. *Blood* 85:2779–2785.

Safford, S.E., Oberley, T.D., Urano, M. and St. Clair, D.K. (1994) Suppression of fibrosarcoma metastasis by elevated expression of manganese superoxide dismutase. *Cancer Res.* 54: 4261–4265.

Sciavolino, P.J., Lee, T.H. and Vilcek, J. (1992) Overexpression of metallothionein confers resistance to the cytotoxic effect of TNF with cadmium in MCF-7 breast carcinoma cells. *Lymphokine and Cytokine Res.* 11:265–270.

Shaffer, J.B., Treanor, C.P. and Del Vecchio, P.J. (1990) Expression of bovine and mouse endothelial cell antioxidant enzymes following TNF-alpha exposure. *Free Radic. Biol. Med.* 8:497–502.

Sheppard, B.C., Fraker, D.L. and Norton, J.A. (1989) Prevention and treatment of endotoxin and sepsis lethality with recombinant human tumor necrosis factor. *Surgery* 106: 156–161.

Simon, M.M., Krone, C., Schwarz, A., Luger, T.A., Jaattela, M. and Schwarz, T. (1995) Heat shock protein 70 overexpression affects to response to ultraviolet light in murine fibroblasts. Evidence for increased cell viability and suppression of cytokine release. *J. Clin. Invest.* 95:926–931.

Slordal, L., Muench, M.O., Warren, D.J. and Moore, M.A.S. (1989) Radioprotection by murine and human tumor necrosis factor: dose-dependent effects on hematopoiesis in the mouse. *Eur. J. Haematol.* 43:428–434.

Slordal, L., Warren, D.J. and Moore, M.A.S. (1990) Protective effects to tumor necrosis factor on murine hematopoiesis during cycle-specific cytotoxic chemotherapy. *Cancer Research* 50:4216–4220.

Talley, A.K., Dewhurst, S., Perry, S.W., Dollard, S.C., Gummuluru, S., Fine, S.M., New D., Epstein, L.G., Gendelman, H.E. and Gelbard, H.A. (1995) Tumor necrosis factor alpha-induced apoptosis in human neuronal cells: protection by the antioxidant N-acetylcysteine and the genes bcl-2 and crmA. *Mol. Cell. Biol.* 15:2359–2366.

Tartaglia, L.A., Goeddel, D.V., Reynolds, C., Figari, I.S., Weber, R.F., Fendly, B.M. and Palladino, M.A., Jr. (1993) Stimulation of human T cell proliferation by specific activation of the 75-kDa tumor necrosis factor receptor. *J. Immunol.* 151:4637–4641.

Tartaglia, L.A., Weber, R.F., Figari, I.S., Reynolds, C., Palladino, J., M.A. and Goeddel, D.V. (1991) The two different receptors for tumor necrosis factor mediate distinct cellular responses. *Proct. Natl. Acad. Sci. USA* 88:9292–9296.

Tewari, M. and Dixit, V.M. (1995) Fas- and tumor necrosis factor-induced apoptosis is inhibited by the poxvirus crmA gene product. *J. Biol. Chem.* 270:3255–3260.

Tewari, M., Quan, L.T., O'Rourke, K., Desnoyers, S., Zeng, Z., Beidler, D.R., Poirier, G.G., Salvesen, G.S. and Dixit, V.M. (1995) Yama/CPP32 beta, a mammalian homolog of CED-3, is a CrmA-inhibitable protease that cleaves the death substrate poly(ADP-ribose) polymerase. *Cell* 81:801–809.

Torti, S.V., Kwak, E.L., Miller, S.C., Miller, L.L., Ringold, G.M., Myambo, K.B., Young, A.P. and Torti, F. (1988) The molecular cloning and characterization of murine ferritin heavy chain, a tumor necrosis factor inducible gene. *J. Biol. Chem.* 263:12638–12644.

Trefzer, U., Brockhaus, M., Loetscher, H., Parlow, F., Kapp, A., Schopf, E. and Krutmann, J. (1991) 55-kd tumor necrosis factor receptor is expressed by human keratinocytes and plays a pivotal role in regulation of human keratinocyte ICAM-1 expression. *J. Investigative Dermatology* 97:911–916.

Tsan, M.F., White, J.E., Treanor, C. and Shaffer, J.B. (1990) Molecular basis for tumor necrosis factor-induced increase in pulmonary superoxide dismutase activities. *Am. J. Phyisol.* 259:L506–L512.

Valentine, J. and Nick, H. (1990) Regulation of manganous and copper-zinc superoxide dismutase messenger RNA in cultured rat intestinal epithelial cells. *FASEB J.* 4:7–11.

Valentine, J.F. and Nick, H.S. (1991) Acute-phase induction of manganese superoxide dismutase in intestinal epithelial cell lines. *Endocrinology* 129:905–912.

van Loon, A.P., Pesold-Hurt, B. and Schatz, G. (1986) A yeast mutant lacking mitochondrial manganese superoxide dismutase is hypersensitive to oxygen. *Proc. Natl. Acad. Sci. USA* 83:3820–3824.

Visner, G.A., Chesrown, S.E., Monnier, J., Ryan, U.S. and Nick, H.S. (1992) Regulation of manganese superoxide dismutase: IL-1 and TNF induction in pulmonary artery and microvascular endothelial cells. *Biochem. Biophys. Res. Commun.* 188:453–462.

Visner, G.A., Dougall, W.C., Wilson, J.M., Bur, I.A. and Nick, H.S. (1990) Regulation of manganese superoxide dismutase by lipopolysaccharide, interleukin-1, and tumor necrosis factor. Role in the acute inflammatory response. *J. Biol. Chem.* 265:2856–2864.

Wallach, D. (1984) Preparations of lymphotoxin induce resistance to their own cytotoxic effect. *J. Immunol.* 132:2464–2469.

Ware, C.F., VanArsdale, T.L., Crowe, P.D. and Browning, J.L. (1995) The ligands and receptors of the lymphotoxin system. *Curr. Top. Microbiol. Immunol.* 198:175–218.

Warner, B.B., Burhans, M.S., Clark, J.C. and Wispe, J.R. (1991) Tumor necrosis factor-alpha increases Mn-SOD expression: protection against oxidant injury. *Am. J. Physiol.* 260:L296–L301.

Weisiger, R.A. and Fridovich, I. (1973) Superoxide dismutase. Organelle specificity. *J. Biol. Chem.* 248:3582–3592.

Wispe, J.R., Clark, J.C., Burhans, M.S., Kropp, K.E., Korfhagen, T.R. and Whitsett, J.A. (1989) Synthesis and processing of the precursor for human mangano-superoxide dismutase. *Biochim. Biophys. Acta.* 994:30–36.

Wong, G.H.W. and Goeddel, D.V. (1986) Tumour necrosis factors alpha and beta inhibit virus replication and synergize with interferons. *Nature* 323:819–822.

Wong, G.H.W. and Goeddel, D.V. (1988a) Induction of manganous superoxide dismutase by tumor necrosis factor: possible protective mechanism. *Science* 242:941–944.

Wong, G.H.W. and Goeddel, D.V. (1988b) Tumor necrosis factors: modulation of synthesis and biological activities. *In:* J.C. Mani and J. Dornaud (eds) *Lymphocyte Activation and Differentiation.* Walter de Gruyter, New York, pp 217–226.

Wong, G.H.W., Elwell, J.H., Oberley, L.W. and Goeddel, D.V. (1989) Manganous superoxide dismutase is essential for cellular resistance to cytotoxicity of tumor necrosis factor. *Cell* 58:923–932.

Wong, G.H.W., McHugh, T., Weber, R. and Goeddel, D.V. (1991) Tumor necrosis factor alpha selectively sensitizes human immunodeficiency virus-infected cells to heat and radiation. *Proc. Natl. Acad. Sci. USA* 88:4372–4376.

Wong, G.H.W., Neta, R. and Goeddel, D.V. (1992a). Protective roles of MnSOD, TNF-alpha, TNF-beta and D-Factor (LIF) in radiation injury. *In:* K. Honn, L. Marnett and T. Walden (eds) *Eicosanoids and Other Bioactive Lipids in Cancer, Inflammation and Radiation Injury.* Kluwer Academic Publishers, Boston, pp 353–357.

Wong, G.H.W., Kamb, A., Tartaglia, L.A. and Goeddel, D.V. (1992b) Possible protective mechanisms of tumor necrosis factors against oxidative stress. *In:* J. Scandalios (ed.) *Molecular Biology of Free Radical Scavenging Systems.* Cold Spring Harbor Press, Cold Spring Harbor, New York, pp 69–96.

Wong, G.H.W., Tartaglia, L.A., Lee, M.S. and Goeddel, D.V. (1992c) Antiviral activity of tumor necrosis factor is signaled through the 55-kDa type I TNF receptor [corrected] [published erratum appears in J. Immunol. 1993 Jan 15; 150(2):705]. *J. Immunol.* 149:3350–3353.

Wong, G.H.W., Kamb, A., Elwell, J.H., Oberley, L.W. and Goeddel, D.V. (1992d) MnSOD induction by TNF and its protective role. *In:* B. Beutler (ed.) *Tumor Necrosis Factors: The Molecules and Their Emerging Role in Medicine.* Raven Press, New York, pp 473–484.

Wong, G.H.W. and Goeddel, D.V. (1994a) Fas antigen and 55-kD TNF receptor signal apoptosis through distinct pathways. *J. Immunol.* 152:1751–1755.

Wong, G.H.W. and Goeddel, D.V. (1994b) One-day northern blotting for detection of mRNA: NDGA inhibits the induction of MnSOD mRNA by agonists of type 1 TNF receptor. *Methods Enzymol.* 234:244–252.

Wong, G.H.W. (1995) Protective roles of cytokines against radiation: induction of mitochondrial MnSOD. *Biochim. Biophys. Acta* 1271:205–209.

Part IV
Paradigms for complex stress responses

Introduction

U. Feige

Department of Pharmacology, AMGEN Center, Mail Stop 15-2-A-224, 1840 De Havilland Drive, Thousand Oaks, CA 91320-1789, USA

The mammalian organism, with its plethora of specialized cells working in concert during life at normal environmental conditions, depends on the ability of all its "member" cells to respond to stressful conditions it faces in harmful conditions such as infection or inflammation. Viral infections induce cells to express hsp70. Prior induction of an hsp70 response renders cells to better resist and survive virus attacks. On the other hand, viral infection results in a change of the stress response of cells, which might contribute of the progression of AIDS. During an infection with bacteria the mammalian organism as well as the infecting bacteria mount stress responses which both generate stress proteins as targets for immune responses and auto-immune responses. The view of the role of HSP in autoimmunity and autoimmune disease has been changing during the last decade. The beneficial role of autoimmunity to HSP in immune homeostasis and maintaining an autoimmune disease resistant status is beginning to be realized. Inflammatory reactions such as those occurring during an autoimmune disease put cells under stress, too, and HSP play important roles in the protection against inflammatory stress and damage. Studies of cells exposed to various stressful stimuli have shown that the stress response is not homogenous but highly dependent on type, severity and duration of stress. The capacity to mount a stress response in response to environmental changes decreases with age in mammalian organisms at the cellular, tissue and organ levels as well as dysfunction of various homeostatic mechanisms. A better understanding of the stress response of individual cells in concert with others may lead to new ways of prevention and treatment of viral and bacterial infections as well as autoimmune disease and inflammatory conditions. However, it appears that we still all have to wait for the "Fountain of Youth", although to "re-juvenate" the stress response of cells seems to be an attractive way to achieve that goal.

Stress-Inducible Cellular Responses
ed. by U. Feige, R.I. Morimoto, I. Yahara and B. Polla
© 1996 Birkhäuser Verlag Basel/Switzerland

Viral infection

M.G. Santoro

Institute of Experimental Medicine, CNR, Vle K. Marx, 15/43 – I-00137 Rome, Italy and Department of Experimental Medicine, University of L'Aquila, I-67200 L'Aquila, Italy

Summary. The relationship between viruses and the cellular stress response is a multifaceted and complex phenomenon which depends on the structural and genetic characteristics of the virus, on the type of infection, as well as on the environmental conditions. It is now well documented that infection of mammalian cells by several types of RNA and DNA viruses often results in alterations of the cellular stress response. Interactions between stress proteins and viral components have been described in a large variety of experimental models at different stages of the viral life cycle, depending on the type of virus and host cell. The presence of heat shock proteins in intact virions has also been described. On the other hand, induction of HSP expression by hyperthermia or other agents results in alterations of the virus replication cycle during acute or persistent infections of mammalian cells, and a possible role of heat shock proteins in the beneficial effect of fever and local hyperthermia during acute infection has been hypothesized. This chapter describes the different aspects of the interaction between viruses and the stress response, and discusses the possible role of stress proteins in the control of virus replication and morphogenesis.

Introduction

Viruses are common pathogens of virtually all forms of life. Due to the simplicity of their structure, viruses have an obligate requirement for intracellular growth and have evolved to exploit the complex metabolic machinery of the host cell and to reprogram it for the synthesis of the macromolecular constituents required for their multiplication and invasion of new cells. Despite their structural simplicity, viruses can adopt a surprising variety of different replicative strategies which will lead to a more or less rapid death of the host cell or to the establishment of a persistent infection.

One of the emerging interesting aspects of the interaction between viruses and their host cells is the induction of heat shock protein expression during infection by different types of RNA and DNA viruses, as well as the finding of a well documented ability of several stress proteins to bind to viral components or to assembly intermediates during virus replication. The presence of hsp70 as a component of mature virions has also been shown.

A different and equally interesting approach to analyze the relationship between viruses and the stress response comes from studies which investigate the effect of high intracellular levels of HSP, induced by either hyperthermia or other agents, on virus replication during acute or persistent infections. While both a positive and a negative role of HSP in the control of virus replication has been hypothesized, the function of heat shock proteins during the virus replication cycle is still not well characterized.

This chapter describes different aspects of the interactions between viruses and stress proteins during infection of mammalian cells, focusing on: (1) the modulation of the stress response by DNA and RNA viruses; (2) the interactions of HSP and other chaperones with viral components during the virus replication cycle; (3) the effect of hyperthermia, HSP inducers and selected antiviral agents on virus replication; (4) AIDS and the cellular stress response. The possible function of stress proteins in the recognition by the immune system when expressed on virus-infected cells is described elsewhere.

Modulation of the cellular stress response by RNA viruses

A small number of known cellular proteins are synthesized at increased rates after infection by RNA viruses. The proteins of the interferon system are the most studied example. Starting from the initial observation by Peluso et al. (1977, 1978) that infection of cultured chick embryo cells by the paramyxoviruses Simian virus 5 (SV5) and Sendai virus stimulated the synthesis of several cellular polypeptides, two of which (97- and 78-kDa proteins) were identified as glucose-regulated proteins (GRP), a growing body of literature has described the induction of stress proteins by different types of RNA viruses in avian and mammalian cells (Fig. 1).

Several studies have focused on paramyxoviruses. In the case of simian virus 5 (SV5), a fivefold increase in the rate of *grp78-BiP* transcription and an increase in grp-BiP protein levels was shown in monkey CV-1 cells at 9 h after infection. When the individual SV5 polypeptides were expressed from cloned cDNAs, the synthesis of the hemagglutinin-neuraminidase (HN) glycoprotein, but not of the fusion (F) glycoprotein or the cytoplasmic proteins P, V and M, led to an increase of *grp78-BiP* accumulation, and it was suggested that the flux of folding-competent HN molecules through the ER of infected cells stimulates grp-BiP synthesis (Watowich at al., 1991). As reported by Collins and Hightower (1982), a different paramyxovirus, Newcastle disease virus (NDV), caused accumulation of heat shock proteins in infected chick embryo cells; avirulent strains were stronger inducers, probably due to the rapid inhibition of host macromolecular synthesis by virulent strains of NDV. A direct association between *in vivo* virus infection and induction of the cellular stress response was shown by Oglesbee and Krakowka (1993) by dual label confocal immunocytochemistry examination of brain tissue from dogs infected with a different member of the *Paramyxoviridae* family, the morbillivirus CDV (canine distemper virus). In this case, elevated levels of a 72 kDa HSP were found in CDV-infected astrocytes as compared to non-infected cells.

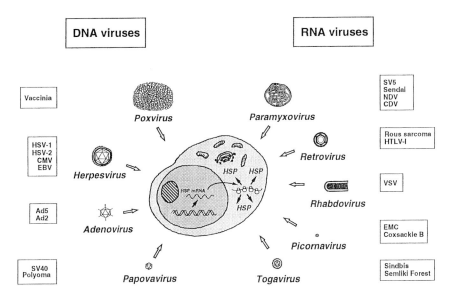

Figure 1. Viruses which induce HSP expression in mammalian cells. HSV = herpes simplex virus; CMV = cytomegalovirus; EBV = Epstein-Barr virus; Ad = adenovirus; SV = simian virus; NDV = Newcastle disease virus; CDV = canine distemper virus; HTLV-1 = human T cell leukemia virus; VSV = vesicular stomatitis virus; EMC = encephalomyocarditis virus.

Induction of stress proteins by the togavirus Sindbis virus and by the rhabdovirus vesicular stomatitis virus (VSV) in chick embryo cells was reported by Garry et al. (1983). The capsid protein (C) of Sindbis virus and nucleocapsid protein (N) of VSV were found to be physically associated with an 89 kDa HSP. Furthermore, the synthesis of virus-induced stress proteins was resistant, relative to the synthesis of most host proteins, to alterations in the intracellular ionic concentration. The capsid proteins of a different member of the *Togaviridae* family, Semliki Forest virus, have been shown to efficiently modulate the stress response of target cells and to confer thermal resistance to HeLa cells (Michel at al., 1994).

Increased levels of hsp70 have detected by Western blot analysis in cultured neonatal myocardial cells from BALB/c mice after infection with two different picornaviruses, encephalomyocarditis (EMC) virus and coxsackievirus B-3 (CVB3) (Huber, 1992). Virus inactivation by ultraviolet irradiation prevented hsp70 induction in these cells. Among the *Picornaviridae,* polioviruses (PV) are known to cause a dramatic shut-off of the host cell protein cap-dependent translation by proteolytically inactivating the cap-binding protein complex (Rueckert, 1990). Poliovirus infection has been shown to inhibit constitutive (Macejak and Sarnow, 1992) as well as heat shock-induced (Muñoz et al., 1984) hsp70 synthesis 2–3 h after infection of human cells. However, the synthesis of heat shock proteins was found to be more resistant to inhibition than normal host proteins after PV infection. On the other hand, as reported by Sarnow (1989), the translation of grp78 is increased in poliovirus-infected HeLa cells, at a time when cap-dependent translation of cellular mRNA is inhibited, suggesting that *grp78* mRNA could be translated by the same "cap-independent" mechanism as the viral RNA. It was then speculated that a virally-induced function could promote the selective translation of both viral and *grp78* mRNA. Whether grp78 is an unnecessary byproduct of the viral infection or has a function in the viral life cycle is yet unclear. A possible role for grp78 in the membrane rearrangements due to proliferation of vesicles generated from ER membranes during PV infection, or its direct participation in the assembly of membrane-bound viral replication complexes by a transient association with viral polypeptides has been hypothesized.

Infection by retroviruses has also been shown to affect the cellular stress response. Stimulation of GRP synthesis has been shown in cells transformed by Rous sarcoma virus by Peluso et al. (1978). A decrease in the transcription and translation of the 47 kDa collagen-binding protein (hsp47) was instead shown in chick embryo fibroblasts after transformation by Rous sarcoma virus (Hirayoshi et al., 1991). Expression of hsp70 has been found on the surface of cells infected with human T cell leukemia virus type I (HTLV-I), suggesting that HSP immunity plays a role in the pathogenesis of HTLV-I infection (Chouchane, 1994). Increased levels of hsp70 synthesis were also reported by D'Onofrio et al. (1995) in a human leukemic T cell line (Molt-4), after exposure to HTLV-I in a cell-to-cell transmission model. Similar data were reported in human cells infected with the human immunodeficiency virus (HIV), as it will be discusses below.

Modulation of the cellular stress response by DNA viruses

Different types of DNA viruses influence the cellular stress response (Fig. 1). As reported by Khandjian and Turler (1983), during the lytic infection of monkey and mouse cells with simian virus 40 (SV40) or polyoma virus is a marked increase in the synthesis of two host heat-inducible proteins of 92 and 72 kDa. In monkey CV-1 cells the main SV40-inducible member of the *hsp70* gene family was recently shown to be the *hsc70* gene, which was only slightly induced by heat (Sainis et al., 1994). An up-regulation of hsp60, hsc70, and hsp90 and a down-regulation of the small heat shock protein hsp28 was recently shown by Honoré et al. (1994) in SV40-transformed keratinocytes as compared to normal human keratinocytes.

Jindal and Young (1992) reported that infection of human monocyte-macrophages by vaccinia virus, while it caused a dramatic decrease in the levels of celular mRNAs, did not cause a significant reduction in the levels of *hsp90* and *hsp60* mRNA and led to substantially increased levels of *hsp70* mRNA, again indicating an increased resistance of HSP transcription and translation during cytopathic virus infection. Interestingly, the expression of stress proteins has been shown to be enhanced during poxvirus infection *in vivo*. A dramatic increase in the expression of the inducible hsp70 and induction of a 32 kDa protein, probably an oxidation-specific stress protein, was shown in vaccinia virus-infected mouse ovaries 6 days post-infection (Sedger and Ruby, 1994).

Infection of human cells with adenovirus has also been found to increase the expression of HSP genes, particularly of the *hsp70* genes, and the earliest virus proteins formed during infection, the E1A 12S and 13S products, were shown to induce *hsp70* gene expression. The E1A 13S product was identified as a specific transcriptional *trans*-activator of the *hsp70* gene (Williams et al., 1989; Simon et al., 1987).

Numerous reports have shown the accumulation of heat shock proteins in several cell lines during herpes simplex virus (HSV) infection. Mutants of HSV type 1 (HSV-1) and 2 (HSV-2) induce HSPs during infection of chick embryo fibroblasts (Notarianni and Preston, 1982) and human neuroblastoma cells (Yura et al., 1987) respectively. The presence of abnormal forms of the HSV-1 immediate early polypeptide Vmw175 was found to be the signal for induction of the stress response in chick embryo fibroblasts infected with the HSV mutant *tsK* (Russel et al., 1987). An altered HSV-1 envelope gB glycoprotein that is retained in the ER of mammalian cells, but not the normal viral envelope protein, was also found to transactivate the *grp78* promoter (Wooden et al., 1991). Lytic infection of BHK cells with several strains of HSV-2 was found to cause intracellular accumulation as well as translocation to the cell surface of a 90 KDa protein (p90) related to the hsp90 family (La Thangue and Latchman, 1988), while increased levels of a heat-inducible 57 kDa phosphoprotein were reported in BHK cells infected with HSV-2 (La Thangue et al., 1984). Kobayashi et al. (1994) have recently reported the presence of elevated *hsp70* mRNA levels in rodent cells within 4 h after infection with HSV types 1 and 2; *hsp70* induction was dependent on protein synthesis but not on viral DNA replication. A different member of the herpesvirus family, the human cytomegalovirus (HCMV), has been shown to transiently (8–48 h post infection, p.i.) induce *hsp70* gene expression in human diploid fibroblasts (Santomenna and Colberg-Poley, 1990). Induction of hsp70 protein was found to occur at later stages of infection, from 16 to 72 h p.i. Finally, infection of human B lymphocytes with Epstein-Barr virus (EBV) was found to induce the synthesis of both hsp70 and hsp90 proteins; induction was dependent on EBV-induced *trans*-membrane Ca^{2+} fluxes and was independent of viral gene expression (Cheung and Dosch, 1993). It was suggested that hsp70 and hsp90 induction is one of the earliest changes in gene expression induced by this virus in a newly infected host cell.

The results described above clearly indicate that the entry of a virus into an eukaryotic cell initiates a cascade of events which, depending on the type of virus, can result in switching on HSP or GRP gene expression. The functional significance of the induction of stress proteins during virus infection is unknown. It appears more and more evident that increased HSP expression by viral infection is not simply part of a non-specific induction of host genes. Studies on the transcriptional regulation of the *hsp70* gene family in monkey and human cells infected with different DNA viruses, adenovirus type 5 (Ad5), HSV-1, SV40 and vaccinia virus, indicated clearly that induction of *hsp70* genes is not a general response to the stress following viral infection (Phillips et al., 1991). The fact that only Ad5 and HSV-1 viruses were able to induce *hsp70* expression in these cells, and that of three 70 kDa heat shock genes examined only *hsp70* was induced, indicated in fact a highly specific cellular response. It is, however, unclear whether induction of HSP by viruses reflects a stress situation of the host cell with no effect on the virus life cycle, or whether HSP could be actively involved in the control of virus replication.

Interactions of HSP with viral proteins

The possibility that stress proteins could be actively involved in the control of virus replication in eukaryotic cells is suggested by the finding that several HSP and GRP often interact with virus proteins as well as host cell proteins (Gething and Sambrook, 1992). Hsp70 was identified as one of the minor virion components in several negative-strand RNA viruses, including Newcastle disease virus, influenza A virus and vesicular stomatitis virus, and was found to be selectively incorporated into rabies virions (Sagara and Kawai, 1992). Association of HSP or GRP with negative-strand RNA viral proteins has been described in several models.

Association of two chaperones, grp78-BiP and calnexin with vesicular stomatitis virus glycoprotein G has been shown in the ER of CHO cells (Hammond and Helenius, 1994). Coimmunoprecipitation of G protein with the chaperones showed that grp78-BiP bound maximally to early folding intermediates of G protein, whereas calnexin bound transiently at a later stage to more folded molecules. Several authors have reported a transient association of grp78-BiP with newly synthesized influenza virus hemagglutinin monomers (Gething et al., 1986), Sendai virus HN glycoprotein (Roux, 1990) and SV5 HN glycoprotein (Ng et al., 1989), and have suggested that the transient formation of these complexes is part of the normal viral glycoprotein maturation pathway. Mulvey and Brown (1995) have recently shown that Sindbis virus membrane glycoprotein E1 interacts with grp78-BiP after synthesis. These authors suggested that ATP-regulated binding and release of BiP has a role in modulating disulphide bond formation during E1 folding.

In the case of positive-strand RNA viruses, the togavirus Semliki Forest virus spike glycoproteins were found to be transiently associated with grp78-BiP (Marquardt and Helenius, 1992). Among the *Retroviridae*, the transforming protein of Rous sarcoma virus pp60[v-src] is known to be transiently complexed with the 90 kDa heat shock protein during or immediately after synthesis in vertebrate cells (Whitelaw et al., 1991); association with hsp90 was shown to stabilize pp60[v-src] protein and to affect both its activity and specificity (Xu and Lindquist, 1993). The association of human immunodeficiency virus HIV-1 with molecular chaperones will be described in a separate section.

In the case of picornaviruses, hsp70 was shown to be associated with newly synthesized capsid precursor P1 of poliovirus and coxsackievirus B1 in infected HeLa cells (Macejak and Sarnow, 1992). The half-life of P1 was increased when bound to hsp70 and hsp70-P1 complexes were uncleavable by the viral protease. The hsp70-P1 complex was found to be part of an assembly intermediate of picornavirus. Since the association of hsp70 with P1 was highly stable, was specific to the P1 precursor and was detected in two different enteroviruses, a role for hsp70 in viral assembly was suggested.

Association of HSP with viral components has also been shown in DNA virus models. Hsp70 was found to be associated with the adenovirus type 5 fiber protein in infected HEp-2 cells (Macejak and Luftig, 1991). Colocalization of cellular hsp70 with adenovirus E1A proteins was shown also in the nucleus of adenovirus type 2 infected HeLa cells (White et al., 1988). In this case, it was shown that colocalization of E1A and hsp70 requires the presence of both functional E1A and a productive virus infection, and is not simply a consequence of the cell expressing the E1A protein.

In Epstein-Barr virus (EBV)-transformed lymphoblastoid cells, the latent membrane transforming protein (LMP-1) was found to colocalize in lysosomes with hsp70 and ubiquitin-protein conjugates (Laszlo et al., 1991), and association of cellular hsp70 with viral components was also detected in vaccinia virus-infected human monocyte-macrophages (Jindal and Young, 1992).

In studies on papovavirus infection, hsp70 was found to associate with polyomavirus medium T antigen (Walter et al., 1987) and with SV40 super T antigen in human, monkey and rat cells (May et al., 1991). Hsp70 was also shown to specifically interact with SV40 large T antigen. The interaction was independent of T antigen/p53 complex formation (Sawai and Butel, 1989), and was necessary for SV40 T antigen nuclear import (Yang and DeFranco, 1994). Besides a chaperone function, it was suggested that hsp70 may interact with T antigen during the formation of replication complexes and may participate in SV40 DNA replication.

Finally, a 60 kDa protein related to the chaperonin t-complex polypeptide 1 (TCP-1) was found to be associated with hepatitis B virus core polypeptides in two different assembly intermediates (but not with the unassembled proteins or with the final mature capsid product) in a cell-free system, suggesting that eukaryotic cytosolic chaperonins may play a distinctive role in virus multimer assembly, apart from their involvement in assisting monomer folding (Lingappa et al., 1994).

As indicated by the results described above, interactions between stress proteins and viral components have been documented at different levels and at different stages of the virus life cycle, depending on the type of virus and host cell; however, their functional significance is still to be defined. Interaction of HSP and virus proteins may simply reflect a "housekeeping" function of HSP in recognizing and facilitating the clearing of foreign and abnormal proteins in the cell. The specificity of the interactions shown in some of the studies described above makes this possibility unlikely. Different types of viruses may instead be able to reprogram the host cell to up- or down-regulate HSP expression, and may utilize different HSP components of the cellular machinery which folds, translocates and assembles proteins, to complete their morphogenesis. Earlier studies in prokaryotic cells have shown that *E. coli* HSP, GroES and GroEL, are involved in the assembly of the head proteins of bacteriophages λ and T4, and the tail proteins of bacteriophage T5 (reviewed by Lindquist and Craig, 1988; Casjens and Hendrix, 1988). In eukaryotic cells, the possibility that HSP or GRP may chaperone virus components and thereby facilitate their replication is supported by numerous evidences in the studies described above.

On the other hand, since the half-lives of some viral proteins are increased as a result of HSP interaction, it has been argued that HSPs could protect the host cell, by precluding the cleavage and subsequent processing of precursors into mature virions (Macejack and Sarnow, 1992). These apparently discordant interpretations, however, do not exclude each other. Due to the amazingly different replicative strategies of different viruses, a unified function of the HSP-virus interaction is difficult to predict. Depending on the more or less rapid take over of the host cell machinery by the virus, which varies for different viruses and is also a function of the infectious dose, HSP could be utilized either by the virus to promote viral particle biogenesis, or by the host cell in an attempt to block or delay virus replication.

It should be underlined that the studies of the interaction between HSP and viruses during infection of cultured cells, may not reflect accurately the events of a viral infection *in vivo* where other conditions, which include changes in body temperature, inflammation, oxidative stress or the increased local concentrations of lipids and cytokines, may influence HSP expression.

Hyperthermia in acute or persistent virus infections

Many viral diseases cause the development of fever in the host. While the possibility that fever could be beneficial during viral infections has been often suggested, a differential effect of hyperthermia during acute or chronic virus infection has been reported.

Fever is a well recognized trigger of the reactivation of latent herpes simplex virus infection in humans. *In vivo* reactivation of HSV-1 in latently infected murine ganglionic neurones was detected as soon as 14 h after transient hyperthermia, while hsp70 synthesis was induced by hyperthermia treatment in trigeminal and lumbosacral ganglia (Sawtell and Thompson, 1992). On the other hand, hyperthermic treatment during acute infection of human neuroblastoma cells with HSV-2 (Yura et al., 1987) or chick embryo fibroblasts with the tsK mutant of HSV-1 (Russel et al., 1987) resulted in a reduction of the synthesis of late viral polypeptides and repression of virus replication.

As reported by Zerbini et al. (1985), heat shock induces the appearance of Epstein-Barr virus (EBV) early antigen in EBV-infected lymphoblastoma cell lines, and enhances cytomegalovirus expression in semipermissive and permissive cells. Enhanced expression of the light nucleocapsid (L-NC) of the morbillivirus canine distemper virus (CDV) was shown in cultured cells persistently infected with CDV, after transient hyperthermia or sodium arsenite treatment (Oglesbee et al., 1993). Postshock increases in viral fusion (F) gene transcripts and F protein levels were associated with dramatic increases in viral cytopathic effect, while only a modest increase in cell-free infectious viral progeny was detected. Finally, as will be discussed below, human cells with stably integrated human immunodeficiency virus can be activated by heat shock to produce virus through an increase in viral gene transcription (Stanley et al., 1990). These results thus indicate a negative role of hyperthermia in chronic virus infection.

On the contrary, exposure of mammalian cells to elevated temperatures during primary virus infection appears to inhibit virus replication. When animals infected with herpesvirus, rabies virus, group B coxsackievirus and other viruses were kept at elevated temperatures, viral infection was reduced in comparison to animals kept at normal temperature (Teisner and Haahr, 1974; Roberts, 1979). Survival of newborn pups after inoculation with canine herpesvirus (CHV) was prolonged and viral growth was greatly reduced when animals were kept in an environment that elevated their body temperature to 38.6°C–39.5°C (Carmichael et al., 1969). Mice receiving 40°C whole body hyperthermia for 30 min on day 5 and 12 after infection with the polycythaemia-inducing strain of Friend virus complex had lower titres of infectious virus in their spleen than untreated controls (Lu et al., 1991). In humans, local hyperthermia (20 to 30 min at 43°C) was found to improve the course of the disease in patients with natural and experimental common colds (Tyrrell et al., 1989).

Continuous hyperthermia at 3 to 4°C above the physiological range can block the replication of several DNA and RNA viruses during primary infection in cultured cells (Bennett and Nicastri, 1960; Panasiak et al., 1990). Togavirus replication was dramatically inhibited in *Aedes albopictus* cells after a 90 min HT at 9°C above the physiological temperature

(Carvalho and Fournier, 1991). Angelidis et al. (1988) reported that a 12 h hyperthermic treatment (HT) at 43 °C selectively inhibited simian virus 40 protein synthesis in CV1 cells. The fact that a similar effect was obtained also after HSP induction by sodium arsenite suggested that inhibition of SV40 replication was a HSP-mediated response. Also in the case of retroviruses, heat shock (45 min at 44 °C, but not 25 at 42.8 °C) was shown to decrease cell-free viral *env* protein and to transiently block the processing of the viral *gag* precursor in mammalian cells infected with Moloney murine leukemia virus (Herman and Yatvin, 1995).

Studies from our laboratory have shown that brief hyperthermic treatment (45 °C for 20 min), if applied during specific stages of the virus cycle, is extremely effective in blocking the replication of the rhabdovirus VSV during primary infection (De Marco and Santoro, 1993). In this model a brief exposure to high temperature did not damage uninfected cells, and only moderately (25 to 35%) inhibited host cell protein synthesis for a period of approximately 4 h. No effect on virus replication was found when heat shock was applied soon after virus entry into the cells, or at later times of infection (8–10 h), when the virus had taken control of the cellular protein synthesis machinery and HSP synthesis was impaired, indicating that the antiviral effect is not due to a general change in membrane fluidity, cell metabolism or stability of virus proteins or RNA. If hyperthermic treatment was applied between 2 and 4 h after virus infection, at the time of genomic amplification when virus proteins start to be synthesized, virus yield was found to be reduced to less than 5% of control. Depending on the temperature used and on the length of treatment, virus replication appeared to be affected at different levels.

These results suggested that synthesis of heat shock proteins at specific stages of the virus cycle could interfere with virus replication. Possibly, in order to achieve an antiviral effect during acute infections, HSP induction has to be carefully timed in relation to virus replication.

AIDS and the cellular stress response

Human immunodeficiency virus (HIV-1) utilizes normal intracellular signaling pathways to regulate its expression, and, like other retroviruses, relies upon host factors for many aspects of its replication cycle. In the last few years, also due to the description of controversial results on the effect of hyperthermia on HIV infection, the attention of several investigators has focused on a possible relationship between HSP and HIV replication.

Furlini et al. (1994) have recently observed an increase in hsp70 synthesis and nuclear translocation early (3 h) after HIV-1 infection of permissive CD4⁺ cells. Also, an increase in the level of *hsp70* mRNA accumulation was shown later (6 h) after infection. Similar results were obtained if cells were exposed to heat-inactivated HIV-1 or to purified recombinant gp120 HIV-1 envelope glycoprotein, suggesting that replication of HIV-1 is not required for the alteration in the synthesis and intracellular translocation of hsp70, and that the simple interaction of the virus with the CD4 receptor on the cell membrane may deliver the signal activating these events. A moderate increase in the level of *hsp70* mRNA accumulation after HIV-1 infection was also shown in CEM-SS cells in our laboratory (De Marco et al., unpublished observations). Finally, an increased expression of hsp70 was shown in human lymphoma cells chronically infected with HIV (Di Cesare et al., 1992), while an hsp60-related protein was found to be associated with purified HIV and SIV virions (Bartz et al., 1994).

HIV envelope protein Env (gp160), which binds to the CD4 receptor and mediates fusion with the plasma membrane of the host cell, has been shown to bind to grp78-BiP in a specific and transient fashion (Earl et al., 1991). Grp78-BiP appears to bind to multiple sites along the protein and to dissociate from it only after the gp160 ectodomain assumes a conformation similar to that of the mature protein.

The interaction between HIV-1 proteins and cyclophilin A, a cytosolic form of the cyclophilin family of peptidyl-prolyl *cis-trans* isomerases, has been recently studied in detail. Luban et al. (1993) reported that cyclophilin A binds specifically to the HIV-1 Gag polyprotein p55^{gag} *in vitro*. This interaction was specific, since other retroviral Gag proteins, including the Gag of the closely related simian immunodeficiency virus (SIV), were unable to bind cyclophilin A. More recently, a similar interaction was reported during HIV-1 replication cycle *in vivo*. Franke et al. (1994), and Thali et al. (1994) reported that high levels of cyclophilin A were specifically packaged into HIV-1 virions, while other retroviruses, including SIV$_{MAC}$, HIV-2, visna virus and Moloney leukemia virus, did not incorporate cyclophilin A into their viral particles. Packaging of cyclophilin A was found to be mediated by a short segment of the HIV-1 capsid protein containing four proline residues, one of which, residue 222 in HIV-1, is required for both cyclophilin A incorporation into the virion and virion infectivity.

The same authors reported that the association of cyclophilin A with HIV-1 virions was inhibited in a dose-dependent manner by cyclosporin A and by a non-immunosuppressive analogue of cyclosporin A (SDZ NIM811). The fact that drugs-induced reductions in virion-associated cyclophilin A levels resulted in reductions in HIV-1 virion infectivity, while SIV virions which do not package cyclophilin A were not affected, indicated that the p55^{gag}-cyclophilin A association is functionally relevant. Both a chaperone-like role of cyclophilin A in HIV-1 morphogenesis, independent of its action as a proline isomerase, and an effect of this protein in facilitating viral polypeptide conformational changes during initial stages of the HIV-1 replication cycle, have been hypothesized (Cullen and Heitman, 1994).

Controversial results on the effect of hyperthermia on HIV replication have been described. Geelen et al. (1988) reported that heat shock (15 min at 42 °C) as well as treatment with chemical inducers of the cellular stress response, including sodium arsenite, induced the expression of the bacterial chloramphenicol acetyltransferase (CAT) gene under the control of the HIV long terminal repeat sequence (LTR) in stably transformed cells. Induction of HIV-1 expression was also reported by Stanley at al. (1990) in chronically infected promonocytic U1 cells (41.9 °C for 2 h) and T-lymphocytic ACH-2 cells (42.7 °C for 1 h), as measured by increases in culture reverse transcriptase activity and detectable viral proteins. HIV-1 LTR contains two NF-κB binding elements, involved in the activation of the promoter during oxidative stress, which are sequence related to the heat shock element HSE. Mutational analysis suggested that the NF-κB consensus sequences of the HIV-LTR are necessary for heat induction. By using transient transfection assays, Kretz-Remy and Arrigo (1994) have recently shown that the kinetics of thermal transcriptional activation of HIV-1 LTR resembles that of the promoter of the gene encoding human hsp70. However, the authors were unable to detect the binding of any protein to κB elements, suggesting that this site is not directly involved in the thermal transcriptional activation of HIV-1 LTR.

HIV infection is characterized by a prolonged period of latency which may be followed by progression to AIDS, associated with enhanced HIV expression. Stanley et al. (1990) observed that physiologic heat can selectively synergize with certain cytokines, including IL-6 and GM-CSF but not TNF-α, in the induction of HIV expression, and concluded that repeated opportunistic infections with fever and associated cytokine production can hasten the clinical progression of disease by accelerating HIV expression.

On the other hand, Wong et al. (1991) have shown that treatment of HUT-78 cells acutely infected with HIV-1 at 42 °C for 3 h reduced the viral titres about three-fold as compared to untreated HIV-infected cells. Moreover, acutely HIV-infected cells were more susceptible to killing by heat than uninfected cells. The authors suggested that the reduced expression of manganous superoxide dismutase (MnSOD) in HIV-infected cells as compared to uninfected cells, could be responsible for increased heat sensitivity. Because of the increased heat sensitivity of HIV-1 infected cells and of the relatively rapid inactivation of HIV at 42 °C–45 °C (Gordon et al., 1988), several authors have suggested that hyperthermic treatment could be potentially beneficial in HIV-1 infection (Yatvin et al., 1993; Pennypacker et al., 1995).

Induction of heat shock proteins by antiviral agents

Several molecules with antiviral activity have been shown to directly induce HSP synthesis or to be able to modulate HSP expression after induction by hyperthermia or other agents. The exposure of human monocytes to tumor necrosis factor (TNF) for 40 min was shown to induce high levels of *hsp70* expression (Fincato et al., 1991). In the same study, interferon-γ was found to have no effect. The relationship between the stress response and interferon (IFN) has been studied in different models. Both IFN-α and -γ were recently shown to transcriptionally up-regulate the expression of the major endoplasmic reticular heat-shock protein gp96 in human cells (Anderson et al., 1994). The enhancement of the expression of gp96 was shown to be mediated at the transcriptional level and to require *de novo* protein synthesis. In the same study, IFN-α or -γ treatment was not able to induce *hsp70* transcription in Daudi cells. Treatment with type I interferon, even though it did not induce HSP synthesis at 37 °C, was shown to potentiate the transcription and translation of HSP mRNA after a mild heat shock (39.5 °C – 40.5 °C) in human promyelocytic HL-60 cells (Chang et al., 1991). Also in murine cells treatment with interferon, which had no effect on HSP expression at 37 °C, was found to enhance and prolong the synthesis of the major inducible hsp70, as well as of the 90-kDa and 110-kDa HSPs, after heat shock (Dubois et al., 1988). IFN was shown to regulate HSP expression at two levels, by enhancing the transcription rate of the heat-shock genes, and by increasing the stability of mRNA coding for HSPs. Results from our laboratory also showed that human natural IFN-α, while it was unable to induce hsp70 synthesis at concentrations ranging from 10 to 1000 I.U./ml, enhanced prostaglandin-induced hsp70 synthesis in cells infected with vesicular stomatitis virus (Pica et al., 1996). The increase in hsp70 synthesis was accompanied by an enhanced antiviral activity.

A well-known inducer of interferon, the double-stranded RNA poly(I)·poly(C$_{12}$U) (dsRNA) was recently shown to induce hsp70 synthesis in human HT-29 cells (Chelbi-Alix and Sripati, 1994). It should be also mentioned that heat shock was found to induce the expression of 2′, 5′-oligoadenylate synthetase, which is usually induced by IFN-α and -β and constitutes one of the pathways of IFN antiviral activity by inhibiting protein synthesis through the activation of a latent endoribonuclease (Chousterman et al., 1987).

Our interest in the relationship between HSPs and virus replication derives from our studies on the antiviral activity of prostaglandins (PGs). Starting from the early observation that prostaglandins of the A type (PGA) inhibit Sendai virus replication and prevent the establishment of persistent infections in cultured cells (Santoro et al., 1980), it is now well established that prostaglandins containing an α, β-unsaturated carbonyl group in the cyclopentane ring structure (cyclopentenone PGs, i.e. PGAs and PGJs) possess a potent antiviral activity against a wide variety of DNA viruses, including herpesviruses and poxviruses, and RNA viruses, including paramyxoviruses, orthomyxoviruses, picornaviruses, rhabdoviruses, togaviruses and retroviruses *in vitro* and *in vivo* (reviewed in Santoro et al., 1990, 1994).

A major characteristic common to prostaglandins with antiviral activity is the ability to function as signal for the induction of heat shock protein synthesis (Santoro et al., 1989a, 1990). Induction of *hsp70* gene transcription by prostaglandin A is mediated by cycloheximide-sensitive activation of heat shock transcription factor (HSF) (Amici et al., 1992). Induction of *hsp70* gene expression has now been described in a large variety of monkey, canine, porcine and human cell lines, as well as in human peripheral blood lymphocytes and macrophages, and primary cells derived from cord blood (reviewed in Santoro et al., 1990). Mouse cells behave differently from other mammalian species in their response to PGA, which is not able to induce hsp70 synthesis, while it increases the synthesis of the constitutive hsc70 in several murine cell lines (Rossi and Santoro, 1994). Synthesis of large amounts of hsp70 or hsc70 proteins can be achieved with concentrations of prostaglandins that do not inhibit nucleic acid or protein synthesis in uninfected cells. Moreover, in contrast to synthesis after brief heat shock, the synthesis of hsp70 continues at high level for several hours (up to 24 h) after induction with cyclopentenone PGs.

The possibility that hsp70 could be involved in the control of virus replication was suggested by the fact that only prostaglandins with antiviral activity were able to induce hsp70 synthesis, and that inhibition of virus replication was always associated with hsp70 induction (reviewed in Santoro et al., 1990). The first evidence for a role of hsp70 in the control of virus replication came from studies in negative-strand RNA virus models. In this case, PGs have been shown to affect two separate events of virus replication, respectively at an early and a late phase of the virus life cycle. Treatment with cyclopentenone PGs in a late phase of the replication cycle of the rhabdovirus VSV or the paramyxovirus Sendai (SV), causes a dramatic block of infectious virus production, which is mediated by alteration in the maturation and intracellular translocation of the VSV glycoprotein G, or the SV glycoprotein HN respectively (Santoro et al., 1983, 1989b). Instead, treatment with the drugs soon after infection results in a selective and dramatic inhibition of virus protein synthesis. These events are associated with hsp70 expression and protection of the host cell from the virus-induced shut-off of cellular protein synthesis (Pica et al., 1993; Amici et al., 1994). While there is no information at the moment about a possible involvement of hsp70 in alterations of virus protein maturation or intracellular translocation, increasingly more experimental evidence is accumulating which indicates that high level of hsp70 synthesis in infected cells antagonize virus protein synthesis in the early phase of acute infections. We have recently shown (Amici et al., 1994) that PGA_1 causes a selective and complete block of Sendai virus protein synthesis, which lasts as long as hsp70 is synthesized by the host cell. The block appears to be at the translational level and is mediated by a cellular component. The possibility that hsp70 itself could be the cellular mediator interfering with SV protein synthesis is suggested by the fact that different hsp70 inducers,

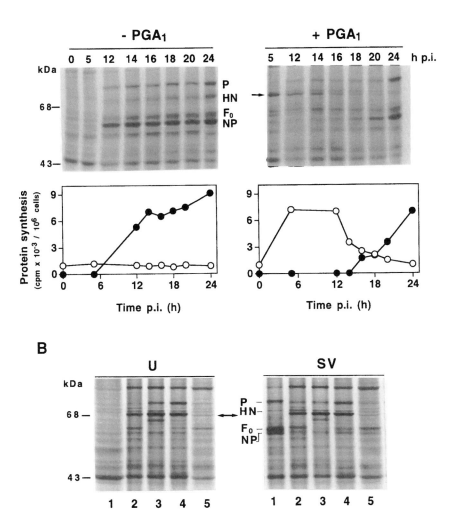

Figure 2. Inhibition of Sendai virus (SV) protein synthesis by prostaglandin A or other hsp70 inducers. A. SV-infected 37RC monkey kidney cells were treated with PGA_1 (4 μg/ml; + PGA_1) or ethanol diluent (-PGA_1) after the 1 h adsorption period, and labeled with ^{35}S-methionine (1 h pulse) at different time intervals (h) after infection (p.i.). Samples containing equal amounts of radio-activity were processed for SDS-PAGE analysis and autoradiography. Virus proteins P, HN, F_0 and NP are indicated. Hsp70 is indicated by the arrow. Densitometric analysis of hsp70 (○) or viral nucleoprotein NP (●) synthesis in SV-infected cells is shown in the panels below. B. Uninfected (U) or SV-infected (SV) 37RC cells were treated with 50 μM $NaAsO_2$ (lane 2), 10^{-5} M $CdCl_2$ (lane 3), 4 μg/ml PGJ_2 (lane 4), or ethanol diluent (lane 1) soon after infection or were subjected to heat shock at 42 °C starting 3 h p.i. for a period of 5 h. Cells were labeled with ^{35}S-methionine (1 h pulse) 9 h p.i. All treatments resulted in the induction of hsp70 (indicated by the arrow) and in the suppression of SV protein synthesis.

including sodium arsenite, cadmium, azetidine and heat shock, all dramatically and selectively inhibited SV protein synthesis as long as heat shock protein synthesis was occurring in infected cells (Fig. 2). Moreover, studies in a murine erythroleukemic cell line (FLC) in which hsp70 is not induced by PGA_1, confirmed that in the absence of hsp70 induction PGA_1 treatment has no effect on SV protein synthesis. Similar results were obtained on the model of rhabdovirus infection previously described. Also in this case, induction of hsp70 synthesis by either PGA_1, sodium arsenite, azetidine or heat shock was associated with a dramatic decrease of VSV protein synthesis in primate cells (Santoro, 1994). During VSV infection of L929 murine fibroblasts, in which PGA is not able to induce hsp70 synthesis, PGA_1-treatment did not inhibit viral protein synthesis, while it caused an alteration of G glycoprotein maturation (Santoro et al., 1983).

Conclusions

Due to the many functions of heat shock proteins and to the many levels of interactions described above, the relationship between viruses and HSP appears to be a complex and multifaceted phenomenon, which depends on the type of virus, the type of infection (acute or persistent), as well as on the environmental conditions. The picture emerging from the large amount of data now available indicates that different viruses can regulate the transcription and/or translation of different HSP, and that these proteins could be either utilized by the virus for its replication and morphogenesis, or by the host cell as an intracellular defense against the invading microorganism. The last possibility is suggested by the studies with hyperthermia, classical HSP inducers or antiviral prostaglandins, which indicate that high levels of HSP synthesis inhibit the expression of virus proteins in several experimental models of acute infection. The mechanism by which HSP can interfere with viral protein synthesis remains to be elucidated. Hsp70 could directly interact with the nascent viral polypeptides, causing a translational block. Schlesinger et al. (1991) have in fact shown that the presence of hsp70 protein during *in vitro* translation of mRNA encoding for the Sindbis virus capsid protein interferes with normal polypeptide synthesis. Our studies on paramyxovirus infection suggest that hsp70 interferes with SV mRNA translation only during its synthesis by the host cell. It could be hypothesized that HSP and virus messages, both of which can be translated in conditions where cellular protein synthesis is impaired, could possess similar mechanisms for preferential translation, and could then compete with each other. For example, both negative strand RNA virus mRNA and hsp70 mRNA are preferentially translated at cytoplasmic ionic concentrations higher than other cellular mRNAs (Nuss and Kock, 1976; Muñoz et al., 1984).

It should be emphasized that during a viral infection *in vivo* HSP expression in the host cell may be influenced not only by the virus, but also by a variety of other conditions which include increased local concentration of cytokines and products of the arachidonate cascade, changes in body temperature, as well as oxidative stress. The description of the effects of oxidative stress on virus replication is beyond the scope of this chapter. However, this is also an interesting aspect of the cellular stress response during virus infection, and is receiving increased attention, especially in relation to the possible use of antioxidants as antiviral drugs in AIDS and other viral diseases (Greenspan, 1993).

In spite of the large amount of data now available, and the increasing interest of virologists in exploring the relationship between viruses and the stress response, many questions remain to be answered. A better understanding of the functional role of heat shock proteins in virus replication is necessary to establish whether HSP or other stress proteins could be part of an intracellular defense strategy against viruses and, in this case, to learn how to manipulate the host cell response for therapeutic gain in the treatment of viral diseases.

Acknowledgements
This work was supported by the Italian Ministry of Public Health, 1996-IX AIDS Research Project.

References

Amici, C., Sistonen, L., Santoro, M.G. and Morimoto, R.I. (1992) Anti-proliferative prostaglandins activate heat shock transcription factor. *Proc. Natl. Acad. Sci. USA* 89:6227–6231.

Amici, G., Giorgi, C., Rossi, A. and Santoro, M.G. (1994) Selective inhibition of virus protein synthesis by prostaglandin A: a translational block associated with HSP70 synthesis. *J. Virol.* 68:6890–6899.

Anderson, D.L., Shen, T., Lou, J., Xing, L., Blachere, N.E., Srivastava, P.K. and Rubin, B.Y. (1994) The endoplasmic reticular heat shock protein gp96 is transcriptionally upregulated in interferon-treated cells. *J. Exp. Med.* 180:1565–1569.

Angelidis, C.E., Lazaridis, I. and Pagoulatos, G.N. (1988) Specific inhibition of simian virus 40 protein synthesis by heat and arsenite treatment. *Eur. J. Biochem.* 172:27–34.

Bartz, S.R., Pauza, C.D., Ivanyi, J., Jindal, S., Welch, W.J. and Malkovsky, M. (1994) An Hsp60 related protein is associated with purified HIV and SIV. *J. Med. Primatol.* 23:151–154.

Bennett, I.L. and Nicastri, A. (1960) Fever as a mechanism of resistance. *Bacteriol. Rev.* 24:16–34.

Carmichael, L.E., Barnes, F.D. and Percy, D.H. (1969) Temperature as a factor in resistance of young puppies to canine herpesvirus. *J. Infect. Dis.* 120:669–678.

Carvalho, M.G.C. and Forunier, M.V. (1991) Effect of heat shock on gene expression of *Aedes albopictus* cells infected with Mayaro virus. *Res. Virol.* 142:25–31.

Casjens, S. and Hendrix, R. (1988) Control mechanisms in dsDNA bacteriophage assembly. *In:* R. Calendar (ed.): *The Bacteriophages,* Vol. 1, Plenum Press, New York, pp 15–91.

Chang, C.C., Konno, S. and Wu, J.M. (1991) Enhanced expression of heat shock protein and mRNA synthesis by type I interferon in human HL-60 leukemic cells. *Biochem. Int.* 24: 369–377.

Chelbi-Alix, M.K. and Sripati, C.E. (1994) Ability of insulin and dsRNA to induce interferon system and hsp70 in fibroblast and epithelial cells in relation to their effects on cell growth. *Exp. Cell Res.* 213:383–390.

Cheung, R.K. and Dosch, H. (1993) The growth transformation of human B cells involves superinduction of hsp70 and hsp90. *Virology* 193:700–708.

Chouchane, L., Bowers, F.S., Sawasdikosol, S., Simpson, R.M. and Kindt, T.J. (1994) Heat-shock proteins expressed on the surface of human T cell leukemia virus type I-infected cell lines induce autoantibodies in rabbits. *J. Infect. Dis.* 169:253–259.

Chousterman, S., Chelbi-Alix, M.K. and Thang, M.N. (1987) 2′, 5′-oligoadenylate synthetase expression is induced in response to heat shock. *J. Biol. Chem.* 262:4806–4811.

Collins, P.L. and Hightower, L. (1982) Newcastle Disease Virus stimulates the cellular accumulation of stress (heat shock) proteins. *J. Virol.* 44:703–707.

Cullen, B.R. and Heitman, J. (1994) Chaperoning a pathogen. *Nature* 372:319–320.

De Marco, A. and Santoro, M.G. (1993) Antiviral effect of short hyperthermic treatment at specific stages of vesicular stomatitis virus replication cycle. *J. Gen. Virol.* 74:1685–1690.

Di Cesare, S., Poccia, F., Mastino, A. and Colizzi, V. (1992) Surface expressed heat-shock proteins by stressed or human immunodeficiency virus (HIV)-infected lymphoid cells represent the target for antibody-dependent cellular cytotoxicity. *Immunol.* 76:341–343.

D'Onofrio, C., Puglianiello, A., Amici, C., Faraoni, I, Lanzilli, G. and Bonmassar, E. (1995) HSP70 production and inhibition of cell proliferation in MOLT-4 T cell after cell-to-cell transmission of HTLV-1: effect of PGA$_1$. *Leukemia Res.* 19:345–356.

Dubois, M.F., Mezger, V., Morange, M., Ferrieux, C., Lebon, P. and Bensaude, O. (1988) Regulation of the heat shock response by interferon in mouse L cells. *J. Cell. Physiol.* 137:102–109.

Earl, P.L., Moss, B. and Doms, R.W. (1991) Folding, interaction with GRP78-BiP, assembly, and transport of the human immunodeficiency virus type 1 envelope protein. *J. Virol.* 65:2047–2055.

Fincato, G., Polentarutti, N., Sica, A., Mantovani, A. and Colotta, F. (1991) Expression of a heat-inducible gene of the hsp70 family in human myelomonocytic cells: regulation by bacterial products and cytokines. *Blood* 77:579–586.

Franke, E.K., Yuan, H.E. and Luban, J. (1994) Specific incorporation of cyclophilin A into HIV-1 virions. *Nature* 372:359–362.

Furlini, G., Vignoli, M., Re, M.C., Gibellini, D., Ramazzotti, E., Zauli, G. and La Placa, M. (1994) Human immunodeficiency virus type 1 interaction with the membrane of CD4$^+$ cells induces the synthesis and nuclear translocation of 70 K heat shock protein. *J. Gen. Virol.* 75:193–199.

Garry, R.F., Emin, T.U. and Bose, H.R. (1983) Induction of stress proteins in Sindbis virus- and vesicular stomatitis virus-infected cells. *Virology* 129:319–332.

Geelen, J.L., Minnaar, R.P., Boom, R., Van der Noordaa, J. and Goudsmit, J. (1988) Heat-shock induction of the human immunodeficiency virus long terminal repeat. *J. Gen. Virol.* 69:2913–2917.

Gething, M.J., McCammon, K. and Sambrook, J. (1986) Expression of wild-type and mutant forms of influenza hemagglutinin. The role of folding in intracellular transport. *Cell* 46:939–950.

Gething, M.J. and Sambrook, J. (1992) Protein folding in the cell. *Nature* 355:33–45.

Gordon, L.M., Jensen, F.C., Curtain, C.C., Mobley, P.M. and Aloia, R.C. (1988) Thermotropic lipid phase separation in the human immunodeficiency virus. *Biochim. Biophys. Acta* 943:331–342.

Greenspan, H.C. (1993) The role of reactive oxygen species, antioxidants and phytopharmaceuticals in human immunodeficiency virus activity. *Med. Hypotheses* 40:85–92.

Hammond, G. and Helenius, A. (1994) Folding of VSV G protein: sequential interaction with BiP and calnexin. *Science* 266:456–458.

Herman, P.P. and Yatvin, M.B. (1995) Effect of heat on viral protein production and budding in cultured mammalian cells. *Int. J. Hyperthermia* 10:627–641.

Hirayoshi, K., Kudo, H., Takechi, H., Nakai, A., Iwamatsu, A., Yamado, K.M. and Nagata K. (1991) HSP47: a tissue specific, transformation sensitive, collagen-binding heat shock protein of chicken embryo fibroblasts. *Mol. Cell. Biol.* 11:4036–4044.

Honoré, B., Rasmussen, H.H., Celis, A., Leffers, H., Madsen, P. and Celis, J. (1994) The molecular chaperones HSP28, GRP78, endoplasmin, and calnexin exhibit strikingly different levels in quiescent keratinocytes as compared to their proliferating normal and transformed counterparts: cDNA cloning and expression of calnexin. *Electrophoresis* 15:482–490.

Huber, S.A. (1992) Heat shock protein induction in adriamycin and picornavirus-infected cardiocytes. *Lab. Invest.* 67:218–224.

Jindal, S. and Young, R.A. (1992) Vaccinia virus infection induces a stress response that leads to association of Hsp70 with viral proteins. *J. Virol.* 66:5357–5362.

Khandjian, E.W. and Turler, H. (1983) Simian virus 40 and polyoma virus induce the synthesis of heat shock proteins in permissive cells. *Mol. Cell. Biol.* 3:1–8.

Kobayashi, K., Ohgitani, E., Tanaka, Y., Kita, M. and Imamishi, J. (1994) Herpes simplex virus-induced expression of 70 kDa heat shock protein (HSP70) requires early protein synthesis but not viral DNA replication. *Microbiol. Immunol.* 38:321–325.

Kretz-Remy, C. and Arrigo, A.P. (1994) The kinetics of HIV-1 long terminal repeat transcriptional activation resemble those of hsp70 promoter in heat-shock treated HeLa cells. *FEBS Lett.* 351:191–196.

La Thangue, N.B., Shriver, K., Dawson, C. and Chan, W.L. (1984) Herpes Simplex virus infection causes the accumulation of a heat shock protein. *EMBO J.* 3:267–277.

La Thangue, N.B. and Latchman, D.S. (1988) A cellular protein related to heat shock protein 90 accumulates during herpes simplex virus infection and is overexpressed in transformed cells. *Exp. Cell Res.* 178:169–179.

Laszlo, L., Tuckwell, J., Self, T., Lowe, J., Landon, M. Smith, S., Hawthorne, J.N. and Mayer, R.J. (1991) The latent membrane protein-1 in Epstein-Barr virus-transformed lymphoblastoid cells is found with ubiquitin-protein conjugates and heat-shock protein 70 in lysosomes oriented around the microtubule organizing centre. *J. Pathol.* 164:203–214.

Lindquist, S. and Craig, E.A. (1988) The heat-shock proteins. *Annu. Rev. Genet.* 22:631–677.

Lingappa, J.R., Martin, R.L., Wong, M.L., Ganem, D., Welch, W.J. and Lingappa, V.R. (1994) A eukaryotic cytosolic chaperonin is associated with a high molecular weight intermediate in the assembly of hepatitis B virus capsid, a multimeric particle. *J. Cell. Biol.* 125:99–111.

Lu, L., Shen, R., Zhou, S.Z., Wu, B., Kim, Y., Lin, Z.H., Ruscetti, S., Ralph, P. and Broxmeyer, H.E. (1991) Efficacy of recombinant human macrophage colony-stimulating factor in combination with whole-body hyperthermia in the treatment of mice infected with the polycythemia-inducing strain of the Friend virus complex. *Exp. Hematol.* 19:804–809.

Luban, J., Bossolt, K.L., Franke, E.K., Kalpana, G.V. and Goff, S.P. (1993) Human immunodeficiency virus type 1 gag protein binds to cyclophilins A and B. *Cell* 73:1067–1078.

Macejak, D.G. and Luftig, R.B. (1991) Association of hsp70 with the adenovirus type 5 fiber protein in infected HEp-2 cells. *Virology* 180:120–125.

Macejak, D.G. and Sarnow, P. (1992) Association of heat shock protein 70 with enterovirus capsid precursor P1 in infected human cells. *J. Virol.* 66:1520–1527.

Marquardt, T. and Helenius, A. (1992) Misfolding and aggregation of newly synthesized proteins in the endoplasmic reticulum. *J. Cell Biol.* 117:505–513.

May, E., Breugnot, C., Duthu, A. and May, P. (1991) Immunological evidence for the association between simian virus 40 115-kDa super T antigen and hsp70 proteins in rat, monkey and human cells. *Virology* 180:285–293.

Michel, M.R., Favre, D., Studer, E., Arrigo, A. and Kempf, C. (1994) Modulation of thermoprotection and translational thermotolerance induced by Semliki Forest virus capsid protein. *Eur. J. Biochem.* 223:791–797.

Mulvey, M. and Brown, D.T. (1995) Involvement of the molecular chaperone BiP in maturation of Sindbis virus envelope glycoproteins. *J. Virol.* 69:1621–1627.

Muñoz, A., Alonso, M.A. and Carrasco, L. (1984) Synthesis of heat shock proteins in HeLa cells: inhibition by virus infection. *Virology* 137:150–159.

Ng, D.T.W., Randall, R.E. and Lamb, R.A. (1989) Intracellular maturation and transport of the SV5 type II glycoprotein hemagglutinin-neuraminidase: specific and transient association with GRP78/BiP in the endoplasmic reticulum and extensive internalization from the cell surface. *J. Cell. Biol.* 109:3273–3289.

Notarianni, E.L. and Preston, C.M. (1982) Activation of cellular stress protein genes by herpes simplex virus temperature-sensitive mutants which overproduce early polypeptides. *Virology* 123:113–122.

Nuss, D.L. and Koch, G. (1976) Differential inhibition of vesicular stomatitis virus polypeptide synthesis by hypertonic initiation block. *J. Virol.* 17:283–286.

Oglesbee, M. and Krakowka, S. (1993) Cellular stress response induces selective intranuclear trafficking and accumulation of morbillivirus major core protein. *Lab. Invest.* 68:109–117.

Panasiak, W., Oraczewska, A. and Luczak, M. (1990) Influence of hyperthermia on experimental viral infections *in vitro*. *In:* H.I. Bicher, J.R. McLaren and G.M. Pigliucci (eds): *Consensus on Hyperthermia,* Plenum Press, New York, pp 471–475.

Peluso, R.W., Lamb, R.A. and Choppin, P.W. (1977) Polypeptide synthesis in simian virus 5-infected cells. *J. Virol.* 23:177–187.

Peluso, R.W., Lamb, R.A. and Choppin, P.W. (1978) Infection with paramyxoviruses stimulates synthesis of cellular polypeptides that are also stimulated in cells transformed by Rous sarcoma virus or deprived of glucose. *Proc. Natl. Acad. Sci. USA* 75:6120–6124.

Pennypacker, C., Perelson, A.S., Nys, N., Nelson, G. and Sessler, D.J. (1995) Localized or systemic *in vivo* heat inactivation of human immunodeficiency virus (HIV): a mathematical analysis. *J. Acquir. Immune Defic. Syndr. Hum. Retrovirol.* 8:321–329.

Phillips, B., Abravaya, K. and Morimoto, R.I. (1991) Analysis of the specificity and mechanism of transcriptional activation of the human hsp70 gene during infection by DNA viruses. *J. Virol.* 65:5680–5692.

Pica, F., De Marco, A., De Cesare, F. and Santoro, M.G. (1993) Inhibition of vesicular stomatitis virus replication by Δ^{12}-prostaglandin J_2 is regulated at two separate levels and is associated with induction of stress protein synthesis. *Antiviral Res.* 20:193–208.

Pica, R., Rossi, A., Santirocco, N., Garaci, E. and Santoro, M.G. (1996) Effect of combined αIFN and prostaglandin A_1 treatment on vesicular stomatitis virus replication and heat shock protein synthesis in epithelial cells. *Antiviral Res.* 29:187–198.

Roberts, N.J. (1979) Temperature and host defense. *Microbiol. Rev.* 43:241–259.

Rossi, A. and Santoro, M.G. (1994) Induction of a 32-kDa stress protein by prostaglandin A_1 in cultured murine cells. *Ann. N.Y. Acad. Sci.* 744:326–329.

Roux, L. (1990) Selective and transient association of Sendai virus HN glycoprotein with BiP. *Virology* 175:161–166.

Rueckert, R.R. (1990) Picornaviridae and their replication. *In:* B.N. Fields, D.M. Knipe, R.M. Chanock, J.L. Melnich, B. Roizman and R.E. Shope (eds): *Virology,* Second Edition. Raven Press, New York, pp 507–548.

Russel, J., Stow, E.C., Stow, N.D. and Preston, C.M. (1987) Abnormal forms of the herpes simplex virus immediate early polypeptide Vmw 175 induce the cellular stress response. *J. Gen. Virol.* 68:2397–2406.

Sagara, J. and Kawai, A. (1992) Identification of heat shock protein 70 in the rabies virion. *Virology* 190:845–848.

Sainis, I., Angelidis, C., Pagoulatos, G. and Lazaridis, I. (1994) The hsc70 gene which is slightly induced by heat is the main virus inducible member of the hsp70 gene family. *FEBS Lett.* 335:282–286.

Santomenna, L.D. and Colberg-Poley, A.M. (1990) Induction of cellular hsp70 expression by human cytomegalovirus. *J. Virol.* 64:2033–2040.

Santoro, M.G., Benedetto, A., Carruba, G., Garaci, E. and Jaffe, B.M. (1980) Prostaglandin A compounds as antiviral agents. *Science* 209:1032–1034.

Santoro, M.G., Jaffe, B.M. and Esteban, M. (1983) Prostaglandin A inhibits the replication of vesicular stomatitis virus: effect on virus glycoprotein. *J. Gen. Virol.* 64:2797–2081.

Santoro, M.G., Garaci, E. and Amici, C. (1989 a) Prostaglandins with antiproliferative activity induce the synthesis of a heat shock protein in human cells. *Proc. Natl. Acad. Sci. USA* 86: 8407–8411.

Santoro, M.G., Amici, C., Elia, G., Benedetto, A. and Garaci, E. (1989 b) Inhibition of virus protein glycosylation as the mechanism of the antiviral action of prostaglandin A_1 in Sendai virus-infected cells. *J. Gen. Virol.* 70:789–800.

Santoro, M.G., Garaci, E. and Amici, C. (1990) Induction of hsp70 by prostaglandins. *In:* M.J. Schlesinger, E. Garaci and M.G. Santoro (eds): *Stress Proteins: Induction and Function,* Springer-Verlag, Berlin, pp 27–44.

Santoro, M.G. (1994) Heat shock proteins and virus replication: hsp70s as mediators of the antiviral effects of prostaglandins. *Experientia* 50:1039–1047.

Sarnow, P. (1989) Translation of glucose-regulated protein 78/immunoglobulin heavy-chain binding protein mRNA is increased in poliovirus-infected cells at a time when cap-dependent translation of cellular mRNAs is inhibited. *Proc. Natl. Acad. Sci. USA* 86: 5795–5799.

Sawai, E. and Butel. J.S. (1989) Association of a cellular heat shock protein with simian virus 40 large T antigen in transformed cells. *J. Virol.* 63:3961–3973.

Sawtell, N.M. and Thompson, R.L. (1992) Rapid *in vivo* reactivation of herpes simplex virus in latently infected murine ganglionic neurons after transient hyperthermia. *J. Virol.* 66: 2150–2156.

Schlesinger, M.J., Ryan, C., Sadis, S. and Hightower, L.E. (1991) *In vitro* inhibition of nascent polypeptide formation by HSP70 proteins. *In:* B. Maresca and S. Lindquist (eds): *Heat Shock Proteins,* Springer-Verlag, Berlin, pp 111–117.

Sedger, L. and Ruby, J. (1994) Heat shock response to vaccinia virus infection. *J. Virol.* 68: 4685–4689.

Simon, M.C., Kitchener, K., Kao, H., Hichey, E., Weber, L., Voellmy, R., Heintz, N. and Nevis, J.R. (1987) Selective induction of human heat shock gene transcription by the adenovirus E1A gene products, including the 12S E1A product. *Mol. Cell. Biol.* 7: 2884–2890.

Stanley, S.K., Bressler, P.B., Poli, G. and Fauci, A.S. (1990) Heat shock induction of HIV production from chronically infected promonocytic and T cell lines. *J. Immunol.* 145: 1120–1126.

Teisner, B. and Haahr, S. (1974) Poikilothermia and susceptibility of suckling mice to coxsackie B1 virus. *Nature* 274: 568–569.

Thali, M., Bukovsky, A., Kondo, E., Rosenwirth, B., Walsh, C.T., Sodroski, J. and Göttlinger, G. (1994) Functional association of cyclophilin A with HIV-1 virions. *Nature* 372: 363–365.

Tyrrell, D., Barrow, I. and Arthus, J. (1989) Local hyperthermia benefits natural and experimental common colds. *Br. Med. J.* 298: 1280–1283.

Walter, G., Carbone, A. and Welch, W.J. (1987) Medium tumor antigen of polyomavirus transformation-defective mutant NG59 is associated with 73-kilodalton heat-shock protein. *J. Virol.* 61: 405–410.

Watowich, S.S., Morimoto, R.I. and Lamb, R.A. (1991) Flux of the paramyxovirus hemagglutinin-neuraminidase glycoprotein through the endoplasmic reticulum activates transcription of the GRP78-BiP gene. *J. Virol.* 65: 3590–3597.

White, E., Spector, D. and Welch, W. (1988) Differential distribution of the adenovirus E1A proteins and colocalization of E1A with the 70-kilodalton cellular heat shock protein in infected cells. *J. Virol.* 62: 4153–4166.

Whitelaw, M.L., Hutchinson, K. and Perdew, G.H. (1991) A 50-kDa cytosolic protein complexed with the 90 kDa heat shock protein (hsp90) is the same protein complexed with pp60[v-src] hsp90 in cells transformed by the Rous sarcoma virus. *J. Biol. Chem.* 266: 16436–16440.

Williams, G.T., McClanahan, T.K. and Morimoto, R.I. (1989) E1A transactivation of the human HSP70 promoter is mediated through the basal transcriptional complex. *Mol. Cell. Biol.* 9: 2574–2587.

Wong, G.H.W., McHugh, T., Weber, R. and Goeddel, D.V. (1991) Tumor necrosis factor α selectively sensitizes human immunodeficiency virus-infected cells to heat and radiation. *Proc. Natl. Acad. Sci. USA* 88: 4372–4376.

Wooden, S.K., Li, L., Navarro, D., Qadri, I., Pereira, L. and Lee, A.S. (1991) Transactivation of the *grp78* promoter by malfolded proteins, glycosylation block, and calcium ionophore is mediated through a proximal region containing a CCAAT motif which interacts with CTF/NF-I. *Mol. Cell. Biol.* 11: 5612–5623.

Xu, Y. and Lindquist, S. (1993) Heat-shock protein hsp90 governs the activity of pp60[v-src] kinase. *Proc. Natl. Acad. Sci. USA* 90: 7074–7078.

Yang, J. and DeFranco, D.B. (1994) Differential roles of heat shock protein 70 in the *in vitro* nuclear import of glucocorticoid receptor and simian virus 40 large tumor antigen. *Mol. Cell. Biol.* 14: 5088–5098.

Yatvin, M.B., Stowell, M.H. and Steinhart, C.R. (1993) Shedding light on the use of heat to treat HIV infections. *Oncology* 50: 380–389.

Yura, Y., Terashima, K., Iga, H., Kondo, Y, Yanagawa, T., Yoshida, H., Hiyashi, Y. and Sato, M. (1987) Macromolecular synthesis at the early stage of herpes simplex type 2 (HSV-2) latency in a human neuroblastoma cell line IMR-32: repression of late viral polypeptide synthesis and accumulation of cellular heat-shock proteins. *Arch. Virol.* 96: 17–28.

Zerbini, M., Musiani, M. and La Placa, M. (1985) Effect of heat shock on Epstein-Barr virus and cytomegalovirus expression. *J. Gen. Virol.* 66: 633–636.

Stress-Inducible Cellular Responses
ed. by U. Feige, R.I. Morimoto, I. Yahara and B. Polla
© 1996 Birkhäuser Verlag Basel/Switzerland

Infection, autoimmunity and autoimmune disease

U. Feige[1] and W. van Eden[2]

[1] Department of Pharmacology, AMGEN Center, Thousand Oaks, CA, USA
[2] Department of Immunology, Faculty of Veterinary Medicine, University of Utrecht,
NL-3508 TD Utrecht, The Netherlands

Summary. Studies of the immune response of mammals to infectious agents have revealed that members of the hsp60 and hsp70 family are highly immunodominant. Given their high conservation during evolution this was surprising, because of the apparent risk of triggering of autoimmunity and autoimmune disease during the defense of a mammal against infection. However, detailed studies of the immune responses to HSP in models of autoimmune diseases in animals resulted in a change of the view that autoimmunity necessarily leads to autoimmune disease. It has been found that modulation of autoimmunity to HSP is one way to prevent autoimmune disease. At least in some cases even treatment of autoimmune diseases by immunization with heat shock protein appears feasible. This was shown in adjuvant arthritis in Lewis rats and insulin dependent diabetes in NOD mice. Hsp60 and hsp70 are ubiquitous proteins. Their involvement in regulatory loops of autoimmunity may serve as basis for the development of strategies to prevent and/or treat autoimmune diseases even without knowledge of the causative (auto-)antigen.

Introduction

The immune system of mammals has evolved while being in continuous contact with bacteria. Likewise, bacteria infecting mammals have adapted to perfect their survival. During infections both mammalian and bacterial cells are under stress and respond accordingly. Quality and quantity of a stress response in cells are dependent on type, severity and duration of stress. In general, a mild or medium stress results in cells being capable to better deal with a stress situation thereafter. In the following, we will discuss the role of heat shock proteins (HSP) as target molecules for the immune system. Bacterial HSP are immunodominant targets for the immune system of an infected mammal. In addition, HSP appear to play a central role at the crossroads of autoimmunity and autoimmune disease. The latter has opened new avenues for prevention and treatment of autoimmune diseases at least in animal models.

Infection

It had been known for quite some time that bacteria are immunogenic and induce antibody responses crossreactive with other bacteria. Using such (anti-)sera, antibody responses to one of the bacterial antigens in particular were found in all bacteria. This antigen was named "common antigen" and later it was identified as hsp60 family member present in bacteria (groEL in *E. coli* or hsp65 in mycobacteria) (Thole et al., 1988; Shinnick et al., 1988).

Heat shock protein synthesis is upregulated in situations stressful for cells. The degree of upregulation can be substantial. For example, when bacteria are killed by heat the content of hsp60 (groEL) can rise from 1.5% to 15% of bacterial dry weight (Neidhardt et al., 1984). Upregulation of HSP synthesis in bacteria during infection of mammals might add to the overall immunogenicity of HSP to the host's immune system. Besides hsp60 bacterial hsp70 (dnaK in *E. coli*) is highly immunogenic (see the chapter by Suzue and Young, this volume).

Homologies of 50% with an additional 20% of conservative amino acid substitutions between HSP family members of bacteria and mammals are the rule. Therefore, it might not be too surprising that immune responses to HSP induced by bacteria crossreact with mammalian HSP (see below). It appears that the benefit for the mammal's immune system to respond to bacterial HSP is large compared to the potential risk of establishing autoimmune disease by means of crossreactivity with self-antigens (Yang and Feige, 1991, 1992; Feige, 1996). However, using ubiquitous proteins such as foreign and self hsp60 as targets might even be an advantageous strategy for the immune system (see paragraph on network in autoimmunity below).

Immune responses to hsp60 in "normal" and infected individuals

Historically, antibody responses were looked at before T cell responses were investigated. However, bacterial hsp60 induces strong cellular immune responses as well. It is not much of a surprise that immune responses to hsp60 are found after a "history" of bacterial infections in mammals.

A search for the antigens detected by γ,δ-T cells began shortly after it had been shown that γ,δ-T cells exist as a second set of T cells besides α,β-T cells. Interestingly, hsp60 were the first antigens recognized to be detected by γ,δ-T cells (Born et al., 1990, 1990a). In fact, a substantial number of γ,δ-T cells recognize mycobacterial hsp60. Research in this area revealed that γ,δ-T cells which recognize hsp60 are not only present after infection, but also can be detected neonatally (O'Brien et al., 1992). It is noteworthy that the response of hsp60 specific γ,δ-T cells is to one epitope in the mycobacterial hsp65, namely 181–187 (Fu et al., 1994). From the fact that cellular immune responses to hsp60 can be detected neonatally (Iwasaki et al., 1991; O'Brien et al., 1992) it may be concluded that autoimmunity to hsp60 exists before the immune system is exposed to foreign (bacterial) HSP. Therefore, the view that there is immunity to bacterial HSP first and that autoimmunity to self-hsp60 develops thereafter might reflect just one of the possibilities when discussing antigenic mimicry between foreign and self HSP.

Antigenic mimicry

As mentioned above, HSP have been highly conserved during evolution. So it is hardly surprising that cross-reactivities can be demonstrated between bacterial and human hsp60. For example, it has been shown that T cells specific for mycobacterial hsp65 can recognize heat-stressed macrophages as well (see, for example, Koga et al., 1989). This is an example of antigenic mimicry. But is this really autoimmunity? It has the basic features of an immune response to self antigen(s), but was induced by bacterial antigens. Where would such an reaction lead if it were to happen during an immune response to bacterial antigens *in vivo*? Analyzing protein or DNA data bases using sequence homology algorythms, identical or homolog linear epitopes can be easily found on mammalian and bacterial proteins (Jones et al., 1994). Whether these homologies are seen by the immune system depends on a multitude of factors among which are (i) the accessibility of the protein, (ii) the processing of the protein during the process of antigen presentation, (iii) the efficiency of this process to produce the crossreactive epitope, (iv) binding affinity of the processed epitope to class II antigen, and (v) other processed peptides around which compete for binding to class II antigen.

From the point of view of HSP such as hsp60 and hsp70 discussed in this article, one might want to distinguish cross-reactivities of epitopes between foreign HSP and self-HSP, between foreign antigens and self-HSP and between foreign HSP and self-non-HSP antigens.

Several possibilities can be postulated as leading to autoimmune disease: (i) A bacterial or viral infection induces an immune response (part of) which is directed against an epitope in the target organ. (ii) Autoimmunity naturally exists to an autoantigen and, in the course of a bacterial infection, antigenic peptides are produced which are cross-reactive with self epitopes and "boost" the autoimmunity present. (iii) Autoimmunity is present and would potentially lead to autoimmune disease, however, the strength of the immune reaction is (just) below treshold. All it needs to induce autoimmune disease is production of stimulatory factors such as cytokines to move the autoimmunity above a certain treshold and autoimmune disease precipitates. The latter would allow for "unspecific" triggering of autoimmune diseases. (iv) Autoimmune disease is not occurring because the immune system is tolerized or tolerant against the autoantigen(s). Tolerance might be broken by an ongoing infection resulting in autoimmune disease.

Involvement of HSP in autoimmune diseases

Adjuvant arthritis in Lewis rats – when involvement of HSP
in autoimmune diseases was detected

Adjuvant arthritis (AdA) in Lewis rats is induced by immunization with a suspension of heat-killed mycobacteria in oil. Involvement of HSP in autoimmune disease drew the attention of the scientific community when it was shown that rat T cell clones which responded to mycobacterial hsp65 peptide 180–188 could induce arthritis (clone A2b) and prevent or treat adjuvant arthritis (clone A2c) in Lewis rats (van Eden et al., 1988). It is noteworthy that both T cell clones have identical epitope specificities and identical T cell receptors (Broeren et al., 1994) indicating that mere epitope specificity is not necessarily bad or good. The biological activities of T cell clones A2b and A2c and identification of their antigen prompted research around the question: does knowledge of an antigen involved in the process of an autoimmune disease enable to define vaccinations and/or therapies against the autoimmune disease? Indeed, this was found to be the case. Preimmunization with mycobacterial hsp65 made Lewis rats resistant to AdA (van Eden et al., 1988). Also pretreatment of Lewis rats with the mycobacterial hsp peptide 180–188, the peptide the arthritogenic T cell clone A2b responds to *in vitro*, led to resistance against adjuvant arthritis (Yang et al., 1990, 1992). Even treatment of AdA with peptide 180–188 after appearance of clinical symptoms is effective (Feige and Gasser, 1994). Furthermore, pretreatment of rats with mycobacterial hsp65 peptide 256–270 resulted in an AdA resistant status (Anderton et al., 1995). Interestingly, epitope 256–270 is homolog on rat hsp60 (see below).

It is important to note that it has been known for a long time that rats could be made resistant to AdA by pretreatment with mycobacteria orally or by subarthritogenic doses of mycobacteria (see Feige and Cohen, 1991). AdA can also be prevented by infecting rats with vaccinia virus which expresses mycobacterial hsp65 (Hogervorst et al., 1991; Lopez-Gurrero et al., 1993, 1994). By analyzing the fine specificity of the cellular immune response to mycobacterial hsp65 in more detail, researchers have found responses to nine peptide epitopes. The pattern of the response was different after immunization with mycobacteria and hsp65, which resulted in AdA or protection from AdA, respectively (Anderton et al., 1994).

What is the autoantigen in AdA? It has been shown that T cell clones A2b and A2c respond *in vitro* to cartilage extracts containing proteoglycan as antigens (Cohen et al., 1985). It appears that the identification of the autoantigen is more difficult than anticipated considering the availability of T cell clones as tools to identify this antigen. Clearly, antibody responses have been found in rats with AdA to a 65 kDa protein which is not an HSP

(Feige et al., 1994; Mollenhauer and Feige, unpublished). Some recent results indicate that the autoantigen is a constitutive chondrocyte protein (CH65) which might belong to the family of cytokeratins. Responses of spleen cells of rats with AdA *in vitro* to CH65 are comparable in magnitude and kinetics to those to hsp65 (U. Feige, unpublished). Interestingly, pretreatment of rats with a preparation of CH65 purified from chicken sterna retarded the onset of adjuvant arthritis comparable to pretreatment with mycobacterial hsp65 (Feige et al., 1994).

Other animal models of arthritis

A first analysis of the fact that a protein as ubiquitous as (myco-)bacterial hsp65 is involved in AdA and can protect rats from this disease led to experiments in which mycobacterial hsp65 pretreatment was tested in other arthritic autoimmune diseases in the same strain of rats. It was found that, indeed, other autoimmune diseases in Lewis rats could be prevented. One example is streptococcal cell wall induced arthritis. Interestingly, avridine induced arthritis was modulated by pretreatment with mycobacterial hsp65 as well. Avridine (CP-20961) is an adjuvant oil inducing a disease in Lewis rats which strongly resembles AdA. The mechanism of disease induction is still unclear in this model and so the protective effect of hsp65 pretreatment in the avidrine model has been a suprise to many (Billingham, 1990). However, the potential of mycobacterial hsp65 to modulate autoimmune disease appears to be even broader. Pristane induced arthritis in mice is another arthritic disease where hsp65 pretreatment was shown to have beneficial effects. It is remarkable that DBA/1 mice which were kept under SPF conditions and, as a consequence, were lacking "natural" immunity to hsp65 did not develop pristane induced arthritis (Thompson and Elson, 1993). Mice which were transferred to conventional housing quickly developed immunity to groEL, hsp65 and autologous hsp60 and also became susceptible to pristane induced arthritis. Another example where the immune system exposed to oil mounts a response to hsp was found when immunization of Balb/c mice with IFA in the footpad resulted in a T cell response to self hsp60 in the draining lymph node seven days later (Anderton et al., 1993).

However, it is important to note that the mere presence of (auto-)immunity to hsp60 does not necessarily indicate involvement of hsp60 in the disease process. When antibody responses to hsp60 were looked at in Lewis rats with arthritis induced by three different arthritogens, antibodies to hsp60 were found only in rats which were housed under conventional conditions, but not in rats housed under germ-free conditions, although incidence and severity of arthritis were identical in both groups (Bjoerk et al., 1994). The authors concluded that the expression of arthritis is a prerequisite for antibody formation to hsp60 and not *vice versa*.

Rheumatoid arthritis (RA)

Immune responses to a panel of antigens including hsp60 have been studied in blood and in synovial fluid (Kogure et al., 1994; see Life et al., 1993, for review). T cell responses were enhanced in synovial fluid (SF) for hsp60 but not for PPD or other antigens. The authors conclude an accumulation of hsp60 specific cells in SF. In this context it is worth mentioning that *in vitro* co-cultures of cartilage explants and mononuclear cells obtained from RA patients mycobacterial hsp65 induced enhanced loss of proteoglycan even in the absence of induction of T cell proliferation (Wilbrink et al., 1993). Although antibody responses to mycobacterial hsp65 have been observed in normal individuals, tuberculosis infected individuals and RA patients, it appears that the antibody response to mycobacterial antigens is skewed towards hsp65 in RA (Ishimoto et al., 1993).

Juvenile rheumatoid arthritis (JRA)

In children with JRA, proliferative responses have been obtained in lymphocytes taken from both the peripheral blood and the synovial compartment. Most significant and reliable responses were seen by stimulating the cells with self-antigen human hsp60 (DeGraeff-Meeder et al., 1991, 1993, 1995; Life et al., 1993). Since most of the patients, however, did respond to mycobacterial hsp60 as well, it seems that in this case, conserved epitopes, equally present in both the human and mycobacterial hsp65, are recognized by patient T cells. From comparison of patients of distinct clinical subgroups, it became evident that responders had oligo-articular (OA) forms of JRA, whereas non-responders had polyarticular or systemic JRA. In other words, those with a remitting from of disease responded, whereas those having a non-remitting form did not (DeGraeff-Meeder et al., 1995). Longitudinal studies indicated that (temporary) remission of OA-JRA coincided with disappearance of responses, and that a clinical relapse coincided with reappearance of responses. Furthermore, it was found that *in vitro* priming of non-responder cells led to positive secondary responses only in case of OA-JRA (remission phase) and not in the case of other clinical subgroups of JRA. Altogether, the data obtained in JRA patients have shown that responses to human hsp60 as a self-antigen do occur and that they are associated with relatively benign forms of arthritis. The presence of responses during the active phase of disease, preceding remission, suggests (in line with the observations made in the AA model) that responses to self-hsp60 may positively contribute to mechanisms leading to disease remission. If so, possibilities for immunological intervention in JRA, and possibly also RA, may be found in strategies aimed at manipulating peripheral tolerance through vaccination with hsp60, or peptides containing defined epitopes. For these purposes hsp60 may well serve as a useful ancillary autoantigen.

Another crossreactivity, namely the susceptibility sequence for RA QKRAA in the third hypervariable region of HLA DRB10401 (formerly HLA Dw4) which is shared with *E. coli* dnaJ was reported. Interestingly, both cellular and humoral immune responses to this epitope have been described in JRA patients (Albani et al., 1994).

Diabetes

Diabetes mellitus in NOD mice is a spontaneous autoimmune disease which becomes overt at 4 to 6 month of age. It is important to note that besides hsp60 several other autoantigens have been reported to be involved in human and mouse IDDM (for review see Gelber et al., 1994). Here, we discuss modulation of disease by autologous and mycobacterial hsp65. A single immunization at 4 weeks of age with an "immunogenic" regimen (hsp65 in incomplete Freund's adjuvant ip) induces a period of transient diabetes after which mice are protected from the spontaneous diabetes which occurs in non hsp60-treated NOD mice (Elias et al., 1990, 1991). It is noteworthy that a single immunization at 4 weeks of age with a "non-immunogenic" regimen (hsp65 in saline ip) renders NOD mice resistant to the spontaneous diabetes (Elias et al., 1991). Importantly, similar effects can be obtained with peptide p277 which is sequence 436 to 460 of mouse hsp60 (Elias et al., 1991). Intraperitoneal treatment of NOD mice with insulitis with 5 μg of p277 in oil prevents diabetes (Elias and Cohen, 1995). The dose response to p277 manifests a threshold effect. Higher doses of p277 are not more effective. Even in NOD mice with overt diabetes at 17 weeks of age (mice not treated will die within another 12 weeks), a single treatment with p277 can stablize or reverse the diabetes (Elias and Cohen, 1994, 1995). It is interesting to relate these results to a study exploring BCG therapy in patients with IDDM (Shehadeh et al., 1994). 17 of 27 newly diagnosed IDDM patients went into remission after treatment with BCG.

In streptozocin-induced diabetes in mice T cell immunity to p277 of mouse hsp60 is found. In fact, existing immunity to hsp60 accelerates the diabetes in these mice (Elias et al., 1994).

Atherosclerosis

During the course of studies undertaken to investigate the role of auto-immunity in the development of atherosclerosis, Xu and Wick observed that immunization of rabbits with complete Freund's adjuvant (CFA) alone resulted in the development of atherosclerotic lesions in the arterial tree (reviewed by Wick et al., 1995). It is interesting to note that only immunization with mycobacterial hsp65 containing adjuvant results in the induction of atherosclerotic lesions. Remarkably, a high cholesterol diet resulted in more severe atherosclerosis in CFA immunized rabbits. However, the diet itself did not induce atherosclerosis. In atherosclerotic lesions high expression of hsp60 and hsp70 as well as T cells reactive to hsp were found. The possibility that immunity to hsp65 is involved in the process of atherosclerosis suggest that vaccination based on hsp60 should be considered as an experimental approach to prevent or treat atherosclerosis in a similar way as other autoimmune diseases (see above) (Wick et al., 1995; Feige and Cohen, 1991).

Immunological mechanisms of prevention of autoimmune disease

The immunological mechanisms which lead to autoimmune disease or to an autoimmune disease resistant state are not yet completely understood. However, quite a bit is known and will be briefly summarized: The AdA resistant state following vaccination with peptide 180–188 in Lewis rats is accompanied by an enhanced immune response to mycobacterial antigens as measured by delayed type hypersensitivity reactions in the ears of rats or proliferation of T cells obtained from rats *in vitro* (Yang and Feige, 1991, 1992; Yang et al., 1992). Studies by Anderton and colleagues have shown that immunization with mycobacterial hsp60 led to responses to a number of distinct epitopes in the molecule and no exclusive dominance of 180–188 was apparent (see above). All epitopes involved were mapped in detail using epitope-specific T cell lines. Upon testing these T cell lines for responses to peptides based on homologous rat hsp60 sequences, only one T cell line also responded to the homologous rat hsp60 peptide. As was expected from the cross-recognition between the mycobacterial and rat sequence, this particular epitope turned out to be strongly conserved (Anderton et al., 1994). The various peptide epitopes were tested for their capacity to protect Lewis rats against AdA. Remarkably, only peptides containing the conserved, rat hsp60 cross-reactive, T cell epitope were protective. These findings suggested that mycobacterial hsp60 was dependent on activity of T cell epitopes in the molecule, which induced responses that cross-reacted with rat self-hsp60.

More recent experiments have shown that non-mycobacterial induced (CP20961 or avridine) arthritis is also effectively prevented by preimmunization with the conserved peptide (Anderton et al., 1995). Teleologically, it seems attractive to propose that the cross-recognition of rat self-hsp60 is the basis of the protection observed. As discussed above, recognition of self-hsp60 may well be subjected to the regulatory pathways of peripheral tolerance. Temporarily compromising such tolerance by forcing the immune system to respond to self-hsp60 could lead to a counteractive regulatory (suppressive) response targeted to the expressed self-hsp60 molecule. Such suppressive regulation may well contribute to the control of local inflammatory processes. Some of the findings made in children suffering from JCA seem to be compatible with these possibilities.

As discussed before AdA can be passively transferred from diseased rats to syngeneic disease free animals by a T cell clone recognizing the non-conserved 180–188 sequence in mycobacterial hsp60. This T cell clone also responded to crude cartilage antigen (Cohen et al., 1985). Recently, it became apparent that induction of tolerance to peptide 180–188 by administering it intranasally or subcutaneously in PBS prior to induction of AdA, led to specific downregulation of responses to peptide 180–188 *in vitro* (Prakken et al., in preparation). Furthermore, this induction of tolerance also resulted in a delay of onset and a reduction of severity of the arthritis. Taking into account that besides the peptide 180–188 epitope additional arthritogenic epitopes may be operative, these findings suggest that mechanisms of dominant bystander suppression are responsible for the arthritis suppressive effect.

Streptococcal cell wall induced arthritis was prevented in Lewis rats by mycobacterial hsp65 applied in a tolerogenic regimen (van den Broek et al., 1989).

In pristane induced arthritis in mice cellular immune responses to hsp65 are lower in mice which did not become arthritic following the pristane injection. However, immune responses to hsp65 appear to be identical in magnitude in arthritic or hsp65 pretreated non-arthritic mice (Thompson et al., 1991). This indicates that immunity to hsp65 must be present in order for DBA/1 mice to be susceptible to arthritis *and* to be protected against arthritis (Thompson and Elson, 1993).

The conclusion from the studies with mycobacterial hsp65 in animal models of autoimmune disease is that tolerance to disease is not necessarily due to tolerance to (auto-)antigen(s) in the immunological sense. Often, a stronger and/or broader immune response is found in protected animals (Yang and Feige, 1991, 1992; Feige and Cohen, 1991; Anderton et al., 1994). However, there are examples where induction of tolerance to (epitopes of) mycobacterial hsp65 results in resistance against autoimmune disease (van den Broek, 1989; Prakken et al., in preparation).

Application to human disease

Immunomodulatory treatment with BCG was used in the past for patients with bladder cancer and malignant melanoma. It is interesting to note that in 0.1 to 1% of treated patients arthritis was observed as a side-effect (Goupille et al. 1994). The latter indicates that, similar to what is believed to be the mechanism leading to reactive arthritis (see Hermann, 1993; Probst et al., 1994), immune responses in defense of bacterial infections may result in arthritis. Applications of hsp70 and hsp60 as adjuvants for vaccines and cancer treatment are reviewed in the chapter by Suzue and Young (this volume).

Treatment implicating the activity of heat-shock proteins may, inadvertently, have made its way to the clinic already. Subreum® (OM-8980) administered orally has shown comparable activity to parenteral gold in clinical trials in RA (Vischer, 1988). It consists of bacterial *(E. coli)* antigens, among them highly immunogenic hsp60 and hsp70. Furthermore, oral administration of Subreum® in rats was found to induce hsp60 and hsp70 T cell responses in the animals. Therefore, it is possible that Subreum® is active, at least in part, by raising immunity to bacterial HSP (Vischer and van Eden, 1994).

Network of autoimmunity

Research in the field of autoimmune diseases and on the role of HSP in induction, prevention, and therapy of autoimmune diseases have put into question the Burnetian view that autoimmunity necessarily leads to autoimmune disease (Cohen, 1991, 1992, 1994; Feige and Cohen, 1991; Yang and Feige, 1991, 1992; Feige, 1996; Segel et al., 1995). On the contrary, autoimmunity appears to be very common and also necessary for immune homeostasis (Cohen and Young, 1991). Autoimmunity will transform into autoimmune disease when self-reactive T cells (autoimmune effector cells) escape the normal regulation or are forced to do so by an immunization with their antigen in a highly immunogenic form. For example, AdA is induced in Lewis rats by an injection of mycobacteria in oil. However, most exposures of Lewis rats to mycobacteria induce an AdA resistant state (see above and Feige and Cohen, 1991). It is important to note that in rats which develop AdA upon immunization with mycobacteria, suppressive mechanisms are activated as well. Antigen-specific and antigen-non-specific suppressor cells have been demonstrated *in vitro* (Karin and Cohen, to be published; see also Atlan and Cohen, 1992). The concepts of the "autoimmune network" (Cohen and Atlan, 1979; Atlan and Cohen, 1992; Cohen, 1994) and the "immunological homunculus" (Cohen, 1989) have been formulated based on these and other data. Autoimmunity is viewed as a complex network of various T cells (such as effector, effector inducer, suppressor inducer, suppressor, etc.) combined with a network of autoantigen(s) and foreign antigen(s). Stimulation of effector cells in this network (e.g., by immunization with an antigen in complete Freund's adjuvant) may lead to autoimmune disease. The perception of autoimmune disease as "disregulation" in a network of T cells opens a much broader spectrum of possibilities to modulate existing autoimmunity, or to prevent autoimmunity from transforming into autoimmune disease, or to revert autoimmune disease (back) to autoimmunity. Modulation of autoimmune disease with foreign or self HSP is just one of these possibilities. Importantly, it seems, that for the sake of treatment the first (auto-)antigen involved at the beginning of the process leading to autoimmune disease does not necessarily have to be known. It appears that hsp60 is a good candidate (auto-)antigen(s) and that safe ways to interfere with autoimmune disease based on hsp60 are feasible (for review see Feige and Cohen, 1991; Cohen, 1991; Yang and Feige, 1991, 1992; Feige, 1996; Segel et al., 1995). It should be noted that autoimmunity is common (normal) and autoimmune disease is rare (not normal?). In other words, the immune system of most individuals does not allow transformation of autoimmunity to autoimmune disease.

The view of autoimmunity as a dynamic network is also underlined by other experiments where autoimmune diseases are controlled by immunomodulation. For example, a single intravenous injection of immunoglobulin (Ig) given at the time of induction of AdA with mycobacteria can prevent AdA in rats (Achiron et al., 1994). Even a single injection of Ig on day 14 inhibited ongoing AdA. In this context, it should be mentioned that Ig treatment of patients with multiple sclerosis (MS) in a 3-year clinical trial resulted in an reduction of the relapse rate (Achiron et al. 1995). The efficacy of Ig treatment in MS patients again indicates the presence of regulatory T cells in ongoing autoimmune disease which can be stimulated by relatively "mild" treatments and in turn can shift the balance in this network in favor of resistance against autoimmune disease. The dynamics of the immune system in dealing with self antigens is highlighted in studies using double-transgenic mice, which are transgenic for the autoantigen (Kb) and a T cell receptor recognizing Kb. In this model, in addition to positive and negative selection of T cells in the thymus, mechanisms were shown to exist in the periphery which enable the immune system to tolerize mature T cells to (self-)antigen(s) although these antigen(s) are newly expressed at adult age (Schönrich et al., 1994; Ferber et al., 1994, and references therein). In fact, various levels of peripheral tolerance have been described, based on consecutive downregulation of T cell receptors induced by low and high concentrations of the "autoantigen" in circulation (Ferber et al., 1994). The findings led to the view of tolerance induction as a multi-step process in mature, peripheral T cells (Schönrich et al., 1994). In analogy, antigens such as cartilage antigens which are not normally "visible" to the immune system could become "visible" autoantigens due to inflammatory reactions during ongoing autoimmune disease and, as a result, modulate ongoing (auto-)immune reactions.

Conclusions

In conclusion, although we still do not know the precise mechanism(s) of the development of autoimmune diseases, research in this area during the last decade has brought forward new possibilities for preventing and treating autoimmune diseases, (therapeutic) vaccination with HSP (peptides) being one of these. "How to define the autoantigen in human disease?" appears no longer to be an absolute prerequisite in order to define immunological treatment(s) for prevention or therapy of autoimmune diseases. Experimental results summarized in this article may help to explain the long observed influence of the so-called environmental factors on autoimmune diseases. In addition, they illustrate that autoimmune disease may be induced by various stimuli. And, if nothing else, research of the role of HSP in autoimmune diseases has (re-)induced a discussion about the immune system's way to distinguish self from non-self. Also, it may be questioned whether application of mere immunosuppressive treatment(s) of autoimmune diseases is appropriate based on the dynamics and the potential of the regulatory processes within the immune system, which appear to be able to regulate (upon stimulation with antigen) towards a state of resistance against autoimmune disease, even after onset of clinical symptoms of disease.

Acknowledgements
We are indebted to Irun R. Cohen in whose laboratory we met in 1984 and started to explore mechanisms of autoimmunity and autoimmune disease under Irun's guidance. U. F. would like to thank Deborah Russell for stimulating discussions and critically reading the manuscript.

References

Achiron, A., Margalit, R., Hershkoviz, R., Markovits, D., Reshef, T., Melamed, E., Cohen, I.R. and Lider, O. (1994) Intravenous immunoglobulin treatment of experimental T cell-mediated autoimmune disease. *J. Clin. Invest.* 93:600–605.

Achiron, A., Cohen, I.R., Lider, O. and Melamed, E. (1995) Intravenous immunoglobulin treatment in multiple sclerosis. *Isr. J. Med. Sci.* 31:7–9.

Albani, S., Ravelli, A., Massa, M., De Benedetti, F., Andree, G., Roudier, J., Martini, A. and Carson, D.A. (1994) Immune responses to the *Escherichia coli* dnaJ heat shock protein in juvenile rheumatoid arthritis and their correlation with disease activity. *J. Pediatr.* 124:561–565.

Anderton, S.M., van der Zee, R. and Goodacre, J.A. (1993) Inflammation activates self hsp60-specific cells. *Eur. J. Immunol.* 23:33–38.

Anderton, S.M., van der Zee, R., Noordzij, A. and van Eden, W. (1994) Differential mycobacterial 65-kDa heat shock protein T cell epitope recognition after adjuvant arthritis inducing or protective immunization protocols. *J. Immunol.* 152:3656–3664.

Anderton, S.M., van der Zee, R., Prakken, B., Noordzij, A. and van Eden, W. (1995) Activation of T cells recognizing self-60-KD heat-shock protein can protect against experimental arthritis. *J. Exp. Med.* 181:943–952.

Atlan, H. and Cohen, I.R. (1992) Paradoxical effects of suppressor T cells in adjuvant arthritis: neural network analysis. *In:* A.S. Perelson and G. Weisbuch (eds): *Theoretical and Experimental Insights into Immunology.* NATO ASI series, Vol. H 66, pp 379–395.

Billingham, M.E., Carney, S., Butler, R. and Colston, M.J. (1990) A mycobacterial 65-kD heat shock protein induces antigen-specific suppression of adjuvants arthritis, but is not itself arthrithogenic. *J. Exp. Med.* 171:339–344.

Bjoerk, J., Kleinau, S., Midtvedt, T., Klareskog, L. and Smedegard, G. (1994) Role of the bowel flora for development of immunity to hsp65 and arthritis in three experimental models. *Scand. J. Immunol.* 40:648–652.

Born, W., Hall, L., Dallas, A., Boymel, J., Shinnick, T., Young, D., Brennan, P. and O'Brien, R. (1990) Recognition of a peptide antigen by heat shock-reactive, γ, δ T lymphocytes. *Science* 249:67–69.

Born, W., Happ, M.P., Dallas, A., Reardon, C., Kubo, R., Shinnick, T., Brennan, P. and Born, W. (1990) Recognition of heat-shock proteins and γ, δ cell function. *Immunol. Today* 11:40–43.

Broeren, C.P.M., Lucassen, M.A., van Stipdonk, M.J.B., van der Zee, R., Boog, C.J.P., Kusters, J.G. and van Eden, W. (1994) CDR1 T cell receptor β-chain peptide induces MHC class-II restricted T-T cell interactions. *Proc. Natl. Acad. Sci. USA* 91:5997–6001.

Cohen, I.R., Holoshitz, J., van Eden, W. and Frenkel, A. (1985) T lymphocyte clones illuminate pathogenesis and affect therapy of experimental arthritis. *Arthrit. Rheum.* 28:841–845.

Cohen, I.R. (1989) Natural id-anti-id networks and the immunological homunculus. *In:* H. Atlan and I.R. Cohen (eds): *Springer Series in Synergetics, Vol. 46*, Springer-Verlag Berlin, Heidelberg, pp 6–12.

Cohen, I.R. and Atlan, H. (1989) Network regulation of autoimmunity: an automaton model. *Autoimmunity* 2:613–625.

Cohen, I.R. (1991) Autoimmunity to chaperonins in the pathogenesis of arthritis. *Annu. Rev. Immunol.* 9:567–589.

Cohen, I.R. and Young, D.B. (1991) Autoimmunity, microbial immunity and the immunological homunculus. *Immunol. Today* 12:105–110.

Cohen, I.R. (1992) Autoimmunity to hsp65 and the immunologic paradigm. *Adv. Intern. Med.* 37:295–311.

Cohen, I.R. (1994) Autoimmunity shifts paradigms. *Isr. J. Med. Sci.* 30:37–38.

DeGraeff-Meeder, E.R., van der Zee, R., Rijkers, G.T., Schuurman, H.J., Kuis, W., Bijlsma, J.W.J., Zegers, B.J.M., van Eden, W. (1991) Recognition of human 60 kD heat shock protein by mononuclear cells from patients with juvenile chronic arthritis. *Lancet* 337: 1368–1372.

DeGraeff-Meeder, E.R., van Eden, W., Rijkers, G.T., Prakken, B.J., Zegers, B.J.M. and Kuis, W. (1993) Heat-shock proteins and juvenile chronic arthritis. *Clin. Exp. Rheumatol.* 11(Suppl. 9):S25–S28.

DeGraeff-Meeder, E.R., van Eden, W., Rijkers, G.T., Kuis, W., Voorhorst-Ogink, M.M., van der Zee, R., Schuurman, H.-J., Helders, P.J.M., Zegers, B.J.M. (1995) Juvenile chronic arthritis: T cell reactivity to human hsp60 in patients with a favourable course of arthritis. *J. Clin. Invest.* 95:934–940.

Elias, D., Markovits, D., Reshef, T., van der Zee, R. and Cohen, I.R. (1990) Induction and therapy of autoimmune diabetes in the non-obese diabetic (NOD/Lt) mouse by a 65-kDa heat shock protein. *Proc. Natl. Acad. Sci. USA* 87:1576–1580.

Elias, D., Reshef, T., Birk, O.S., van der Zee, R., Walker, M.D. and Cohen, I.R. (1991) Vaccination against autoimmune mouse diabetes with a T cell epitope of the human 65-kDa heat shock protein. *Proc. Natl. Acad. Sci. USA* 88:3088–3091.

Elias, D. and Cohen, I.R. (1994) Peptide therapy for diabetes in NOD mice. *Lancet* 343: 704–706.

Elias, D., Prigozin, H., Polak, N., Rapoport, M., Lohse, A.W. and Cohen, I.R. (1994) Auto-immune diabetes induced by the β-cell toxin STZ. Immunity to the 60-kDa heat shock protein and to insulin. *Diabetes* 43:992–998.

Elias, D. and Cohen, I.R. (1995) Treatment of autoimmune diabetes and insulitis in NOD mice with heat shock protein 60 peptide p277. *Diabetes* 44:1132–1138.

Feige, U. and Cohen, I.R. (1991) The 65-kDa heat-shock protein in the pathogenesis, prevention and therapy of autoimmune arthritis and diabetes mellitus in rats and mice. *Springer Semin. Immunopathol.* 13:99–113.

Feige, U. and Gasser, J. (1994) Therapeutic intervention with mycobacterial heat shock protein peptide 180–188 in adjuvant arthritis in Lewis rats. *Mediat. Inflammat.* 3:304.

Feige, U., Schulmeister, A., Mollenhauer, J., Brune, K. and Bang, H. (1994) A constitutive 65 kDa chondrocyte protein as a target antigen in adjuvant arthritis in Lewis rats. *Auto-immunity* 17:233–239.

Feige, U. (1996) Heat-shock proteins and arthritis. *In:* M. Zierhut, J. Saal and H.J. Thiel (eds): *Immunology of the Joint and the Eye*. Aeolus Press, Buren, The Netherlands, pp 47–61.

Ferber, I., Schönrich, U., Schenkel, J., Mellor, A.L., Hämmerling, G.J. and Arnold, B. (1994) Levels of tolerance induced by different doses of tolerogen. *Science* 263:674–676.

Fu, Y.X., Vollmer, M., Kalataradi, H., Heyborne, K., Reardon, C., Miles, C., O'Brien, R. and Born, W. (1994) Structural requirements for peptides that stimulate a subset of γ, δ T cells. *J. Immunol.* 152:1578–1588.

Gelber, C., Paborsky, L., Singer, S., McAteer, D., Tisch, R., Jolicoeur, C., Buelow, R., McDevitt, H. and Fathman, C.G. (1994) Isolation of nonobese diabetic mouse T cells that recognize novel autoantigens involved in the early events of diabetes. *Diabetes* 43:33–39.

Goupille, P., Poet, J.L., Jattiot, F., Mattei, J.P., Vedere, V., Tonolli-Seabian, I., Roux, H. and Valat, J.P. (1994) Three cases of arthritis after BCG therapy for bladder cancer. *Clin. Exp. Rheumatol.* 12:195–197.

Hermann, E. (1993) T cells in reactive arthritis. *APMIS* 101:177–186.

Hogervorst, E.J., Schouls, L., Wagenaar, J.P., Boog, C.J., Spaan, W.J., van Embden, J.D. and van Eden, W. (1991) Modulation of experimental autoimmunity: treatment of adjuvant arthritis by immunization with a recombinant vaccinia virus. *Infect. Immun.* 59: 2029–2035.

Ishimoto-T., Sato-K., Higaki-M., Nomaguchi-H., Osumi-K., Kashiwazaki-S. (1993) Specific increase of IgG antibody to 65 kDa heat shock protein but not to crude mycobacterial extract in RA [letter]. *J. Rheumatol.* 20:1089–1090.

Iwasaki-A., Yoshikai-Y, Yuuki-H., Takimoto-H., Nomoto-K. (1991) Self-reactive T cells are activated by the 65-kDa mycobacterial heat-shock protein in neonatally thymectomized mice. *Eur. J. Immunol.* 21:597–603.

Jones, D.B., Coulson, A.F.W. and Duff, G.W. (1994) Sequence homologies between hsp60 and autoantigens. *Immunol. Today* 14:115–118.

Koga, T., Wand-Württemberger, A., DeBruyn, J., Munk, M.E., Schoel, B. and Kaufmann, S.H.E. (1989) T cells against a bacterial heat-shock protein recognize stressed macrophages. *Science* 245:1112–1115.

Kogure, A., Miyata, M., Nishimaki, T. and Kasukawa, R. (1994) Proliverative response of synovial fluid mononuclear cells of patients with rheumatoid arthritis to mycobacterial 65 kDa heat shock protein and its association with HLA-DR+·γδ+ T cells. *J. Rheumatol.* 21:1403–1408.

Life, P., Hassell, A., Williams, K., Young, S., Bacon, P., Southwood, T. and Gaston, J.S.H. (1993) Responses to Gram negative enteric bacterial antigens by synovial T cells from patients with juvenile chronic arthritis: Recognition of heat shock protein 60. *J. Rheumatol.* 20: 1388–1396.

Lopez-Guerrero, J.A., Lopez-Bote, J.P., Ortiz, M.A., Gupta, R.S., Paez, E. and Bernabeu, C. (1993) Modulation of adjuvant arthritis in Lewis rats by recombinant vaccinia virus expressing the human 60-kilodalton heat shock protein. *Infect. Immun.* 61:4225–4231.

Lopez-Guerrero, J.A., Ortiz, M.A., Paez, E., Bernabeu, C. and Lopez-Bote, J.P. (1994) Therapeutic effect of recombinant vaccinia virus expressing the 60-kd heat-shock protein on adjuvant arthritis. *Arthrit. Rheum.* 37:1462–1467.

Neidhardt, F.C., VanBogelen, R.A. and Vaughn, V. (1984) The genetics and regulation of heat-shock proeins. *Ann. Rev. Genet.* 18:295–329.

O'Brien, R.L., Fu, Y.X., Cranfill, R., Dallas, A., Ellis, C., Reardon, C., Lang, J., Carding, S.R., Kubo, R. and Born, W. (1992) Heat shock protein hsp60-reactive gamma delta cells: a large, diversified T-lymphocyte subset with highly focused specificity. *Proc. Natl. Acad. Sci. USA* 89:4348–4352.

Probst, P., Hermann, E. and Fleischer, B. (1994) Role of bacteria-specific T cells in the immunopathogenesis of reactive arthritis. *Trends Microbiol.* 2:329–332.

Schönrich, U., Alferink, J., Klevenz, A., Kühlbeck, G., Auphan, N., Schmitt-Verhulst, A.-M., Hämmerling, G. and Arnold, B. (1994) Tolerance induction as a multistep process. *Eur. J. Immunol.* 24:285–293.

Segel, L.A., Jäger, E., Elias, D. and Cohen, I.R. (1995) A quantitative model of autoimmune disease and T cell vaccination: does more mean less? *Immunol. Today* 16:80–84.

Shehadeh, N., Calcinaro, F., Bradley, B.J., Bruchlim, I., Vardi, P. and Lafferty, K.J. (1994) Effect of adjuvant therapy on development of diabetes in mouse and man. *Lancet* 343: 706–707.

Shinnick, T.M., Vodkin, M.H. and Williams, J.L. (1988) The *Mycobacterium tuberculosis* 65 kDa protein is a heat shock protein which corresponds to common antigen and to the *E. coli* groEL protein. *Infect. Immun.* 56:446–451.

Thole, J.E.R., Hindersson, P., de Bruyn, J., Cremers, F., van der Zee, R., de Cock, H., Tommassen, J., van Eden, W. and van Embden, J.D. (1988) Antigenic relatedness of a strongly immunogenic 65 kDa mycobacterial protein antigen with a similarly sized ubiquitous bacterial common antigen. *Microbiol. Pathogenesis* 4:71–83.

Thompson, S.J., Hitsumoto, Y., Ghoraishian, M., van der Zee, R. and Elson, C.J. (1991) Cellular and humoral reactivity pattern to the mycobacterial heat shock protein hsp65 in pristane induced arthritis susceptible and hsp65 protected DBA/1 mice. *Autoimmunity* 11:89–95.

Thompson, S.J. and Elson, C.J. (1993) Susceptibility to pristane-induced arthritis is altered with changes in bowel flora. *Immunol. Lett.* 36:227–232.

Van den Broek, M.F., Hogervorst, E.J., Van Bruggen, M.C., Van Eden, W., van der Zee, R. and van den Berg, W.B. (1989) Protection against streptococcal cell wall-induced arthritis by pretreatment with the 65-kD mycobacterial heat shock protein. *J. Exp. Med.* 170:449–466.

van Eden, W., Thole, J.E., van der Zee, R., Nordzij, A., van Embden, J.D., Hensen, E.J. and Cohen, I.R. (1988) Cloning of the mycobacterial epitope recognized by T lymphocytes in adjuvant arthritis. *Nature* 331:171–173.

van Eden, W., Anderton, S.M., van der Zee, R., Prakken, A.B.J. and Rijkers, G.T. (1995) Specific immunity as a critical factor in the control of autoimmune arthritis – the example of hsp60 as an ancilliary and protective autoantigen. *Scand. J. Rheumatol. Suppl.* 101:141–145.

Vischer, T.L. (1988) A double blind multicentre study of OM-8980 and auronofin in rheumatoid arthritis. *Ann. Rheum. Dis.* 47:482–487.

Vischer, T.L. and van Eden, W. (1994) Oral desensibilization in rheumatoid arthritis (RA). *Ann. Rheum. Dis.* 53:708–710.

Wick, G., Schett, G., Amberger, A., Kleindienst, R. and Xu, Q. (1995) Is atherosclerosis an immunologically mediated disease? *Immunol. Today* 16:27–33.

Wilbrink, B., Hoilewijn M., Bijlsma, J.W.J., van Roy, J.A.L.M., den Otter, W. and van Eden, W. (1993) Suppression of human cartilage proteoglycan synthesis by rheumatoid synovial fluid mononuclear cells activated with mycobacterial 60 kD heat-shock protein. *Arthrit. Rheum.* 36:514–518.

Yang, X.-D., Gasser, J., Riniker, B. and Feige, U. (1990) Prevention of adjuvant arthritis in rats by a nonapeptide from the 65-kD mycobacterial heat shock protein. *Clin. Exp. Immunol.* 81:189–194.

Yang, X.-D. and Feige, U. (1991) The 65 kDa heat shock protein: a key molecule mediating the development of autoimmune arthritis? *Autoimmunity* 9:83–88.

Yang, X.-D. and Feige, U. (1992) Heat shock proteins in autoimmune disease. From causative antigen to specific therapy? *Experientia* 48:650–656.

Yang, X.-D., Gasser, J. and Feige, U. (1992) Prevention of adjuvant arthritis in rats by a nonapeptide from the 65-kD mycobacterial heat shock protein: specificity and mechanism. *Clin. Exp. Immunol.* 87:99–104.

Stress-Inducible Cellular Responses
ed. by U. Feige, R.I. Morimoto, I. Yahara and B. Polla
© 1996 Birkhäuser Verlag Basel/Switzerland

Stress proteins in inflammation

B. S. Polla[1] and A. Cossarizza[2]

[1] *Laboratoire de Physiologie Respiratoire, UFR Cochin Port-Royal, Université Paris V,
24, rue du Faubourg Saint-Jacques, F-75014 Paris, France and* [2] *Department of Biomedical
Sciences, University of Modena, Via Campi 287, I-41100 Modena, Italia*

Summary. Inflammation provides those searching in the field with a number of "models" allowing them to study, *in vivo*, in humans and in animals, the regulation and the functions of HSP, which are being considered as a new and promising marker for the severity and the prognosis of inflammatory diseases. HSP are differentially regulated according to the type of inflammation, whether acute or chronic, whether self-limiting (inflammatory cell elimination by apoptosis) or self-perpetuating (inflammatory cell death by necrosis). We propose that mitochondria are a key organelle in determining the outcome of inflammation, because they are both the cellular "switchboard" for apoptosis and a selective target for the protective effects of HSP against the cytotoxic effects of TNFα and ROS. On the other hand, HSP exert multiple protective effects in inflammation, including self/non-self discrimination, enhancement of immune responses, immune protection, thermotolerance and protection against the cytotoxicity of inflammatory mediators. The latter protective effects against the deleterious effects of the mediators of inflammation, including ROS and cytokines, open new avenues for the development of original anti-inflammatory therapies, such as non-toxic inducers of a complete HS response. It may well be that the "beneficial effects of fever" already described by Hippocrates actually relate to increased HSP expression during fever, and to their protective effects…

Introduction

Inflammatory reactions result from the recruitment, to the site of inflammation, of phagocytic cells (monocytes-macrophages [mφ], neurophils, eosinophils), from their activation and their interactions with specific immune cells (T- and B-lymphocytes) in a given tissue. Inflammation is associated with the increased production of a number of mediators which have been shown to modulate the expression of stress proteins, including reactive oxygen species (ROS), the lipid mediators of inflammation, and cytokines. On the other hand, heat shock (HS) and subsequent overexpression of stress proteins may protect *in vitro* and *in vivo* cells and animals from the toxic effects of some of these mediators. This chapter will deal with the expression and regulation of stress proteins in inflammation and the protective effects of HS/stress proteins (HSP) in inflammation.

Expression and regulation of stress proteins in inflammation

In vitro regulation of stress proteins by the mediators of inflammation

Reactive oxygen species
As far as the regulation of HSP is concerned, ROS are the most important mediators of inflammation. In infection and subsequent inflammation, ROS act primarily as defense against pathogens and secondarily as toxic factor leading to tissue damage and amplification of the inflammatory reaction. ROS may induce HSP as part of the host defenses against infection, or as an indicator of tissue damage. When produced at low concentration, ROS may act as second messengers, for example, for the effects of cytokines. They also are involved in cell metabolism (such as the arachidonic acid cascade) and may signal cell proliferation.

In inflammation, the main producers of ROS are the phagocytes (for reviews, Badwey and Karnovsky, 1980; Beaman and Beaman, 1984). These cells are the host's first line of defense against a wide variety of pathogens, and their ability to kill invading microorganisms depends to a large extent on the respiratory burst, a sequence of events whereby the activated phagocyte reduces molecular oxygen into toxic metabolites, the so-called ROS.

The primary biochemical event of the respiratory burst is the univalent reduction of oxygen to superoxide (O_2^-) by the complex enzymatic system, NADPH oxidase. Subsequent reductions lead to the formation of other toxic oxygen species including hydrogen peroxide (H_2O_2) and hydroxyl radical ($\cdot OH$) the latter reaction being catalyzed by iron. Stimulated neutrophils release myeloperoxidase (MPO) which catalyzes the conversion of H_2O_2 and Cl^- to hypochlorous acid (HOCl) and singlet oxygen ($^1O_2$). Rodent mϕ, when activated with phagocytic stimuli, lipopolysaccharides (LPS), or cytokines, also generate nitric oxide ($NO\cdot$) (Nussler and Biliar, 1993). The nitrogen atom of $NO\cdot$ is derived from the N-guanidino-terminal group of the amino acid, L-arginine, whereas the oxygen atom is provided by molecular oxygen (O_2). $NO\cdot$ reacts with other ROS: as an example, $NO\cdot$ and O_2^- interact with each other yielding the peroxynitrite anion ONOO-, an unstable species at physiological pH, protonating to peroxynitrous acid ONOOH which spontaneously decomposes to NO_2 and $\cdot OH$. Peroxynitrites have strong prooxidant properties and contribute to free radical-dependent toxicity. The generation and metabolism of major ROS are illustrated in Figure 1.

Some, but not all of these ROS are involved in HSP regulation. $NO\cdot$, for example, although mediating the anti-tumor and anti-microbial effects of rodent mϕ, appears, as an essentially redundant and inefficient mediator in human mϕ (Albina, 1995) and is unlikely to be directly involved in the activation of HS genes. Indeed, inhibitors of $NO\cdot$ synthases do not affect, in human cells, the induction of the HS response by HS or during

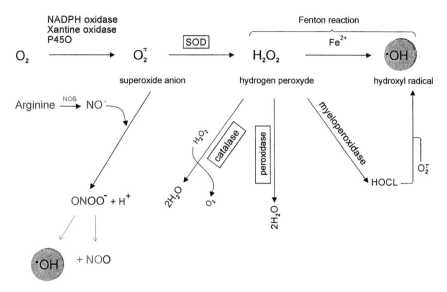

Figure 1. Generation and metabolism of major ROS.

phagocytosis (M.S. Richard and B.S. Polla, unpublished data). However, even if NO· is not an efficient mediator of human mφ responses, it is involved in sepsis, in adult respiratory distress syndrome (ARDS), and in diabetes, and these three conditions have been shown to benefit from HS treatment, either *in vitro* or *in vivo* (Villar et al., 1993, 1994; Bellmann et al., 1995) (see section on *Protective effects of HS/stress protein over-expression in inflammation*).

In order to determine which are the specific ROS and possible other relevant inducers of stress proteins during inflammation, we used phago-cytosis as a convenient *in vitro* model. Indeed, phagocytosis is active during inflammation and is associated with ROS production and with the intra-cellular accumulation of non-self (phagocytosed) material. Such non-self material could serve as an additional trigger, together with ROS, for activating the HS response. We established that erythrophagocytosis by human mφ induces in the latter a high expression of HSP (Clerget and Polla, 1990). Neither accumulation of intracellular non-self material nor O_2^- alone (even when produced at high levels and whether extracellularly, intracellularly or both) are by themselves sufficient to induce HSP: more toxic ROS are necessary for the upregulation of HSP. Furthermore, in con-trast to H_2O_2 and/or ·OH, which lead to the activation of the HS transcrip-tion factor (HSF), O_2^-, is unable to do so (Jacquier-Sarlin et al., 1995; Jacquier-Sarlin and Polla, 1996). The molecular mechanisms involved in oxidative stress-regulated HS gene activation and the transcriptional regu-lator(s) involved are reviewed in more detail in the chapter by Storz and Polla (this volume).

If one assumes that alteration in protein conformation is one of the final common stressors leading to HSF activation, one could propose that H_2O_2 and $\cdot OH$, but nor O_2^- or $NO\cdot$, lead to sufficient alterations in protein structure and/or function so as to serve as signals for HSF activation. During phagocytosis, while O_2^- production is not directly responsible of HSP induction, its resulting transformation into $\cdot OH$, in the presence of iron, may be essential to lead to oxidative alterations of nonself (or self) proteins and the subsequent induction of HSP synthesis.

Lipid mediators

The effects of the lipid mediators of inflammation, i.e., the products of the arachidonic acid cascade, on the HS response, are described in detail in the chapter by Santoro (this volume).

Cytokines

Cytokines such as Interleukin 1 (IL-1) and tumor necrosis factor α (TNFα) contribute to acute and chronic inflammation by activating phospholipase A_2 (PLA$_2$) thus leading to the generation of the lipid mediators of inflammation and further ROS, and by modulating the production of extracellular matrix proteins and activating a number of other cells. Binding of TNFα to its receptor results in a rapid rise in intracellular (mitochondrial) ROS (O_2^-) which does not occur in TNFα-resistant cells (Hennet et al., 1993) while IL-1 selectively imposes an oxidative stress to pancreatic β cells (Argilés et al., 1992; Rabinovitch et al., 1992). Both TNFα and IL-1 may prime the respiratory burst enzyme NADPH oxidase for increased O_2^- production. However, though modulating cellular oxidative metabolism, most proinflammatory cytokines, as a rule, do not activate a HS response.

Nonetheless, there are a number of reports on cytokine-mediated regulation of HSP expression. IL-1, for example, induces hsp70 in the β cells of the pancreas, but not in other cells (Helqvist et al., 1991). Interestingly, this cytokine is cytotoxic in these cells only, while it is not in the many other cells in which it does not induce HSP expression. IL-2 has been shown to induce hsp70 in IL-2-receptor-bearing cells and mitogens activate HSP in lymphocytes (Haire et al., 1988). There is a single report indicating that TNFα induces HSP (Fincato et al., 1991) while endotoxin, the strongest activator of TNFα, does not (Rinaldo et al., 1990). On the other hand, transforming growth factor β (TGFβ), which is a tissue repair rather than a proinflammatory cytokine, induces hsp70 and hsp90 in chicken embryo cells (Takenaka and Hightower, 1992). This effect does not relate to cytotoxicity but rather to an increased requirement in chaperoning subsequent to the activation of protein synthesis by this cytokine. IL-6 induces metallothioneins (reviewed in Polla and Kantengwa, 1991), along with other acute phase proteins, and might be important in the regulation of hsp90 expression, since the promoter of the hsp90α gene contains the response element for the IL-6 nuclear transactivator factor (V. Jérôme and M. Catelli, personal communication).

Yet another approach has been taken to the study of interactions between HS and cytokines. A number of cell types such as lymphocytes, display increased responsiveness to cytokines when incubated at elevated temperatures (reviewed in Polla and Kantengwa, 1991). Furthermore, HS might modulate the production of proinflammatory cytokines: while it decreases the production of TNFα (see below), HS induces the release of fibroblast growth factor 1 (FGF1), however in an inactive form (Jackson et al., 1992).

Role of protein kinase c activation in stress protein regulation
The role of kinases, and in particular that of mitogen activated protein kinases (MAPkinases), in stress signaling, is reviewed in detail in the chapter by Holbrook at al. (this volume). Thus we will discuss here solely protein kinase C (PKC), which contributes to the post-transcriptional regulation of HSP expression.

Phorbol myristate acetate (PMA) is a potent activator of PKC and of NADPH oxidase (Maridonneau-Parini et al., 1986). PMA induces in human mϕ hsc70 and hsp90, and, to a lesser extent, hsp70, an upregulation independent from O_2^- production and mediated by the activation of PKC (Jacquier-Sarlin et al., 1995). While HS- and oxidant-induced HSP synthesis is essentially controlled at the transcriptional level (activation of HSF), the PMA-induced HSP synthesis results from a stabilization of HSP mRNA, most likely by activation and binding of AU-binding factors to the AU rich sequences within the 3' regions of the relevant RNA, as has been described for cytokines (Malter and Hong, 1991; Gorospe et al., 1993; Moseley et al., 1993).

Thus, during inflammation, HSP expression is likely regulated by both transcriptional and posttranscriptional mechanisms (Fig. 2). Under conditions associated with a sustained HSP expression such as chronic inflammation, stabilization of mRNA may be the essential mechanism for HSP regulation, as it is in cells stimulated with PMA. In contrast, stresses such as HS or ·OH which lead to rapid alterations in cellular proteins induce a more rapid, transient, transcriptional regulation of HS genes and subsequent HSP synthesis.

Infection as a paradigm for inflammation
Infection represents a stress for both the host and the pathogen. Infection realizes an interesting model of "cellular sociology," the interactions between host and pathogen possibly leading to the death of the one or the other, or alternatively, to prolonged cohabitation.

The host cell stress response is differentially regulated: according to the type of infected cell and the type of pathogen, to the micro-environ-

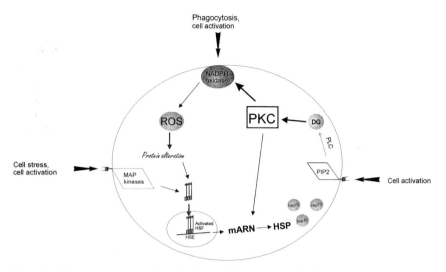

Figure 2. HSP expression is regulated in human inflammatory cells by both transcriptional and posttranscriptional mechanisms. Protein kinase c (PKC) activation occurs secondary to the interaction of ligands with receptors activating inositol phosphate metabolism (PIP2), and increases HSP expression by posttranscriptional mechanisms. In contrast, ROS-mediated protein alteration leads to transcriptional activation of the heat shock factor (HSF), DNA binding and subsequent HSP synthesis. Mitogen activated protein kinases (MAPK) are also involved in the regulation of HSP expression.

ment and to the specific intercellular or intracellular second messengers, host-pathogen interactions will lead, or not, to a stress response within the host cell and/or the pathogen (Hevin et al., 1993; Kantengwa and Polla, 1993; Kantengwa et al., 1995; Polla et al., 1995b, c; Barazzone et al., 1996b).

Since all of the HS genes share similar promoter regions, regulatory mechanisms involving a distinct unique secondary messenger or a cascade of messengers must exist to explain this differential HSP expression. For example, it appears that during phagocytosis of *Staphylococcus aureus*, NADPH oxidase activation is essentially mediated by PKC, thus leading to a posttranscriptional upregulation of host HSP, while during phagocytosis of *Pseudomonas aeruginosa*, NADPH oxidase activation is mediated by second messengers not involved in HSP upregulation (Polla et al., 1995b; Barazzone et al., 1996b). Furthermore, infection induces other selective responses within human mφ: for example, infection with gram negative bacteria induces high levels of intracellular IL-1, while infection with *S. aureus* induces other protective responses within the mφ, including increased activity of superoxide dismutase (Kantengwa and Polla, 1993; Barazzone et al., 1996b).

HSP synthesis during phagocytosis thus requires either PKC activation or the production of ROS toxic enough so as to lead to protein alteration (Fig. 2), or both.

In vivo regulation of stress proteins expression in inflammation

This section will deal with a selected number of animal models for inflammation and focus on human diseases such as rheumatoid arthritis (RA), asthma, and Duchenne Muscular Dystrophy (DMD). RA is a most classical chronic inflammatory disease, involving essentially mononuclear cells, while asthma is the clinical counterpart of eosinophilic inflammation. In contrast, DMD is a genetic disease resulting from a lack in dystrophin, while the biochemical defects leading to muscle necrosis include oxidative stress, alterations in intracellular calcium, and local inflammation.

The immunological functions of HSP in autoimmune diseases are presented in detail in the chapter by Feige and van Eden (this volume). In RA, HSP have been proposed both as marker for inflammation and/or as (auto)antigens, an opinion which has recently been challenged (Sharif et al., 1992). They appear today rather to represent a potential therapy, or prevention of RA. Indeed, clinical trials in patients with RA have shown that bacterial extracts (Subreum®) which contain hsp65 (Polla et al., 1995a), when administered orally, have comparable clinical efficiency to parenteral gold (Vischer, 1988). The effects of hsp65 (if indeed hsp65 mediates this therapeutic effect of Subreum®) may be secondary to a modulation of the immune response towards suppression. Hsp65 may modify the balance between TH_1 and TH_2 lymphocytes in favor of TH_2 cells and, as a consequence, in favor of the anti-inflammatory cytokines IL-4 and IL-10. The fact that HSP may serve not only as antigens, but also as allergens (Aki et al., 1994), supports the possibility that HSP might, under certain immunological conditions, preferentially activate TH_2 cells. Alternatively, hsp65 may raise suppressive cells able to secrete IL-4, IL-10 and transforming growth factor β (TGFβ) ("bystander suppression"), the latter being able, in turn, to upregulate HSP expression (Takenaka and Hightower, 1992; Polla et al., 1995a).

Alveolar macrophages (AM) from patients with a number of inflammatory lung diseases produce excess amounts of ROS. Based on the observation that exogenous ROS induce HSP, we hypothesized that endogenous production of ROS during disease may induce HSP in AM as an autoprotective mechanism (Kantengwa et al., 1991). We thus investigated the *ex vivo* HSP synthesis of human AM recovered during broncholaveolar lavage performed for diagnostic purposes and found high levels of spontaneous HSP synthesis by AM of two particular patients, whose pulmonary inflammation was characterized by an important number of alveolar eosinophils (Polla et al., 1993). Eosinophilic inflammation is the hallmark of a number of specific diseases, including asthma and, as mentioned above, certain parasitic infections, the eosinophils being activated and attracted at the site of inflammation by IL-5, which is, along with IL-4, preferentially produced by TH_2 cells. The epithelial cells and AM of patients with severe asthma, but not with mild asthma (Vignola et al., 1995; Fajac et al., submitted), overexpress HSP, and HSP expression correlates with broncholaveolar lavage eosinophilia. The regulation and function of HSP in eosinophilic inflammation are reviewed in more detail in Christie et al. (1996).

Corticosteroids still represent the most widely used and potent anti-inflammatory agents. They are routinely applied for the treatment of asthma and other inflammatory diseases such as systemic lupus. However, a number of patients are resistant to the anti-inflammatory effects of steroids. An interesting possibility is that corticoresistance relates to alterations in hsp90 expression or function in inflammation. Indeed, hsp90 selectively chaperones steroid receptors (see chapter by Pratt et al., this volume). Hsp90 is required to render these receptors competent for steroid binding, but must then be released in order for the steroid hormone-receptor complex to enter the nucleus and to bind and activate the corre-sponding hormone-response elements on the steroid-responsive genes. Thus, increases in the expression of hsp90 or modifications of its binding affinity for the steroid receptors (either related to inflammation or to poly-morphisms of the hsp90 genes) may determine steroid responsiveness of steroid resistance.

In DMD, an increased expression of both the HSP and the glucose-regu-lated proteins (GRP) has been found (Bornman et al., 1995, 1996). HSP expression relates to inflammation, oxidative stress, cell activation or regeneration, while GRP expression relates to increased intracellular calci-um levels in the dystrophin-deficient muscle fibers, thus supporting the idea that the expression of HSP/GRP corresponds, *in vivo* as *in vitro*, to distinct intracellular events. Furthermore, the regulation of the HSP them-selves is distinct, as hsp70 is increased secondarily to cell damage, while hsp90 upregulation rather reflects cell regeneration (myogenesis) (Born-man et al., 1996; Bornman and Polla, 1996). It is interesting, in this respect, to recall that the transcription factor MyoDI, which plays a major role in myogenesis, is conformationally activated to its DNA binding form by hsp90 (Shaknovitch, et al., 1992).

Protective effects of HS/stress protein over-expression in inflammation

In vitro protective effects of HS/stress protein overexpression

The first function to be attributed to HSP was thermotolerance, i.e., pro-tection from further exposure to HS (Welch and Mizzen, 1988; Li et al., 1991). One of the major reasons for the subsequent interest of biomedical research in HSP has been the hypothesis that HSP would provide protec-tion not only from HS, but from a wide array of other stresses. Initially this possibility has been investigated with respect to the types of stresses or molecules themselves capable of inducing a HS response, such as arsenite. More recently, the interest in the protective effects of HSP has developed in other fields, including their possible contribution to protective immunity. Here, we will focus on the protective potential of HSP against the deleterious effects of the toxic mediators of inflammation, i.e., cytokines and ROS, on cells and organisms.

We propose that mitochondria might play a key role in controlling inflammation. Indeed, mitochondrial function is profoundly affected by ROS and TNFα (Schulze-Osthoff et al., 1992; Hennet et al., 1993) while in turn, mitochondria are a selective target for the protective effect of HS against ROS (Polla et al., 1996).

Protection against cytokines
IL-1. The protective effects of HS/hsp70 against IL-1 particularly apply to diabetes. Insulin-dependent diabetes mellitus (IDDM) is an autoimmune disorder ultimately resulting in the destruction of the pancreatic β cells, hence causing insulin deficiency. Among immune cells, mϕ are the first to invade pancreatic islets during the development of the disease, mϕ-derived IL-1 and other products leading to oxidative stress. Since β cells are characterized by particularly low levels of endogenous antioxidant enzymes (scavengers), hsp70 induced by IL-1 may compensate this defect by exerting a novel antioxidant mechanism (Helqvist et al., 1991). In agreement with this hypothesis, purified hsp70 introduced into β cells *via* liposomes renders these cells fully resistant to subsequent exposure to IL-1 (Margulis et al., 1991). In rat islet cells, HS is capable of reducing damage and mortality provoked by ROS or streptozotocin and rat islet cell lysis induced by the NO$\cdot$ donor sodium nitroprusside (Bellmann et al., 1995). The mechanism for protection is linked to the inhibition of poly (ADP-ribose)polymerase (PARP), an enzyme activated by the DNA strand breaks caused by NO$\cdot$ and ROS (Radons et al., 1994) which plays a crucial role in provoking intracellular NAD+ depletion (Cleaver et al., 1987) and cell death (Franceschi et al., 1988).

TNFα. TNFα is a multifunctional cytokine with a crucial role in immune and inflammatory reactions (Beutler and Cerami, 1989). By analogy with ROS, TNFα exerts both beneficial effects by contributing to anti-infectious defenses and killing tumor cells and deleterious effects by leading to tissue damage and organ wasting. Apart from its toxic effects on mitochondria (Schulze-Osthoff et al., 1992). TNFα causes NAD depletion through the activation of PARP (Agarwal et al., 1988), the activity of which is inhibited by HS (see above; Bellmann et al., 1995).

The protective role of HSP against TNFα-mediated cytotoxicity has been particularly studied in tumor cells but could be extended to other systems. A short pre-treatment of mouse fibrosarcoma cells (WEHI) at 39–42 °C decreases TNFα-mediated lysis by approximately 50% (Jäättelä et al.,

1989). This protective effect roughly coincides with the kinetics of HSP induction.

The human hsp70 gene was then stably transfected into the highly TNFα-sensitive WEHI-S tumor line both in sense and in antisense orientation. Cells overexpressing hsp70 are protected from the lytic effects of TNFα (Jäättelä et al., 1989, 1992; Jäättelä, 1993). Similar results were obtained using hsp27 (Mehlen et al., 1995). Indeed, when genes encoding hsp27 from different species are placed under the control of a constitutive promoter and stably expressed in the TNFα-sensitive L929 fibrosarcoma cells, they rescue cells from TNFα-mediated killing (Mehlen et al., 1995). Given the multiplicity of HSP and their coordinate upregulation during diverse stresses, there likely is a requirement for cooperation between several HSP for maximal protection.

An interesting correlation has been observed between TNFα-sensitivity, TNFα-induced arachidonic acid metabolism and hsp70 synthesis. Indeed, an inhibition of arachidonic acid metabolism was observed in tumor cells overexpressing the hsp70 (Jäättelä, 1993). Whatever the precise mechanism of hsp70-associated TNFα-resistance, it was found to interfere with the TNFα-induced signal transduction pathway at a step after receptor binding but before activation of PLA_2. *In vivo*, most of TNFα is produced by activated mφ. Thus, hsp70, induced in mφ by TNFα, may provide these cells with a protective mechanism against the potential autotoxicity of TNFα.

Protection from ROS-mediated toxicity and cell death

HSP may protect prokaryotes as well as eukaryotes from the toxic effects of ROS, thus serving similar protective functions in both. As an example, mutant *Salmonella* resistant to oxidative stress overexpress a number of HSP, whereas bacteria unable to overexpress HSP (normally induced during the infection of mφ and the resulting oxidative stress) although able to survive within mφ, lose their virulence (Buchmeier and Heffron, 1990). Pre-exposure of avirulent *Leishmania* promastigotes to HS display increased hsp70 transcription paralleled by resistance to H_2O_2 (Zarley et al., 1991). In the human premonocytic line U937, preexposure to HS induces protection from subsequent exposure to H_2O_2 (Polla et al., 1988) and there are similar reports from many different laboratories showing HS-induced resistance to oxidants (reviewed in Polla et al., 1991).

HSP-induced eukaryotic cellular protection against ROS may be targetted to any of the following: membranes (lipid peroxidation), proteins, DNA and mitochondria. The protective effects of hsp70 towards lipid peroxidation and DNA damage have been recently reviewed (Jacquier-Sarlin et al., 1994). Mitochondria are both an important source of oxidative stress, including NO· (Sung and Dietert, 1994), and a selective target for HS-induced cellular protection (Patriarca and Maresca, 1990; Patriarca et al., 1992; Gabai and Kabakov, 1993; Lithgow et al., 1993; Kabakov and Gabai,

1994; Polla et al., 1996). Consequently, and because mitochondria profoundly affect the cellular "choice" between necrosis (amplification of inflammation) or apoptosis (limitation of inflammation) (Mangan et al., 1993), they may represent a key organelle in the control of inflammation.

Apoptosis, or programmed cell death, is used by almost all living organisms to self-destruct cells by an active "suicide" program when they are no longer needed, or have become seriously damaged (reviewed by Steller, 1995; Thompson, 1995). The morphological characteristics of apoptosis include nuclear and cytosolic condensation, and the formation of membrane-bound apoptotic bodies that are rapidly engulfed and digested by neighboring cells, particularly by mφ. Apoptotic cells do not release their potentially noxious contents and phagocytosis of apoptotic bodies occurs without activation of the mφ. On the contrary, cells undergoing necrosis (as a consequence of overwhelming injuries or loss in their capacity to adequately cope with stress) swell, lyse, and release cytoplasmic material which can trigger an inflammatory response. Furthermore, phagocytosis of necrotic cells or debris activates the mφ, thus leading to amplification of damage. As a consequence, whether a given cell undergoes apoptosis or necrosis is crucial in the control of inflammation. Since mitochondria are a "switchboard for apoptosis" (Richter et al., 1996), they also may be the switchboard defining, under the control of HSP, the outcome of acute inflammation, whether complete healing or chronic inflammation.

Previous morphological observations on mitochondria in apoptotic cells suggested that these organelles remain intact during the earliest phases of the process. The lack of fully reliable techniques available to study their functionality in a reasonable number of cells prompted us to set up a cytofluorimetric method that allows the measure of mitochondrial membrane potential ($\Delta\psi$) either at the single cell (Cossarizza et al., 1993) or at the single organelle level (Cossarizza et al., 1996). The method is based upon the use of a lipophilic, cationic, fluorescent dye, 5,5', 6,6'-tetrachloro-1,1', 3,3'-tetraethylbenzymidazol carbocyanine iodide (JC-1), which is able to selectively enter the mitochondria and indicate their $\Delta\psi$. This method allowed us not only to ascertain the presence of functional heterogeneity in isolated mitochondrial populations after treatment with different depolarizing agents, but also to study the function and responsiveness of mitochondria within their physiological environment, i.e., within the cell, under several physiopathological conditions including apoptosis of rat thymocytes, of U937 treated with TNFα, or of peripheral blood monocytes treated with 2-deoxyribose (Cossarizza et al., 1994, 1995 a, b).

Kinetic studies on rat thymocytes treated with dexamethasone showed that cells first undergo the typical morphological changes characteristic of this process, including physical parameters and DNA content, then lose their $\Delta\psi$. In TNFα-treated U937 cells or lymphocytes incubated with 2-D-deoxyribose, the rescue of $\Delta\psi$ results in protection from apoptosis. These observations indicate that functional mitochondria are crucial for apoptosis to occur. Accordingly, it has been proposed that mitochondrial functional integrity and maintenance of ATP levels are important during the early phases of apoptosis, when a variety of energy-requiring intracellular processes occur (Richter et al., 1996). In contrast, if intracellular ATP levels are very low, and if mitochondrial function is impaired or damaged, necrosis eventually occurs. Given the above-mentioned differences between apoptosis and necrosis in terms of amplification or limitation of inflammation, the maintenance of mitochondrial function may protect cells and tissues from inflammatory damages, as the result of the cell's "choice" to undergo the active apoptotic program rather than passive necrosis.

Another link between mitochondria and the protective cellular responses is detailed in the chapter by Wong et al. (this volume). These authors review the evidence that pretreatment with TNFα can induce the synthesis of protective mitochondrial proteins such as MnSOD, thus protection cells and animals from the lethal effects of radiation or chemotherapeutic drugs. Mitochondria may thus play a key role in radio- and chemoprotection as well as in apoptosis and inflammation.

Mitochondria may also be the central site of the overproduction of ROS in sepsis, and mitochondrial oxidative alterations play a key role in multiple organ failure (MOF) (Taylor et al., 1995). The fact that septic animals benefit from the protective effects of HSP overexpression (see below) provides one more link between ROS, mitochondria, and HSP-induced protection against oxidative injury and inflammation.

In vivo protective effects of HS/stress protein overexpression in animal models for inflammation

Adult respiratory distress syndrome (ARDS)
One model disease to consider with respect to the *in vivo* protective effects of HSP is ARDS. ARDS is an acute pulmonary inflammation which follows a number of diverse triggers and is characterized by the massive generation of ROS within the lung, along with activation of PLA$_2$ and production of TNFα. Despite major efforts to develop new therapeutic approaches, ARDS remains lethal for about 50% of the patients. The implications of HSP in ARDS have been recently reviewed (Jolliet et al., 1994) as well as the protective effects of HSP in animal models of this disease (Barazzone et al., 1996a). Briefly, it should be mentioned that a rodent model for ARDS, in which intratracheal administration of PLA$_2$ leads to a reproducible mortality of 30%, has been used to evaluate the protective effects of HS. Preexposure of these animals to HS, leading to massive induction of hsp70 within the lung, is associated with a remarkable decrease in pulmonary inflammation and the complete prevention of lethality (Villar et al., 1993). In human ARDS, circulating inflammatory cells of some

of the patients express increased levels of hsp70 (Kindas-Mügge et al., 1993).

Sepsis
Acute heat stress protects rats and mice from the lethal effects of endotoxins and entotoxic shock (Ryan et al., 1992; Hotchkiss et al., 1993). Experimental hyperthermia and arsenite administration (both of which induce HSP *in vivo*) also enhance survival in septic animals (Villar et al., 1994; Ribeiro et al., 1994). It is proposed that hsp70 may prevent the secretion of TNFα by selective intracellular chaperoning of the cytokine (S. P. Ribeiro et al., personal communication; Pugin and Polla, unpublished data).

Protection in transgenic mice overexpressing hsp70
Plumier et al. recently generated hsp70-transgenic mice using the human inducible hsp70 gene linked to an actin promoter (Plumier et al., 1995). In these animals, overexpression of human hsp70 does not affect normal protein synthesis or the stress response, in comparison with control, non-transgenic animals. However, upon reperfusion and following ischemia, transgenic hearts show significantly improved recovery of contractile force (more than twice that of controls), rate of contraction, and rate of relaxation. Creatine kinase, taken as an index of cellular damage, is released in transgenic animals about 50 times less than in controls. Thus, murine myocardium can be significantly protected from ischemia and reperfusion injury by the overexpression of a single HSP, the human inducible hsp70, suggesting a critical role of this molecule in preventing cellular damages.

Immune protection
Finally, HSP contribute to protection from autoimmune disease and from inflammation, as well as to immune defenses against pathogens or tumors. These aspects are covered in detail in the chapters by Suzue and Young (this volume) and by Feige and van Eden (this volume), respectively.

References

Aki, T., Fujikawa, A., Wada, T., Jyo, T., Shigeta, S., Murooka, Y., Oka, S. and Ono, K. (1994) Cloning and expression of cDNA coding for a new allergen from the house dust mite, *Dermatophagoides farinae*: homology with human heat shock cognate proteins in the heat shock protein 70 family. *J. Biochem.* 115:435–440.

Albina, J.E. (1995) On the expression of nitric oxide synthase by human macrophages. Why no NO? *J. Leukoc. Biol.* 58:643–649.

Agarwal, S., Drysdale, B.E. and Shin, H.S. (1988) Tumor necrosis factor-mediated cytotoxicity involves ADP-ribosylation. *J. Immunol.* 140:4187–4192.

Argilés, J.M., Lopez-Soriano, J., Angel Ortiz, A., Pou, J.M. and Lopez-Soriano, F.J. (1992) Interleukin-1 and β-cell function: more than one second messenger? *Endocr. Rev.* 13:515–524.

Badwey, J.A. and Karnovsky, M.L. (1980) Active oxygen species and the functions of phagocytic leukocytes. *Ann. Rev. Biochem.* 49:695–726.

Barazzone, C., Christie, P. and Polla, B.S. (1996a) Heat-shock proteins in the cell defence mechanisms of the lung. *In:* J. Chrétien, D. Dusser (eds): *Airways and Environment: From Injury to Repair.* Marcel Dekker Inc., New York; *in press.*

Barazzone, C., Kantengwa, S., Suter, S. and Polla, B.S. (1996b) Phagocytosis of *Pseudomonas aeruginosa* fails to elicit a heat shock response in human monocytes. *Inflammation* 20:243–262.

Beaman, L. and Beaman, B.L. (1984) The role of oxygen and its derivatives in microbial pathogenesis and host defense. *Annu. Rev. Microbiol.* 38:27–48.

Bellmann, K., Wenz, A., Radons, J., Burkart, V., Kleemann, R. and Kolb, H. (1995) Heat shock induces resistance in rat pancreatic islet cells against nitric oxide, oxygen radicals and streptozotocin toxicity in vitro. *J. Clin. Invest.* 95:2840–2845.

Beutler, B. and Cerami, A. (1989) The biology of cachectin/TNFa – a primary mediator of the host response. *Annu. Rev. Immunol.* 7:625–655.

Bornman, L., Polla, B.S., Lotz, B.P. and Gericke, G.S. (1995) Expression of heat shock proteins in muscle from Duchenne muscular dystrophy patients. *Muscle Nerve* 18:23–31.

Bornman, L. and Polla, B.S. (1996) Heat shock proteins in Duchenne's muscular dystrophy and other muscular diseases. *In:* W. van Eden, D. Young (eds): *Stress Proteins in Medicine.* Marcel Dekker Inc. New York, Basel, Hong Kong, pp 495–507.

Bornman, L., Polla, B.S. and Gericke, G.S. (1996) Heat shock protein 90 and ubiquitin: developmental regulation during myogenesis. *Muscle Nerve* 19:574–580.

Buchmeier, N.A. and Heffron, F. (1990) Induction of Salmonella stress proteins upon infection of macrophages. *Science* 248:730–732.

Christie, P., Janin, A., Bousquet, J. and Polla, B.S. (1996) Heat shock proteins in eosinophilic inflammation. *In:* W. van Eden, D. Young (eds): *Stress Proteins in Medicine.* Marcel Dekker Inc. New York, Basel, Hong Kong, pp 479–493.

Cleaver, J.E., Borek, C., Milam, K. and Morgan, W.F. (1987) The role of poly(ADP-ribose) synthesis in toxicity and repair of DNA damage. *Pharmacol. Ther.* 31:269–293.

Clerget, M. and Polla, B.S. (1990) Erythrophagocytosis induces heat shock protein synthesis in human monocytes-macrophages. *Proc. Natl. Acad. Sci. USA* 87:1081–1085.

Cossarizza, A., Baccarani, Contri, M., Kalashnikova, G. and Franceschi, C. (1993) A new method for the cytofluorimetric analysis of mitochondrial membrane potential using the J-aggregate forming lipophilic cation 5,5', 6,6'-tetrachloro-1,1', 3,3'-tetraethylbenzimidazol-cabocyanine iodide (JC-1). *Biochem. Biophys. Res. Comm.* 197:40–45.

Cossarizza, A., Kalashnikova, G., Grassili, E., Chiappelli, F., Salvioli, S., Capri, M., Barbieri, D., Troiano, L., Monti, D., Franceschi, C. (1994) Mitochondrial modifications during rat thymocyte apoptosis: a study at the single cell level. *Exp. Cell Res.* 214:323–330.

Cossarizza, A., Franceschi, C., Monti, D., Salvioli, S., Bellesia, E., Rivabene, R., Biondo, L., Rainaldi, G., Tinari, A. and Malorni, W. (1995a) Protective effect of N-acetylcysteine in tumor necrosis factor-α-induced apoptosis in U937 cells: the role of mitochondria. *Exp. Cell Res.* 220:232–240.

Cossarizza, A., Salvioli, S., Franceschini, M.G., Kalashnikova, G., Barbieri, D., Monti, D., Grassilli, E., Tropea, F., Troiano, L. and Franceschi, C. (1995b) Mitochondria and apoptosis: a cytofluorimetric approach. *Fundam. Clin. Immunol.* 3:67–68.

Cossarizza, A., Ceccarelli, D. and Masini, A. (1996) Functional heterogeneity of an isolated mitochondrial population revealed by cytofluorimetric analysis at the single organelle level. *Exp. Cell. Res.* 222:84–94.

Fincato, G., Polentarutti, N., Sica, A., Mantovani, A. and Colotta, F. (1991) Expression of a heat-shock inducible gene of the HSP70 family in human myelomonocytic cells: regulation by bacterial products and cytokines. *Blood* 77:579–586.

Franceschi, C., Marini, M., Zunica, G., Monti, D., Cossarizza, A., Bologni, A., Gatti, C. and Brunelli, M.A. (1988) Effect of ADP-ribosyl transferase (ADPRT) on the survival of human lymphocytes after exposure to different DNA damaging agents. Ann. N.Y. Acad. Sci. 551:446–447.

Gabai, V.L. and Kabakov, A.E. (1993) Rise in heat-shock protein level confers tolerance to energy deprivation. *FEBS Lett.* 327:247–250.

Gorospe, M., Kumar, S. and Baglioni, C. (1993) Tumor necrosis factor increases stability of interleukin-1 mRNA activating protein kinase C. *J. Biol. Chem.* 268:6214–6220.

Haire, R.N., Peterson, M.S. and O'Leary, J.J. (1988) Mitogen activation induces the enhanced synthesis of two heat-shock proteins in human lymphocytes. *J. Cell Biol.* 106:883–891.

Helqvist, S., Polla, B.S. Johannasen, J. and Nerup, J. (1991) Heat shock protein induction in rat pancreatic islets by recombinant human interleukin 1β. *Diabetologia* 34:150–156.

Hennet, T., Richter, C. and Peterhans, E. (1993) Tumour necrosis factor-α induces superoxide anion generation in mitochondria of L929 cells. *Biochem. J.* 289:587–592.

Hevin, B., Morange, M. and Fauve, R.M. (1993) Absence of an early detectable increase in heat-shock protein synthesis by *Listeria monocytogenes* within mouse monuclear phagocytes. *Res. Immunol.* 144:679–689.

Hotchkiss, R., Nunnally, I. Lindquist, S., Taulien, J., Perdrizet, G. and Karl, I. (1993) Hyperthermia protects mice against the lethal effects of endotoxin. *Am. J. Physiol.* 265 (*Regulatory Integrative Comp. Physiol.* 31): R1447–R1457.

Jäättelä, M., Saksela, K. and Saksela, E. (1989) Heat shock protects WEHI-164 target cells from the cytolysis by tumour necrosis factor α and β. *Eur. J. Immunol.* 19:1413–1417.

Jäättelä, M., Wissing, D., Bauser, P.A. and Li, G.C. (1992) Major heat shock protein hsp70 protects tumor cells from tumor necrosis factor cytotoxicity. *EMBO J.* 11:3507–3512.

Jäättelä, M. (1993) Overexpression of major heat shock protein hsp70 inhibits tumour necrosis factor-induced activation of phospholipase A2. *J. Immunol.* 151:4286–4294.

Jackson, A., Friedman, S., Zhan, X., Engleka, K.A., Forough, R. and Maciag, T. 81992) Heat shock induces the release of fibroblast growth factor 1 from NIH 3T3 cells. *Proc. Natl. Acad. Sci. USA* 89:10691–10695.

Jacquier-Sarlin, M.R. and Polla, B.S. (1996) Dual regulation of heat-shock transcription factor (HSF) activation and DNA-binding activity by H₂O₂: role of thioredoxin. *Bio Chem.* 318:187–193.

Jacquier-Sarlin, M.R., Fuller, K., Dinh-Xuan, A.T., Richard, M.J. and Polla, B.S. (1994) Protective effects of hsp70 in inflammation. *Experientia* 50:1031–1038.

Jacquier-Sarlin, M.R., Jornot, L. and Polla, B.S. (1995) Differential expression and regulation of hsp70 and hsp90 by phorbol esters and heat shock. *J. Biol. Chem.* 270:14094–14099.

Jolliet, Ph., Slosman, D.O. and Polla, B.S. 81994) Heat shock proteins in critical illness: markers of cellular stress or more? *In:* J.L. Vincend (ed.): *Yearbook of Intensive Care and Emergency Medicine.* Springer-Verlag, Berlin, pp 24–34.

Kabakow, A.E. and Gabai, V.L. (1994) Heat-shock proteins maintain the viability of ATP-deprived cells: what is the mechanisms? *Trends Cell. Biol.* 4:193–195.

Kantengwa, S., Donati, Y.R., Clerget, M., Maridonneau-Parini, I., Sinclair, F., Mariethoz, E., Perin, M., Rees, A.D., Slosman, D.O. and Polla, B.S. (1991) Heat shock proteins: an autoprotective mechanism for inflammatory cells? *Sem. Immunol.* 3:49–56.

Kantengwa, S. and Polla, B.S. (1993) Phagocytosis of *Staphylococcus aureus* induces a stress response in human monocytes-macrophages (MF): modulation by MF differentiation and by iron. *Infect. Immun.* 61:1281–1287.

Kantengwa, S., Müller, I., Louis, J. and Polla, B.S. (1995) Infection of human and murine macrophages with *Leishmania major* is associated with early parasite heat shock protein synthesis but fails to induce host stress response. *Immunol. Cell Biol.* 73:73–80.

Kindas-Mügge, I., Hammerle, A.H., Fröhlich, I., Oismüller, C., Micksche, M. and Trautinger, F. (1993) Granulocytes of critically ill patients spontaneously express the 72 kD heat shock protein. *Circulatory Shock*, 39:247–252.

Li, G.C., Li, L.G., Liu, Y.K., Mak, J.Y., Chen, L.L. and Lee, W.M. (1991) Thermal response of rat fibroblasts stably transfected with the human 70-kDa heat shock protein-encoding gene. *Proc. Natl. Acad. Sci. USA* 88:1681–1685.

Lithgow, T., Ryan, M., Anderson, R.L., Hoj, P.B. and Hoogenraad, N.J. (1993) A constitutive form of heat-shock protein 70 is located in the outer membranes of mitochondria from rat liver. *FEBS Lett.* 332:277–281.

Malter, J. and Hong, Y. (1991) A redox switch and phosphorylation are involved in the post-translational up-regulation of the adenosine-uridine binding factor by phorbol ester and ionophore. *J. Biol. Chem.* 266:3167–3171.

Mangan, D.F., Mergenhagen, S.E. and Wahl, S.M. (1993) Apoptosis in human monocytes: possible role in chronic inflammatory diseases. *J. Periodontol.* 64:461–466.

Margulis, B.A., Sandler, S., Eizirik, D.L., Welsh, N. and Welsh, M. (1991) Liposomal delivery of purified heat shock protein hsp70 into rat pancreatic islets as protection against interleukin 1β-inducted impaired β-cell function. *Diabetes* 40:1418–1422.

Maridonneau-Parini, I., Tringale, S.M. and Tauber, A.I. (1986) Identification of distinct activation pathways of human neutrophil NADPH-oxidase. *J. Immunol.* 137:2925–2929.

Mehlen, P., Preville, X., Chareyron, P., Briolay, J., Klemenz, R. and Arrigo, P. (1995) Constitutive expression of human hsp27, *Drosophila* hsp27, or human αB-crystallin confers resistance to TNF- and oxidative stress-induced cytotoxicity in stably transfected murine L929 fibroblasts. *J. Immunol.* 154:363–374.

Moseley, P.L., Wallen, E.S., McCafferty, J.D., Flanagan, S. and Kern, J.A. (1993) Heat stress regulates the human 70-kDa heat-shock gene through the 3'-untranslated region. *Am. J. Physiol.* 264 (*Lung Cell. Mol. Physiol.* 8):L533–L537.

Nussler, A.K. and Billiar, T.R. (1993) Inflammation, immunoregulation, and inducible nitric oxide. *J. Leukoc. Biol.* 54:171–178.

Patriarca, E.J. and Maresca, B. (1990) Acquired thermotolerance following heat shock protein synthesis prevents impairment of mitochondrial ATPase activity at elevated temperatures in *Saccharomyces cerevisae. Exp. Cell Res.* 190:57–64.

Patriarca, E.J., Kobayashi, G.S. and Maresca, B. (1992) Mitochondrial activity and heat-shock response during morphogenesis in the pathogenic fungus *Histoplasma capsulatum. Biochem. Cell Biol.* 70:207–214.

Plumier, J-C.L., Ross, B.M., Currie, R.W., Angelidis, C.E., Kazlaris, H., Kollias, G. and Pagoulatos, G.N. (1995) Transgenic mice expressing the human heat shock protein 70 have improved post-ischemic myocardial recovery. *J. Clin. Invest.* 95:1854–1860.

Polla, B.S., Bonventre, J.V. and Krane, S.M. (1988) 1,25-dihydroxyvitamin D3 increases the toxicity of hydrogen peroxide in the human monocytic line U937: the role of calcium and heat shock. *J. Cell Biol.* 107:373–380.

Polla, B.S., and Kantengwa, S. (1991) Heat shock proteins and inflammation. *Curr. Top. Microbiol. Immunol.* 167:93–104.

Polla, B.S., Mili, N. and Kantengwa, S. (1991) Heat shock and oxidative injury in human cells. *In:* B. Marresca and S. Lindquist (eds): *Heat Shock.* Springer-Verlag, Berlin, Heidelberg, New York, pp 279–290.

Polla, B.S., Kantengwa, S., Gleich, G.J., Kondo, M., Reimert, C.M. and Junod, A.F. (1993) Spontaneous heat shock protein synthesis by alveolar macrophages in interstitial lung disease associated with phagocytosis of eosinophils. *Eur. Respir. J.* 6:483–488.

Polla, B.S., Baladi, S., Fuller, K. and Rook, G. (1995a) Species-specific hsp65 in bacterial extracts (OM-89): a possible mediator or orally-induced tolerance? *Experientia* 51:775–779.

Polla, B.S., Mariéthoz, E., Hubert, D. and Barazzone, C. (1995b) Heat-shock proteins in host-pathogen interactions: implications for cystic fibrosis. *Trends Microbiol.* 10:392–396.

Polla, B.S., Stubbe, H., Kantengwa, S., Maridonneau-Parini, I. and Jacquier-Sarlin, M.R. (1995c) Differential induction of stress proteins and functional effects of heat shock in human phagocytes. *Inflammation* 19:363–378.

Polla, B.S., Kantengwa, S., François, D., Salvioli, S., Franceschi, C., Marsac, C. and Cossarizza, A. (1996) Mitochondria are selective targets for the protective effects of heat shock against oxidative injury. *Proc. Natl. Acad. Sci. USA; in press.*

Rabinovitch, A., Suarez, W.L., Thomas, P.D., Strynadka, K. and Simpson, I. (1992) Cytotoxic effects of cytokines on rat islets: evidence for involvement of free radicals and lipid peroxidation. *Diabetologia* 35:409–413.

Radons, J., Heller, B., Bürkle, A., Hartmann, B., Rodriguez, M-L., Kröncke, K-D., Burkart, V. and Kolb, H. (1994) Nitric oxide toxicity in islet cells involves poly(ADP-ribose)polymerase activation and concomitant NAD depletion. *Biochem. Biophys. Res. Commun.* 199:1270–1277.

Ribeiro, S.P., Villar, J., Downey, G.P., Edelson, J.D. and Slutsky, A.S. (1994) Sodium arsenite induces heat shock protein-72 kilodalton expression in the lungs and protects rats against sepsis. *Crit. Care Med.* 22:922–929.

Richter, C., Schweizer, M., Cossarizza, A. and Franceschi, C. (1996) Hypothesis: Control of apoptosis by the cellular ATP level. *FEBS Lett.* 378:107–110.

Rinaldo, J.E., Gorry, M., Strieter, R., Cowan, H., Abdolrasulnia, R. and Sheperd, V. (1990) Effect of endotoxin-induced cell injury on 70-kD heat shock proteins in bovine lung endothelial cells. *Am. J. Respir. Cell. Mol. Biol.* 3:207–216.

Ryan, A.J., Flanagan, S.W., Moseley, P.L. and Gisolfi, C.V. (1992) Acute heat stress protects rats against endotoxin shock. *J. Appl. Physiol.* 73:1517–1522.

Shaknovich, R., Shue, G. and Kohtz, D.S. (1992) Conformational activation of a basic helix-loop-helix-protein (MyoD1) by the C-terminal region of murine hsp90 (hsp84). *Mol. Cell Biol.* 12:5059–5068.

Schulze-Osthoff, K., Bakker, A.C., Vanhaesebroeck, B., Beyaert, R., Jacob, W.A. and Fiers, W. (1992) Cytotoxic activity of tumor necrosis factor is mediated by early damage of mitochondria function. Evidence for the involvement of mitochondria radical generation. *J. Biol. Chem.* 267:5317–5323.

Sharif, M., Worrall, J.G., Sing, B., Gupta, R.S., Lydyard, P.M., Lambert, C., McCulloch, J. and Rook, G.A. (1992) The development of monoclonal antibodies to the human mitochondrial 60-kd heat-shock protein, and their use in studying the expression of the protein in the rheumatoid arthritis. *Arthritis Rheum.* 35:1427–1433.

Steller, H. (1995) Mechanisms and genes of cellular suicide. *Science* 267:1445–1449.

Sung, Y-J. and Dietert, R.R. (1994) Nitric oxide (·NO)-induced mitochondrial injury among chicken ·NO-generating and target leukocytes. *J. Leukoc. Biol.* 56:52–58.

Takenaka, I.M. and Hightower, L.E. (1992) Transforming growth factor-β1 rapidly induces Hsp70 and Hsp90 molecular chaperones in cultured chicken embryo cells. *J. Cell. Physiol.* 152:568–577.

Taylor, D.E., Ghio, A.J. and Piontadosi, C.A. (1995) Reactive oxygen species produced by liver mitochondria of rats in sepsis. *Arch. Biochem. Biophys.* 316:70–76.

Thompson, C.B. (1995) Apoptosis in the pathogenesis and treatment of disease. *Science* 267:1456–1462.

Vignola, A.M., Chanez, P., Polla, B.S., Vic, P., Godard, P. and Bousquet, J. (1995) Increased expression of heat shock protein 70 on airway cells in asthma and chronic bronchitis. *Am. J. Respir. Cell Mol. Biol.* 13:683–691.

Villar, J., Edelson, J.D., Post, M., Brendan, J., Mullen, M. and Slutsky, A.S. (1993) Induction of heat stress proteins is associated with decreased mortality in an animal model of acute lung injury. *Am. Rev. Respir. Dis.* 147:177–181.

Villar, J., Ribeiro, S.P., Brendan, J., Mullen, M., Kuliszewski, M., Post, M. and Slutsky, A.S. (1994) Induction of the heat shock response reduces mortality rate and organ damage in a sepsis-induced acute lung injury model. *Crit. Care Med.* 22:914–921.

Vischer, T.L. (1988) A double blind multicentre study on OM-8980 and auronofin in rheumatois arthritis. *Ann. Rheum. Dis.* 47:482–489.

Welch, W.J. and Mizzen, L.A. (1988) Characterization of the thermotolerant cell. Effects on intracellular distribution of heat-shock protein 70, intermediate filaments and small nuclear ribonucleoprotein complexes. *J. Cell. Biol.* 106:1117–1130.

Zarley, J.H., Britigan, B.E. and Wilson, M.E. (1991) Hydrogen peroxide-mediated toxicity for *Leishmania donovani chagesi* promastigotes. Role of hydroxyl radical and protection by heat shock. *J. Clin. Invest.* 88:1511–1521.

Stress-Inducible Cellular Responses
ed. by U. Feige, R.I. Morimoto, I. Yahara and B. Polla
© 1996 Birkhäuser Verlag Basel/Switzerland

Attenuated heat shock transcriptional response in aging: Molecular mechanism and implication in the biology of aging

A.Y.-C. Liu, Y.-K. Lee, D. Manalo and L.E. Huang

Department of Biological Sciences, Rutgers, The State University of New Jersey, Piscataway, NJ 08855-1059, USA

Summary. A characteristic feature of aging is a progressive impairment in the ability to adapt to environmental challenges. The purpose of this review is to present the experimental evidence of an attenuated heat shock transcriptional response to heat and physiological stresses in a number of aging mammalian model systems. These include the human diploid fibroblasts in culture, whole animals and animal derived cells and cell cultures, as well as peripheral blood mononuclear cells obtained from human donors. The possibility that age-dependent changes in cellular redox status, as exemplified by the increased production of reactive oxygen intermediates and accumulation of oxidatively-modified proteins, affects the regulation and function of the heat shock factor 1 (HSF1) and contributes to the attenuated heat shock transcriptional response in aging cells and organisms is discussed. Given the fundamentally important role of HSPs in many aspects of protein homeostasis and signal transduction, it seems likely that the inability, or compromised ability, of aging cells and organisms to produce HSPs in response to stress would contribute to the well known increase in morbidity and mortality of the aged when challenged.

Introduction

A characteristic feature of aging is a progressive impairment in the ability to adapt to environmental changes. Manifestation of this in the aging human may include decreased responsiveness to cold and heat stress, reduced carbohydrate tolerance, declined immune function, adrenocortical abnormalities, and impaired ability of a number of enzymes to undergo adaptive regulation (Driscoll, 1971; Oechsli and Bueckley, 1970; Schneider and Rowe, 1990; Shock, 1977). The biochemical and molecular bases of these age-related changes are not understood, in part due to the multicomponent nature and complex circuits involved in the regulation of these processes at the organismic level. Nonetheless, it is generally believed that functional decline at the cellular, tissue, and organ levels as well as dysfunction of various homeostatic mechanisms contribute to the increased morbidity and mortality of the aged.

---

[1] *Present address:* Brigham and Women's Hospital, Division of Hematology/Oncology, Harvard Medical School, Boston, MA 02115, USA.

The heat shock response, defined as the rapid induction of heat shock proteins (HSP), was first described in 1962 by Ritossa in *Drosophila* (Ritossa, 1992). It has since been shown that the response is ubiquitous in all organisms examined, and that it can be elicited by a wide range of noxious stimuli (for reviews, see Lis and Wu, 1993; Morimoto, 1993; Morimoto et al., 1990). Thus the terms stress response and stress proteins were introduced to recognize the more general nature of the response and the implicit function of the proteins. The HSP are highly conserved, and they are important in maintaining protein homeostasis (Gething and Sambrook, 1992; Hendrick and Hartl, 1993; Hartmann and Gething and Cyr and Neupert, this volume) as well as in conferring survival under adverse conditions (Li and Nussenzweig, this volume). Together, these observations underscore the biological importance of the heat shock response and the HSP.

Given this background of the biology of aging and heat shock response, it would seem likely that changes in the regulation and function of the heat shock transcriptional response would occur during aging, and that these changes would contribute to the functional decline of the aged. The purpose of this review is present the experimental evidence in support of such a suggestion.

Definition of terms and the use of the human diploid fibroblast as a model system for studying aging

Definition

For those unfamiliar with the field of aging research, a word on terminology is in order. Aging designates the development sequence of changes that begin before birth and continue throughout the life span. Senescence is most commonly used to imply a generalized functional deterioration with aging. Although aging and senescence are often used interchangeably, there is interpretative differences in the two: aging refers more appropriately to the process of growing old regardless of chronologic age, whereas senescence is restricted to the state of old age characteristic of the later years of lifespan.

Human diploid fibroblasts as a model for studying cellular aging

The finite replicative lifespan of normal diploid cells in tissue culture was first reported more than 30 years ago by Hayflick and Moorhead (1961). They observed that cells from a variety of normal human tissues proliferate for various lengths of time but eventually cease to divide, and attributed this limited dividing potential of normal diploid cells in tissue culture as an expression of senescence at the cellular level (Hayflick and Moorhead, 1961). The initial study of Hayflick and Moorhead has since been extensively confirmed, expanded and refined by work from many other laboratories using cells from a variety of tissues from different species (for reviews, see Goldstein, 1990; Hayflick 1980; Norwood et al., 1990). We now know that: (1) the replicative potential of human diploid fibroblast culture *in vitro* is inversely related to the age of the donor from which the cells were derived; (2) cells from individuals with genetic disorders associated with some features of premature aging (e.g., progeroid syndromes) display a reduced replicative potential, and (3) there is a relationship between the replicative potential of fibroblast-like cultures and the maximum life span of the species from which the cell cultures are derived. For example, mouse fetal fibroblasts usually senesce after 10–15 population doublings, human cells after ~ 60 doublings, whereas cells from the Galapagos tortoise can proliferate for more than 100 population doublings. Collectively, these results reveal a critical connection between replicative senescence of fibroblast-like culture *in vitro* and aging at the organismic level, and suggest that the fibroblast culture system can be a useful model system for studying aging at the cellular level.

However, the use of fibroblast-like culture – notably the Human Diploid Fibroblast (HDF) – as a model system for studying aging has not been without controversy. It has been argued that many tissues of the body consist largely or entirely of post-mitotic, non-dividing cells from infancy onward. These are usually highly specialized cells such as neurons and muscle cells, and they live and function through the life span of the organism without undergoing cell division. It has also been argued that for cell populations that can be stimulated to divide *in vivo*, even though there is a limit of how many times these cells can divide, the limit is so large (~ 2^{60}) that it is rarely if ever reached by cells *in vivo* but is demonstrable *in vitro*. The corollary is that it is most unlikely that animals age and die because one or more important cell populations – including cell populations that are periodically or continuously renewed (such as fibroblasts) – lose their proliferative capacity. Finally, it has been noted that there are discrete examples whereby aging and age-related pathological conditions are associated with an over-active cell division process, benign prostate hyperplasia being one such example.

Perhaps some of the difficulties in appreciating the utility of HDF as a model system for studying aging have to do with an overemphasis of the loss of dividing potential as a working definition of the senescent HDF, to the extent that other functional parameters are often overlooked. The fact that "cellular senescence" and "replicative senescence" are interchangeable terms further perpetuates the notion that a loss of replicative potential is *the* parameter that defines a senescent cell. This is unfortunate, because it has long been recognized that many functional changes occur in HDF grown *in vitro*, and these functional changes are expressed well before the cells lose their capacity to replicate (Hayflick, 1979). In the context of cellular adaptive/defensive responses, we note, in addition to the attenuated response to heat shock which will be discussed below, that both the basal and the induced expression of interleukin-6 have been shown to decline progressively with age in HDF (Goodman and Stein, 1994). Similarly, the induction of metallothionein expression by cadmium has been shown to be down-regulated in aging HDF (Luce et al., 1993).

The inevitable conclusion of these studies is that important functional declines occur in the aging HDF. Some of these functional declines are likely to compromise the cell's ability to adjust to and survive adverse conditions. We suggest that these functional deficits observed at a cellular level contribute to the increased morbidity and mortality of the aging human. We further suggest that a better understanding of the molecular mechanism(s) of these functional deficits in the fibroblast culture system – including the attenuated heat shock transcriptional response – would provide the necessary framework for a better understanding of the biology of aging as well as in the development of pharmacological agents to moderate the aging process.

Aging and the attenuated heat shock transcriptional response

Attenuated heat shock transcriptional response in aging human diploid fibroblasts

Research in our laboratory has focused on the regulation and function of the heat shock transcriptional response in cell aging. Our intent is to use the human diploid fibroblast culture as a model system to gain insights into the molecular and cellular basis of aging. Analysis of the effects of heat shock (42 °C) and canavanine (an arginine analogue) treatment in the induction of HSP synthesis in young and old IMR-90 cells, a diploid cell strain derived from human embryonic lung tissue, showed that while similar series of HSPs were induced in the young and old cells, there was an inverse correlation of the inducibility of HSP synthesis and the replicative potential of cells, such that the magnitude of induction of HSP synthesis in the old cells was significantly reduced when compared to that of the young cells (Liu et al., 1989 a, b). Quantitation of the amount of translatable and hybridizable mRNA by the methods of *in vitro* translation and Northern blot hybridization, and of the basal and heat-induced transcription rate of the hsp70 gene by nuclear run-on assay, as well as analysis of the expression of an hsp70 promoter-driven reporter gene construct in transient transfection assay provided clear evidence that the difference in expression of HSPs in young and old cells is attributable to a transcriptional mechanisms (Liu et al., 1989 a, b). This conclusion is corroborated by analysis of the heat inducible heat shock transcription factor (HSF) DNA-binding activity (Choi et al.,

1990). Quantitation of the affinity and capacity of the HSF DNA-binding activity by Scatchard analysis showed that while the affinity of the binding does not vary as a function of age, the capacity of this heat-inducible DNA-binding activity is inversely related to the age of the cells (Choi et al., 1990). Furthermore, we showed, by varying the temperature and time of heat shock, that as cells age a progressively higher temperature and longer time of incubation at the heat shock temperature are required to induce the response (Liu et al., 1991).

Qualitatively similar results have been reported by other laboratories using a different cell strain, the WI-38 human diploid lung fibroblasts (Campanini et al., 1990; Luce and Christofalo, 1992). Furthermore, it was noted that the decreased induction of HSPs in senescent HDF is correlated with increased thermosensitivity (Luce and Christofalo, 1992), a result consistent with the suggestion that HSPs confer thermotolerance to cells (Li and Nussenzweig, this volume). Whether an attenuated heat shock transcriptional response is a signature feature of *all* aging human diploid fibroblast culture systems is not entirely clear. We note that in an earlier study, Tsuji et al., (1986) reported little age-dependent difference in either the basal or the heat-induced HSP pattern in the TIG-1 strain of human diploid fibroblasts. Whether this difference in result is attributable to the use of different diploid cell strain or to some other reason has yet to be determined.

In considering the possible molecular mechanism of the attenuated heat shock response as well as the implication of this observation in the biology of aging, we note that this functional change in response to heat shock is progressive and can be detected well before the HDF lose their capacity to replicate (Liu et al., 1991). For example, the ability of HDF to mount a response to heat shock of 39 °C, a physiologically relevant temperature in the human, is already substantially compromised after 35 cumulative population doublings (Liu et al., 1991). Given the average limit of 50–60 cumulative population doublings of HDF *in vitro*, this result shows that a deficient response to heat shock is not restricted to cells at the very end of their replicative life-span. This finding is particularly noteworthy from a gerontological perspective for the following reason. The replicative potential of the human diploid fibroblast is quite large, such that even cells from the oldest donor still have some 20 population doublings left (Martin et al., 1970). Our observation of a significant functional decline of the heat shock response of cells at this stage of their life history would suggest that this phenotype at the cellular level is relevant to the biology of human aging.

Attenuated heat shock transcriptional response in aging animal systems

Studies done in other laboratories showed that an attenuated response to heat shock is also observed in aging animal model systems and in primary cell cutures derived from these animals. Thus, work from Holbrook's laboratory showed a decreased expression of heat shock protein 70 mRNA and protein upon heat shock of lung- or skin-derived fibroblasts of aged Wistar rats when compared to cells derived from young/adult animals (Fagnoli et al., 1990). Using hepatocytes isolated from young/adult and old Fischer F334 rats as the experimental system, studies from Richardson's laboratory showed a 40–50% decrease in the induction of hsp70 synthesis, mRNA levels, transcription rate of the hsp70 gene, and HSF DNA-binding activity by heat shock in cells isolated from old animals when compared to that of cells isolated from young/adult animals (Heydari et al., 1993). The relevance of this attenuated heat shock transcriptional response to the biology of aging was further strengthened by the observation that caloric restriction, the only regimen know to retard aging and increase survival of mammals, reverses the age-related decline in the induction of hsp70 transcription in hepatocytes (Heydari et al., 1993).

Experiments using whole body hyperthermia as a paradigm of heat shock showed that while there is clearly the trend of an attenuated response to heat shock in the aging animals, the extent of the response may vary depending on the particular cell type, tissue, organ, and the particular individual animal under consideration (Blake et al., 1990, 1991a; Pardue et al., 1992). For example, analysis of the effects of thermal stress in the regulation of heat shock gene expression in the central nervous system of the rat demonstrated that the induction of hsp70 mRNA was highest in a subpopulation of glia, intermediate in granule cells of the dentate gyrus, and lowest in pyramidal cells of Ammon's horn (Pardue et al., 1992). Further, the levels of this heat-induced mRNA were several-fold lower in the dentate gyrus granule cells of aged rats as compared to that of young/adult animals, and were also reduced in many pyramidal cells of the hippocampus but not in hippocampal glia (Pardue et al., 1992). Studies of whole body hyperthermia treatment on the induction of hsp70 mRNA in various organ systems also showed clear age-related decline in the induction in brain, lung and skin; however, hsp70 expression in liver of these animals was usually low and displayed no apparent correlation with age or heat-stress conditions (Blake et al., 1990, 1991a).

Induction of the heat shock transcriptional response is dependent on the temperature and duration of heat stress. This being given, the possibility that there may be an age-dependent difference in body temperature regulation in response to the hyperthermic treatment needs to be addressed. In the study on the regulation of HSP expression in the CNS, there is no difference in the mean colonic temperature versus time profiles for heat shocked young/adult and aged rats; the rate of increase of colonic temperature as well as the rate of recovery from the heat shock regimen are identical (Pardue et al., 1992). On the other hand, Blake et al. (1991a) observed an attenuated increase in colonic temperature of old rats when compared to that of the young animals subjected to the same regimen of hyperthermia, leading the authors to suggest that the observed age-related differences in hsp70 mRNA expression may be due to a difference in core body temperature in response to the heat stress rather than an intrinsic impairment in the regulation of hsp70 gene expression. The reason for this difference in result reported in the two studies is not clear.

The effects of age on heat shock gene expression has also been determined in a fourth model system: the peripheral blood mononuclear cells obtained from human donors. Results showed that induction of hsp expression in response to heat- and mitogen-stimulation is inversely related to the age of the donor from which the cells were derived (Deguchi et al., 1988; Faassen et al., 1989). Furthermore, the decreased expression of HSPs following mitogen stimulation of lymphocytes from aged donors is associated with a decreased proliferative response of the cells (Faassen et al., 1989). Since peripheral blood lymphocytes can be obtained from human donors relatively easily (especially when compared to other tissues or cell types), this system may provide an especially attractive avenue for exploring age-related changes in human.

Together, these results show that an attenuated heat shock response is a feature common to a number of aging mammalian cells and organisms. Whether this phenotype is applicable to other non-mammalian aging model systems – including the fruit fly (*D. melanogaster*), the worm (*C. elegans*), and yeast (*S. cerevisiae*) – remains to be seen. In fact, studies done in *Drosophila melanogaster* showed an unusual pattern of expression of the HSPs in senescent flies (Fleming et al., 1988; Niedzwiecki et al., 1991). Using two-dimensional gel electrophoresis as the tool of analysis of protein synthesis, the authors reported that while a 20-min heat shock ($25 \rightarrow 37\,°C$) results in the expression of 14 new proteins in the young-adult flies, the same treatment of old flies leads to the expression of at least 50 new or highly up-regulated proteins. Furthermore, feeding young flies with canavanine, a proceduce that results in the production of analogue substituted and presumably mis-folded proteins in the flies, gives a pattern of HSP response similar to that observed in senescent flies (Fleming et al., 1988; Niedzwiecki et al., 1991). The authors suggest that the accumulation of conformationally altered proteins in the senescent flies plays a role in the regulation of the heat shock protein response.

Aging, stress and the stress response

Virtually any significant challenge, physical or psychological, to an organism's well being results in a neurally mediated release of catecholamines from the adrenal medulla and initiates the neuroendocrine cascade of events that results in the secretion of glucocorticoids from the adrenal cortex. The release of these hormones in turn triggers a well-characterized array of stress or emergency responses that involve a number of organ systems and metabolic processes (McCubbin et al., 1991). There is overwhelming evidence that this cascade of events is important in maintaining the well being of the organism, and that impairment in this and other homeostatic control mechanisms contributes to the increased morbidity and mortality of the aged (Ferrari et al., 1990; Greden et al., 1986; Spencer et al., 1990).

In this context, it becomes clear that an important question that we need to ask in order to appreciate the importance of the attenuated heat shock transcriptional response in the biology of aging is whether other forms of physiological and behavioral stress of the organism – not just heat shock – induce the expression of HSPs, and whether this induction attenuates with the age of the organism. Using physical restraint of the animal as a means of activating the hypothalamic-pituitary-adrenal axis, studies from Holbrook's laboratory showed that restraint stress induces the expression of hsp70 mRNA in both the adrenal cortex and vasculature (notably the smooth muscle layer of the thoracic and abdominal aortas as well as the vena cava) (Blake et al., 1991b; Fawcett et al., 1994; Udelsman et al., 1993). The response is rapid and is mediated by activation of the HSF1 DNA-binding activity. Furthermore, while the adrenal response is adreno-corticotropin-dependent, the vascular response is under adrenergic control. Thus, the ability of restraint to induce hsp70 expression in the adrenal cortex is virtually eliminated in hypophysectomized animals and is restored by the exogenous administration of ACTH, whereas the vasculatur response is selectively blocked by either $\alpha 1$- or β-adrenergic blocking agents. It may be noted that qualitatively similar results were obtained in animals subjected to surgical stress (ether anesthesia, laparotomy, hemorrhage, and recovery) (Udelsman et al., 1991). The elegance of these findings is that the induction of hsps is initiated by the animal, is tissue-specific, and is under neurohormonal control, suggesting that induction of the heat shock transcriptional response can provide a molecular and biochemical index of stress of the animal.

Significantly, both the adrenal as well as the vascular response to restraint stress down-regulates with age. Thus, induction of hsp70 expression and activation of the HSF DNA-binding activity were dramatically reduced in older animals when compared to that of the young/adult rats (Blake et al., 1991b; Fawcett et al., 1994; Udelsman et al., 1993). Whether this attenuated expression of HSP70 in the aging rats is due to a functional change of the hypothalamus-pituitary-adrenal activity or an alternation in the target cells/tissues of the mechanisms involved in hsp gene expression is not entirely clear. It is also noteworthy that it has thus far not been possible to mimic the *in vivo* response to ACTH in *in vitro* cultured adrenocortical cells using ACTH or other stimulators of cAMP, an intracellular mediator of the effects of ACTH (Fawcett et al., 1994). This result highlights the complex neuro-endocrine response to stress, and suggests a requirement for some unknown factor, perhaps acting in concert with ACTH, in eliciting the induction of HSPs in a target tissue specific manner.

In summary, these studies – using a number of different mammalian models of aging as well as different means to elicit the heat shock response – show that an attenuated response to stress in the induction of HSPs is a biomarker of aging both *in vivo* (whole organisms) and *in vitro*.

We would like to suggest that the inability or diminished ability of the aging cells and organisms to produce HSPs when challenged could have a broad range of negative functional consequences. Analysis of the function of HSPs provided clear evidence that they serve as molecular chaperones in cells and are involved in many aspects of protein homeostasis, from the folding and assembly of newly synthesized proteins to their trafficking, repair or degradation (Gething and Sambrook, 1992; Hendrick and Hartl, 1993; Hartman and Gething, Sherman and Goldberg, this volume). Not surprisingly, induction of HSPs has been shown to confer thermotolerance and is necessary for the survival of cells under adverse conditions (see chapter by Li and Nussenzweig, this volume). An equally important but often overlooked function of HSPs is that these protein chaperones, by facilitating protein conformation change, have fundamentally important roles in signal transduction and regulation of gene expression. It is well known that HSPs have essential roles in regulating the function, folding, and trafficking of the family of steroid receptor proteins and the transforming tyrosine kinase $pp60^{v-src}$ (Pratt et al., 1993; Rutherford and Zuker, 1994; Pratt et al., this volume). It has been suggested that the HSP molecular chaperones that promote the folding and assembly of nascent proteins also play a pivotal role in a variety of regulatory processes by facilitating structural transitions of signaling molecules and transcription factors to switch between "on" and "off" states (Rutherford and Zuker, 1994). Clearly, this has important implications concerning the role of HSPs in a variety of complex biological processes such as cell growth regulation, cell transformation and aging.

Changes in the regulation and function of the heat shock factor1 (HSF1) in aging

HSF1 is a constitutively expressed protein and is stored in a latent monomeric form in the cytosol of non-stressed cells. Activation of HSF1, which mediates the effects of heat shock, involves the conversion of the HSF1 monomer to a sequence-specific-DNA-binding homotrimer, nuclear translocation, as well as phosphorylation events that may modulate the transactivating activity of the protein (Morimoto et al., this volume). Given that the attenuated heat shock response in aging cells and animals is directly attributable to a decrease in the amount/activity of activated HSF1, it reasonably follows that a better understanding of the regulation and function of HSF1 would provide important insights into the mechanistic basis of this attenuation.

In evaluating the possible changes in the regulation and function of HSF1 as a basis for the attenuated heat shock response in aging cells and organisms, the two most obvious possibilities are: (1) a decrease in the expression and steady state level of HSF1, and (2) a compromised activation mechanism of HSF1, perhaps due to defects in the sensing and signal transduction mechanism of heat shock, or changes in the biochemistry and/or subcellular localization of HSF1 that attenuates its ability to undergo trimerization and activation.

In their analysis of the effects of neurohormonal stress and aging on the activation of HSF1 in rats, studies from Holbrook's laboratory showed that while the levels of HSF1 remain unchanged as a function of age, it nevertheless exhibits a decreased ability to bind DNA (Fawcett et al., 1994). This result is corroborated by studies done in our laboratory on heat shocked young- and old-IMR-90 human diploid lung fibroblast. Furthermore, the decreased HSF DNA-binding activity is directly attributable to a reduced ability of heat shock to promote the trimerization of HSF1 (Y.-K. Lee and A.Y.-C. Liu, unpublished data). Together, these results show that while the expression of HSF1 does not change with age, its ability to sense the heat shock signal and to undergo trimerization attenuates with age.

Redox regulation, the heat shock response, and aging: Unanswered questions and future directions

We do not yet know how cells sense changes in the temperature of their environment. There is little information concerning the nature of the cellular "thermometer" or the signal transduction mechanism involved. Suffice it to say that the process is exquisitely sensitive (e.g., a change of $2\,^{\circ}C$ of incubation temperature in IMR-90 cells is sufficient to elicit the response; Liu et al., 1991), rapid (within minutes if not seconds; Zimarino and Wu, 1987; O'Brien and Lis, 1991, 1993), and most likely detects a relative as opposed to an absolute temperature change (Rabindran et al., 1993; Liu et al., 1994).

A hallmark of the heat shock response is the broad variety of stressors besides heat that induce the response, and in this context, the term stress response serves a useful purpose. A survey of the reagents/conditions that elicit the heat shock response indicates that many of them are capable of either generating reactive oxygen intermediates in cells or depleting cellular SH function; these include reoxygenation following hypoxia, oxidizing agents (H_2O_2), uncouplers of oxidative phosphorylation, heavy metal ions, and SH-reactive reagents (Burdon et al., 1990; Loven, 1988; Nover, 1984). The observations that heat shock increases the rate of cyanide-resistant respiration and induces the expression of superoxide dismutase (Privalle and Fridovich, 1987); that inhibition of antioxidant defenses increases susceptibility to killing by heat shock (Mitchell and Russo, 1983); that overexpression of HSPs confers resistance to oxidative stress (Spitz et al., 1987); and that deletion of the rpoH gene (which encodes the heat shock σ factor in *E. coli*) results in sensitization of the cells to oxidative stress (Kogoma and Yura, 1992) further suggest that heat and oxidative stress have common cellular effects.

These considerations summarized above prompted us to ask the following two questions: (1) whether a redox mechanism is involved in the heat shock signal transduction pathway? and (2) whether the attenuated heat shock response of aging cells may be causally related to age-dependent changes in the redox status of cells? In an attempt to address the first question, we evaluated if induction of the heat shock transcriptional response may be modulated by sulfhydryl-reducing reagents (Huang et al., 1994). Our results showed that dithiothreitol inhibits the heat-induced increase in the synthesis of HSPs, abundance of mRNA of hsp70, hsp70 gene promoter activity, and the heat shock factor DNA-binding activity. This effect of DTT is specific and rapidly reversible; further, oxidized DTT is ineffective whereas other thiol-reducing compounds have the same effect as DTT. Furthermore, analysis of the effects of DTT on the regulation and function of HSF suggests that DTT blocks an early and important step in the activation process without having a direct effect on the HSF protein. These results, together with the observation that activation of HSF DNA-binding activity is attenuated under an anoxic condition and that hydrogen peroxide can activate the HSF DNA-binding activity suggest the involvement of a redox mechanism as an early and important step in the heat shock signal transduction pathway (Huang et al., 1994).

The notion that redox is an important step in the sequence of events leading to the activation of HSF1 may also have important implications in our analysis of the biochemical and molecular basis of the attenuated heat shock transcriptional response in aging cells. A variety of evidence suggests that oxygen free radicals (i.e., ROI) and oxidant-induced damages to cellular macromolecules are important contributing factors in the aging process and age-related pathological condition (for reviews, see Ames et al., 1993; Emerit and Chance, 1992; Harman, 1992; Shigenaga et al., 1994; Stadtman, 1992). These evidence include: (i) Increases in endogenous levels of ROI in the latter part of the life span in insects as well as mammals (Sohal, 1993; Sohal and Brunk, 1992; Sohal and Sohal, 1991). (ii) The amount of oxidized cellular proteins increases as a function of age in a number of *in vivo* and *in vitro* aging models (Stadtman, 1992). (iii) Manipulation of the physical activity (metabolism) of houseflies uncovered an inverse correlation between oxidative molecular damage and life expectancy (Agarwal and Sohal, 1994). (iv) Protection against oxidative damages by over-expression of antioxidant enzymes (superoxide dismutase and catalase) results in a significant extension of life-span, a lower amount of protein oxidative damage and a delayed loss in physical performance in *Drosophila melanogaster* (Orr and Sohal, 1994). (v) Genetic and developmental conditions that extend the life span of *C. elegans* also up-regulate superoxide dismutase and catalase activities and confer resistance against oxidative damage (Larsen, 1993). (vi) The common oxidizing reagent, H_2O_2, can induce an irreversible senescent-like state in human diploid fibroblasts (Chen and Ames, 1994).

We note that changes in the redox status of cells can have significant and broad implication in a number of important cellular processes. Reactive oxygen intermediates (ROI; or oxygen radicals) have been suggested to serve as a second messenger in the activation of NFκB (Schreck and Baeuerle, 1991; Schreck et al., 1991). There is also ample evidence that active oxygen can induce the expression of *c-Fos and c-Jun*, and this induction may contribute to the mitogenic and/or tumor promotion effects of oxidants (Amstadt et al., 1992; Crawford et al., 1988; Devary et al., 1992; Shibanuma et al., 1988). *In vitro* analysis of the regulation and function of *c-Fos, c-Jun* and NF-kB showed that preservation of the SH-function of a conserved cysteine residue in the DNA-binding domain of the proteins is essential for DNA-binding activity; modification (alkylation or oxidation) of this cys-SH group blocks DNA-binding (Abate et al., 1990; Hayashi et al., 1993; Matthews et al., 1990). The recent isolation of a ubiquitous nuclear redox factor (Ref-1) which reduces and stimulates the *in vitro* DNA-binding activity of *Fos/Jun* and NFκB further underscores the importance of redox in the regulation of transcription factor activity (Xanthoudakis and Curran, 1992; Xanthoudakis et al., 1992).

These considerations suggest that age-dependent changes in the cellular redox status may impact on a number of signal transduction processes as well as the function of a variety of transcription factors – perhaps the heat shock response and the HSF1 included. We suggest that a better understanding of the attenuated heat shock transcriptional response in cell aging would contribute to our understanding of how changes in the cellular environment (redox in particular) modulates gene activity and cell function, and would provide useful insights into the molecular basis of impaired adaptive functions in senescence.

Acknowledgement
This work was supported by grants from the American Cancer Society and the National Science Foundation. We thank Dr. David T. Denhardt for a critical reading of the manuscript.

References

Abate, C., Patel, L., Rauscher, F.J. and Curran, T. (1990) Redox regulation of *Fos* and *Jun* DNA-binding activity *in vitro*. *Science* 249:157–1161.

Agarwal, S. and Sohal, R.S. (1994) DNA oxidative damage and life expectancy in houseflies. *Proc. Natl. Acad. Sci. USA* 91:12332–12335.

Ames, B.N., Shigenaga, M.K. and Hagen, T.M. (1933) Oxidants, antioxidants, and the degenerative diseases of aging. *Proc. Natl. Acad. Sci. USA* 90:7915–7922.

Amstad, P.A., Kupitza, G. and Gerutti, P.A. (1992) Mechanism of *c-fos* induction by active oxygen. *Cancer Res.* 52:3952–3960.

Blake, M.J., Gershon, D., Fargnoli, J. and Holbrook, N.J. (1990) Discordant expression of heat shock protein mRNAs in tissues of heat stressed rats. *J. Biol. Chem.* 265:15275–15279.

Blake, M.J., Fargnoli, J. Gershon, D. and Holbrook, N.J. (1991a) Concomitant decline in heat induced hyperthermia and HSP 70 mRNA expression in aged rats. *Amer. J. Physiol.* 260:R663–R667.

Blake, M.J., Udelsman, R., Feulner, G.J., Norton, D.D. and Holbrook, N.J. (1991b) Stress-induced heat shock protein 70 expression in adrenal cortex: an adrenocorticotropic hormone-sensitive, age-dependent response. *Proc. Natl. Acad. Sci. USA* 88:9873–9877.

Burdon, R.H., Gill, V. and Evans, C.R. (1990) Active oxygen species and heat shock protein induction. *In:* M.J. Schlesinger, M.G. Santaro and E. Garaci (eds): *Stress Proteins: Induction and Function.* Springer-Verlag, Berlin, New York, pp 19–25.

Campanini, C., Petronini, P.G., Marmiroli, N. and Borghette, A.F. (1990) Heat shock response in senescent human fibroblasts. *In:* H.L. Sega, M. Rothsein and E. Gergamini (eds): *Protein Metabolism in Aging.* Wiley-Liss, New York, pp 239–244.

Chen, Q. and Ames, B.N. (1994) Senescence-like growth arrest induced by hydrogen peroxide in human diploid fibroblast F 65 cells. *Proc. Natl. Acad. Sci. USA* 91:4130–4134.

Choi, H.S., Lin, Z., Li, B. and Liu, A.Y.-C. (1990) Age-dependent decrease in the heat-inducible DNA-sequence-specific binding activity in human diploid fibroblasts. *J. Biol. Chem.* 265:18005–18011.

Crawford, D., Zbinden, R., Amstad, P. and Cerutti, P. (1988) Oxidant stress induces the proto-oncogenes *c-fos* and *c-myc* in mouse epidermal cells. *Oncogene* 3:27–32.

Deguchi, Y., Negoro, S. and Kishimoto, S. (1988) Age-related changes of heat shock protein gene transcription in human peripheral blood mononuclear cells. *Biochem. Biophys. Res. Commun.* 157:580–584.

Devary, Y., Gottlieb, R.A., Smeal, T. and Karin, M. (1922) The mammalian ultraviolet response is triggered by activation of *Src* tyrosine kinase. *Cell* 71:1081–1091.

Driscoll, D.M. (1971) The relationship between weather and mortality in the major metropolitan areas in the United States, 1962–1965. *Int. J. Biometeorol.* 15:23–39.

Emerit, I. and Chance, B. (eds) (1992) *Free Radicals and Aging.* Birkhäuser Verlag, Basel.

Faassen, A.E., O'Leary, J.J., Rodysill, K.J., Bergh, N. and Hallgren, H.M. (1989) Diminished heat shock protein synthesis following mitogen stimulation of lymphocytes from aged donors. *Expt. Cell Res.* 183:326–334.

Fagnoli, J., Kunisada, T., Fornace, A.J., Schneider, E.L. and Holbrook, N.J. (1990) Decreased expression of heat shock protein 70 mRNA and protein after heat treatment in cells of aged rats. *Proc. Natl. Acad. Sci. USA* 87:846–850.

Fawcett, T.W., Sylvester, S.L., Sarge, K.D., Morimoto, R.I. and Holbrook, N.J. (1994) Effects of neurohormonal stress and aging on the activation of mammalian heat shock factor 1. *J. Biol. Chem.* 269:32272–32278.

Ferrari, E., Solerte, S.B., Magri, F., Fioravanti, M., Magni, P., Dori, D. and Bossolo, P. (1990) Age-related changes of the pituitary-adrenal axis in humans. *In:* G. Nappi, E. Martignoni, A.R. Genazzani and F. Petraglia (eds): *Stress and the Aging Brain: Integrative Mechanisms.* Raven Press, New York, pp 39–52.

Fleming, J.E., Walton, J.K., Dubitsky, R. and Bench, K.G. (1988) Aging results in an unusual expression of *Drosophila* heat shock proteins. *Proc. Natl. Acad. Sci. USA* 85:4099–4103.

Gething, M.J. and Sambrook, J. (1992) Protein folding in the cell. *Nature* 355:33–45.

Goldstein, S. (1990) Replicative senescence: The human fibroblast comes of age. *Science* 249:1129–1133.

Goodman, L. and Stein, G.H. (1994) Basal and induced amounts of interleukin-6 mRNA decline progressively with age in human fibroblasts. *J. Biol. Chem.* 269:19250–19255.

Greden, J.F., Flegel, P., Haskett, R., Dilsaver, S., Carroll, B.J., Grunhaus, L. and Genero, N. (1986) Age effect in serial hypothalamic-pituitary-adrenal monitoring. *Psychoneuroendocrinology* 11:195–204.

Harman, D. (1992) Free radical theory of aging: History. *In:* I. Emerit and B. Chance (eds): *Free Radicals and Aging.* Birkhäuser Verlag, Basel, pp 1–10.

Hayashi, T., Ueno, Y. and Okamoto, T. (1993) Oxidoreductive regulation of nuclear factor kB: Involvement of a cellular reducing catalyst thioredoxin. *J. Biol. Chem.* 268:11380–11388.

Hayflick, L. and Moorhead, P.S. (1961) The serial cultivation of human diploid cell strains. *Exp. Cell. Res.* 25:585–621.

Hayflick, L. (1979) Cell aging. *In.:* A. Cherkin, C. Finch, N. Kharasch, F.L. Scott, T. Makinodan and B. Strehler (eds): *Physiology and Cell Biology of Aging.* Raven Press, New York, pp 3–19.

Hayflick, L. (1980) The cell biology of human aging. *Scientific American* 242:58065.

Hendrick, J.P. and Hartl, F.-U. (1993) Molecular chaperone functions of heat shock proteins. *Ann. Rev. Biochem.* 62:349–384.

Heydari, A.R., Wu, B., Takahashi, R., Strong, R. and Richardson, A. (1993) Expression of heat shock protein 70 is altered by age and diet at the level of transcription. *Mol. Cell. Biol.* 13:2902–2918.

Huang, L.E., Zhang, H., Bae, S.W. and Liu, A.Y.-C. (1994) Thiol reducing reagents inhibit the heat shock response: Involvement of a redox mechanism in the heat shock signal transduction pathway. *J. Biol. Chem.* 269:30718–30725.

Kogoma, T. and Yura, T. (1992) Sensitization of *E. coli* cells to oxidative stress by deletion of the rpoH gene, which encodes the heat shock sigma factor. *Mol. Cell. Biol.* 174:630–632.

Larsen, P.L. (1993) Aging and resistance to oxidative damage in *C. elegans. Proc. Natl. Acad. Sci. USA* 90:8905–8909.

Lis, J. and Wu, C. (1993) Protein traffic on the heat shock promoter: parking, stalling and trucking along. *Cell* 74:1–4.

Liu, A.Y.-C., Bae-Lee, M.S., Choi, H.S. and Li, B. (1989a) Heat shock induction of HSP89 is regulated in cellular aging. *Biochem. Biophys. Res. Commun.* 162:1302–1310.

Liu, A.Y.-C., Lin, Z., Choi, H.S., Sorhage, F. and Li., B. (1989b) Attenuated induction of heat shock gene expression in aging diploid fibroblasts. *J. Biol. Chem.* 264:12037–12045.

Liu, A.Y.-C., Choi, H.S., Lee, Y.K. and Chen, K.Y. (1991) Molecular events involved in transcriptional activation of heat shock genes become progressively refractory to heat stimulation during aging of human diploid fibroblasts. *J. Cell. Physiol.* 149:560–566.

Liu, A.Y.-C., Bian, H., Huang, L.E. and Lee, Y.C. (1994) Transient cold shock induces the heat shock response upon recovery at 37°C in human cells. *J. Biol. Chem.* 269:14768–14775.

Loven, D.P. (1988) A role for reduced oxygen species in heat induced cell killing and the induction of thermotolerance. *Medical Hypothesis* 26:39–50.

Luce, M.W. and Cristofalo, V.J. (1992) Reduction in heat shock gene expression correlates with increased thermosensitivity in senescent human fibroblasts. *Expt. Cell. Res.* 202:9–16.

Luce, M.C., Schyberg, J.P. and Bunn, C.L. (1933) Metallothionein expression and stress responses in aging human diploid fibroblasts. *Expt. Gerontology* 28:17–38.

Martin, G.M., Sprague, C.A., Epstein, C.J. (1970) Replicative life-span of cultivated human cells. *Lab. Invest.* 23:86–92.

Matthews, J.R., Wakasugi, N., Virelizier, J., Yodoi, J. and Hay, R.T. (1992) Thioredoxin regulates the DNA binding activity of NFκB by reduction of a disulphide bond involving cysteine 62. *Nucleic Acid Res.* 20:3821–3830.

McCubbin, J.A., Kaufmann, P.G. and Nemeroff, C.B. (1991) *Stress, Neuropeptides and Systemic Disease.* Academic Press, Harcourt Brace Jovanovich Publishers, New York.

Mitchell, J.B., Russo, A. (1983) Thiols, thiol depletion, and thermosensitivity. *Rad. Res.* 95:471–485.

Morimoto, R.I. (1993) Cells in stress: transcriptional activation of heat shock genes. *Science* 259:1409–1410.

Morimoto, R.I., Tissieres, A. and Georgopoulos, C. (1990) *Stress Proteins in Biology and Medicine,* Cold Spring Harbor Laboratory Press, Cold Spring Harbor, New York, p 449.

Niedzwiecki, A., Kongpachith, A.M. and Fleming, J.E. (1991) Aging affects expression of 70 kDa heat shock proteins in *Drosophila*. *J. Biol. Chem.* 266:9332–9338.

Norwood, T., Smith, J.R. and Stein, G.H. (1990) Aging at the cellular level: the human fibro-blastlike cell model. *In:* E.L. Schneider and J.W. Rowe (eds): *Handbook of the Biology of Aging,* 3rd Edition. Academic Press, Harcourt Brace Jovanovich Publishers, New York, pp 131–154.

Nover, L., (1984) Inducers of heat shock protein synthesis. *In:* L. Nover (ed): *Heat Shock Response of Eukaryotic Cells.* Springer-Verlag, Berlin, pp 7–11.

O'Brien, T. and Lis, J.T. (1991) RNA polymerase II pauses at the 5′ end of the transcriptionally induced *Drosophila* hsp70 gene. *Mol. Cell. Biol.* 11:5285–5290.

O'Brien, T. and Lis, J.T. (1993) Rapid changes in *Drosophila* transcription after an instanta-neous heat shock. *Mol. Cell. Biol.* 13:3456–3463.

Oechsli, F.W. and Bueckley, R.W. (1970) Excess mortality associated with three Los Angeles September hot spells. *Environ. Res.* 3:277–284.

Orr, W.C. and Sohal, R.S. (1994) Extension of life-span by overexpression of superoxide dismutase and catalase in *Drosophila melanogaster. Science* 263:1128–1130.

Pardue, S., Groshan, K., Raese, J.D. and Morrison-Bogorad, M. (1992) Hsp 70 mRNA induc-tion is reduced in neurons of aged rat hippocampus after thermal stress. *Neurobiol. Aging* 13:661–672.

Pratt, W.B. (1993) The role of heat shock proteins in regulating the function, folding and trafficking of the glucocorticoid receptor. *J. Biol. Chem.* 268:21455–21458.

Privalle, C.T. and Fridovich, I. (1987) Induction of superoxide dismutase in *E. coli* by heat shock. *Proc. Nat. Acad. Sci. USA* 84:2723–2726.

Rabindran, S.K., Haroun, R.I., Clos, J., Wisniewski, J. and Wu, C. (1993) Regulation of heat shock factor trimer formation: role of a conserved leucine zipper. *Science* 259:230–234.

Ritossa, F. (1962) A new puffing pattern induced by temperature shock and DNP in *Drosophila. Experientia* 18:571–573.

Rutherford, S.L. and Zuker, C.S. (1994) Protein folding and the regulation of signaling path-ways. *Cell* 79:1129–11132.

Schneider, E.L. and Rowe, J.W. (1990) *Handbook of the Biology of Aging,* 3rd Edition, Academic Press, Harcourt Brace Jovanovich Publishers, New York.

Schreck, R. and Baeuerle, P. (1991) A role for oxygen radicals as second messengers. *Trends in Cell Biol.* 1:39–42.

Schreck, R., Rieber, P. and Baeuerle, P.A. (1991) Reactive oxygen intermediates as apparently widely used messengers in the activation of the NF-κB transcription factor and HIV-1. *EMBO J.* 10:2247–2258.

Shibanuma, M., Kuroki, T. and Nose, K. (1988) Induction of DNA replication and expression of proto-oncogene *c-myc* and *c-fos* in quiescent Balb/3T3 cells by xanthine/xanthine oxidase. *Oncogene* 3:17–21.

Shigenaga, M.K., Hagen, T.M. and Ames, B.N. (1994) Oxidative damage and mitochondrial decay in aging. *Proc. Natl. Acad. Sci. USA* 91:10771–10778.

Shock, N.W. (1977) Systems integration. *In:* C.E. Finch, L. Hayflick, H. Brody, I. Rossman and F. Sinex (eds): *Handbook of the Biology of Aging.* van Nostrand Reinhold Co. New York, 639–665.

Sohal, R.S. and Sohal, B.H. (1991) Hydrogen peroxide release by mitochondria increases during aging. *Mechanisms of Ageing and Dev.* 57:187–202.

Sohal, R.S. and Brunk, U.T. (1992) Mitochondrial production of pro-oxidants and cellular senescende. *Mutation Res. Lett.* 275:295–304.

Sohal, R.S. (1993) Aging, cytochrome oxidase activity, and hydrogen peroxide release by mitochondria. *Free Radical Biol. and Med.* 14:583–588.

Spencer, R.L., Miller, A.H., Young, E.A. and McEwen, B.S. (1990) Stress-induced changes in the brain: Implications for aging. *In:* G. Nappi, E. Martignoni, A.R. Genazzani and F. Petraglia (eds): *Stress and the Aging Brain: Integrative Mechanisms.* Raven Press, New York, pp 17–29.

Spitz, D.R., Dewey, W.L. and Li, G.C. (1987) Hydrogen peroxide or heat shock induces resistance to hydrogen peroxide in Chinese hamster fibroblasts. *J. Cell. Physiol.* 131:364–373.

Stadtman, E.R. (1992) Protein oxidation and aging. *Science* 257:1220–1224.

Tsuji, Y., Ishibashi, S. and Ide, T. (1986) Induction of heat shock proteins in young and senes-cent human diploid fibroblasts. *Mechanisms Ageing and Dev.* 36:155–160.

Udelsman, R., Blake, M., Holbrook, N.J. (1991) Molecular response to surgical stress: specific and simultaneous heat shock protein induction in the adrenal cortex, aorta and vena cava. *Surgery* 110:1125–1130.

Udelsman, R., Blake, M.J., Stagg, C.A., Li, D., Putney, J. and Holbrook, N.J. (1993) Vascular heat shock protein expression in response to stress. Endocrine and autonomic regulation of this age-dependent response. *J. Clin. Invest.* 91:465–473.

Xanthoudakis, S. and Curran, T. (1992) Identification and characterization of Ref-1, a nuclear protein that facilitates AP-1 DNA-binding activity. *EMBO J.* 11:653–665.

Xanthoudakis, S., Miao, G., Wang, F., Pan, Y.-C. and Curran, T. (1992) Redox activation of Fos-Jun DNA-binding activity is mediated by a DNA-repair enzyme. *EMBO J.* 11: 3323–3335.

Zimarino, V. and Wu, C. (1987) Induction of sequence-specific binding of *Drosophila* heat shock activator protein without prior protein synthesis. *Nature* 327:727–730.

Part V
Applications of stress responses in toxicology and pharmacology

Part V: Applications of stress responses in toxicology and pharmacology

I. Yahara

The Tokyo Metropolitan Institute of Medical Science, Department of Cell Biology, Honkomagome 3-18-22, Bunkyo-Ku, Tokyo 113, Japan

Stress responses affect physiological and pathological features of various organisms including the human. Since the main event associated with stress responses is the synthesis of stress proteins, these proteins can be used for medical diagnosis and prognosis. In the same sense, stress proteins induced in microorganisms, plants, invertebrates and vertebrates may be useful for monitoring environmental pollution. Expression of stress proteins in organs to a sufficient level confers resistance on them against injury caused by stresses. This phenomenon may give a basis for thera-peutic and preventive treatments of certain diseases such as ischemia and parasite infection. On the contrary, it also gives rise to thermotolerance of tumor cells that prevents hyperthermic therapy of tumors. Stress proteins and their derived-peptides have particular antigenicity and are, therefore, able to modulate immune response. All of the above applications of stress responses are dealt with in this part.

Stress-Inducible Cellular Responses
ed. by U. Feige, R.I. Morimoto, I. Yahara and B. Polla
© 1996 Birkhäuser Verlag Basel/Switzerland

Stress proteins as molecular biomarkers for environmental toxicology

J. A. Ryan and L. E. Hightower

Department of Molecular and Cell Biology, University of Connecticut, 75 N. Eagleville Road, Storrs, CT 06269-3044, USA

Summary. Biomarkers are increasingly being used in environmental monitoring to provide evidence that organisms have been exposed to, or affected by, xenobiotic chemicals. Usually, these biomarkers rely on biochemical, histological, morphological, and physiological changes in whole organisms; however, changes at the cellular and molecular levels of organization, especially in nucleic acids and proteins, are increasingly being used to supplement these more traditional biomarkers. This chapter starts by giving a brief overview of biomarkers and some of the basic requirements for their effective use. Then stress-inducible proteins that are potentially useful as environmental biomarkers are explored, and some examples of their application as biomarkers and methods of detecting them are presented.

Introduction

Although stress protein research is just over 20 years old, potential applications in medicine, pharmacology, and toxicology are already beginning to appear. One major application currently the focus of much attention is the use of stress proteins as molecular biomarkers for environmental monitoring, for both estimating environmental exposures to toxic chemicals, and detecting damage resulting from such exposures. By using biomarkers at the molecular level of organization, it is hoped that problems related to chemical exposures can be identified more efficiently and with greater sensitivity, allowing earlier detection than measuring biomarkers from higher levels of organization (tissue pathologies, reproductive failure, mass mortality, loss of species diversity within ecological communities, etc.). This newer monitoring approach can then be integrated with the more traditionally used physical, biological, and chemical monitoring methods, and provides for more effective environmental monitoring and management. This chapter will give a brief overview of molecular biomarkers, and give some examples of how they are used and studied.

Overview of biomarkers

Biomarkers are measurements at any level of biological organization, in either wild populations from contaminated habitats or organisms (including cultured cells and tissues) experimentally exposed to physical or chemical stressors, that provide a sensitive index of exposures to, or adverse sublethal effects of toxic chemicals. At higher levels of biological organization (tissue, organ, and whole organism), these biomarkers include changes in metabolism, physiology, morphology, histology, and immunology. At the lower molecular and subcellular levels of organization, biomarkers rely primarily on changes in nucleic acids and proteins (reviewed in Fossi and Leonzio, 1994; Huggett et al., 1992; McCarthy and Shugart, 1990).

For biomarkers to be effectively and reliably used, their requirements and limitations must be understood.

(1) Biomarkers must have biological significance: They should correlate in a time or dose/concentration-dependent manner with stress-induced damage or other effects to the organism. This can be accomplished with wild populations by comparing differences in biomarkers in populations from contaminated sites with the biomarkers in corresponding populations from pristine or minimally contaminated sites. In the laboratory, carefully controlled dose-response studies can be performed in whole animals or cell cultures to directly link dose-dependent changes in a biomarker with a specific stressor.

(2) They should be more sensitive than conventional endpoints such as growth, survival, or reproductivity, but still linkable to population, community, and ecosystem-level effects. It is very important to be able to link molecular level changes to both cellular and higher level effects.

(3) Biomarkers should be sensitive to many different contaminants and be easy and inexpensive to reliably measure or detect in a wide variety of organisms. Enough must be known about the biomarker to differentiate between natural (developmental, reproductive, dietary, seasonal, etc.) variations and the effects of anthropogenic stressors (see also Depledge, 1994; Stegeman et al., 1992).

Since all environmental pollutants ultimately induce or begin their harmful or pathological processes at the molecular or subcellular levels, changes in potential biomarkers at these levels should be more sensitive than at higher levels, and thus provide earlier signals that exposure or injury has occurred (Moore et al., 1994). To determine the biological significance of these changes at the molecular and subcellular levels, and thus their potential usefulness as biomarkers, the alterations in cellular physiology that accompany these changes must be clearly understood and directly linked to the stressor(s) that induced them. Both organismal and cellular physiological responses to stressors can be separated into three categories: (1) adaptive responses to prevent damage; (2) responses to sublethal damage, and (3) responses to lethal damage. The first two categories can be used conceptually to divide biomarkers into two subgroups: biomarkers of effect and biomarkers of exposure (Sanders, 1990).

(1) Adaptive responses resulting from contact with potentially toxic stressors are useful as biomarkers of exposure, also called Tier 2 biomarkers. At the molecular level, adaptive responses are designed to prevent or protect against cellular damage by detoxifying, binding, or excreting stressors before they can cause damage (see Fig. 1). Most traditional biomarkers fall into this category, showing exposure to stressors rather than their effects (Depledge, 1994). Cytochrome P450s are good examples of molecular biomarkers of exposure.

(2) Physiological responses to sublethal damage, from which an organism or cell can recover, are useful as biomarkers of effect, also called Tier 1 biomarkers. At the cellular level, sublethal damage results when defenses are insufficient, or incapable of handling the source of stress; then systems for mediating, repairing, or removing the stressor-induced damage respond (see Fig. 1). Traditional heat shock proteins (HSPs), especially the hsp70 family, are good examples of molecular level Tier 1 biomarkers; proteotoxic damage to the cell elicits a rapid increase in their cellular levels.

(3) The last category, responses to lethal damage from which the organism or cell cannot recover (apoptotic or necrotic changes, for example), result when all protective and corrective measures (including both Tier 1 and Tier 2 biomarkers) have failed. These responses have little or no value as biomarkers since they occur too late to be useful.

Thus it is very important to distinguish into which category a response falls, so its potential use as a biomarker can be accurately determined (Depledge, 1994; Mayer et al., 1992).

There is already a large body of information on the induction of stress proteins by environmentally relevant pollutants or stressors (reviewed in Sanders, 1993; Stegeman et al., 1992). All organisms, from bacteria and yeast to humans, respond to many physical and chemical stressors at the cellular level by increasing their synthesis of a small group of stress proteins. These stress proteins generally have two major roles: the first is to prevent stress-induced damage; the second, if the first is overwhelmed, is to repair it, limit further damage and prevent future damage. The most abundant and widely studied group of stress-related proteins, the hsp70 family, was originally identified as HSPs, but they are now known to be induced by a wide variety of agents and conditions that either damage proteins directly, or indirectly by causing the cellular production of abnormal proteins (Hightower, 1991). These stressors, in addition to heat, include anoxia, heavy metals, and oxidizing agents (reviewed in Lindquist and Craig, 1988; Morimoto et al., 1990).

There are several major families of HSPs; the following are three of the major eukaryotic HSP families that show good promise as environmental biomarkers (Sanders, 1990).

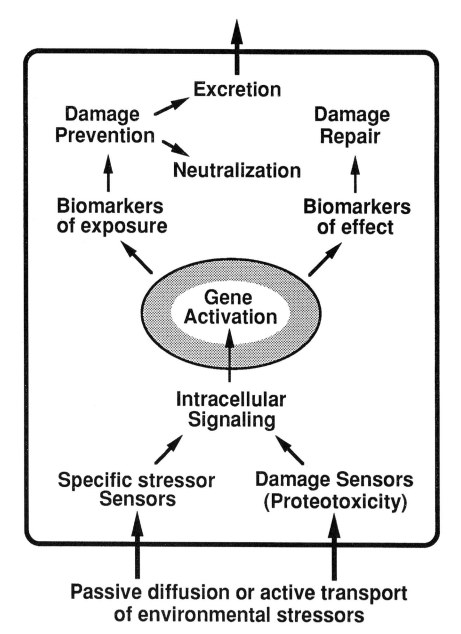

Figure 1. Based on their functional roles in the cell, stress-inducible molecular biomarkers can be subdivided into two classes: biomarkers of exposure that attempt to prevent damage, and biomarkers of effect that attempt to repair damage. Cytochrome P4501A1 is an example of a biomarker of exposure; hsp70 is a biomarker of effect.

hsp70

This protein, the most widely studied HSP, belongs to a multigene family with at least one form that is constitutively expressed (hsc70 or cognate) in unstressed cells, and one or more isoforms (hsp70) that are only stress inducible. In unstressed cells the major roles of hsc70 are facilitating the folding of nascent proteins by preventing their aggregation, and then chaperoning them to sites of membrane translocation. In addition, both hsc70 and hsp70 help to refold denatured proteins that occur in cells following heat shock or exposure to other proteotoxic stressors. Because these proteins are so highly conserved and have a very strong response to protein damage, they are ideal candidates for Tier 1 biomarkers, showing that cells have suffered sublethal damage as the result of an environmental stress.

hsp60

This highly conserved HSP belongs to the cpn60 family of chaperonins involved in protein folding, of which the hsp60 member is localized in the mitochondria. Because of this subcellular localization, hsp60 may be very useful as an organelle-specific biomarker of effect for chemicals that perturb the mitochondria.

Small HSPs

This HSP family is less highly conserved than the hsp70 and hsp60 families, with molecular weights ranging from 15 to 30 kDa. However, all members show substantial homology with α-crystallins found in vertebrate eye lens tissue. Their cellular roles are not as well understood as other HSPs but they may be involved in actin binding to stabilize microfilaments. They are also phosphorylated when signal transduction pathways, involving protein kinase cascades, are stimulated by stress, or serum, phorbol esters and other compounds with mitogenic activity. They are developmentally regulated and may function as molecular chaperones (Arrigo and Landry, 1994). These HSPs have good potential as biomarkers for the effects of stressors on specific signaling pathways, or targets such as the cytoskeleton; however, a better understanding of their cellular roles is essential to fully evaluate their use as biomarkers.

In addition to the above heat shock proteins, these other stress-inducible proteins also show potential as environmental biomarkers (reviewed in Sanders, 1990; Stegeman et al., 1992).

Cytochrome P450 mixed function oxidases (MFOs)

This large heme-containing enzyme family (Phase I conjugating enzymes) biotransforms many relatively insoluble organic compounds, including drugs, pesticides, and polycyclic aromatic or halogenated hydrocarbons, into more water soluble forms (often with assistance from Phase II conjugating enzymes) so that they can be excreted in bile or urine. Unfortunately they can also biotransform some compounds (dimethylbenzanthracene and benzo(a)pyrene for example) into chemically reactive forms more toxic than the parent compounds, capable of damaging DNA and causing mutations or cancer. Induction of MFOs by their substrates or related compounds leads to significant increases in both protein levels and enzymatic activity. Field and laboratory studies have repeatedly shown the usefulness of changes in MFO induction and enzymatic activity in vertebrates as sensitive indicators of exposure for environmental monitoring. MFO induction is currently the most widely studied molecular biomarker. Additional research to determine the sensitivities of the different members of MFO gene families, their regulation and expression, and any interspecies differences, will be required before their potential as biomarkers can be fully evaluated (reviewed in Payne et al., 1987).

Metallothioneins

These are a class of small (61 amino acids in most mammals) cysteine-rich metal-binding proteins, found throughout the animal kingdom. One of the major roles of mammalian metallothioneins (MTs), the chelation of heavy metal ions (including Cd^{2+}, Hg^{2+}, Zn^{2+}, and Cu^{2+}), is believed to protect cells against free heavy metal ion-induced damage. However, the cellular roles of MTs are complex and not yet fully understood. They also help regulate zinc and copper ion metabolism, making these ions available to the cell as necessary, and may act as sulfhydryl-rich scavengers to prevent damage from stress-induced free radicals. Less is known about MTs from other species. Because MT levels can increase significantly following exposures to a variety (depending on species) of metal ions, they are strong candidates for biomarkers of toxic metal ion exposure. However, they are also induced by other agents including glucocorticoids; physiological conditions, including changes in nutritional and reproductive status, may also affect their levels. As a result, before MTs can be used as biomarkers, additional information is necessary to determine the normal functions of MTs and the mechanisms by which their levels are controlled (see chapter by Zhu and Thiele, this volume; Garvey, 1990).

P-glycoproteins

When some human tumors are treated with an anti-cancer drug, the tumor cells often acquire resistance not only to the drug they were treated with, but also to other unrelated anti-cancer drugs. This phenomenon, referred to as multidrug resistance (MDR), is commonly linked directly with the overexpression of P-glycoproteins (PGPs). PGPs are large (170 kDa in mammals) membrane-associated transport proteins found throughout the animal kingdom that apparently function as ATP-dependent efflux pumps to remove many functionally and structurally unrelated compounds from the cell. PGPs are inducible by heat and arsenite in some human tumor-derived cell lines, increasing both their protein and mRNA levels 7–8 fold as a result of these exposures (Chin et al., 1990). A similar system operates against environmental pollutants in animal cells (reviewed in Kurelec, 1992). It is called multixenobiotic resistance and appears to be a major defense mechanism against both natural and man-made toxins. Mono-clonal antibodies against human PGPs cross-react with PGPs from all animal species tested, including several species of marine invertebrates (Minier et al., 1993). Because of their role in protecting cells against environmental stressors, and their apparent ability to be induced following stressor exposures, PGPs appear to be good potential candidates for molecular biomarkers of exposure. However, very few field studies have been done to demonstrate their usefulness as environmental biomarkers.

Phase II conjugating enzymes

These important proteins include glutathione transferases and UDP-glu-curonosyltransferases, two families of enzymes that continue the process of solubilizing potentially toxic chemicals begun by MFOs so that they can be detoxified or more readily excreted by the cell. Both of these enzymes respond to some environmental contaminants by increasing their levels or activities, however with other contaminants they have yielded conflicting results. Further studies will be needed to determine their usefulness as biomarkers.

Antioxidant proteins

These proteins help prevent cellular damage resulting from oxidative stress, primarily caused by activated oxygen species (free-radicals) generated in the cell during normal metabolic processes (MFO function-ing, for example) or as the result of exposure to environmental contamin-ants. When cellular antioxidant defenses are inadequate or overwhelmed, excess free-radicals in the cell can oxidize nucleic acids, lipids, and pro-teins leading to extensive cellular damage, and in the case of DNA oxida-tion, potentially leading to carcinogenesis. The most important antioxidant proteins are superoxide dismutases, catalases, peroxidases, and gluta-thione-related enzymes. They have all been be shown to increase their levels of activity following exposures to agents that increase the production of free-radicals in the cell. While antioxidant proteins offer potential as bio-markers of exposure, very few environmental studies have been done to demonstrate their utility as biomarkers (Wong and Kaspar; Storz and Polla; this volume).

Biomarker applications

Molecular biomarkers are potentially useful in a number of important environmental applications. As sensitive indicators of exposure or effect, they have been used in the wild for comparing organisms from sites known or suspected of being polluted with those from sites known to be pristine. In the laboratory, biomarkers have been used with *in vitro* bioassays for evaluating the potential environmental impact (effects) of individual com-pounds, and for identifying the presence of pollutants in samples (ex-posures) as part of environmental monitoring programs. *In vitro* and *in vivo* systems have also been used in the laboratory as models to screen for, or evaluate potential biomarkers under carefully controlled conditions before beginning field studies, and to elucidate the underlying mechanisms that control their expression at the cellular level. As biomarker-based test systems become more developed, they will also play an important role in evaluating progress in remediating polluted areas, helping provide answers to the question: When is it clean enough? The following examples of some biomarker applications and methods demonstrate the potential for bio-markers in both field and laboratory research.

Field evaluation of potential biomarkers in organisms in the wild is an important step for validating their usefulness in environmental monitoring. Collier and colleagues (1995) monitored changes in hepatic cytochrome P4501A (CYP1A) enzyme activity and protein levels in three species of flat fish (*Pleuronectes vetulus, Lepidopsetta bilineata, and Platichthys stellatus*) over the course of one year at two sites with significantly diffe-rent levels of chemical contamination. Two commonly used catalytic activi-ty assays were performed: the first measured aryl hydrocarbon hydrolase (AHH) activity, the second ethoxyresorufin *O*-deethylase (EROD) activity. In addition, an enzyme-linked immunoassay (ELISA) was performed (using rabbit anti-cod CYP1A) to measure changes in CYP1A protein levels. All three methods showed consistent site differences: the more con-taminated site had consistently higher levels of CYP1A induction for all three methods in all three species of fish. They concluded that although CYP1A can be used by itself as a biomarker of exposure, it would be more useful when combined with a suitable biomarker of effect.

Another method for monitoring contaminated sites is to maintain caged animals for short periods in the contaminated sites and then compare them with identical animals maintained under clean conditions in the laboratory. This approach was used (Haasch et al., 1993) to compare hepatic cytochrome P4501A1 (CYP1A1) activity in caged catfish (*Ictalurus punctatus*) and largemouth bass (*Micropterus salmoides*), as well as in wild killifish (*Fundulus heteroclitus*) collected from both pristine and contaminated sites. They measured CYP1A1 induction at three levels: accumulation of mRNA by nucleic acid hybridization (Northern and dot blots), protein accumulation by immunodetection (Western blotting), and finally enzyme (EROD) activity. Measures at all three levels of expression showed increases in CYP1A1 induction in all three species of fish exposed at the contaminated sites. They found measuring induction at each of the three levels of expression to be a more reliable approach for biomonitoring since it eliminates the possibility of interferences that could occur at individual levels. These interferences have been shown to occur when measuring stressor-induced changes in CYP1A1 enzymatic activities (see the discussion below of the paper by Hahn et al., 1993).

An alternative approach for environmental monitoring is to use cell cultures in addition to whole animals. Using cell cultures provides a rapid, efficient, and economical way for both screening potential environmental stressors, as well as testing samples from contaminated sites to evaluate levels or types of contamination. One such *in vitro* system uses a rat hepatoma-derived cell line, H4IIE, as a bioassay to determine levels of planar halogenated hydrocarbons (PHH) in environmental extracts by measuring their ability to induce CYP1A1-associated EROD activity in the cells (as compared to a standard, 2,3,7,8,-tetrachlorodibenzo-*p*-dioxin [TCDD] and expressed as TCDD-equivalents).This system is useful for evaluating the relative biological and toxicological properties of complex mixtures of PHHs in environmental and biological samples. Tillitt and colleagues (1991) used this bioassay system to evaluate the overall potency of extracts from bird eggs from contaminated areas. The highest levels of TCDD equivalents were found in birds from the most polluted sites; and these levels could be linked with the reproductive success rates of the birds from these areas.

As an alternative to using difficult enzyme assays to detect or quantify biomarkers, cell cultures have been genetically engineered to express a variety of reporter genes whose products are easier to measure. Postlund and colleagues (1993) placed the 5′-flanking region of the human CYP1A1 gene, containing three dioxin responsive enhancers (DREs), immediately upstream of the firefly luciferase gene to design a reporter construct. This reporter gene was then stably transfected into the human hepatoma-derived HepG2 cell line to create a bioassay that is sensitive to TCDD and other halogenated aromatic hydrocarbons (HAHs). This system is similar to the H4IIE bioassay (described above) but is easier to use and more sensitive due to the use of a bioluminescent assay system. A similar reporter gene system (El-Fouly et al., 1994) using human placental alkaline phosphatase as the reporter has also been developed. This reporter system was transfected into a mouse hepatoma-derived cell line and used for the detection of dioxin-like compounds.

In vitro methods have also been developed that show the effects of exposures to environmental contaminants, and thus have potential as Tier 1 biomarkers. Fischbach and colleagues (1993) developed an *in vitro* test system by stably transfecting mouse NIH/3T3 cells with a plasmid construct carrying the human hsp70 promoter coupled to the human growth hormone (hGH) gene. The resulting cell line only expressed the reporter gene product when subjected to stress. Induction of the hsp70 promoter by stressor-induced proteotoxicity resulted in the dose-dependent production of the hormone which was then measured using a very sensitive, commercially available enzyme immunoassay. This system was used to successfully rank 31 metal compounds as inducers of this cellular stress response (hGH production); these results showed good agreement with other *in vivo* and *in vitro* toxicity data.

When studying biomarkers of effect it is very important to link stressor-induced changes in biomarker levels with corresponding stressor-induced changes in physiological state or health of the cell or organism. To make this link in laboratory reared bay mussels (*Mytilus edulis*), Steinert and Pickwell (1993) measured bistributyltin-(TBT) induced changes in hsp70 levels by immunoblotting and radiolabelling, and compared them with changes in their feeding (filtration) rates. There was a linear, dose-dependent increase in hsp70 levels detected by immunoblotting with a greater than 10-fold increase over controls at the highest TBT concentrations. As the hsp70 levels increased there was a corresponding linear decrease in filtration (feeding) rates, effectively linking changes in hsp70 levels to a physiological change in the organism (filtration rate).

Stringham and Candido (1994) have developed a rapid and cost-effective *in vivo* bioassay system in *Caenorhabditis elegans*, a small free-living soil nematode, by fusing a copy of two of its small 16 kDa HSP genes (also containing their heat shock elements and TATA boxes) to the *Escherichia coli lacZ* reporter gene (β-galactosidase). The resulting transgene construct was strictly stress inducible in these worms. These transgenic nematodes were incubated with a variety of stressors, including Cd^{2+}, Cu^{2+}, Hg^{2+}, Pb^{2+}, Zn^{2+}, AsO_3^{2-}, and heat, and then assayed either qualitatively by staining whole animals with the histochemical stain 5-bromo-4-chloro-3-indolyl-galactoside (Xgal) to show tissue distribution of the stress response, or quantitatively by photometrically measuring the conversion of *o*-nitrophenyl-β-D-galactoside (ONPG, a substrate for β-galactosidase) to a colored end product, to determine the magnitude of the stress response. The patterns of expression of these genes was shown to be stressor-dependent: arsenite and heat shock resulted in transgene expression in the entire organism; mercuric chloride only caused expression in the gut; cadmium chloride in the pharynx; lead nitrate in the posterior bulb of the pharynx; cupric chloride in the anterior end of the pharynx; and zinc chloride in the hypodermal cells of the body cuticle. In addition, transgene expression for a given stressor always occurred below its LC50, showing that their test system was sensitive to sublethal stressor levels and linked with survival, important characteristics for biomarkers of effect.

Cell cultures can also be used as models to screen for and evaluate potential biomarkers under carefully controlled conditions before beginning field studies. Ryan and Hightower (1994) developed a simple *in vitro* system to rapidly screen fish cell cultures exposed to environmental stressors to identify good biomarker candidates. They used SDS-PAGE to detect stressor-induced, concentration-dependent changes in stress protein levels while simultaneously using the neutral red cytotoxicity assay to determine the effects of these stressors on cellular physiology. They found a concentration-dependent relationship between sublethal cytotoxic effects and increases in stress proteins levels (hsp70 and hsp27) in both a hepatoma-derived cell line (PLHC-1, a cell line from the desert topminnow (*Poeciliopsis lucida*) that is functionally similar to the human HepG2 and rat H4IIE cell lines described above) and primary winter flounder (*Pleuronectes americanus*) kidney cultures. These increases in protein levels were reversible; when the stressors were removed from the cultures and the cells allowed to recover for several days, stress protein levels rapidly returned to near normal levels.

In vitro systems can also be used in the laboratory to study the underlying mechanisms that control stressor-induced gene expression at the cellular level. Hahn and colleagues (1993) used the PLHC-1 cell line (described above) as a model system to characterize and study the regulation of CYP1A. They studied changes in both enzyme (EROD) activity and protein levels (by immunoblotting) induced by exposure to increasing doses of 3,3′,4,4′-tetrachlorobiphenyl (TCB). CYP1A protein increased with increasing TCB concentration, EROD activity initially increased but was inhibited at higher concentrations of TCB. This apparent inhibition of EROD activity by TCB did not appear to be related to cytotoxicity and has been observed both *in vivo* and *in vitro* in other systems from several species. Thus a low level of EROD activity can result from either a high or low level exposure to TCB. They concluded that PLHC-1 cells are a good fish cell model for studying the aryl hydrocarbon (Ah) receptor mediated regulation of gene expression. It also has potential as a bioassay system that is more representative of stress-induced toxicity in fish than the rat H4IIE or human HepG2 cell line-based bioassays described above.

Detecting biomarkers

The traditional approach for detecting changes in the levels of stressor-induced protein induction has been to expose organisms or cells to a stressor in the presence of radiolabelled amino acids and assay for their incorporation into newly synthesized proteins. However this approach, with its short time frame and requirements for radiolabelling, while useful for some short-term laboratory studies, is not suitable for examining the effects of chronic, longer term xenobiotic exposures. Protein radiolabelling, while it can clearly show an acute, transient response to a stressor, cannot be directly related to longer term changes in cellular protein levels. Also, there are no practical methods for radiolabelling wild populations in the field (Steinert and Pickwell, 1993).

Measuring RNA levels by hybridization using Northern or dot blotting (Haasch et al., 1993) or by using quantitative Reverse Transcriptase Polymerase Chain Reaction (RT-PCR) (Jessen-Eller, et al., 1994) can be useful, especially for mechanistic studies on gene regulation of the stress response. Unfortunately, stressor-induced increases in mRNA levels are often more transient than the corresponding changes in protein levels, may not be directly linked to their corresponding protein levels and, like protein radiolabelling, are not suited for longer field studies (Sanders, 1993). Both Collier and colleagues (1995) and Hahn and colleagues (1993) used both protein levels and enzymatic activities when comparing induction levels of cytochrome P4501A. They advocated using both enzyme activity and immunological assays (Western blotting, dot blotting, and ELISA) to quantify protein levels, since enzyme activity may be altered during processing or inhibited by stressors. Ultimately, it is the amount and activity of stress proteins in the cell, not their rates of synthesis and degradation, or the amount of their mRNA, that determine their physiological impact.

One of the most interesting and latest developments in biomarker detection and measurement has been the application of transgenic reporter systems in both cell cultures (El-Fouly et al., 1994; Fischbach et al., 1993; Postlund et al., 1993) and whole organisms (Stringham and Candido, 1994). The use of appropriate transgenic reporter systems should provide inexpensive, and both more convenient and practical, approaches for detecting changes in molecular biomarkers than some of the methods currently employed. However field evaluations to validate this approach remain to be done.

Conclusion

Stress-inducible proteins appear to be excellent candidates for molecular biomarkers. Most of the stress proteins currently being evaluated in laboratory and field studies are biomarkers of exposure. However, these biomarkers have been difficult to directly link with the physiological health of the cell or organism. Preliminary studies with hsp70 and other HSPs (Ryan and Hightower, 1994; Steinert and Pickwell, 1993) have shown that it is also possible to link biomarkers of effect directly with adverse physiological events at at both the cellular and organismal levels. A combined biomonitoring approach, using a suite of both types of biomarkers integrated together into a single system, needs to be developed and utilized. This would provide for sensitive *in vivo* or *in vitro* assays that could both detect exposures to environmental stressors and show whether these exposures are directly linked with sublethal injuries or other adverse biological effects. These could then be used to supplement traditional biological, chemical, and physical methods for monitoring the environment, effluents from industrial streams, and progress at contaminated sites undergoing remediation; or used for screening man-made chemicals or basic studies of the mechanism of toxicity for chemicals.

References

Arrigo, A.-P. and Landry, J. (1994) Expression and function of the low-molecular-weight heat shock proteins. *In:* R.I. Morimoto, A. Tissieres and C. Georgopoulos (eds): *The Biology of Heat Shock Proteins and Molecular Chaperones,* Cold Spring Harbor Laboratory Press, Plainville, NY, pp 335–373.

Chin, K.-V., Tanaka, S., Darlington, G., Pastan, I. and Gottesman, M.M. (1990) Heat shock and arsenite increase expression of the multidrug resistance (*MDR1*) gene in human renal carcinoma cells. *J. Biol. Chem.* 265(1):221–226.

Collier, T.K., Anulacion, B., Stein, J.E., Goksoyr, A. and Varnashi, U. (1995) A field evaluation of Cytochrome P4501A as a biomarker of contaminant exposure in three species of flat fish. *Environ. Toxicol. Chem.* 14(1):143–152.

Depledge, M.H. (1994) The rational basis for the use of biomarkers as ecotoxicological tools. *In:* M.C. Fossi and C. Leonzio (eds): *Nondestructive Biomarkers in Vertebrates,* Lewis Publishers, Boca Raton, FL, pp 271–295.

El-Fouly, M.H., Richter, C., Giesy, J.P. and Denison, M.S. (1994) Production of a novel recombinant cell line for use as bioassay system for detection of 2, 3, 7, 8,-tetrachlorodibenzo-*p*-dioxin-like chemicals. *Environ. Toxicol. Chem.* 13(10):1581–1588.

Fischbach, M., Sabbioni, E. and Bromley, P. (1993) Induction of the human growth hormone gene placed under human hsp70 promoter control in mouse cells: A quantitative indicator of metal toxicity. *Cell Biol. Toxicol.* 9(2):177–188.

Fossi,M.C. and Leonzio, C. (1994) *Nondestructive Biomarkers in Vertebrates.* Lewis Publishers, Boca Raton, FL.

Garvey, J.S. (1990) Metallothionein: A potential biomonitor of exposure to environmental toxins. *In:* J.F. McCarthy and L.R. Shugart (eds): *Biomarkers of Environmental Contamination,* Lewis Publishers, Boca Raton, FL, pp 267–288.

Haasch, M.L., Prince, R., Wejksnora, P.J., Cooper, K.R. and Lech, J.J. (1993) Caged and wild fish: induction of hepatic Cytochrome P-450 (CYP1A1) as an environmental biomarker. *Environ. Toxicol. Chem.* 12:885–895.

Hahn, M.E., Lamb, T.M., Schultz, M.E., Smolowitz, R.M. and Stegeman, J.J. (1993) Cytochrome P4501A induction and inhibition by 3, 3', 4, 4'-tetrachlorobiphenyl in an Ah receptor-containing fish hepatoma cell line (PLHC-1). *Aquat. Toxicol.* 26:185–208.

Hightower, L.E. (1991) Heat shock, stress proteins, chaperones, and proteotoxicity. *Cell* 66: 191–197.

Huggett, R.J., Kimerle, R.A., Mehrle, Jr. P. M. and Bergman, H.L. (1992) *Biomarkers – Biochemical, Physiological, and Histological Markers of Anthropogenic Stress.* Lewis Publishers, Boca Raton, FL.

Jessen-Eller, K., Picozza, E. and Crivello, J.F. (1994) Quantitation of metallothionein mRNA by RT-PCR and chemiluminescence. *Biotechniques* 17 (5):962–973.

Kurelec, B. (1992) The multixenobiotic resistance mechanism in aquatic organisms. *Crit. Rev. Toxicol.* 22(1):23–43.

Lindquist, S. and Craig, E.A. (1988) The heat-shock proteins. *Annu. Rev. Genet.* 22: 631–677.

Mayer, F.L., Versteeg, D.J., McKee, M.J., Folmer, L.C., Graney, R.L., McCume, D.C. and Rattner, B.A. (1992) Physiological and nonspecific biomarkers. *In:* R.J. Huggett, R.A. Kimerle, P.M. Mehrle Jr. and H.L. Bergman (eds): *Biomarkers – Biochemical, Physiological, and Histological Markers of Anthropogenic Stress.* Lewis Publishers, Boca Raton, FL, pp 5–85.

McCarthy, J.F. and Shugart, L.R. (1990) *Biomarkers of Environmental Contamination.* Lewis Publishers, Boca Raton, FL.

Minier, C., Akcha, F. and Galgani, F. (1993) P-glycoprotein expression in *Crassostrea gigas* and *Mytilus edulis* in polluted sea water. *Comp. Biochem. Physiol.* 106B(4):1029–1036.

Moore, M.N., Kohler, A., Lowe, D.M. and Simpson, M.G. (1994) An integrated approach to cellular biomarkers in fish. *In:* M.C. Fossi and C. Leonzio (eds): *Nondestructive Biomarkers in Vertebrates,* Lewis Publishers, Boca Raton, FL, pp 171–197.

Morimoto, R.I., Tissieres, A. and Georgopoulos, C. (1990) *Stress Proteins in Biology and Medicine.* Cold Spring Harbor Laboratory Press, Cold Spring Harbor, N. Y.

Payne, J.F., Fancey, L.L., Rahimtula, A.D. and Porter, E.L. (1987) Review and perspective on the use of mixed-function oxygenase enzymes in biological monitoring. *Comp. Biochem. Physiol.* 86C (2):233–245.

Postlund, H., Vu, T.P., Tukey, R.H. and Quattorochi, L.C. (1993) Response of human CYP1-luciferase plasmids to 2,3,7,8-tetrachlorodibenzo-*p*-dioxin and polycyclic aromatic hydrocarbons. *Toxicol. Appl. Pharmacol.* 118:255–262.

Ryan, J.A. and Hightower, L.E. (1994) Evaluation of heavy-metal ion toxicity in fish cells using a combined stress protein and cytotoxicity assay. *Environ. Toxicol. Chem.* 13(8):1231–1240.

Sanders, B. (1990) Stress proteins: potential as multitiered biomarkers. *In:* J.F. McCarthy and L.R. Shugart (eds): *Biomarkers of Environmental Contamination*, Lewis Publishers, Boca Raton, FL, pp 165–192.

Sanders, B.M. (1993) Stress proteins in aquatic organisms – an environmental perspective. *Crit. Rev. Toxicol.* 23(1):49–75.

Stegeman, J.J., Brouwer, M., Di Giulio, R.T., Forlin, L., Fowler, B.A., Sanders, B.M. and Van Veld, P.A. (1992) Enzyme and protein synthesis as indicators of contaminant exposure and effect. *In:* R.J. Huggett, R.A. Kimerie, P.M. Mehrle, Jr. and H.L. Bergman (eds): *Biomarkers – Biochemical, Physiological, and Histological Markers of Anthropogenic Stress.* Lewis Publishers, Boca Raton, FL, pp 235–335.

Steinert, S.A. and Pickwell, G.V. (1993) Induction of hsp70 proteins in mussels by ingestion of tributyltin. *Mar. Environ. Res.* 35:89–93.

Stringham, E.G. and Candido, E.P.M. (1994) Transgenic hsp16-*lacZ* strains of the soil nematode *Caenorhabditis elegans* as biological monitors of environmental stress. *Environ. Toxicol. Chem.* 13(8):1211–1220.

Tillitt, D.A., Ankley, G.T., Verbrugge, D.A., Giesy, J.P., Ludwig, J.P. and Kubiak, T.J. (1991) H4IIE rat hepatoma cell bioassay-derived 2,3,7,8-tetrachlorodibenzo-*p*-dioxin equivalents in colonial fish-eating waterbird eggs from the Great Lakes. *Arch. Environ. Contam. Toxicol.* 21:91–101.

Stress-Inducible Cellular Responses
ed. by U. Feige, R.I. Morimoto, I. Yahara and B. Polla
© 1996 Birkhäuser Verlag Basel/Switzerland

Thermotolerance and heat shock proteins: Possible involvement of Ku autoantigen in regulating Hsp70 expression

G.C. Li and A. Nussenzweig

Department of Medical Physics and Department of Radiation Oncology, 1275 York Avenue, Memorial Sloan-Kettering Cancer Center, New York, NY 10021, USA

Summary. Here we characterize and compare the phenomenon of thermotolerance and permanent heat resistance in mammalian cells. The biochemical and molecular mechanisms underlying the induction of thermotolerance, and the role that heat shock proteins play in its development and decay are discussed. Finally, we describe a novel constitutive HSE-binding factor (CHBF/Ku) that appears to be involved in the regulation of the heat shock response.

Introduction

One of the most interesting aspects of thermal biology in the mammalian system is the response of heated cells to subsequent heat challenges. Mammalian cells, when exposed to a non-lethal heat shock, have the ability to acquire a transient resistance to one or more subsequent exposures at elevated temperatures. This phenomenon has been termed thermotolerance (Gerner, 1983; Gerner and Schneider, 1975; Henle and Dethlefsen, 1978; Henle and Leeper, 1976). The molecular mechanism(s) by which cells develop thermotolerance is not well understood, but early experimental evidence suggested that protein synthesis is required for its manifestation. On the molecular level, heat shock activates a specific set of genes, so-called heat shock genes, and results in the preferential synthesis of heat shock proteins (Lindquist, 1986; Lindquist and Craig, 1988; Morimoto et al., 1990). The heat shock response has been extensively studied in the past decade, and has attracted the attention of a wide spectrum of investigators ranging from molecular and cell biologists to radiation and hyperthermia oncologists. There is much data supporting the hypothesis that heat shock proteins play a key role in modulating the cellular response to environmental stress, and are involved in the development of thermotolerance.

We begin this review by summarizing our current knowledge about the induction of thermotolerance by heat shock and other environmental stresses. We will characterize and compare the phenomena of thermotolerance and permanent heat resistance. We then describe the biochemical and molecular mechanisms underlying the induction of thermotolerance and the role that heat shock proteins play in its development and decay. Finally, we discuss the regulation of the heat shock response and the involvement of a novel constitutive HSE-binding factor (CHBF/Ku) in this process.

Transient thermotolerance and permanent heat resistance

Exposure of mammalian cells in culture to temperatures above 40 °C leads to reproductive death. Survival curves, when plotted as a function of duration of heating, resemble X-ray survival curves and are characterized by an initial shoulder region followed by exponential decrease in surviving fractions (Dewey et al., 1980). The survival of cells after hyperthermia depends on both the applied temperature and the duration of exposure. The time required to reduce the survival on the exponential region to 37% of its initial value is defined as D_0. The kinetics of heat killing for a wide variety of mammalian cells has been analyzed in terms of Arrhenius plots, where $1/D_0$ is plotted as the inverse of the absolute temperature. For most cell lines studied, the Arrhenius plots of cell inactivation seem to be composed of two segments, with a break at or near 43 °C (Bauer and Henle, 1979; Dewey et al., 1977; Westra and Dewey, 1971). Above that temperature, the activation energy is between 110 and 150 kcal/mol, a value consistent with the view that protein damage is responsible for cell death. At lower temperatures, the activation energy is relatively higher, i.e., around 300 to 400 kcal/mol. On the basis of the difference in activation energies, several authors have suggested that there may be different primary modes of heat-induced cell death, one dominant above 43°C and the other below 43 °C. On the other hand, the change in slope of Arrhenius plots below 43 °C could be a manifestation of the ability of the cells to develop thermotolerance (Li and Hahn, 1980; Sapareto et al., 1978). Mammalian cells vary appreciably in their intrinsic thermal sensitivity (Raaphorst et al., 1979). A possible interpretation of these findings is that permanent heat resistance is simply a genetic alteration of constitutive levels of macromolecules transiently induced in thermotolerant cells.

Henle and Leeper (Henle and Leeper, 1976) and Gerner and Schneider (Gerner and Schneider, 1975) first showed that cultured mammalian cells exposed to a nonlethal heat treatment have the ability to develop resistance to subsequent heat challenge. During this tolerance state, cells are much more heat resistant than cells which have never been preexposed to elevated temperatures (Field and Anderson, 1982; Gerner, 1983; Henle and Dethlefsen, 1978). *In vitro*, thermotolerance can be induced by a short initial heat treatment at temperatures above 43 °C followed by a 37 °C incubation before the second heat challenge. The survival curves after the second heat exposure show an increase in D_0, as well as an increase in the width of the shoulder. Thermotolerance can also be induced during continuous heating at temperatures below 43 °C (Gerweck, 1977; Harisiadis et al., 1977; Palzer and Heidelberger, 1973; Sapareto, et al., 1978). The degree of thermotolerance developed can be dramatic as evidenced by increases in survival levels by several orders of magnitude.

The thermal history, heat fractionation interval, and recovery conditions all significantly modify the kinetics of thermotolerance (Gerner, 1983; Henle and Dethlefsen, 1978). The proliferative and nutritional status of mammalian cells also significantly affect their thermal sensitivity (Hahn, 1982). Nielsen and Overgaard (1979) and Goldin and Leeper (1981) examined the effect of the extracellular pH on thermotolerance induction during fractionated hyperthermia; their data indicate that reducing the extracellular pH can partially inhibit the induction and expression of split-dose thermotolerance. Gerweck (1977) has shown that the development of thermotolerance during continuous heating can be delayed below physio-logical pH values.

The effects of varying temperature and the duration of the first heat treatment on the subsequent expression of thermotolerance in Chinese hamster HA-1 cells in culture were studied in detail by Li and coworkers (1982). These sets of experiments indicated that temperatures of 43 °C or higher did not permit the development of thermotolerance during this first heat exposure. A subsequent incubation at 37 °C was found to be required for its manifestation. In contrast, if the priming dose was at 41 °C, thermo-tolerance was almost fully expressed by the end of the initial treatment. On the basis of these and other data, Li and Hahn (1980) proposed an opera-tional model of thermotolerance. The authors suggest that thermotolerance can be divided into three complementary and sometimes competing pro-cesses: an initial event ("trigger"), the expression of resistance ("develop-ment"), and the gradual disappearance of resistance ("decay"). Each of these components may have its own temperature dependence as well as dependence on other factors such as pH and nutrients. Conceptually, the three components of thermotolerance may be considered to be independent processes. However, independent measurements of each component are not always possible.

In this simplified model, thermotolerance develops in at least two steps. First, the triggering event converts normal cells to the triggered state with a rate constant k_1. This process very likely involves the activation of the heat shock transcription factor, HSF1 (Lis and Wu, 1993; Morimoto, 1993). Second, these triggered cells are converted to thermotolerant cells with a rate constant k_2. Above 42.5 °C, $k_2 = 0$; the triggered cells remain sensitive, and if transferred to 37 °C become converted to thermotolerant cells. This thermotolerant state is manifested by the elevated expression of heat shock proteins, enhanced protection and faster recovery from thermal damage. Finally, thermotolerant cells all reconvert to their sensitive state at a slow rate governed by rate constant k_3. An Arrhenius plot of the induc-tion of thermotolerance as measured by the cell survival assay is shown in

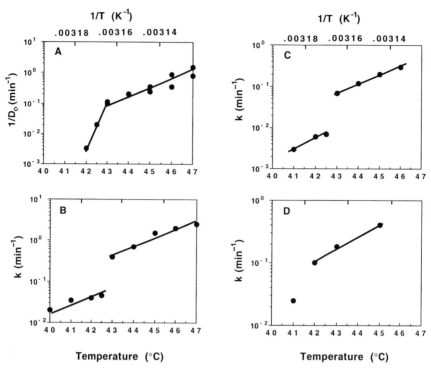

Figure 1. Arrhenius plots for heat-killing, induction of thermotolerance, induction of heat shock protein hsp70, and heat shock factor (HSF1) activation. (A) Plot of the inverse of the log of the slope of the survival curves obtained at various temperatures against the inverse of the absolute temperature; (B) Rate at which thermotolerance is induced as a function of the duration of the triggering. This rate is plotted vs. the inverse of the absolute temperature of the appropriate treatment; (C) Similar to (B), except that the time to induce a maximum amount of hsp70 was used to determine the ordinate. (D) The rate of maximum HSF1 binding activity to the heat shock element (HSE) of the rat heat shock promoter, determined by quantifying bands of HSE-HSF1 complexes that were obtained using the gel-mobility shift assay. Extracts were prepared immediately after cells were heat shocked for different times at a given temperature.

Figure 1 B. There it is compared to the Arrhenius plot for cell killing (Fig. 1 A), for the induction of synthesis of one of the heat shock proteins (hsp70) (Fig. 1 C), and for the activation of the heat shock transcription factor HSF1 (Fig. 1 D). The slopes of the straight lines can be used to calculate the activation energy for the rate-limiting reaction involved. The calculated value of activation energy is $\sim$ 120 kcal/mole, suggesting protein unfolding or protein denaturation. The similarity in the four graphs (at temperatures above 43 °C) strongly suggests that cell killing by heat, the induction of thermotolerance, the induction of heat shock response both in terms of hsp70 synthesis and HSF1 activation all have similar origin.

Characterization of thermotolerant cells

So far, thermotolerance has been discussed in terms of an increased survival after a heat treatment. However cell killing is the result of heat-induced damage to cellular structures/functions, and viability is determined by the cells' capacity to repair this damage. When cells are heated, the absorption of energy is evenly distributed throughout the cell resulting in damage to virtually all cellular structures/functions (Laszlo, 1992). In thermotolerant cells, however, many of these structures/functions show an increased resistance against heat-induced modifications. The data in Table 1 show that thermotolerance develops for nuclear, cytoplasmic and membrane components. Thus, there seems to be no spatial preference in the cell for the expression of tolerance. However, an increasing amount of data suggest that the development of thermotolerance can be more specific than previously assumed, depending on the agents used for induction. For example, when thermotolerance is induced by sodium arsenite, diamide or ethanol treatment, thermotolerance only develops in cellular fractions that are damaged during the thermotolerance-inducing treatment (Tab. 2, Kampinga, 1993). Thus the subcellular localization of the induced resistance appears to correspond with the localization of the damage induced during the tolerance-inducing treatment.

Several reports support the idea of a (spatial) correlation between damage and the expression of tolerance: (i) Lunec and Cresswell (1983) observed an enhanced resistance of the ATP-synthesis in thermotolerant cells, but only in a cell line (L5178Y-S) in which the ATP-synthesis was impaired by the tolerance-inducing heat treatment. In another cell line (Ehrlich ascites) this treatment had no effect on the ATP-synthesis, and in the thermotolerant cells no enhanced resistance of this process was observed; (ii) Nguyen et al., (1989), using firefly luciferase-transfected cells, found that heat treatments that decreased the luciferase activity but not protein synthesis in these cells, induced an increased heat resistance of the luciferase activity but not of the protein synthesis. After a priming treatment at a higher temperature ($45\,^{\circ}$C instead of $42\,^{\circ}$C), both activities were impaired, and in the thermotolerant cells resistance against a subsequent heat treatment was observed for both the luciferase activity and protein synthesis; and (iii) Data by Anderson and Hahn (1985) indicate that the correlation between damage and induced resistance might even be true for multidomain proteins in which the domains differ in heat sensitivity. These authors, working on Na^+, K^+, ATPase, reported that thermotolerance could be induced in a heat sensitive domain of the protein (the ouabain binding domain) without changing the heat sensitivity of a more heat resistant domain (the ATP-hydrolyzing domain).

For all cellular proteins/structures/functions that have been studied it has been shown that thermotolerance can be induced. However, there are indications that thermotolerance is not a non-specific protection of the entire

Table 1. Characterization of the thermotolerant mammalian cell: molecules, cellular structures or functions that are protected after a mild heat shock.

Macromolecules, cellular structure and function	References
– *DNA*	
Polymerase activity[a]	Chu and Dewey, 1987; Dikomey et al., 1987
Synthesis	Van Dongen et al., 1984
– *RNA*	
rRNA synthesis	Burdon, 1986; Nover et al., 1986
Splicing	Yost and Lindquist, 1986
– *Protein*	
Synthesis	Hahn et al., 1985; Mizzen and Welch, 1988; Sciandra and Subjeck, 1984
Denaturation	Lepock et al., 1990; Nguyen, et al., 1989; Pinto et al., 1991
Removal of agregates	Kampinga et al., 1989; Wallen and Landis, 1990
– *Membrane*	
con A capping	Stevenson et al., 1981
Na[+], K[+], ATPase[b]	Anderson and Hahn, 1985; Burdon and Cutmore, 1982
Permeability	Maytin et al., 1990
Insulin receptors	Calderwood and Hahn, 1983
– *Cytoplasm*	
Cytoskeletal reorganization	Wiegant et al., 1987
cAMP levels	Calderwood et al., 1985
ATP levels[c]	Lunec and Cresswell, 1983
– *Cell cycle progression*	Van Dongen et al., 1984

[a] Not observed by Kampinga et al., 1985 and Jorritsma et al., 1986.
[b] Anderson and Hahn (1985) observed thermotolerance development for the ouabain binding capacity only, the heat sensitivity of the ATP hydrolyzing activity was unchanged in thermotolerant cells.
[c] Only in L5178Y-S cells, no resistance was observed in Ehrlich ascites cells.

cell, but is restricted to those targets in the cell that were damaged during the thermotolerance-inducing treatment. This direct correlation between damage and tolerance opens the possibility to study the role of inactivation of different proteins/structures in hyperthermic cell killing.

Role of heat shock proteins in thermotolerance and heat resistance

The mechanism underlying thermotolerance is not well understood, although several studies suggest that the heat shock proteins (HSP) are involved in its development (Landry et al., 1982; Li and Werb, 1982;

Table 2. Comparison of the damage and resistance induced in the membrane fraction (PF) and the nuclear fraction of HeLa S3 cells by different resistance inducing agents

	TTR_{10}	Damage to membranes (by agent)	Enhanced resistance of membranes	Damage to nuclei (by agent)	Enhanced resistance of nuclei
C	(1.0)				
HTT	2.3	yes	yes	yes	yes
ATT	1.8	yes[a]	yes	no	no
DTT	2.5	n.d.	no	yes	yes
ETT	2.3	no	no	yes	yes

As a measure for the induced thermotolerance, the thermotolerance ratio is given (TTR_{10}: ratio of heating times required to reduce survival to 10%). Thermotolerance is induced as follows: HeLa S3 cells were pretreated with heat (15 min, 44 °C + 5 h, 37 °C: HTT), sodium arsenite (1 h, 100 μM + 5 h, 37 °C: ATT), diamide, (1 h, 500 μM + 5 h, 37 °C: DTT) or ethanol (1 h, 6% + 4 h, 37 °C: ETT) (Data from Burgman et al., 1993 and Kampinga et al., 1994). C: control, non-tolerant cells. n.d. = not determined. [a] taken from Yih et al., 1991. (Reproduced with permission from Dr. Paul Burgman's 1993 Ph.D. thesis).

Subjeck et al., 1982). Qualitative evidence exists for a causal relationship between HSP synthesis and thermotolerance (Landry et al. 1982; Laszlo and Li, 1985; Li, 1985; Li and Werb, 1982; Subjeck et al., 1982): (i) heat shock induces transiently enhanced synthesis of HSP that correlates temporally with the development of thermotolerance, (ii) the persistence of thermotolerance correlates well with the stability of HSP; (iii) agents known to induce HSP induce thermotolerance; (iv) conversely, agents known to induce thermotolerance induce HSP (see Tab. 3); and (v) stable heat-resistant variant cells express high levels of HSP constitutively.

One notable exception to the correlation summarized above is that amino acid analogues have been shown to induce HSP but not thermotolerance; Chinese hamster HA-1 cells treated with such compounds are more sensitive to elevated temperatures (Li and Laszlo, 1985). This apparent lack of correlation can be attributed, however, to the dysfunction of analogue-substituted HSP: the nonfunctional, analogue-substituted HSP would not be expected to protect cells from thermal stress. In support of this notion, it was found that the incorporation of amino acid analogues into cellular proteins inhibits the development of thermotolerance and that thermo-tolerant cells or permanently heat-resistant cells are more resistant to the thermal-sensitizing action of amino acid analogues (Laszlo and Li, 1993; Li and Laszlo, 1985).

Quantitatively, of the many HSP preferentially synthesized after heat shock, the concentration of the 70-kDa heat shock protein (hsp70) appears to correlate best with heat resistance, either permanent or transient (Laszlo and Li, 1985; Li, 1985; Li and Werb, 1982). However, a good correlation between levels of a 27-kDa heat shock protein (hsp27) and thermal resistance also has been reported (Landry et al., 1989).

Table 3. Relation between induction of thermotolerance and induction of HSP synthesis

Thermotolerance inducing treatment	HSP synthesis	References
Heat	+	Laszlo, 1988; Li, 1983
Heavy metals	+	Li and Mivechi, 1986
Ethanol	+	Boon-Niermeijer et al., 1988, Burgman et al., 1993; Henle et al., 1986; Li 1983; Li and Hahn, 1978
Sodium arsenite	+	Crete and Landry, 1990; Kampinga et al., 1992; Li, 1983
Procaine, lidocaine	+	Hahn et al., 1985
Aliphatic alcohols (C_5-C_8)	+	Hahn et al. 1985
Dinitrophenol[a]	0	Boon-Niermeijer et al., 1986; Haveman et al., 1986
	+[c]	Ritossa, 1962, 1963
CCP[a]	+	Haveman et al., 1986
Puromycin[b]	+	Lee and Dewey, 1987
Prostaglandin A	+	Amici et al., 1993

+: increased, 0: unaffected.
[a] Not observed by Rastogi et al. (1988); this might however be due to the long interval between DNP or CCP treatment and test heating, and the low concentrations of CCP and DNP used.
[b] Only at intermediate concentrations of puromycin (3–30 µg/ml) that inhibit protein synthesis by 15–80%; not by higher concentrations.
[c] Induction of new puffs in *Drosophila* salivary gland giant chromosomes was observed, a phenomenon shown to be involved in the induction of synthesis of new proteins (Tissieres et al., 1974). (Reproduced with permission from Dr. Paul Burgman's 1993 Ph.D. thesis).

In mammalian cells, three types of experiments were performed before 1990 to vary the intracellular concentration of hsp70 and to correlate this change with thermal-stress response. (i) Microinjection of affinity-purified anti-hsp70 antibodies into rat cells appeared to prevent the nuclear and nucleolar accumulation of hsp70 after a test heat shock and greatly increased the lethality of a 45 °C, 30-min heat treatment (Riabowol et al., 1988). (ii) The 5'-control region of the hsp70-encoding gene was inserted into a plasmid containing the dihydrofolate reductase gene; this recombinant plasmid was then introduced into a Chinese hamster ovary (CHO) cell line, and a 20000-fold elevation in its copy number was achieved by selection of cells with methotrexate. These copies of the *hsp70* regulatory region presumably competed with the endogenous *hsp70* promoter sequence for factors that activate hsp70 expression to reduce heat-inducible expression from the intact endogenous gene for hsp70 by at least 90%. It was found that cells containing the amplified regulatory sequences display increased thermosensitivity (Johnston and Kucey, 1988). (iii) Human hsp70 microinjected directly into CHO cells increased the resistance of cells to 45 °C heating (Li, 1989).

The expression of hsp70 under heterologous promoters has yielded additional insight into its structure and function. Thus, transient expression of
Drosophila hsp 70 in monkey COS cells demonstrated that hsp70 accelerates the recovery of cell nucleoli after heat shock (Munro and Pelham,
1984). Similarly, the domains of human hsp70 responsible for nucleolar
localization and for ATP-binding were dissected (Milarski and Morimoto,
1989). Using retroviral-mediated gene transfer technique, Li et al. have
established rat cell lines stably and constitutively expressing a cloned
human hsp70 gene (Li et al., 1992). These cell lines provide a direct means
of studying the effects of selected hsp70 expression on cell survival after
heat shock.

Survival curves for pooled populations of Rat-1 cells constitutively
expressing either intact or mutant human hsp70 are plotted in Figure 2.
Figure 2A shows that MV6-infected Rat-1 cells (vector control) exhibited
similar clonogenic survival compared to Rat-1 cells after 45, 60, or 75 min
of heat shock at 45°C. By contrast, six independent pools of MVH-infected Rat-1 cells (expressing human hsp70) exhibit approximately 100-fold
higher survival after 60 and 75 min of 45°C heat treatment. Similar results
were obtained when the 45°C cell survival experiments were performed
with individual clones of Rat-1 cells expressing human hsp70 (Li, et al.,
1992). To evaluate the protective effect of different levels of human hsp70
expression, the level of human hsp 70 at 37°C was measured in individual
clones by flow cytometry (Li et al., 1992). In a parallel experiment the
clonogenic survival of these cells after 75 min at 45°C heat shock was
determined. Figure 2C shows that clones expressing more human hsp70
generally survive thermal stress better than clones expressing lower levels,
and there appears to be a good correlation between levels of exogenous
human hsp70 expressed and the degree of thermal resistance. Interestingly,
increases in human hsp70 expression beyond a certain point do not yield
greater thermal protection. Rat-1 cells expressing human hsp70 missing its
ATP-binding domain are also thermal resistant as shown in Figure 2B.
These data provide direct evidence for a causal relation between expression of a functional form of mammalian hsp70 and survival of cells at
elevated temperatures. However, one should note that production of
hsp70 is only part of the program of protein biosynthesis initiated after
heat shock, and other components of this response might also enhance cell
survival.

The role played by hsp70 in thermotolerance development was also
examined by studying the heat shock response in a cell line that constitutively expresses antisense *hsc70* RNA (Nussenzweig and Li, 1993).
The antisense RNA, which is complementary to 712 nucleotides of the
coding strand of the rat *hsc70* gene, was designed to target both *hsp70*
(p72) and *hsc70* (p73), which share greater than 75% identity in nucleotide
sequence. As shown in Figure 3, the expression of hsp70 is significantly
reduced and delayed after heat shock in the transfected cells compared with

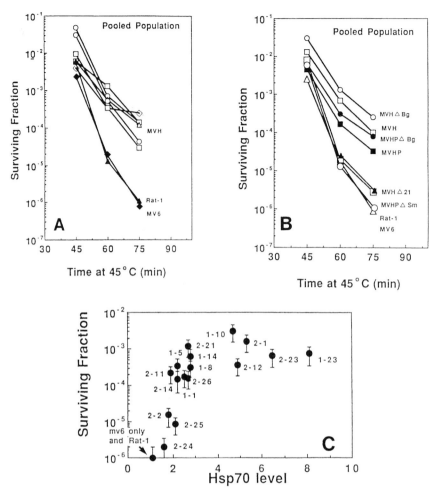

Figure 2. Expression of human hsp70 gene confers thermal resistance to Rat-1 cells. (A) Survival at 45 °C of pooled MVH-infected (expressing human hsp70), MV6-infected (vector control), and uninfected Rat-1 cells. Survivals from six pooled populations of MVH-infected cells independently derived from separate infection experiments are shown (open symbols). Each pool is derived by pooling 200–600 colonies. Survival values after 30-min heating at 45 °C are clustered around 10% for all cells and are, therefore, omitted for clarity. (B) Thermal protection offered by mutant hsp70 missing its ATP-binding domain. Survival after 45 °C heat shock of pooled population of Rat-1 expressing intact or mutant human hsp70. MVH, and MVHP cells express intact untagged human hsp70 and intact human hsp70 tagged with substance P, respectively. MVHΔBg and MVHPΔBg (tagged with substance P) cells express mutant hsp70 missing its ATP-binding domain. MVHPΔSm cells express mutant hsp70 missing its nucleolar localization domain. MVHΔ21 represents Rat-1 cells transfected with a 4-base pair out-of-frame deletion mutant. (C) Monolayers of exponentially growing individual clones of MVH-infected cells were exposed to 45 °C for 75 min, and survival was determined. In parallel experiments, relative levels of human hsp70 in these MVH clones were measured by flow cytometry (Becton Dickinson, FACS 440), using mAb C92F3A-5, specifically against human hsp70. Relative levels of hsp70 were estimated by mean FITC fluorescence intensity of cell populations. At least 20000 cells were analyzed for each flow cytometric measurement. Thermal survivals of various infected cell lines are plotted against the relative level of human hsp70.

the parental Rat-1 cells. For example, the hsp70 band appears as early as 4 h after a 45 °C, 15 min heat shock in Rat-1, while its appearance is 2 to 4 h delayed in the transfected cells. In contrast, there were no significant changes in hsc70 concentration in the different cell lines. It is plausible that higher levels of antisense RNA must be present to produce significant inhibition of *hsc70* transcription, which constitutes approximately 0.1 % of total mRNA in Rat-1 cells. This is consistent with previous attempts to regulate gene expression by antisense RNA in which a large excess of antisense over sense transcripts is often necessary to produce significant results. Figures 3 B and 3 C show that the antisense transfected cells are more heat-sensitive than Rat-1, and impaired in thermotolerance development. While greater than 20 % of the parental cells survive a 30 min exposure at 45 °C, only 5 % of A12 cells and 0.1 % of A13 cells survive such a heat treatment. It is clear from Figure 3 that the cell line exhibiting the greatest delay in hsp70 synthesis (A13) is also the most heat sensitive. Cells expressing antisense RNA also show a reduced maximum tolerance level, as well as a delay in thermotolerance development. Clearly, thermotolerance does develop in cells in which hsp70 levels are reduced, but to a much less degree than in the parental cells.

It is well established that heat shock inhibits RNA and protein synthesis. This inhibition is reversible, and the transcriptional and translational activity recovers gradually when heated cells are returned to 37 °C incubation. The role of hsp70 in these processes was examined using Rat-1 fibroblasts expressing a cloned human hsp70 gene, designated M21 (Li et al., 1992). The constitutive expression of the human hsp70 gene in Rat-1 cells confers heat resistance as evidenced by the enhanced survival of heat-treated cells and resistance against heat-induced translational inhibition (termed translational tolerance) (Liu et al., 1992). In addition, after a 45 °C, 25-min heat treatment, the time required for RNA and protein synthesis to recover was considerably shorter in M21 cells than control Rat-1 cells. These data demonstrate that the expression of human hsp70 in Rat-1 cells, by itself, not only confers heat resistance and translational tolerance, but also facilitates the ability of cells to recover from translational inhibition after thermal stress.

In parallel experiments, the effects that mutations of hsp70 have on the cellular transcriptional/translational activity after heat shock were studied. Both a 4-bp out-of-frame deletion ($\Delta 21$) and in-frame deletion of the nucleolar localization domain (ΔSm) of human hsp70 had no significant affect on the cells' intrinsic heat sensitivity, heat-induced transcriptional/translational inhibition, nor on the subsequent recovery at 37 °C. On the other hand, cells expressing a mutant human hsp70 missing the ATP-binding domain (ΔBg) were found to be heat resistant. However, when the recovery kinetics of protein synthesis were evaluated, no significant differences were observed between Rat-1 cells and cells expressing any of these mutated forms of hsp70.

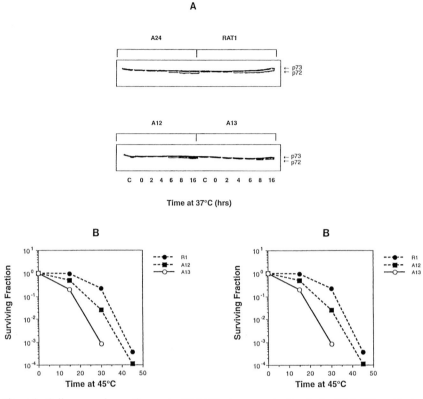

Figure 3. Cells expressing antisense *hsc70* (p73) show a reduction in hsp70 (p72) synthesis, increased thermal sensitivity and a delay in the development of thermotolerance. (A) Kinetics of hsp70 accumulation following heat shock is demonstrated by Western analysis. Antisense transfected cells are designated A12, A13 and A24. Note that the band for p72 appears later in the lines expressing antisense RNA. Equal amounts of protein were separated by one-dimensional polyacrylamide gel electrophoresis, transferred to nitrocellulose membranes, and probed with a mixture of mAb C92F3A-5 (specific for hsp70 (p72)) and N27F3-4 (recognizes both *hsc70* (p73) and hsp70 (p72)); (B) Cellular survivals after heat shock at 45°C, determined by the colony formation assay; (C) Development of thermotolerance. Exponentially growing cultures were first exposed to a 45°C, 15-min pre-treatment and then incubated at 37°C for various times. After such incubations cells were challenged with a second heat treatment, 45°C for 45-min, and survival curves are plotted as a function of the 37°C incubation time between the two heat treatments. Compared to the wild-type, Rat-1, cells transfected with antisense *hsc70* show a greatly reduced maximum tolerance level as well as a delay in thermotolerance development.

These studies provide strong evidence for a direct link between the expression of a functional form of mammalian hsp70 and protection of cells' translational machinery at elevated temperature. They suggest that ATP-binding and/or hydrolysis by hsp70 are dispensable in the hsp70-mediated protection against thermal killing and translational inhibition. It is plausible that hsp70 lacking its ATP-binding domain can still bind to cellular proteins (such as RNPs), stabilize them and prevent their aggregation at elevated temperatures.

On the other hand, because the mutant hsp70 missing its ATP-binding domain does not facilitate the recovery from translational inhibition, ATP-binding and/or hydrolysis may be important in enabling the dissociation of hsp70 from its substrates, or in facilitating the dissociation of aggregated protein complexes to restore their functional integrity.

The ability of heat shock proteins to protect various cellular processes and enzymatic activities during stress has been demonstrated in a series of experiments. *In vitro* studies by Skowyra et al. (Skowyra et al., 1990) have shown that the *E. coli* hsp70 homologue, DnaK, can protect RNA polymerase from inactivation during heat treatment, and that thermally inactivated RNA polymerase can be reactivated by DnaK in a process that requires ATP hydrolysis. Furthermore, the thermal protection efficiency of DnaK is enhanced by the action of its partner proteins DnaJ and GrpE (Hendrick and Hartl, 1993). Similarly, mitochondrial hsp60 has been shown to have thermal protective functions through its ability to prevent the thermal inactivation of dihydrofolate reductase imported into the mitochondria (Martin et al., 1992). The *in vivo* enzymatic activities of luciferase and β-galactosidase were also demonstrated to be protected in thermotolerant mouse and *Drosophila* cells given a heat challenge (Nguyen et al., 1989).

Heat shock causes denaturation and aggregation of various cellular proteins which become insoluble and functionally inactive. It has been previously shown that in mammalian cells, foreign reporter enzymes, such as luciferase, are rapidly inactivated and become insoluble during heat stress (Nguyen et al., 1989). Heat inactivation and insolubilization of luciferase are dramatically attenuated in thermotolerant cells.

As mentioned above, it has been suggested that hsp70 may play a role in protecting cells from thermal stress either by stabilization and prevention of thermal denaturation of normal proteins, and/or by facilitating the dissociation of protein aggregates. Studies were initiated to test this hypothesis by using firefly luciferase as a model protein. Luciferase was introduced into mammalian cells by transfecting Rat-1 cells or M21 cells (a clone isolated from Rat-1 cells expressing high level of human hsp70) with plasmids containing the coding sequence of luciferase gene. The clone M21-L isolated from successful transfectants expresses high levels of both luciferase and human hsp70 at 37°C, while clone LCG expresses only luciferase, but not human hsp70. Monolayers of M21-L or LCG cells were heated at 43°C for various times, and the effects of heat shock on luciferase activity in these cells were determined by luminescence detection using standard scintillation counting; in parallel, the solubility of luciferase in heat-shocked cells were monitored by immunoblot analysis of the soluble and insoluble protein fractions using antisera against luciferase, respectively. Our data demonstrate that in cells expressing a high level of intact human hsp70, the heat inactivation of luciferase enzyme activity is attenuated (Fig. 4). Similarly, a decrease in the insoluble fraction of luciferase molecules in these cells is observed.

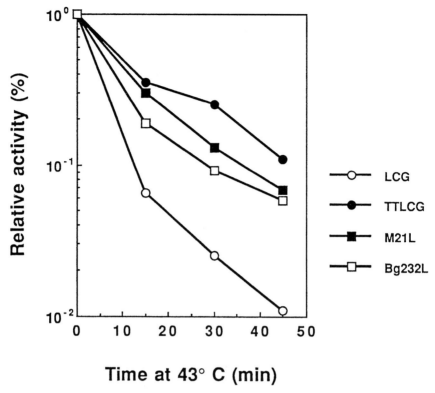

Figure 4. Heat shock inactivation of luciferase in LCG cells, thermotolerant LCG cells and in cells overexpressing intact or mutant human hsp70 gene. Cells were exposed to 43°C for 0–45 min. The luciferase activity was measured using a scintillation counter. Extracts from control cells or cells heat-shocked for 15, 30 and 45 min at 43°C were analyzed. M21L, cells express both human hsp70 and luciferase; BgL, cells express luciferase and mutant hsp70 missing its ATP-binding domain; LCG, cells express luciferase; TT-LCG, thermotolerant LCG cells (thermotolerance was induced by heating LCG cells at 45°C for 15 min and followed with 16 h incubation at 37°C).

It has been shown earlier that the ATP-binding domain of human hsp70 appears to be dispensable in the hsp70-mediated protection of cells against thermal killing (Li et al., 1992). The importance of the ATP-binding domain of human hsp70 in the protection of luciferase against heat-induced denaturation was also evaluated. For this purpose, the luciferase gene was stably transfected in Rat-1 cells expressing a mutant human hsp70 gene missing its ATP-binding domain (Bg232-L). It was found that in these cells, both the enzyme activity and the solubility of luciferase were protected against heat shock nearly as well as in M21-L cells (Fig. 4). It seems that similar to the hsp70-mediated protection of cells from thermal killing, the ATP-binding domain of hsp70 is dispensable in the hsp70-mediated protection against heat inactivation of luciferase enzymes. An attractive interpretation for this finding is that hsp70 nonspecifically stabilizes soluble protein structures, in this case the firefly luciferase, thus preventing the formation of heat-induced insoluble protein aggregates, and maintaining their functional integrity.

Heat shock proteins have also been implicated in protecting the splicing machinery from disruption by heat shock (Yost and Lindquist, 1986; Yost and Lindquist, 1988; Yost and Lindquist, 1991): high temperatures are found to result in the accumulation of mRNA precursors due to the block in splicing. However, in cells made thermotolerant, splicing is no longer disrupted and mature mRNAs accumulate.

In addition to hsp70, the small heat shock protein, hsp27, has been shown to have thermal protective functions. Landry and co-workers have over-expressed hsp27 from various species into different cell lines and have observed that resistance to heat shock correlates with levels of hsp27 (Huot et al., 1991; Lavoie et al., 1993). In addition it appears that phosphorylation of hsp27 may play an important role in its thermal protective function. In support of this, Crete and Landry 1990 (Crete and Landry, 1990) observed that chemical agents such as cycloheximide, A23187 and EGTA, which induce phosphorylation but not the accumulation of hsp27, resulted in a significant degree of thermal protection. Furthermore, phosphorylation mutants of hsp27 failed to protect cells from heat stress (Landry et al., 1993). However, this point remains controversial because of recent evidence that the chaperone properties of the small heat shock protein contributes to the increased cellular thermoresistance in a phosphorylation-independent manner (Knauf et al., 1994). This study suggests that the phosphorylation dependent function of hsp27 is distinct from its thermo-resistance-mediating functions.

Regulation of heat shock response: Possible involvement of Ku autoantigen

Although the importance of the heat shock transcription factor HSF1 in the regulation of mammalian heat shock genes is well-established (Morimoto et al., 1990), recent data indicate that activation of HSF1 by itself is not sufficient for the induction of hsp70 mRNA synthesis, suggesting the existence of additional regulatory factors (Jurivich et al., 1992; Liu et al., 1993; Mathur et al., 1994). In rodent cells, we have found that a constitutive heat shock element (HSE)-binding factor (CHBF) appears to be involved in the regulation of hsp70 transcription. In order to study the role of CHBF in the regulation of heat shock gene expression, we have purified this protein and found that CHBF is identical, or closely related to the Ku-auto-antigen (Kim et al., 1995).

The Ku autoantigen is a heterodimer of 70-kDa and 86-kDa polypeptides (Mimori et al., 1981; Reeves and Sthoeger, 1989). *In vitro* studies have shown that this abundant nuclear protein binds to the termini of double-stranded DNA and DNA ending in stem-loop structures, via the 70-kDa subunit (Mimori and Hardin, 1986). By virtue of its DNA binding activity, Ku serves as a regulatory component of the mammalian DNA-dependent protein kinase, DNA-PK which is activated by DNA ends (Gottlieb and Jackson, 1993). The other component of DNA-PK is a 450-kDa catalytic subunit (DNA-PK$_{cs}$). Recent evidence suggests that the Ku DNA-end binding protein complex is essential for the repair of double strand breaks. The 86-kDa subunit of Ku has been identified as the XRCC5 gene that complements the DNA double-strand break rejoining deficiency and V(D)J recombination defect in radiosensitive CHO cell lines (Taccioli et al., 1994). In another report, the 450-kDa catalytic subunit of DNA-PK was linked to the V(D)J recombination and DNA repair machinery, and shown to be the product of the murine gene SCID (Kirchgessner et al., 1995). These biochemical characteristics suggest a role for the Ku-protein in basic metabolic processes such as DNA replication, repair, recombination and transcription.

The fact that CHBF is identical to Ku suggests that Ku/CHBF may play a role in the heat shock response. To investigate this possibility, rodent cell lines were established that stably and constitutively express the human Ku-70 protein. In Figure 5, monolayers of Rat-1 and R70-15 cells expressing human Ku-70 were subject to a heat shock, and the induction of *hsp70* mRNA and hsp70 protein were analyzed. Following a 15 min exposure to 45 °C, *hsp70* mRNA is rapidly induced in Rat-1 cells, reaching a maximal level about 6 h after the heat shock (Fig. 5A); in R70-15 cells; however, the induction of *hsp70* mRNA by the same treatment is barely detectable during the same time period. Similar patterns were observed when the synthesis of hsp70 protein was monitored (Fig. 5B). After a 45 °C, 15 min heat shock, there is a marked reduction in the level of hsp70 protein in R70-15 cells relative to Rat-1 cells. However, the cellular level of hsc70, the constitutive cognate of hsp70, is essentially unaffected by the expression of human Ku-70. The repression of the heat-induced gene expression by a high cellular level of human Ku-70 is spe-

---

▶

Figure 5. Northern and Western blot analysis of hsp70 protein from heat-shocked cells during post-heat-shock recovery at 37 °C. Rat-1 cells and R70-15 cells overexpressing human Ku-70 protein were exposed to 45 °C for 15 min, and returned to 37 °C for 0–10 h. (A) Equal amounts of cytoplasmic RNA (10 µg) were loaded per lane, size fractionated on a 1% agarose gel, transferred to Hybond-N membrane (Amersham), and probed with the human hsp70 gene (*upper* panel). C: no heat shock, 0, 2, 4, 6, 8, 10, recovery time (h) at 37 °C. The heat-inducible hsp70 mRNA is indicated by an arrowhead on the left margin. Note that the level of hsp70 mRNA in heat-shocked R70-15 cells is significantly reduced relative to that in Rat-1 cells. The same membrane was subsequently probed with a β-actin gene. The levels of actin mRNA at all time points are not significantly different for both cell lines (*lower* panel). (B) The heat-induced accumulation of hsp70 protein was analyzed by Western blot using a mixture of antibodies which recognized both rat *hsc70* and rat hsp70 (N27F3-4 and C92F3A-5, StressGen). Note that in R70-15 cells, the hsp70 protein is barely detectable after the 45 °C, 15 min treatment.

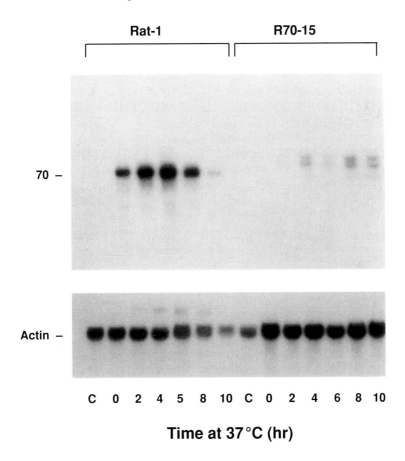

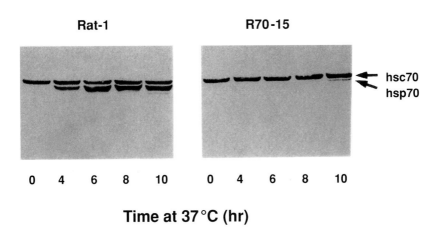

Time at 37°C (hr)

cific to hsp70. The thermal induction of other heat shock proteins is not affected in these Ku-overexpressing cells (Li et al., 1995).

It is plausible that expression of human Ku-70 at a level significantly higher than that of the endogenous rat Ku-70 might affect the cellular level of HSF1, or its state of oligomerization or phosphorylation, which in turn might affect the thermal induction of hsp70. To test this possibility, R70-15 cells and Rat-1 cells were exposed to 45 °C for 15 min, and cell extracts were obtained from these cells at various times after the heat shock (Li et al., 1995). As shown in Figure 6, the overexpression of human Ku-70 protein in R70-15 cells does not affect the heat-induced HSF1-HSE binding activity. In both cell lines, HSF1 rapidly converts to its HSE-binding form upon heat shock. The high cellular level of this form of HSF1 persists for at least 4 h after the 45 °C, 15 min heat shock, and then decreases to a barely detectable level by 6 h (in Rat-1) or 8 h (in R70-15) after heat shock (Fig. 6, *upper* panels). To examine whether HSF1 phosphorylation might

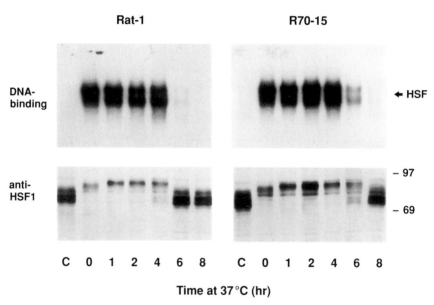

Figure 6. Analysis of the HSE-binding activity and the state of phosphorylation of HSF1 in extracts from Rat-1 and R70-15 cells during a 45 °C, 15 min heat shock and the subsequent recovery phase at 37 °C. Monolayers of Rat-1 cells and R70-15 cells (Rat-1 cells overexpressing human Ku-70 protein) were exposed to 45 °C for 15 min, and returned to 37 °C for 0–8 h. Equal amounts of whole cell extracts were subjected to gel mobility shift analysis (*upper* panel), or immunoblot analysis (*lower* panel) using HSF1-specific antiserum. The level of phosphorylation of HSF1 during heat shock and subsequent recovery was examined by immunoblot analysis with antisera specific to HSF1, taking advantage of the known reduction in the electrophoretic mobility of HSF1 in SDS-polyacrylamide gel upon its phosphorylation. The increase in the apparent size of HSF1 observed upon heat shock is due to phosphorylation, as demonstrated by a decrease in molecular size upon phosphatase treatment (data not shown). The position of HSF1-HSE binding (HSF) complex is indicated by an arrow. The molecular sizes are indicated in kilodaltons.

be modified by the expression of human Ku-70, the state of phosphorylation of HSF1 during and after heat shock was monitored in the same cell
extracts that were used for the gel mobility shift assay. As shown in
Figure 6 (*lower* panel), immunostaining with antiserum against HSF1 indicates that upon heat shock the gel electrophoretic mobility of HSF1 from
either R70-15 or Rat-1 cells is reduced, signifying enhanced phosphorylation. The level of phosphorylation of HSF1 and the appearance of the HSE-
binding form of HSF1 rise and fall in near synchrony, and no significant
difference was observed between R70-15 and control Rat-1 cells.

The fact that Ku is involved in gene regulation is not without precedent.
Both the 70- and 86-kDa subunits of Ku contain leucine/serine repeats
which are reminiscent of similar "zipper" motifs found in a variety of
transcription factors. Furthermore, Ku is a component of DNA-PK which
phosphorylates several transcription factors, such as SP1 (Gottlieb and
Jackson, 1993), c-Jun (Bannister et al., 1993), p53 (Lees-Miller et al.,
1990), c-Myc, Oct-1, and Oct-2 (Lees-Miller and Anderson, 1991) *in vitro*.
Studies also indicate that the Ku/DNA-PK$_{cs}$ complex modulates RNA
polymerase I transcription (Hoff and Jacob, 1993; Knuth et al., 1990; Kuhn
et al., 1995; Kuhn et al., 1993). Additionally, Ku has been found to be
localized on certain transcriptionally active loci of chromosomal DNA
(Amabis et al., 1990; Reeves, 1992). Although these data support a role for
DNA-PK/Ku in transcription, specific genes regulated by DNA-PK/Ku
have not yet been identified.

The mechanisms by which the overexpression of human Ku-70 specifically suppresses the heat induction of hsp70, and the role of the 86-kDa
subunit of the Ku protein in this regulatory process, remains unknown. It is
likely that one function of Ku-70 is to repress hsp70 expression as we have
previously proposed for CHBF. An alternative interpretation is that the
Ku-70/Ku-80 heterodimer or Ku-80 alone has a positive regulatory role
that is disrupted by the overexpression of a single, heterologous Ku-70
subunit.

Discussion

Research from the past few years has made it clear that heat shock proteins
are required for the development of thermotolerance. The fact that mammalian cells become permanently thermoresistant when transfected with
hsp70, and conversely, that cells become thermal sensitive when hsp70
levels are reduced, directly demonstrates that hsp70 plays a vital role in
thermotolerance. Hsp70 acts by stabilizing and preventing thermal denaturation of proteins, and by facilitating the dissociation of protein aggregates
that are formed during conditions of stress. Other HSPs, such as hsp27,
may also have thermal protective functions, and there may even be
cooperative action among members of the heat shock protein family during

thermotolerance development, as has been shown for the hsp70 and hsp60 chaperones during the folding of denatured proteins (Hendrick and Hartl, 1993).

Studies both *in vitro* and *in vivo* have shown that chaperones have the ability to recognize structures that are exposed in unfolded proteins, protect polypeptides from aggregation or premature folding, and promote the refolding of proteins after denaturation. These folding and unfolding reactions are critical both during normal cellular biogenesis and during metabolic stress. Under various conditions of stress the increased need to repair denatured proteins and protect vital structures is fulfilled by the enhanced and preferential synthesis of chaperones.

The importance of heat shock factor HSF1 in the regulation of mammalian heat shock gene expression is well-established. However, recent data indicate that activation of HSF1, by itself, is not sufficient for the induction of *hsp70* mRNA synthesis. We have previously proposed a dual control mechanism for the regulation of *hsp70* gene expression in mammalian cells: a positive control mechanism mediated by HSF1, and a negative control mechanism mediated by CHBF. We have purified CHBF to apparent homogeneity and found that CHBF is identical to the Ku-autoantigen. To examine the *in vivo* role(s) of Ku in the modulation of heat shock response, we have established rodent cell lines that stably and constitutively express the 70-kDa subunit of the human Ku-autoantigen. Our results show that expression of the 70-kDa polypeptide of human Ku in rodent cells specifically suppresses the induction of hsp70 upon heat shock (Li et al., 1995). Thermal induction of the other heat shock proteins appears not to be affected, nor is the state of phosphorylation or the DNA-binding ability of HSF1 affected. A challenge for future studies will be to elucidate how Ku is functionally involved in the regulation of heat-induced hsp70 expression.

References

Amabis, J.M., Amabis, D.C., Kahuraki, J. and Stollar, B.O. (1990) The presence of an antigen reactive with a human autoantibody in *Trichosia pubescens* (Diptera: sciaridae) and its association with certain transcriptionally active regions of the genome. *Chromosoma* 99: 102–110.

Amici, C., Palamara, T. and Santoro, M.G. (1993) Induction of thermotolerance by prostaglandin A in human cells. *Exp. Cell Res.* 207: 230–234.

Anderson, R.L. and Hahn, G.M. (1985) Differential effects of hyperthermia on the Na$^+$, K$^+$-ATPase of Chinese hamster ovary cells. *Radiat. Res.* 102: 314–323.

Bannister, A.J., Gottlieb, T.M., Kouzarides, T. and Jackson, S.P. (1993) c-Jun is phosphorylated by the DNA-dependent protein kinase in vitro; definition of the minimal kinase recognition motif. *Nucleic Acids Res.* 21: 1289–1295.

Bauer, K.D. and Henle, K.J. (1979) Arrhenius analysis of heat survival curves from normal and thermotolerant CHO cells. *Radiation Res.* 78: 251–263.

Boon-Niermeijer, E.K., Tuyl, M. and Van der Scheur, H. (1986) Evidence for two states of thermotolerance. *Int. J. Hyperthermia* 2: 93–105.

Boon-Niermeijer, E.K., Souren, J.E.M., De Waal, A.M. and Van Wijk, R. (1988) Thermotolerance induced by heat and ethanol. *Int. J. Hyperthermia* 4: 211–222.

Burdon, R.H. and Cutmore, C.M.M. (1982) Human heat shock gene expression and the modulation of Na⁺K⁺-ATPase activity. *FEBS Lett.* 140:45–48.

Burdon, R.H. (1986) Heat shock and the heat shock proteins. *Biochem. J.* 240:313–324.

Burgman, P.W.J.J. (1993) *Denaturation of membrane proteins and hyperthermic cell killing.* Ph. D. Thesis, University of Groningen, The Netherlands.

Burgman, P.W.J.J., Kampinga, H.H. and Konings, A.W.T. (1993) Possible role of localized protein denaturation in the induction of thermotolerance by heat, sodium-arsenite, and ethanol. *Int. J. Hyperthermia* 9:151–162.

Calderwood, S.K. and Hahn, G.M. (1983) Thermal sensitivity and resistance of insulin-receptor binding. *Biochim. Biophys. Acta* 756:1–8.

Calderwood, S.K., Stevenson, M.A. and Hahn, G.M. (1985) Cyclic AMP and the heat shock response in Chinese hamster ovary cells. *Biochem. and Biophys. Res. Comm.* 126:911–916.

Chu, G.L. and Dewey, W.C. (1987) Effect of cycloheximide on heat-induced cell killing, radiosensitization, and loss of cellular DNA polymerase activities in Chinese hamster ovary cells. *Radiat. Res.* 112:575–580.

Crete, P. and Landry, J. (1990) Induction of hsp27 phosphorylation and thermoresistance in Chinese hamster cells by arsenite, cycloheximide, A23187, and EGTA. *Radiation Res.* 121:320–327.

Dewey, W.C., Hopwood, L.E., Sapareto, S.A. and Gerweck, L.E. (1977) Cellular responses to combination of hyperthermia and radiation. *Radiology* 123:464–477.

Dewey, W.C., Freeman, M.L., Raaphorst, G.P., Clark, E.P., Wong, R.S., Highfield, D.P., Spiro, I.J., Tomasovic, S.P., Denman, D.L. and Coss, R.A. (1980) Cell biology of hyperthermia and radiation. *In:* R.E. Meyn and H.R. Withers (eds): *Radiation Biology in Cancer Research,* Raven Press, New York, pp 589–623.

Dikomey, E., Becker, W. and Wielckens, K. (1987) Reduction of DNA-polymerase β activity of CHO cells by single and combined heat treatments. *Int. J. Radiat. Biol.* 52:775–785.

Field, S.B. and Anderson, R.L. (1982) Thermotolerance: a review of observations and possible mechanisms. Cancer therapy by hyperthermia, drugs and radiation. *National Cancer Institute Monograph* 61:193–201.

Gerner, E.W. (1983) Thermotolerance. *In:* F.K. Storm (ed): *Hyperthermia in Cancer Therapy,* G.K. Hall, Boston, pp 141–162.

Gerner, E.W. and Schneider, M.J. (1975) Induced thermal resistance in HeLa cells. *Nature* 256:500–502.

Gerweck, L.E. (1977) Modification of cell lethality at elevated temperatures. The pH effect. *Radiat. Res.* 70:224–235.

Goldin, E.M. and Leeper, D.B. (1981) The effect of low pH on thermotolerance induction using fractionated 45°C hyperthermia. *Radiat. Res.* 85:472–479.

Gottlieb, T.M. and Jackson, S.P. (1993) The DNA-dependent protein kinase: requirement for DNA ends and association with Ku antigen. *Cell* 72:131–142.

Hahn, G.M., Shiu, E.C., West, B., Goldstein, L. and Li, G.C. (1985) Mechanistic implications of the induction of thermotolerance in Chinese hamster cells by organic solvents. *Cancer Res.* 45:4138–4143.

Harisiadis, L., Sung, D.I. and Hall, E.J. (1977) Thermal tolerance and repair of thermal damage by cultured cells. *Radiology* 123:505–509.

Haveman, J., Li, G.C., Mak, J.Y. and Kipp, J.B. (1986) Chemically induced resistance to heat treatment and stress protein synthesis in cultured mammalian cells. *Int. J. Radiat. Biol.* 50:51–64.

Hendrick, J.P. and Hartl., F.-U. (1993) Molecular chaperone functions of heat-shock proteins. *Ann. Rev. Biochem.* 62:349–384.

Henle, K.J. and Leeper, D.B. (1976) Interaction of hyperthermia and radiation in CHO cells: recovery kinetics. *Radiat. Res.* 66:505–518.

Henle, K.J. and Dethlefsen, L.A. (1978) Heat fractionation and thermotolerance: a review. *Cancer Res.* 38:1843–1851.

Henle, K.J., Moss, A.J. and Nagle, W.A. (1986) Temperature-dependent induction of thermotolerance by ethanol. *Radiat. Res.* 66:505–518.

Hoff, C.M. and Jacob, S.T. (1993) Characterization of the factor EIBF from a rat hepatoma that modulates ribosomal RNA gene transcription and its relationship to the human Ku antoantigen. *Biochem. Biophys. Res. Comm.* 190:747–753.

Huot, J., Roy, G., Lambert, H., Chretien, P. and Landry, J. (1991) Increased survival after treatments with anticancer agents of Chinese hamster cells expressing the human Mr 27000 heat shock protein. *Cancer Res.* 51:5245–5252.

Johnston, R.N. and Kucey, B.L. (1988) Competitive inhibition of hsp70 gene expression causes thermosensitivity. *Science* 242:1551–1554.

Jorritsma, J.B.M., Burgman, P., Kampinga, H.H. and Konings, A.W.T. (1986) DNA polymerase activity in heat killing and hyperthermic radiosensitization of mammalian cells as observed after fractionated heat treatments. *Radiat. Res.* 105:307–319.

Jurivich, D.A., Sistonen, L., Kroes, R.A. and Morimoto, R.I. (1992) Effect of sodium salicylate on the human heat shock response. *Science* 255:1243–1245.

Kampinga, H.H., Jorritsma, J.B.M. and Konings, A.W.T. (1985) Heat-induced alterations in DNA polymerase activity of HeLa cells and of isolated nuclei. Relation to cell survival. *Int. J. Radiat. Biol.* 47:29–40.

Kampinga, H.H., Turkel-Uygur, N., Roti Roti, J.L. and Konings, A.W.T. (1989) The relationship of increased nuclear proteins content induced by hyperthermia to killing of HeLa S3 cells. *Radiat. Res.* 117:511–522.

Kampinga, H.H., Brunsting, J.F. and Konings, A.W.T. (1992) Acquisition of thermotolerance induced by heat and arsenite in HeLA cells: multiple pathways to induce tolerance? *J. Cell. Physiol.* 150:406–415.

Kampinga, H.H. (1993) Thermotolerance in mammalian cells: protein-denaturation, -aggregation and stress proteins. *J. Cell. Sci.* 104:11–17.

Kampinga, H.H., Brunsting, J.F., Stege, G.J.J., Konings, A.W.T. and Landry, J. (1994) Cells overexpressing hsp27 show accelerated recovery from heat-induced nuclear protein aggregation. *Biochem. Biophys. Res. Comm.* 204:1170–1177.

Kim, D., Ouyang, H., Yang, S.-H., Nussenzweig, A., Burgman, P. and Li, G.C. (1995) A constitutive heat shock element-binding factor is immunologically identical to the Ku-autoantigen. *J. Biol. Chem.* 270:15277–15284.

Kirchgessner, C.U., Patil, C.K., Evans, J.W., Cuomo, C.A., Fried, L.M., Carter, T., Oettinger, M.A. and Brown, J.M. (1995) DNA-dependent kinase (p350) as a candidate gene for the murine SCID defect. *Science* 267:1178–1183.

Knauf, U., Jakob, U., Engel, K., Buchner, J. and Gaestel, M. (1994) Stress- and mitogen-induced phosphorylation of the small heat shock protein hsp25 by MAPKAP kinase 2 is not essential for chaperone properties and cellular thermoresistance. *EMBO J.* 13:54–60.

Knuth, M.W., Gunderson, S.I., Thompson, N.E., Strasheim, L.A. and Burgess, R.R. (1990) Purification and characterization of proximal sequence element-binding protein 1, a transcription activating protein related to Ku and TREF that binds the proximal sequence element of the human U1 promoter. *J. Biol. Chem.* 265:17911–17920.

Kuhn, A., Stefanovsky, V. and Grummt, Z. (1993) The nucleolar transcription activator UBF relieves Ku-antigen-mediated repression of mouse ribosomal gene transcription. *Nucleic. Acid. Res.* 21:2057–2063.

Kuhn, A., Gottlieb, T.M., Jackson, S.P. and Grummt, I. (1995) DNA-dependent protein kinase: a potent inhibitor of transcription by RNA polymerase I. *Genes & Dev.* 9:193–203.

Landry, J., Bernier, D., Chretien, P., Nicole, L.M., Tanguay, R.M. and Marceau, N. (1982) Synthesis and degradation of heat shock proteins during development and decay of thermotolerance. *Cancer Res.* 42:2457–2461.

Landry, J., Chretien, P., Lambert, H., Hickey, E. and Weber, L.A. (1989) Heat shock resistance conferred by expression of the human HSP27 gene in rodent cells. *J. Cell. Biol.* 109:7–15.

Laszlo, A. and Li, G.C. (1985) Heat-resistant variants of Chinese hamster fibroblasts altered in expression of heat shock protein. *Proc. Natl. Acad. Sci. USA* 82:8029–8033.

Laszlo, A. (1988) The relationship of heat-shock proteins, thermotolerance, and protein synthesis. *Exp. Cell. Res.* 178:401–414.

Laszlo, A. (1992) The effects on hyperthermia on mammalian cell structure and function. *Cell Proliferation* 25:59–87.

Laszlo, A. and Li, G.C. (1993) Effect of amino acid analogs on the development of thermotolerance and on thermotolerant cells. *J. Cell. Physiol.* 154:419–432.

Lavoie, J.N., Bingras-Breton, G. and Landry, J. (1993) Induction of Chinese hamster hsp27 gene expression in mouse cells confers resistance to heat shock. *J. Biol. Chem.* 268:3420–3429.

Lee, Y.J. and Dewey, W.C. (1987) Induction of heat shock proteins in Chinese hamster ovary cells and development of thermotolerance by intermediate concentrations of puromycin. *J. Cell. Physiol.* 132:1–11.

Lees-Miller, S.P., Chen, Y.-R. and Anderson, C.W. (1990) Human cells contain a DNA-activated protein kinase that phosphorylates simian virus 40 T antigen, mouse p53, and the human Ku antoantigen. *Mol. Cell. Biol.* 10:6472–6481.

Lees-Miller, S.P. and Anderson, C.W. (1991) The DNA-activated protein kinase, DNA-PK: a potential coordinator of nuclear events. *Cancer Cells* 3:341–346.

Lepock, J.R., Frey, H.E., Heynen, M.P., Nishio, J., Waters, B., Ritchie, K.P. and Kruuv, J. (1990) Increased thermostability of thermotolerant CHL V79 cells as determined by differential scanning calorimetry. *J. Cell. Physiol.* 142:628–634.

Li, G.C. and Hahn, G.M. (1978) Ethanol-induced tolerance to heat and to adriamycin. *Nature* 274:699–701.

Li, G.C. and Hahn, G.M. (1980) A proposed operational model of thermotolerance based on effects of nutrients and the initial treatment temperature. *Cancer Res.* 40:4501–4508.

Li, G.C. and Werb, Z. (1982) Correlation between synthesis of heat shock proteins and development of thermotolerance in Chinese hamster fibroblasts. *Proc. Natl. Acad. Sci. USA* 79:3218–3222.

Li, G.C., Fisher, G.A. and Hahn, G.M. (1982) Modification of the thermal response by D2O. II. Thermotolerance and the specific inhibition of development. *Radiat. Res.* 92:541–551.

Li, G.C. (1983) Induction of thermotolerance and enhanced heat shock protein synthesis in Chinese hamster fibroblasts by sodium arsenite and by ethanol. *J. Cell. Physiol.* 115:116–122.

Li, G.C. (1985) Elevated levels of 70000 dalton heat shock protein in transiently thermotolerant Chinese hamster fibroblasts and in their stable heat resistant variants. *Int. J. Radiat. Oncol. Biol. Physics* 11:165–177.

Li, G.C. and Laszlo, A. (1985) Amino acid analogs while inducing heat shock proteins sensitize CHO cells to thermal damage. *J. Cell. Physiol.* 122:91–97.

Li, G.C. and Mivechi, N.F. (1986) Thermotolerance in mammalian systems: a review. *In:* Anghileri, L.J. and Robert, J. (eds): *Hyperthermia in Cancer Treatment,* CRC Press, Inc., Boca Raton, pp 59–77.

Li, G.C., Li, L., Liu, R.Y., Rehman, M. and Lee, W.M.F. (1992) Heat shock protein hsp70 protects cells from thermal stress even after deletion of its ATP-binding domain. *Proc. Natl. Acad. Sci. USA* 89:2036–2040.

Li, G.C., Yang, S.-H., Kim, D., Nussenzweig, A., Ouyang, H., Wei, J., Burgman, P. and Li, L. (1995) Suppression of heat-induced hsp70 expression by the 70-kDa subunit of the human Ku-autoantigen. *Proc. Natl. Acad. Sci. USA* 92:4512–4516.

Lindquist, S. (1986) The heat-shock response. *Ann. Rev. Biochem.* 55:1151–1191.

Lindquist, S. and Craig, E.A. (1988) The heat shock proteins. *Ann. Rev. Gen.* 22:631–677.

Lis, J. and Wu, C. (1993) Protein traffic on the heat shock promoter: parking, stalling, and trucking along. *Cell* 74:1–4.

Liu, R.Y., Li, X, Li, L. and Li, G.C. (1992) Expression of human hsp70 in rat fibroblasts enhances cell survival and facilitates recovery from translational and transcriptional inhibition following heat shock. *Cancer Res.* 52:3667–3673.

Liu, R.Y., Kim, D., Yang, S.-H. and Li, G.C. (1993) Dual control of heat shock response: Involvement of a constitutive heat shock element-binding factor. *Proc. Natl. Acad. Sci. USA* 90:3078–3082.

Lunec, J. and Cresswell, S.R. (1983) Heat-induced thermotolerance expressed in the energy metabolism of mammalian cells. *Radiat. Res.* 93:588–597.

Martin, J., Horwich, A.L. and Hartl, F.-U. (1992) Prevention of protein denaturation under heat stress by the chaperonin Hsp60. *Science* 258:995–998.

Mathur, S.K., Sistonen, L., Brown, I.R., Murphy, S.P., Sarge, K.D. and Morimoto, R.I. (1994) Deficient induction of human hsp70 heat shock gene transcription in Y79 retinoblastoma cells despite activation of heat shock factor. 1. *Proc. Natl. Acad. Sci. USA* 91:8695–8699.

Maytin, E.V., Wimberly, J.M. and Anderson, R.R. (1990) Thermotolerance and the heat shock response in normal human keratinocytes in culture. *J. Invest. Dermatol.* 95:635–642.

Milarski, K.L. and Morimoto, R.I. (1989) Mutational analysis of the human HSP70 protein: distinct domains for nucleolar localization and adenosine triphosphate binding. *J. Cell Biol.* 109:1947–1962.

Mimori, T., Akizuki, M., Yamagata, H., Inada, S., Yoshida, S. and Homma, M. (1981) Characterization of a high molecular weight acidic nuclear protein recognized by auto-antibodies in sera from patients with polymyositis-scleroderma overlap. *J. Clin. Invest.* 68: 611–620.

Mimori, T. and Hardin, J.A. (1986) Mechanism of interaction between Ku protein and DNA. *J. Biol. Chem.* 261: 10375–10379.

Mizzen, L.A. and Welch, W.J. (1988) Characterization of the thermotolerant cell. I. Effects on protein synthesis activity and the regulation of heat-shock protein 70 expression. *J. Cell Biol.* 106: 1105–1116.

Morimoto, R.I., Tissieres, A. and Georgopoulos, C. (1990) *Stress Proteins in Biology and Medicine.* Cold Spring Harbor Laboratory Press, Cold Spring Harbor, NY.

Morimoto, R.I. (1993) Cells in stress: Transcriptional activation of heat shock genes. *Science* 259: 1409–1410.

Munro, S. and Pelham, H.R.B. (1984) Use of peptide tagging to detect proteins expressed from cloned genes: deletion mapping functional domains of *Drosophila* hsp70. *EMBO J.* 3: 3087–3093.

Nguyen, V.T., Morange, M. and Bensaude, O. (1989) Protein denaturation during heat shock and related stress. *J. Biol. Chem.* 264: 10487–10492.

Nielsen, O.S. and Overgaard, J. (1979) Effect of extracellular pH on thermotolerance and recovery of hyperthermic damage *in vitro. Cancer Res.* 39: 2772–2778.

Nover, L., Munsche, D., Neumann, D., Ohme, K. and Scharf, K.D. (1986) Control of ribosome biosynthesis in plant cell cultures under heat-shock conditions. Ribosomal RNA. *Eur. J. Biochem.* 160: 297–304.

Palzer, R.J. and Heidelberger, C. (1973) Studies on the quantitative biology of hyperthermic killing of Hela cells. *Cancer Res.* 33: 415–421.

Pinto, M., Morange, M. and Bensaude, O. (1991) Denaturation of proteins during heat shock. *J. Biol. Chem.* 266: 13941–13946.

Raaphorst, G.P., Romano, S.L., Mitchell, J.B., Bedford, J.S. and Dewey, W.C. (1979) Intrinsic differences in heat and/or X-ray sensitivity of seven mammalian cell lines cultured and treated under identical conditions. *Cancer Res.* 39: 396–401.

Rastogi, D., Nagle, W.A., Henle, K.J., Moss, A.J. and Rastogi, S.P. (1988) Uncoupling of oxidative phosphorylation does not induce thermotolerance in cultured Chinese hamster cells. *Int. J. Hyperthermia* 4: 333–344.

Reeves, W.H. and Sthoeger, Z.M. (1989) Molecular cloning of cDNA encoding the p70 (Ku) lupus autoantigen. *J. Biol. Chem.* 264: 5047–5052.

Reeves, W.H. (1992) Antibodies to p70/p80 (Ku) antigens in systemic lupus erythematosus. *Rheumatic Disease Clinics of North America* 18: 391–414.

Riabowol, K.T., Mizzen, L.A. and Welch, W.J. (1988) Heat shock is lethal to fibroblasts micro-injected with antibodies against hsp70. *Science* 242: 433–436.

Ritossa, F.M. (1962) A new puffing pattern induced by heat shock and DNP in *Drosophila. Experientia* 18: 571–573.

Ritossa, F.M. (1963) New puffs induced by temperature shock, DNP and salicylate in salivary chromosomes of *Drosophila melanogaster. Drosophila Inf. Service* 37: 122–123.

Sapareto, S.A., Hopwood, L.E., Dewey, W.C., Raju, M.R. and Gray, J.W. (1978) Effects of hyperthermia on survival and progression of Chinese hamster ovary cells. *Cancer Res.* 38: 393–400.

Sciandra, J.J. and Subjeck, J.R. (1984) Heat shock proteins and protection of proliferation and translation in mammalian cells. *Cancer Res.* 44: 5188–5194.

Skowyra, D., Georgopoulos, C. and Zylicz, M. (1990) The *E. coli* dnaK gene product, the hsp70 homolog, can reactivate heat-inactivated RNA polymerase in an ATP hydrolysis-dependent manner. *Cell* 62: 939–944.

Stevenson, M.A., Minton, K.W. and Hahn, G.M. (1981) Survival and concavalin-A induced capping in CHO-fibroblasts after exposure to hyperthermia, ethanol and x-irradiation. *Radiat. Res.* 86: 467–478.

Subjeck, J.R., Sciandra, J.J. and Johnson, R.J. (1982) Heat shock proteins and thermotolerance: comparison of induction kinetics. *British J. Radiol.* 55: 579–584.

Taccioli, G.E., Gottlieb, T.M., Blunt, T., Priestly, A., Demengeot, J., Mizuta, R., Lehmann, A.R., Alt, F.A., Jackson, S.P. and Jeggo, P.A. (1994) Ku80: product of the *XRCC5* gene and its role in DNA repair and V(D)J recombination. *Science* 265: 1442–1445.

Tissieres, A., Mitchell, H.K. and Tracy, U. (1974) Protein synthesis in the salvary glands of *Drosophila melanogaster.* Relation to chromosome puffs. *J. Mol. Biol.* 84:389–398.

Van Dongen, G., Van de Zande, L., Schamhart, D. and Van Wijk, R. (1984) Comparative studies on the heat-induced thermotolerance of protein synthesis and cell cycle division in synchronized mouse neuroblastoma cells. *Int. J. Radiat. Biol.* 46:759–769.

Wallen, C.A. and Landis, M. (1990) Removal of excess nuclear protein from cells heated in different physiological states. *Int. J. Hyperthermia* 6:87–95.

Westra, A. and Dewey, W.C. (1971) Variation in sensitivity to heat shock during the cell-cycle of Chinese hamster cells *in vitro*. *Int. J. Radiat. Biol.* 19:467–477.

Wiegant, F.A.C., Van Bergen en Henegouwen, P.M.P., Van Dongen, G. and Linnemans, W.A.M. (1987) Stress-induced thermotolerance of the cytoskeleton of mouse neuroblastoma N2A cells and rat Reuber H35 hepatoma cells. *Cancer Res.* 47:1674–1680.

Yih, L.-H., Huang, H., Jan, K.Y. and Lee, T.-C. (1991) Sodium arsenite induces ATP depletion and mitochondrial damage in HeLa cells. *Cell Biol. Int. Rpts.* 46:253–264.

Yost, H.J. and Lindquist, S. (1986) RNA splicing is interrupted by heat shock and is rescued by heat shock protein synthesis. *Cell* 45:185–193.

Yost, H.J. and Lindquist, S. (1988) Translation of unspliced transcripts after heat shock. *Science* 242:1544–1548.

Yost, H.J. and Lindquist, S. (1991) Heat shock proteins affect RNA processing during the heat shock response of *Saccharomyces cerevisiae. Mol. Cell. Biol.* 11:1062–1068.

Stress-Inducible Cellular Responses
ed. by U. Feige, R.I. Morimoto, I. Yahara and B. Polla
© 1996 Birkhäuser Verlag Basel/Switzerland

Heat shock proteins as immunological carriers and vaccines

K. Suzue and R. A. Young

Whitehead Institute for Biomedical Research, Nine Cambridge Center, Cambridge, MA 02142, USA and Department of Biology Massachusetts Institute of Technology, Cambridge, MA 02139, USA

Summary. HSPs are among the major targets of the immune response to bacterial, fungal and parasitic pathogens. The antigenic nature of HSPs is emphasized by evidence that mammals are capable of recognizing multiple B- and T cell epitopes in these proteins. The powerful immunological features of HSPs have led to their experimental use as immunomodulators and as subunit vaccine candidates. Mycobacterial hsp70 and hsp60 have been found to be excellent immunological carriers of molecules against which an immune response is desired; in the absence of adjuvants, the HSPs can stimulate strong and long-lasting immune responses against molecules which have been covalently attached to the HSPs. When used as subunit vaccines, HSPs derived from a variety of bacterial and fungal pathogens have been found to stimulate protective immunity in animal models. These studies suggest that HSPs might be used as immunomodulators or subunit vaccines against infectious disease in man.

Introduction

Humans are exposed to a large number and variety of pathogenic and non-pathogenic microorganisms. Our bodies provide niches for a multitude of microbes that do not typically cause disease. For example, approximately 10^6 cocci and diptheroids reside on one cubic centimeter of human skin (Leyden et al., 1991) and 1 ml of saliva contains about 10^8 microbes (Rosebury, 1962). We are also continuously confronted by pathogenic organisms of substantial diversity.

To protect us from these microorganisms, the immune system has evolved to recognize a remarkable variety of antigenic determinants. Despite this capacity, the immune system appears to devote considerable attention to the members of one family of proteins, the ubiquitous heat shock proteins (HSPs) (Young, 1990; Kaufmann, 1990; Young et al., 1990; Murray and Young, 1992; Kaufmann and Schoel, 1994). The HSPs are among the major targets of the immune response to most bacterial and parasitic pathogens. Humoral and cellular immune responses to HSPs have been observed following exposure to a broad spectrum of infectious agents, including gram positive and gram negative bacteria, fungi, helminths and protozoa (Tab. 1).

The immune responses to HSPs elicited by mycobacterial pathogens have been particularly well-studied. Exposure to *Mycobacterium tuber-*

Table 1. Pathogens induce immune responses to HSPs

Infectious agent	Disease	HSP	References
Bacteria			
Bordetella pertussis	pertussis	hsp70, hsp60	Del Giudice et al., 1993
Borrelia burgdorferi	Lyme disease	hsp60	Hansen et al., 1988
Brucella abortus	brucellosis	hsp60	Roop, et al., 1992
Chlamydia trachomatis	blinding trachoma	hsp70, hsp60	Taylor et al., 1990; Cerrone et al., 1991; Zhong and Brunham, 1992
Coxiella burnetti	Q fever	hsp60	Vodkin and Williams, 1988
Helicobacter pylori	gastritis	hsp60, hsp10	Suerbaum, 1994, Ferrero et al., 1995
Legionella pneumophila	Legionnaires' disease	hsp60	Hoffman et al., 1990
Mycobacterium leprae	leprosy	hsp70, hsp60, small hsps	Mehra et al., 1992; Young et al., 1988
Mycobacterium tuberculosis	tuberculosis	hsp70, hsp60, small hsps	Young et al., 1988, Shinnick et al., 1988; Baird et al., 1988
Treponema pallidum	syphilis	hsp60	Hindersson et al., 1987
Fungi			
Aspergillus fumigatus	aspergillosis	hsp60	Kumar et al., 1993
Candida albicans	candidasis	hsp90	Matthews et al., 1987; Matthews and Burnie 1989
Histoplasma capsulatum	histoplasmosis	hsp60, hsp70	Gomez et al, 1991a; 1992; 1995
Helminths			
Brugia malayi	lymphatic filariasis	hsp70	Selkirk et al., 1989
Onchocerca volvus	ocular filariasis	hsp70	Rothstein et al., 1989
Schistosoma mansoni	schistosomiasis	hsp90, hsp70, small hsps	Johnson et al., 1989; Hedstrom et al., 1987; Nene et al., 1986
Schistosoma japonicum	schistosomiasis	hsp70	Scallon et al., 1987; Hedstrom et al., 1988
Protozoa			
Leishmania braziliensis	leishmaniasis	hsp70	Levy Yeyati et al., 1992
Leishmania donovani	visceral leishmaniasis	hsp90, hsp70	MacFarlane et al., 1990; de Andrade et al., 1992
Plasmodium falciparum	malaria	hsp70	Mattei et al., 1989; Renia et al., 1990
Trypanosoma cruzi	Chagas' disease	hsp70	Levy Yeyati et al., 1992; Requena et al., 1993

culosis or *Mycobacterium leprae* leads to humoral and cellular immune responses to hsp70, hsp60 and small hsps (18kD, 14kD, 10kD) (Adams et al., 1990; Husson and Young, 1987; Young et al., 1988; Nerland et al., 1988; Verbon et al., 1992; Mehra et al., 1992). The cellular responses to mycobacterial HSPs are profound; limiting dilution analysis indicates that 20% of the murine CD4$^+$ T lymphocytes that recognize mycobacterial antigens are directed against hsp60 alone (Kaufmann et al., 1987). The high frequency with which human CD4$^+$ T cells directed against mycobacterial hsp70 and hsp60 have been detected suggest that these HSPs are also major targets of the cellular response in humans (Munk et al., 1988). Limiting dilution analysis of human T lymphocytes from a tuberculoid leprosy patient as well as a patient contact revealed that one-third of *M. leprae* reactive T cells were directed against hsp10 (Mehra et al., 1992).

The powerful antigenic nature of HSPs is emphasized by evidence that mammals are capable of recognizing multiple B and T cell epitopes in these proteins. Murine and human B cell epitopes have been mapped in HSPs from *M. tuberculosis, M. leprae, Trypanosoma cruzi* and *Plasmodium falciparum* (Mehra et al., 1986; Thole et al., 1988; Richman et al., 1989; Mattei et al., 1989; Requena et al., 1993). These mapping data indicate that B cells can recognize many portions of the hsp70 and hsp60 protein molecules. Murine and human T cell epitopes have been mapped most extensively for mycobacterial hsp60 and hsp70 (Lamb et al., 1987; Munk et al., 1990; Van Schooten et al., 1988; Adams et al., 1994; Oftung et al., 1994). This evidence indicates that mycobacterial HSPs can be presented in the context of multiple MHC haplotypes, and that T cell epitopes can be found throughout these HSPs.

The powerful immunological features of HSPs have led to their experimental use as immunomodulators and as subunit vaccine candidates. Mycobacterial hsp70 and hsp60 have been found to be excellent immunological carriers of molecules against which an immune response is desired; in the absence of adjuvants, the HSPs can stimulate strong and long-lasting immune responses against molecules which have been covalently attached to the HSPs. When used as subunit vaccines, HSPs derived from a variety of pathogens have been found to stimulate protective immunity in animal models. We describe below the concepts behind immunological carrier proteins and adjuvants, and then discuss the adjuvant-free carrier effects of hsp60 and hsp70. We then review evidence supporting the use of HSPs as vaccines against specific pathogens.

Immunological carrier proteins and adjuvants

Many polysaccharides and simple chemical compounds are inherently non-immunogenic and fail to elicit strong antibody responses. Landsteiner (1936) observed that these substances, "haptens", react *in vitro* with antibodies but do not have the capacity to elicit antibodies *in vivo*. However, if the hapten was administered in combination with a "carrier" protein, antibodies to the hapten could be generated.

Ovary and Benacerraf (1963) demonstrated that the same carrier protein used in the primary immunization must be used in the subsequent immunization in order to elicit a secondary immune response to the hapten. The hapten and carrier had to be physically linked and within this conjugate molecule, cells recognized one antigenic determinant on the hapten and a second determinant on the carrier (Mitchison, 1971a; Rajewsky et al., 1969). It was determined that in the generation of an antibody response, two distinct types of cells were involved: bone marrow derived lymphocytes (B cells) and thymus derived lymphocytes (T cells) (Claman et al., 1966; Davies et al., 1967; Miller and Mitchell, 1967). Involvement of B and T cell cooperation in the "carrier effect" was demonstrated by adoptive transfer experiments in which one mouse was injected with the hapten, a second mouse was injected with the carrier, and a third mouse was irradiated and received hapten-primed B cells and carrier-primed T cells (Mitchison, 1971b; Raff, 1970). The B and T cells could collaborate to generate an antibody response only if they were obtained from syngeneic mice (Kindred and Shreffler, 1972; Katz et al., 1973).

The "carrier effect" is believed to occur in the following manner. When animals are primed with a hapten-carrier preparation, and then exposed to a second dose, hapten-specific B cells recognize, internalize, and process the hapten-carrier conjugate. The B cell can then present peptides derived from the carrier molecule on its surface in the context of an MHC molecule. This MHC/peptide complex is bound by a carrier-primed T cell, leading to the direct release of cytokines by the T cell to the B cell. These soluble factors stimulate the B cell to proliferate, differentiate, and secrete antibodies. In this review, a "carrier" will refer to a molecule containing T cell epitopes which, when covalently linked to a second molecule, help to elicit and enhance immune responses against the second molecule.

Carriers are an important component of some human vaccines. Tetanus toxoid (TT), diphtheria toxoid (DT) and neisseria outer membrane proteins are the carrier molecules used in the various *Haemophilus influenza* vaccines licensed for use in humans (Smith et al., 1989). In these vaccines, the principal virulence determinant of *H. influenza* type b, a repeating polymer of ribose, ribitol and phosphate, had to be conjugated to a carrier in order to elicit the high levels of anti-polysaccharide antibodies necessary for protective immunity in young children (Robbins and Schneerson, 1990).

The immune response to an antigen of interest can also be enhanced by use of an adjuvant. Indeed, adjuvants are often necessary to elicit desired immune responses (Geerligs 1989; Kenney et al., 1989). In contrast to a carrier, an adjuvant does not need to be covalently coupled to the antigen to perform its function. Instead, the adjuvant and antigen are absorbed or mixed together and are co-administered (Nicklas, 1992). In general, adjuvants function by slowly releasing the antigen, thereby acting as a long-lived antigen reservoir (called the depot effect), and by causing general inflammation at the injection site and thus recruiting immunological mediators such as macrophages. Many adjuvants containing bacterial components, oils and various chemicals have been described (Edelman and Tacket, 1990).

Alum, which contains the aluminum salts $Al(OH)_3$ and $AlPO_4$, is currently the only adjuvant licensed for use in humans and is included in vaccines against diphtheria, tetanus, pertussis, *H. influenza* and hepatitis B. However, alum is poor at stimulating cell-mediated responses and is not effective in the induction of humoral responses against certain antigens (Altman and Dixon, 1989). Thus, a considerable effort is being made to develop new safe and effective adjuvants for use in man (Lussow et al., 1990a; Dintzis, 1992; Audibert and Lise, 1993).

Adjuvant-free carrier effect of hsp60 and hsp70 proteins

Purified protein derivative (PPD), prepared from mycobacterial culture supernatant, is a protein mix which contains hsp60 and hsp70 and elicits a delayed type hypersensitivity reaction in individuals previously exposed to mycobacteria. The powerful immunostimulatory properties of PPD also suggested that it might have some utility as an immunological carrier. Indeed, when PPD was cross-linked to small chemical haptens or peptides, the conjugates elicited a strong antibody response against the attached molecules (Lachmann and Amos, 1970; Lachmann et al., 1986). In addition, a PPD-tumor cell conjugate has been found to enhance the immune response to tumor cells (Lachmann and Sikora, 1978). For PPD to be an effective carrier, physical linkage of PPD to the antigen was crucial and carrier priming with bacille Calmette-Guerin (BCG) was necessary. The powerful carrier effect of PPD was also evident in comparative studies using various conjugates administered with Freund's adjuvant; these studies demonstrated that PPD was a more effective carrier than bovine serum albumin or keyhole limpet hemacyanin (Lachman et al., 1986). PPD has been shown to be an effective carrier in the absence of adjuvant. When the synthetic malarial peptide $(NANP)_{40}$, an epitope from the *Plasmodium falciparum* major surface protein, was conjugated to PPD and administered in Freund's adjuvant or in saline to BCG primed mice, the anti-$(NANP)_{40}$ antibody titers were equivalent (Lussow et al., 1990b).

Recombinant, mycobacterial hsp60 and hsp70 proteins can substitute for PPD in a $(NANP)_{40}$ conjugate (Lussow et al., 1991), suggesting that these were among the components of PPD responsible for the adjuvant-free carrier effect. Hsp60-$(NANP)_{40}$ and hsp70-$(NANP)_{40}$ conjugates were found to elicit anti-$(NANP)_{40}$ antibodies in mice and squirrel monkeys in the absence of adjuvants (Lussow et al., 1991; Barrios et al., 1992; Perraut et al., 1993). Moreover, mycobacterial hsp60 and hsp70 were also found to be effective adjuvant-free carriers when conjugated to the poorly immunogenic meningococcal group C oligosaccharide (MenC) (Barrios et al., 1992).

Other results from the (NANP)$_{40}$ studies support the notion that HSPs can be powerful carriers, but indicate that not all HSPs behave identically. (NANP)$_{40}$ alone acts as a hapten in most strains of mice (Good et al., 1986; Del Giudice et al., 1986), and the presence of covalently linked myco-bacterial hsp60 or hsp70 carrier molecules was essential to obtain antibody responses against (NANP)$_{40}$ (Lussow et al., 1990b). The corresponding heat shock proteins from *E. coli*, GroEL and DnaK, could also function as adjuvant-free carriers to elicit anti-(NANP)$_{40}$ antibodies (Barrios et al., 1994). However the hsp70 conjugates exhibited one useful feature that was not observed with hsp60 conjugates. While priming with recombinant hsp60 in Freund's adjuvant or with living BCG was necessary to obtain the carrier effect with the hsp60 conjugates, but priming was unnecessary with the hsp70 conjugates. Moreover, previous exposure to BCG or to hsp70 neither augmented nor suppressed the antibody response against the antigen attached to hsp70.

To further investigate the adjuvant-free carrier effect of mycobacterial hsp70, an hsp70 fusion vector system was created (Suzue and Young, 1996). This system enabled the production of proteins composed of one mole of antigen fused to one mole of the hsp70 carrier protein in which the number and position of potential epitopes were identical for each mole-cule. In contrast, hsp70 conjugates generated by glutaraldehyde cross-lin-king are a pool of nonidentical molecules with variable epitope density. Immune responses to an antigen can be strongly affected by differences in the molar ratio of antigen and carrier, in the mode of linkage of hapten and carrier and in the position of B and T cell epitobes (Hanna et al., 1972; Klaus and Cross, 1974; Snippe et al., 1975; Anderson et al., 1989; Dintzis, 1992). Thus, hsp70 fusion proteins reduce the variables associated with the study of HSPs as immunological carriers.

The hsp70 fusion vector system was used to produce an HIV Gag-hsp70 fusion protein and to investigate the humoral and cellular immune respon-se to it (Suzue and Young, 1996). The mycobacterial hsp70 moiety was found to dramatically increase the immunogenicity of the Gag p24 antigen. Mice immunized with the p24-hsp70 fusion protein in phosphate buffer generated a vigorous humoral immune response against p24, whereas administration of p24 elicited low levels of anti-p24 antibody wich was not long lasting. The high level of anti-p24 antibody elicited by the p24-hsp70 fusion protein was long-lived; high titres of anti-p24 antibodies were de-tected up to 68 weeks after immunization. The p24-hsp70 fusion protein also stimulated T cell responses to p24 in the absence of adjuvant. Splen-ocytes from mice immunized with the fusion protein exhibited p24 antigen-dependent proliferation and production of the cytokines IFN-γ, IL5 and IL-2. Covalent linkage of hsp70 to p24 was essential in order for hsp70 to exert an adjuvant-free carrier effect with p24. Thus, under these conditions, hsp70 is not a conventional adjuvant, in that its administration together with an unlinked antigen does not stimulate immune responses to that antigen. Instead, hsp70 functions as an exceptional carrier, since it stimulates vigorous immune responses when covalently linked to the antigen of interest. The reason that hsp70 is an exceptional carrier, and the reason that the hsp70 carrier does not require priming, is perhaps due to previous and ongoing exposure to substantial amounts of diverse microbial HSPs. It is also possible that the protein chaperone function of hsp70 contributes to the adjuvant-free carrier effect.

The above studies demonstrate that HSPs can act as adjuvant-free carriers when attached to a synthetic peptide, oligosaccharide or full-length protein. HSP fusion proteins and conjugates can induce strong humoral and cellular immune responses to a linked antigen. Thus, HSPs appear to be an important addition to the small repertoire of carriers available for the development of new vaccines.

HSPs as vaccines

HSPs from bacterial and fungal pathogens are among the major targets of the immune response to infection and have been found to be capable of inducing protective immunity in experimental animal models (Tab. 2). We

Table 2. HSPs can elicit protective immune responses

Infectious agent	HSP	Animal model	References
Bacteria			
Helicobacter pylori	hsp60, hsp10	mouse	Ferrero et al., 1995
Legionella pneumophila	hsp60	guinea pig	Blander and Horwitz, 1993
Mycobacterium leprae	hsp60, hsp10	mouse	Gelber et al., 1992; 1994
Mycobacterium tuberculosis	hsp70	mouse, guinea pig	Pal and Horwitz, 1992; Hubbard et al., 1992; Andersen 1994; Horwitz et al., 1995
Mycobacterium tuberculosis	hsp60	mouse	Silva et al., 1994a, 1994b; Lowrie et al., 1994
Yersinia enterocolitica	hsp60	mouse	Noll et al., 1994
Fungi			
Candida albicans	hsp90	mouse	Matthews et al., 1991
Histoplasma capsulatum	hsp60, hsp70	mouse	Gomez et al., 1991a, 1991b; 1992, 1995

review below examples of pathogenic microorganisms for which there is evidence that the immune response to HSPs contributes to protection against infection and disease.

The ability of *Legionella pneumophila* hsp60 to protect against Legionnaires' disease has been examined in a guinea pig model (Blander and Horwitz, 1993). Immunization of guinea pigs with purified *L. pneumophila* hsp60 was found to protect the animals from a lethal aerosol challenge with the organism. Cell-mediated immunity is known to be critical to the host defense against intracellular *L. pneumophila* (Horwitz, 1983) and animals immunized with purified *L. pneumophila* hsp60 exhibited delayed-tpye hypersensitivity (DTH) to hsp60, indicating that cellular responses had been elicited by the experimental vaccine.

M. tuberculosis hsp70 can contribute to protective immunity in animal models of infection. Immunizing mice and guinea pigs with *M. tuberculosis* culture filtrate proteins prior to challenge with the bacterium significantly reduced the number of organisms in the lung and spleen (Pal and Horwitz, 1992; Hubbard et al., 1992; Andersen, 1994). The colony forming units of mycobacteria recovered from the spleen after challenge was reduced by more than 95%, and even 22 weeks after vaccination, T cells from immunized mice could transfer protection to recipient mice (Andersen, 1994). Subsequently, abundant proteins in the *M. tuberculosis* culture filtrate were purified and analyzed for immunoprotective activity (Horwitz et al., 1995). Immunization of guinea pigs with purified hsp70 alone induced protection in some instances while a combination of five purified mycobacterial proteins, which included hsp70, consistently protected guinea pigs against weight loss and lung destruction when the animals were challenged with *M. tuberculosis*.

In contrast to hsp70, *M. tuberculosis* hsp60 does not elicit protective responses when administered as a soluble protein. However, vaccination of mice with syngenic J774 macrophage cells expressing mycobacterial hsp60 afforded remarkable protection against *M. tuberculosis* (Silva and Lowrie 1994). In mice vaccinated with J774-hsp60, 100 times fewer *M. tuberculosis* CFUs could be recovered from the liver 5 weeks after challenge, compared to unvaccinated mice. Hsp60 specific T cells cloned from the vaccinated mice could adoptively transfer protection to non-vaccinated mice (Silva et al., 1994). Vaccination of mice with mycobacterial hsp60 was also effective when administered to mice as a naked DNA vaccine (Lowrie et al., 1994). These studies indicate that eliciting protective immune responses can depend on the method of administration of an antigen. The *ex vivo* cell vaccine and the naked DNA vaccine approaches may better mimic an aspect of normal macrophage infection by *M. tuberculosis* than inoculation with a soluble antigen.

HSPs are also among the antigens of *Mycobacterium leprae* that appear to contribute to immunological protection. In order to identify protective components of *M. leprae*, mice were injected with various fractions derived from the organism (Gelber et al., 1992). Vaccination of mice with a soluble protein fraction afforded significant protection against *M. leprae* infection and was more effective than administering killed *M. leprae* organisms. Studies were then carried out with recombinant *M. leprae* hsp60 and hsp10 to determine whether the highly protective nature of the soluble protein fraction could be attributed to a particular protein (Gelber et al., 1994). By challenging mice with *M. leprae* at different time intervals following vaccination, the extent of protection provided by each protein was determined. The soluble protein fraction provided protection and inhibited *M. leprae* growth up to one year after vaccination; hsp10 provided protection up to 2 months later and hsp60, up to 4 months. These experiments suggest that a cocktail containing multiple *M. leprae* proteins may be necessary to confer long-lasting protective immunity.

The inclusion of *Helicobacter pylori* HSPs in experimental vaccine preparations has been found to contribute to protection against gastroduodenal disease in a mouse model of *H. pylori* infection (Ferrero et al., 1995). Oral administration of recombinant *H. pylori* hsp60 or hsp10 protected a significant percentage of mice from infection. Immunization with both hsp10 and UreB, a urease subunit protein, conferred complete protection to the mice, and was found to be as effective as administering sonicated heliobacterial whole-cell extract.

HSPs have also been implicated in immunological protection against major fungal diseases. Immunization with purified *Histoplasma capsulatum* HSPs has been found to induce protective immunity against the pathogenic fungus in a mouse model of infection. Early experiments revealed that a detergent cell wall/cell membrane (CW/M) extract of *H. capsulatum* injected into mice could confer protection against a lethal challenge of yeast cells (Gomez et al., 1991a). Further analysis revealed that a 62 kD protein component of the CW/M extract (HIS-62) was a target of *H. capsulatum*-specific T cells and was sufficient to engender protection against a lethal challenge (Gomez et al., 1991b). Cloning of HIS-62 led to its identification as *H. capsulatum* hsp60 (Gomez et al., 1995). In contrast to the fortuitous identification of hsp60 as a protective antigen, hsp70 from *H. capsulatum* was identified and isolated with the intent of studying its involvement in immunity. Purified *H. capsulatum* hsp70 could induce DTH responses and protect mice from a sublethal injection of yeast cells (Gomez et al., 1992). Immunization with hsp60 confers protection against a higher challenge dose of *H. capsulatum* than does immunization with hsp70. It will be interesting to determine whether a vaccine consisting of both hsp60 and hsp70 might afford even better protection against *H. capsulatum*.

The significance of anti-HSP immune responses in some bacterial and fungal diseases has been affirmed in passive immunization studies. For example, the bacterial pathogen *Yersinia enterocolitica* elicits humoral and cellular responses against hsp60 in mice (Noll et al., 1994). The relevance of HSP specific T cells in protective immunity against murine yesiniosis was examined by intravenously transferring 10^7 hsp60 specific T cell clones into naive mice. Three different T cell clones were examined and all of the passively immunized mice were protected when challenged with a lethal dose of pathogen. In studies with *Candida albicans*, hsp90 has been identified as an immunodominant antigen, and passive administration of an hsp90 monoclonal antibody was found to confer protection against systemic candidosis (Matthews et al., 1991).

The immune responses against HSPs confer protection against a broad range of pathogens. Although there is some concern with using HSPs in vaccine formulations due to their highly conserved nature and homology with self-HSPs, it must be emphasized that healthy individuals are routinely stimulated to respond to HSPs without causing autoimmunity. For example, the trivalent vaccine against tetanus, diphtheria and pertussis, which is routinely administered to infants, induces anti-hsp70 immune responses (Del Giudice et al., 1993). Live BCG, which contains substantial amounts of hsp70 and hsp60, has been used to immunize 80% of the world's children against tuberculosis. Thus, current knowledge suggests to us that the inclusion of HSPs in vaccines against a broad spectrum of infectious diseases would be both safe and beneficial.

References

Adams, E., Britton, W., Morgan, A., Sergeantson, S. and Basten, A. (1994) Individuals from different populations identify multiple and diverse T cell determinants on mycobacterial HSP70. *Scand. J. Immunol.* 39:588–596.

Altman, A. and Dixon, F.J. (1989) Immunomodifiers in vaccines. *In:* J.L. Bittle and F.L. Murphy (eds): *Advances in Veterinary Science and Comparative Medicine.* Academic Press, San Diego, pp 313–316.

Andersen, P. (1994) Effective vaccination of mice against Mycobacterium tuberculosis infection with a soluble mixture of secreted mycobacterial proteins. *Infect. Immun.* 62: 2536–2544.

Anderson, P.W., Pichichero, M.E., Stein, E.C., Porcelli, S., Betts, R.F., Connuck, D.M., Korones, D., Insel, R.A., Zahradnik, J.M. and Eby, R. (1989) Effect of oligosaccharide chain length, exposed terminal group, and hapten loading on the antibody response of human adults and infants to vaccines consisting of Haemophilus influenzae type b capsular antigen unterminally coupled to the diphtheria protein CRM197. *J. Immunol.* 142:2464–2468.

Audibert, F.M. and Lise, L.D. (1993) Adjuvants: current status, clinical perspectives and future prospects. *Immunol. Today* 14:281–284.

Baird, P.N., Hall, L.M. and Coates, A.R. (1988) A major antigen from Mycobacterium tuberculosis which is homologous to the heat shock proteins groES from *E. coli* and the htpA gene product of Coxiella burneti. *Nucleic Acids Res.* 16:9047.

Barrios, C., Lussow, A.R., Van Embden, J., Van der Zee, R., Rappuoli, R., Costantino, P., Louis, J.A., Lambert, P.H. and Del Giudice, G. (1992) Mycobacterial heat-shock proteins as carrier molecules. II: The use of the 70-kDa mycobacterial heat-shock protein as carrier for conjugated vaccines can circumvent the need for adjuvants and Bacillus Calmette Guerin priming. *Eur. J. Immunol.* 22:1365–1372.

Barrios, C., Georgopoulos, C., Lambert, P.H. and Del Giudice, G. (1994) Heat shock proteins as carrier molecules: *in vivo* helper effect mediated by *Escherichia coli* GroEL and DnaK proteins requires cross-linking with antigen. *Clin. Exp. Immunol.* 98:229–233.

Blander, S.J. and Horwitz, M.A. (1993) Major cytoplasmic membrane protein of *Legionella pneumophila*, a genus common antigen and member of the hsp60 family of heat shock proteins, induces protective immunity in a guinea pig model of Legionnaires' disesase. *J. Clin. Invest.* 91:717–23.

Cerrone, M.C., Ma, J.J. and Stephens, R.S. (1991) Cloning and sequence of the gene for heat shock protein 60 from *Chlamydia trachomatis* and immunological reactivity of the protein. *Infect. Immun.* 59:79–90.

Claman, H.N., Chaperon, E.A. and Triplett, R.F. (1966) Thymus-marrow cell combinations. Synergism in antibody production. *Proc. Soc. Exp. Biol. Med.* 122:1167–1171.

Davies, A.J.S., Leuchars, E., Wallis, V., Marchant, R. and Elliott, E.V. (1967) The failure of thymus-derived cells to produce antibody. *Transplantation* 5:222–231.

de Andrade, C.R., Kirchhoff, L.V., Donelson, J.E. and Otsu, K. (1992) Recombinant Leishmania Hsp90 and Hsp70 are recognized by sera from visceral leishmaniasis patients but not Chagas' disease patients. *J. Clin. Microbiol.* 30:330–335.

Del Giudice, G., Cooper, J.A., Merino, J., Verdini, A.S., Pessi, A., Togna, A.R., Engers, H.D., Corradin, G. and Lambert, P.H. (1986) The antibody response in mice to carrier-free synthetic polymers of Plasmodium falciparum circumsporozoite repetitive epitope is I-Ab-restricted: possible implications for malaria vaccines. *J. Immunol.* 137:2952–2955.

Del Giudice, G., Gervaix, A., Costantino, P., Wyler, C.A., Tougne, C., de Graeff-Meeder, E.R., Van Embden, J., Van der Zee, R., Nencioni, L., Rappuoli, R., Suter, S. and Lambert, P.H. (1993) Priming to heat shock proteins in infants vaccinated against pertussis. *J. Immunol.* 150:2025–2032.

Dintzis, R.Z. (1992) Rational design of conjugate vaccines. *Pediatr. Res.* 32:376–385.

Edelman, R. and C.O. Tacket (1990) Adjuvants. *Int. Rev. Immunol.* 7:51–66.

Ferrero, R.L., Thiberge, J.-M., Kansau, I., Wuscher, N., Huerre, M. and Labigne, A. (1995) The GroES homolog of *Helicobacter pylori* confers protective immunity against musosal infection in mice. *Proc. Natl. Acad. Sci. USA* 92:6499–6503.

Geerligs, H.J., Weijer, W.J., Welling, G.W. and Welling-Wester, S. (1989) The influence of different adjuvants on the immune response to a synthetic peptide comprising amino acid residues 9-21 of herpes simplex virus type 1 glycoprotein D. *J. Immunol. Methods* 124:95–102.

Gelber, R.H., Murray, L., Siu, P. and Tsang, M. (1992) Vaccination of mice with a soluble protein fraction of Mycobacterium leprae provides consistent and long-term protection against M. leprae infection. *Infect. Immun.* 60:1840–1844.

Gelber, R.H., Mehra, V., Bloom, B., Murray, L.P., Siu, P., Tsang, M. and Brennan, P.J. (1994) Vaccination with pure Mycobacterium leprae proteins inhibits M. leprae multiplication in mouse footpads. *Infect. Immun.* 62:4250–4255.

Gomez, A.M., Rhodes, J.C. and Deepe, Jr., G. (1991a) Antigenicity and immunogenicity of an extract from the cell wall and cell membrane of Histoplasma capsulatum yeast cells. *Infect. Immun.* 59:330–336.

Gomez, F.J., Gomez, A.M. and Deepe, Jr., G. (1991b) Protective efficacy of a 62-kilodalton antigen, HIS-62, from the cell wall and cell membrane of Histoplasma capsulatum yeast cells. *Infect. Immun.* 59:4459–4464.

Gomez, F.J., Gomez, A.M. and Deepe, Jr. G. (1992) An 80-kilodalton antigen from Histoplasma capsulatum that has homology to heat shock protein 70 induces cell-mediated immune responses and protection in mice. *Infect. Immun.* 60:2565–2571.

Gomez, F.J., Allendoerfer, R. and Deepe, G.S. (1995) Vaccination with recombinant heat shock protein 60 from *Histoplasma capsulatum* protects mice against pulmonary histoplasmosis. *Infect. Immun.* 63:2587–2595.

Good, M.F., Berzofsky, J.A., Maloy, W.L., Hayashi, Y. Fujii, N., Hockmeyer, W.T. and Miller, L.H. (1986) Genetic control of the immune response in mice to a Plasmodium falciparum sporozoite vaccine. Widespread nonresponsiveness to single malaria T epitope in highly repetitive vaccine. *J. Exp. Med.* 164:655–660.

Hanna, N., Jarosch, E. and Leskowitz, S. (1972) Altered immunogenicity produced by change in mode of linkage of hapten to carrier. *Proc. Soc. Exp. Biol. Med.* 140:89–92.

Hansen, K., Bangsborg, J.M., Fjordvang, H., Pedersen, N.S. and Hindersson, P. (1988) Immunochemical characterization of and isolation of the gene for a *Borrelia burgdorferi* immunodominant 60-kilodalton antigen common to a wide range of bacteria. *Infect. Immun.* 56:2047–2053.

Hedstrom, R., Culpepper, J., Harrison, R.A., Agabian, N. and Newport, G. (1987) A major iummunogen in *Schistosoma mansoni* infections is homologous to the heat-shock protein Hsp70. *J. Exp. Med.* 165:1430–1435.

Hedstrom, R., Culpepper, J., Schinski, V., Agabian, N. and Newport, G. (1988) Schistosome heat-shock proteins are immunologically distinct host-like antigens. *Mol. Biochem. Parasitol.* 29: 275–282.

Hindersson, P., Knudsen, J.D. and Axelsen, N.H. (1987) Cloning and expression of treponema pallidum common antigen (Tp-4) in *Escherichia coli* K12. *J. Gen. Microbiol.* 133: 587–596.

Hoffman, P.S., Houston, L. and Butler, C.A. (1990) *Legionella pneumophila* htpAB heat shock operon: nucleotide sequence and expression of the 60-kilodalton antigen in *L. pneumophila*-infected HeLa cells. *Infect. Immun.* 58:3380–3387.

Horwitz, M.A. (1983) Cell-mediated immunity in Legionnaires' disease. *J. Clin. Invest.* 71: 1686–1697.

Horwitz, M.A., Lee, B.W., Dillon, B.J. and Harth, G. (1995) Protective immunity against tuberculosis induced by vaccination with major extracellular proteins of Mycobacterium tuberculosis. *Proc. Natl. Acad. Sci. USA* 92:1530–1534.

Hubbard, R.D., Flory, C.M. and Collins, F.M. (1992) Immunization of mice with mycobacterial culture filtrate proteins. *Clin. Exp. Immunol.* 87:94–98.

Husson, R.N. and Young, R.A. (1987) Genes for the major protein antigens of Mycobacterium tuberculosis: the etiologic agents of tuberculosis and leprosy share an immunodominant antigen. *Proc. Natl. Acad. Sci. USA* 84:1679–1683.

Johnson, K.S., Wells, K., Bock, J.V., Nene, V., Taylor, D.W. and Cordingley, J.W. (1989) The 86-kilodalton antigen from *Schistosoma mansoni* is a heat-shock protein homologous to yeast HSP-90. *Mol. Biochem. Parasitol.* 36:19–28.

Katz, D.H., Hamaoka, T. and Benacerraf, B. (1973) Cell interactions between histoincompatible T and B lymphocytes. II. Failure of physiologic cooperative interactions between T and B lymphocytes from allogeneic donor strains in humoral response to hapten-protein conjugates. *J. Exp. Med.* 137:1405–1418.

Kaufmann, S.H., Vath, U., Thole, J.E., Van Embden, J.D. and Emmrich, F. (1987) Enumeration of T cells reactive with Mycobacterium tuberculosis organisms and specific for the recombinant mycobacterial 64-kDa protein. *Eur. J. Immunol.* 17:351–357.

Kaufmann, S.H. (1990) Heat shock proteins and the immune response. *Immunol. Today* 11:129–136.

Kaufmann, S.H.E. and Schoel, B. (1994) Heat shock proteins as antigens in immunity against infection and self. *In:* R.I. Morimoto, A. Tissieres and C. Georgopoulos (eds): *The Biology of Heat Shock Proteins and Molecular Chaperones.* Cold Spring Harbor Laboratory Press, Cold Spring Harbor, NY, pp 495–531.

Kenney, J.S. Hughes, B.W., Masada, M.P. and Allison, A.C. (1989) Influence of adjuvants on the quantity, affinity, isotype and epitope specificity of murine antibodies. *J. Immunol. Methods* 121:157–166.

Kindred, B. and Shreffler, D.C. (1972) H-2 dependence of co-operation between T and B cells *in vivo*. *J. Immunol.* 109:940–943.

Klaus, G.G. and Cross, A.M. (1974) The influence of epitope density on the immunological properties of hapten-protein conjugates. I. Characteristics of the immune response to hapten-coupled albumen with varying epitope density. *Cell. Immunol.* 14:226–241.

Kumar, A., Reddy, L.V., Sochanik, A. and Kurup, V.P. (1993) Isolation and characterization of a recombinant heat shock protein of *Aspergillus fumigatus*. *J. Allergy Clin. Immunol.* 91: 1024–1030.

Lachmann, P.J. and Amos, H.E. (1970) Soluble factors in the mediation of the cooperative effect. *In:* P.A. Miescher (ed.): *Immunopathology.* Basel, Schwabe, p. 65.

Lachman, P.J. and Sikora, K. (1978) Coupling PPD to tumour cells enhances their antigenicity in BCG-primed mice. *Nature* 271:463–464.

Lachmann, P.J., Strangeways, L., Vyakarnam, A. and Evan, G. (1986) Raising antibodies by coupling peptides to PPD and immunizing BCG-sensitized animals. *Ciba Found Symp.* 119: 25–57.

Lamb, J.R., Ivanyi, J., Rees, A.D., Rothbar, J.B., Howland, K., Young, R.A. and Young, D.B. (1987) Mapping of T cell epitopes using recombinant antigens and synthetic peptides. *EMBO J.* 6:1245–1249.

Landsteiner, K. (1936) *The Specificity of Serological Reactions.* Dover Publications, New York.

Levy Yeyati, P., Bonnefoy, S., Mirkin, G., Debrabant, A., Lafon, S., Panebra, A., Gonzalez-Cappa, E., Dedet, J.P., Hontebeyrie-Joskowicz, M. and Levin, M.J. (1992) The 70-kDa heat-shock protein is a major antigenic determinant in human *Trypanosoma cruzi/Leishmania braziliensis braziliensis* mixed infection. *Immunol. Lett.* 31:27–33.

Leyden, J.J., Nordstrom, K.M. and McGinley, K.J. (1991) Cutaneous Microbiology. *In:* L.A. Goldsmith (ed.): *Physiology Biochemistry and Molecular Biology of the Skin.* Oxford University Press, New York, pp 1403–1421.

Lowrie, D.B., Tascon, R.E., Colston, M.J. and Silva, C.L. (1994) Towards a DNA vaccine against tuberculosis. *Vaccine* 12:1537–1540.

Lussow, A.R., Aguado, M.T., Del Giudice, G. and Lambert, P.H. (1990a) Towards vaccine optimisation. *Immunol. Lett.* 25:255–263.

Lussow, A.R., Del Giudice, G., Renia, L., Mazier, D., Verhave, J.P., Verdini, A.S., Pessi, A., Louis, J.A. and Lambert, P.H. (1990b). Use of a tuberculin purified protein derivative-Asn-Ala-Asn-Pro conjugate in bacillus Calmette-Guerin primed mice overcomes H-2 restriction of the antibody response and avoids the need for adjuvants. *Proc. Natl. Acad. Sci. USA* 87:2960–2964.

Lussow, A.R., Barrios, C., van Embden, J., Van der Zee, R., Verdini, A.S., Pessi, A., Louis, J.A., Lambert, P.H. and Del Giudice, G. (1991) Mycobacterial heat-shock proteins as carrier molecules. *Eur. J. Immunol.* 21:2297–2302.

MacFarlane, J., Blaxter, M.L., Bishop, R.P., Miles, M.A. and Kelly, J.M. (1990) Identification and characterisation of a Leishmania donovani antigen belonging to the 70-kDa heat-shock protein family. *Eur. J. Biochem.* 190:377–384.

Mattei, D., Scherf, A., Bensaude, O. and da Silva, L.P. (1989) A heat shock-like protein from the human malaria parasite *Plasmodium falciparum* induces autoantibodies. *Eur. J. Immunol.* 19:1823–1828.

Matthews, R.C., Burnie, J.P. and Tabaqchali, S. (1987) Isolation of immunodominant antigens from sera of patients with systemic candidiasis and characterization of serological response to Candida albicans. *J. Clin. Microbiol.* 25:230–237.

Matthews, R.C. and Burnie, J.P. (1989) Cloning of a DNA sequence encoding a major fragment of the 47 kilodalton stress protein homologue of *Candida albicans. FEMS Microbiol. Letts.* 60:25.

Matthews, R.C., Burnie, J.P., Howat, D., Rowland, T. and Walton, F. (1991) Autoantibody to heat-shock protein 90 can mediate protection against systemic candidosis. *Immunology* 74:20–24.

Mehra, V., Sweetser, D. and Young, R.A. (1986) Efficient mapping of protein antigenic determinants. *Proc. Natl. Acad. Sci USA* 83:7013–7017.

Mehra, V., Bloom, B.R., Bajardi, A.C., Grisso, C.L., Sieling, P.A., Alland, D., Convit, J., Fan, X.D., Hunter, S.W., Brennan, P.J., Rea, T.H. and Modlin, R.L. (1992) A major T cell antigen of *Mycobacterium leprae* is a 10-kd heat-shock cognate protein. *J. Exp. Med.* 175:275–284.

Miller, J.F. and Mitchell, G.F. (1967) The thymus and the precursors of antigen reactive cells. *Nature* 216:659–663.

Mitchison, N.A. (1971a) The carrier effect in the secondary response to hapten-protein conjugates. I. Measurement of the effect with transferred cells and objections to the local environment hypothesis. *Eur. J. Immunol.* 1:10–17.

Mitchison, N.A. (1971b) The carrier effect in the secondary response to hapten-protein conjugates. II. Cellular cooperation. *Eur. J. Immunol.* 1:18–27.

Munk, M.E., Schoel, B. and Kaufmann, S.H. (1988) T cell responses of normal individuals toward recombinant protein antigens of Mycobacterium tuberculosis. *Eur. J. Immunol.* 18:1835–1838.

Munk, M.E., Shinnick, T.M. and Kaufmann, S.H. (1990) Epitopes of the mycobacterial heat shock protein 65 for human T cells comprise different structures. *Immunobiology* 180:272–277.

Murray, P.J. and Young, R.A. (1992) Stress and immunological recognition in host-pathogen interactions. *J. Bacteriol.* 174:4193–4196.

Nene, V., Dunne, D.W., Johnson, K.S., Taylor, D.W. and Cordingley, J.S. (1986) Sequence and expression of a major egg antigen from *Schistosoma mansoni.* Homologies to heat shock proteins and alpha-crystallins. *Mol. Biochem. Parasitol.* 21:179–188.

Nerland, A.H., Mustafa, A.S., Sweetser, D., Godal, T. and Young, R.A. (1988) A protein antigen of Mycobacterium leprae is related to a family of small heat shock proteins. *J. Bacteriol.* 170:5919–5921.

Nicklas, W. (1992) Aluminum salts. *Res. Immunol.* 143:489–494.

Noll, A., Roggenkamp, A., Heesemann, J. and Autenrieth, I.B. (1994) Protective role for heat shock protein-reactive alpha beta T cells in murine yersiniosis. *Infect. Immun.* 62:2784–2791.

Oftung, F., Geluk, A., Lundin, K.E., Meloen, R.H., Thole, J.E., Mustafa, A.S. and Ottenhoff, T.H. (1994) Mapping of multiple HLA class II-restricted T cell epitopes of the mycobacterial 70-kilodalton heat shock protein. *Infect. Immun.* 62:5411–5418.

Ovary, Z. and Benacerraf, B. (1963) Immunological specificity of the secondary response with dinitrophenylated proteins. *Proc. Soc. Exp. Biol. Med.* 114:72–76.

Pal, P.G. and Horwitz, M.A. (1992) Immunization with extracellular proteins of Mycobacterium tuberculosis induces cell-mediated immune responses and substantial protective immunity in a guinea pig model of pulmonary tuberculosis. *Infect. Immun.* 60:4781–4792.

Perraut, R., Lussow, A.R., Gavoille, S., Garraud, O., Matile, H., Tougne, C., Van Embden, J., Van der Zee, R., Lambert, P.H., Gysin, J. and Del Giudice, G. (1993) Successful primate immunization with peptides conjugated to purified protein derivative or mycobacterial heat shock proteins in the absence of adjuvants. *Clin. Exp. Immunol.* 93:382–386.

Raff, M.C. (1970) Role of thymus-derived lymphocytes in the secondary humoral immune response in mice. *Nature* 226:1257–1258.

Rajewsky, K., Schirrmacher, V., Nase, S. and Jerne, N.K. (1969) The requirement of more than one antigenic determinant for immunogenicity. *J. Exp. Med.* 129:1131–1143.

Renia, L., Mattei, D., Goma, J., Pied, S., Dubois, P., Miltgen, F., Nussler, A., Matile, H., Menegaux, F., Gentilini, M. and Mazier, D. (1990) A malaria heat-shock-like determinant expressed on the infected hepatocyte surface is the target of antibody-dependent cell-mediated cytotoxic mechanisms by nonparenchymal liver cells. *Eur. J. Immunol.* 20:1445–1449.

Requena, J.M., Soto, M., Guzman, F., Maekelt, A., Noya, O., Patarroyo, M.E. and Alonso, C. (1993) Mapping of antigenic determinants of the *T. cruzi* hsp70 in chagasic and healthy individuals. *Mol. Immunol.* 30:1115–1121.

Richman, S.J., Vedvick, T.S. and Reese, R.T. (1989) Peptide mapping of conformational epitopes in a human malarial parasite heat shock protein. *J. Immunol.* 143:285–292.

Robbins, J.B. and Schneerson, R. (1990) Polysaccharide-protein conjugates: a new generation of vaccines. *J. Infect. Dis.* 161:821–832.

Roop, R., Price M.L., Dunn, B.E., Boyle, S.M., Sriranganathan, N. and Schurig, G.G. (1992) Molecular cloning and nucleotide sequence analysis of the gene encoding the immunoreactive *Brucella abortus* Hsp60 protein, BA60K. *Microb. Pathog.* 12:47–62.

Rosebury, T. (1962) *Microorganisms Indigenous to Man.* McGraw-Hill Book Co., Inc., New York. pp 310–350.

Rothstein, N.M., Higashi, G., Yates, J. and Rajan, T.V. (1989) *Onchocerca volvulus* heat shock protein 70 is a major immunogen in amicrofilaremic individuals from a filariasis-endemic area. *Mol. Biochem. Parasitol.* 33:229–235.

Scallon, B.J., Bogitsh, B.J. and Carter, C.E. (1987) Cloning of a *Schistosoma japonicum* gene encoding a major immunogen recognized by hyperinfected rabbits. *Mol. Biochem. Parasitol.* 24:237–245.

Selkirk, M.E., Denham, D.A., Partono, F. and Maizels, R.M. (1989) Heat shock cognate 70 is a prominent immunogen in Brugian filariasis. *J. Immunol.* 143:299–308.

Shinnick, T.M., Vodkin, M.H. and Williams, J.C. (1988) The Mycobacterium tuberculosis 65-kilodalton antigen is a heat shock protein which corresponds to common antigen and to the *Escherichia coli* GroEL protein. *Infect. Immun.* 56:446–451.

Silva, C.L. and Lowrie, D.B. (1994) A single mycobacterial protein (hsp65) expressed by a transgenic antigen-presenting cell vaccinates mice against tuberculosis. *Immunology* 82:244–248.

Silva, C.L., Silva, M.F., Pietro, R.C. and Lowrie, D.B. (1994) Protection against tuberculosis by passive transfer with T cell clones recognizing mycobacterial heat-shock protein 65. *Immunology* 83:341–346.

Smith, D.H., Madore, D.V., Eby, R.J., Anderson, P.W., Insel, R.A. and Johnson, C.L. (1989) Haemophilus b oligosaccharide-CRM197 and other Haemophilus b conjugate vaccines: a status report. *Adv. Exp. Med. Biol.* 251:65–82.

Snippe, H., Graven, W.G. and Willems, P.J. (1975) Antibody formation in the mouse induced by hapten-carrier complexes. *Immunology* 28:885–895.

Suerbaum, S., Thiberge, J.M., Kansau, I., Ferrero, R.L. and Labigne, A. (1994) Helicobacter pylori hspA-hspB heat-shock gene cluster: nucleotide sequence, expression, putative function and immunogenicity. *Mol. Microbiol* 14:959–974.

Suzue, K. and Young, R.A. (1996) Adjuvant-free hsp70 fusion protein system elicits humoral and cellular immune responses to HIV-1 p24. *J. Immunol.* 156:873–879.

Taylor, H.R., Maclean, I.W., Brunham, R.C., Pal, S. and Wittum-Hudson, J. (1990) Chlamydial heat shock proteins and trachoma. *Infect. Immun.* 58:3061–3063.

Thole, J.E., van Schooten, W.C., Keulen, W.J., Hermans, P.W., Janson, A.A., de Vries, R.R., Kolk, A.H. and van Embden, J.D. (1988) Use of recombinant antigens expressed in *Escherichia coli* K-12 to map B cell and T cell epitopes on the immunodominant 65-kilodalton protein of mycobacterium bovis BCG. *Infect. Immun.* 56:1633–1640.

Van Schooten, W.C., Ottenhoff, T.H., Klatser, P.R., Thole, J., De Vries, R.R. and Kolk, H. (1988) T cell epitopes on the 36K and 65K Mycobacterium leprae antigens defined by human T cell clones. *Eur. J. Immunol.* 18:849–854.

Verbon, A., Hartskeerl, R.A., Schuitema, A., Kolk, A.H., Young, D.B. and Lathigra, R. (1992) The 14000-molecular-weight antigen of Mycobacterium tuberculosis is related to the alpha-crystallin family of low-molecular-weight heat shock proteins. *J. Bacteriol.* 174:1352–1359.

Vodkin, M.H. and Williams, J.C. 81988) A heat shock operon in *Coxiella burnetti* produces a major antigen homologous to a protein in both mycobacteria and *Escherichia coli. J. Bacteriol.* 170:1227–1234.

Young, D., Lathigra, R., Hendrix, R., Sweetser, D. and Young, R.A. (1988) Stress proteins are immune targets in leprosy and tuberculosis. *Proc. Natl. Acad. Sci. USA* 85:4267–4270.

Young, R.A. (1990) Stress proteins and immunology. *Annu. Rev. Immunol.* 8:401–420.

Young, D.B., Mehlert, A. and Smith, D.F. (1990) Stress proteins and infectious diseases. *In:* R.I. Morimoto, A. Tissieres and C. Georgopoulos (eds): *Stress Proteins in Biology and Medicine.* Cold Spring Harbor Laboratory, Cold Spring Harbor, NY, pp 131–165.

Zhong, G. and Brunham, R.C. (1992) Antibody responses to the chlamydial heat shock proteins hsp60 and hsp70 are H-2 linked. *Infect. Immun.* 60:3143–3149.

Stress-Inducible Cellular Responses
ed. by U. Feige, R.I. Morimoto, I. Yahara and B. Polla
© 1996 Birkhäuser Verlag Basel/Switzerland

Regulation of thermotolerance and ischemic tolerance

K. Nagata

Department of Cell Biology, Chest Disease Research Institute, Kyoto University, 53 Kawaharacho, Shogo-inn, Sakyo-ku, Kyoto 606-01, Japan

Summary. Thermotolerance and ischemic tolerance are two major biological aspects where heat shock (stress) proteins exert essential roles for survival in cells as well as in various tissues. Bioflavonoids prevent the cells from acquiring thermotolerance after stresses through specific inhibition in the induction of heat shock proteins. The mechanism of this inhibition is revealed to be due to the prevention of the activation of heat shock factor 1 after heat shock. The induction of stress proteins during the ischemic stress is then described in global as well as focal cerebral ischemic model in rats. The activation of heat shock factor 1 after ischemia is first shown to induce various stress proteins in the central nervous system.

Introduction

When cells or organisms are exposed to heat shock, they respond to it by synthesizing a group of proteins called heat shock proteins (HSPs) (Morimoto et al., 1994). HSPs are among the most highly conserved proteins during the evolution, and the important roles of HSP have been clarified not only in stressed cells but also in cells under normal growth conditions.

The induction of HSP is observed under various adverse circumstances including heat shock, hypoxia, glucose starvation, and the addition of sodium arsenite, heavy metals, amino acid analogs and so on. Thus, HSPs are more generally termed as stress proteins. Drugs which positively and negatively modulate the induction of HSP after heat shock are now reported (Morimoto et al., 1994; Hosokawa et al., 1990). The induction of HSP have been reported to be directly involved in the acquisition of the resistance against subsequent cellular stresses including heat shock and ischemia (Li et al., 1991; Landry et al., 1989; Nowak and Abe, 1994; Ohtsuki et al., 1992). Although HSPs have been shown to be involved in the acquisition of the tolerance against the subsequent stress, it is still uncertain how HSPs participate in the acquisition, maintenance, and decay of thermotolerance and ischemic tolerance in cultured cells as well as in the whole body.

In this chapter, I will summarize two aspects of the phenomena, thermotolerance and ischemic tolerance, both of which are developed by the induction of HSP *in vitro* and *in vivo*. Since thermotolerance is discussed in detail by Li and Nussenzweig in this volume, I will focus on the specific

inhibition of the induction of HSP as well as the inhibition of thermotolerance with specific drugs exogenously added to the cells. Then, I will
move to another aspect of the HSP-related tolerance in the cells, ischemia,
using an *in vivo* model of cerebral ischemia as well as ischemia in the
isolated heart. Here, I will show the evidence that the induction of stress
proteins in the tissues is mediated by the activation of the heat shock
factors as was reported in the cultured cells.

Inhibition of thermotolerance by bioflavonoids

Inhibition of the induction of HSPs

Although a lot of stresses that induce HSP have been reported, there are few
reports about the drugs or the treatments which inhibit the induction of
HSP. The incubation of the cells in heavy water (D_2O) is reported to repress
the induction of HSP after heat shock, due to the inefficient transduction of
the heat (Edington et al., 1989). We have reported that the induction of
various HSPs after heat shock is uniformly inhibited by incubation of
the cells with several kinds of flavonoids, such as quercentin, genistein,
luteorin and so on (Hosokawa et al., 1990).

COLO 320DM cells, derived from a human colon cancer, were treated at
the elevated temperature in the presence and absence of quercetin. Two-
dimensional gel electrophoresis revealed that the synthesis of major HSPs
including HSP110, HSP90, HSP70, HSP47 (Nagata et al., 1986) and
HSP40 (Ohtsuka et al., 1990) were clearly induced after heat shock. When
quercetin at the concentration of 100 μM was present during heat shock, it
inhibited the induction of all of these HSPs while the synthesis of other proteins was not affected (Fig. 1). Similar results were obtained for other cell
lines. The specific inhibition of the synthesis of major HSPs, including
HSP90, HSP70s (HSC70 and HSP70), HSP47 and HSP27 was confirmed
by immunoprecipitation with specific antibodies. The synthesis of these
HSPs after heat shock was essentially at the same level without heat shock
in the presence of quercetin. The degree of the inhibition was observed to
be dependent on the concentration of quercetin added to the culture medium.

The synthesis of HSP70 induced by exposing the cells to sodium arsenite
or azetidine was also inhibited by quercetin treatment. Thus, the effect of
quercetin on the induction of HSP is not restricted to the heat shock, but
may be observed for more general stresses. We also examined the effects
of various flavonoids other than quercetin; flavone and luteolin are fla-
vones, quercetin and kaempferol are flavonols, and genistein is an iso-
flavone. Rutin is a 3-rhamno-glucoside of quercetin. These flavonoids were
added to the medium at the concentration (25–300 μM) where the cell
number did not increase during 3 days incubation. In the presence of
flavone, kaempferol or genistein, the induction of HSP70, HSP110 and

basic   **acidic**

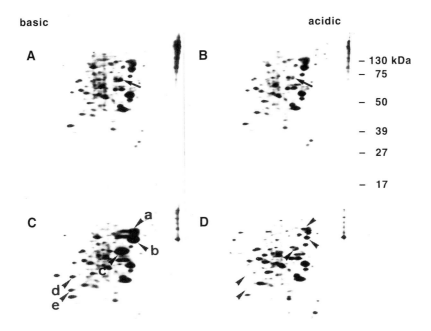

Figure 1. Inhibition in the induction of heat shock proteins in COLO 320DM cells by treatment with quercetin. Cells were pretreated with 100 µM quercetin or 0.25% DMSO as a vehicle for 6 h, followed by heat shock at 43°C for 1.5 h. After incubation at 37°C for 2 h, cells were labeled with [^{35}S]methionine for 1 h. Quercetin or vehicle was present during heat shock and recovery as well as the pretreatment periods. (A) vehicle (0.25% DMSO) without heat treatment; (B) 100 µM quercetin without heat treatment; (C) vehicle with heat shock, and (D) 100 µM quercetin with heat shock. Arrows indicate heat shock cognate protein 70 (HSC70) and arrow heads show inducible-type HSPs (a: HSP110, b: HSP90, d: HSP47 and e: HSP40).

HSP40, which was easily observed on one dimensional gel electrophoresis, was inhibited nearly to the level of that without heat shock. Luteolin caused a weak inhibition of HSP70 induction, and a moderate inhibition of other HSPs. Rutin hardly influenced the induction of all these HSPs.

Inhibitory mechanisms of HSP induction after heat shock by flavonoids

By Northern blot analysis, the effect of quercetin on the levels of HSP70 mRNA of COLO 320DM cells was examined by parallel analysis of β-actin mRNA as an internal control. The level of HSP70 mRNA was markedly induced after heat shock, and this induction was completely inhibited by quercetin both immediately after heat shock and after a 2-h recovery period, while β-actin mRNA remained virtually unchanged under all the conditions examined (Hosokawa et al., 1990). An RNase protection assay quantitatively showed the decrease of HSP70 mRNA by flavonoids. Quercetin also completely inhibited the CAT activity of the cells transfect-

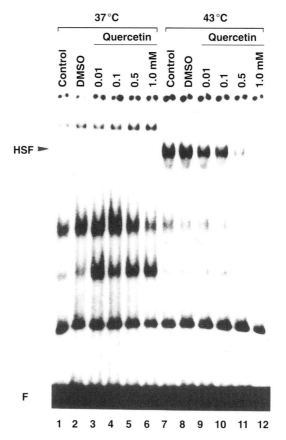

Figure 2. Inhibition of HSF activation *in vitro* by heat in the presence of quercetin. The DNA-binding activity of HSF was examined by the gel mobility shift assay, with the endlabeled HSE oligonucleotide as a probe. Cytoplasmic extracts were prepared from non-heat-shocked COLO 320 DM cells and incubated at 37°C or 43°C for 45 min in the presence or absence of various concentrations of quercetin.

ed with the plasmid containing HSP70 promoter after heat shock (Hoso-kawa et al., 1992). These results suggest that the inhibition of HSP70 synthesis by quercetin is due to the inhibition at the transcriptional level. Elia and Santoro (1994) have shown that quercetin affect HSP70 synthesis at more than one level, depending on the conditions used: post-transcriptional level at non-toxic concentration after severe heat shock and transcriptional level after prolonged exposure at mild temperature.

To examine the effect of quercetin on HSF activation after heat shock, we next performed the gel mobility shift assay with the end-labeled HSE oligonucleotide as a probe (Hosokawa et al., 1992). The HSF-HSE complex was detected after heat shock as a specific shifted band which disappeared in the presence of a 50-fold molar excess of nonradioactive HSE oligonucleotide. Quercetin inhibited the appearance of HSE-binding activity of HSF activated *in vivo* by heat shock. Dose-dependent inhibition of HSF activation by quercetin was observed.

HSF is also activated *in vitro* by heat shock. The cytoplasmic extracts mixed with various concentrations of quercetin were incubated for 45 min at 37 °C or 43 °C. *In vitro* activation of HSF was suppressed by the presence of 0.5–1.0 mM quercetin (Fig. 2). The DNA-binding activity of octamer-binding proteins (OBPs), SP1, and AP1 was not inhibited by the presence of quercetin, which indicated that the effect of quercetin is specific for the interaction of HSF to HSE. When the quercetin was added only during the binding reaction after HSF was activated by *in vitro* heat shock, it did not inhibit the DNA binding of this activated HSF to HSE. This result suggests that quercetin inhibits the HSF activation itself caused by heat shock without affecting the binding of activated HSF to HSE. On the other hand, Lee et al. (1994) recently reported that the dissociation of HSF-HSE complex is markedly delayed during treatment with a combination of low pH and quercetin and that this treatment also causes a suppression of initiation and elongation of HSP70 mRNA.

In *Drosophila* and vertebrate, HSF exists as an inactive monomer which is rapidly converted to trimer form after heat shock resulting in binding to DNA (Westwood et al., 1991). In addition, heat-shocked human HSF has been reported to be phosphorylated (Jurivich et al., 1994). It has been shown that HSF1 is the primary component of the HSE-binding activity in cells exposed to heat shock and many other physiological stresses (Barler et al., 1993; Rabindran et al., 1993; Sarge et al., 1993). Recently, it has been reported that quercetin does not affect HSF1 oligomerization, but reduces the amount of HSF1, especially of the phosphorylated form of HSF1 (Nagai et al., 1995). HSF1 exists as two closely sized forms either in normal or in stressed conditions, although heat shock causes a marked increase in the size of HSF1 owing to its hyper-phosphorylation. Quercetin causes the decrease in this upper band either under normal or under stressed condition. In addition to this effect, quercetin inhibits the amount of HSF1 in the cell. Similar effects of quercetin are reported for p53 in human breast carcinoma cells, where the amount of p53 is translationally down-regulated by treatment with quercetin (Avila et al., 1994). The conformational alterations in HSF1 due to the inhibition of phosphorylation may cause the degradation of HSF1, especially in heat shocked cells.

Quercetin inhibits the acquisition of thermotolerance in cancer cells

Quercetin inhibits the heat shock-induced HSP synthesis at the transcriptional level through the inhibition of HSF activation. So, the bioflavonoids including quercetin could be expected to inhibit the acquisition of thermotolerance induced by sublethal heat shock. Cells became resistant to heat treatment at 45 °C when they were preheated at sublethal temperature. This induction of thermotolerance was dose-dependently inhibited by treatment with quercetin during the first, and second heating as well as the interval between (Koishi et al., 1992). Quercetin was also found to inhibit the acquisition of thermotolerance induced by sodium arsenite. Since quercetin inhibits the induction of all the major HSPs reported to date and has a relatively low toxicity *in vivo*, it would seem possible to use it as a heat-sensitizer in hyperthermic cancer therapy.

As quercetin inhibits the activation of HSF by cellular stresses, it might be expected that quercetin can inhibit the expression of HSE-containing genes other than HSPs. This possibility was examined using the multidrug resistance (*MDR1*) gene (Chin et al., 1990). Quercetin has been shown to inhibit the expression of *MDR1* gene by treatment with arsenite at the transcriptional level (Kioka et al., 1992). The product of *MDR1* gene (P-glycoprotein), is also inhibited by the incubation of the cells with quercetin.

Many reagents have been screened for reversing multidrug resistance and some calcium channel blockers, such as verapamil and quinidine, and other chemicals have been reported to have this property. These reversing agents interact with P-glycoprotein and inhibit its function competitively. However, the most promising way to conquer multidrug resistance should be to suppress the emergence of cancer cells with acquired resistance. Thus, flavonoids including quercetin might be most promising agents that can be applied for the suppressor of multidrug resistance in the course of cancer chemotherapy as well as for the sensitizer in hyperthermic cancer therapy.

Ischemic tolerance in brain and heart

Recent studies of cerebral pathological events such as ischemia, epilepsy and trauma and of ischemic heart disease caused by coronary thrombosis have generated great interest in the activation of specific genetic programs including the induction of some immediate early genes as well as the altered protein synthesis including stress proteins. These altered genetic programs *in vivo* may correlate with the ability of individual cells to survive stressful conditions such as ischemia. Since the progression to cellular necrosis is not, however, instantaneous, efforts to understand the mechanisms by which cells are damaged during ischemia and to identify compensatory or adaptive responses that may augment cell survival are of paramount importance.

There has been reported an intriguing result of "ischemic tolerance" following global ischemia in rodents. Certain neuronal cells pretreated with sublethal ischemic stress acquire tolerance against subsequent lethal ischemic stress. Similarly, preconditioning stimulus such as hyperthermia or brief ischemia induces the tolerance against the subsequent prolonged ischemia in isolated heart and open-chest animal models. An inverse correlation between expression of major stress proteins induced by ischemic preconditioning and the severity of subsequent injury resulting from prolonged ischemia has been reported. While the biological importance of these phenomena is well recognized, the molecular mechanisms regulating the stress response in the mammalian nervous and myocardiac systems are not well understood. In this brief review, we will show and discuss the regulatory mechanism for the stress response in the mammalian nervous and myocardiac systems are not well understood. In this brief review, I will show and discuss the regulatory mechanism for the stress response in rat cerebral and heart ischemia models.

Induction of HSPs in global and focal cerebral ischemia

Induction of HSP70 in established models of global cerebral ischemia was first reported in gerbil (Nowak 1985), and then in rat (Dienel et al., 1986). The increase of HSP70 mRNAs after global ischemia in the hippocampus of the gerbil was also shown by in situ hybridization using specific oligonucleotide probe (Nowak, 1991). After 5 min ischemia, the accumulation of HSP70 mRNA is initially expressed in all of the major hippocampal neuron populations but is sequentially lost from the less vulnerable dentate granule cells and CA3 pyramidal neurons, whereas the signal is persistently expressed in CA1 neurons for up to 48 h. This time-dependent expression and attenuation of HSP70 is more clearly observed in the protein level.

Immunological localization of HSP70 at 48 h of recirculation after 5 min ischemia in the gerbil is observed only in CA3 but not in CA1 neurons that show prolonged expression of the HSP70 mRNA (Vass et al., 1988). On the contrary, brief 2-min ischemia results in the clear staining of HSP70 in CA1 neurons. Similarly, CA1 neurons are strongly stained with anti-HSP70 antibody at 2 days after 5-min ischemia in the rat, while only CA3 and CA4 neuronal cells are stained after 30-min ischemia (Nishi et al., 1993). These observations are thought to be due to different vulnerability of the translational activity in different areas of hippocampus after global ischemia (Thilmann et al., 1986).

In the focal ischemic model is reported a similar phenomena showing restricted expression patterns of HSP70 mRNA during the progression of ischemia. HSP70 mRNA is reported to be induced in the "penumbra" region of the infarct, the border zone surrounding the ischemic core, with relatively little mRNA within the core region in rat focal ischemic model (Welsh et al., 1992). The accumulation of HSP70 mRNA has been more carefully examined and shown to be localized in the core region of the infarct at the early period of ischemia (2 h), and then move to the penumbra region at a relatively late period (4−8 h) in the spontaneously hypertensive rat (Higashi et al., 1994). These observations suggest that the

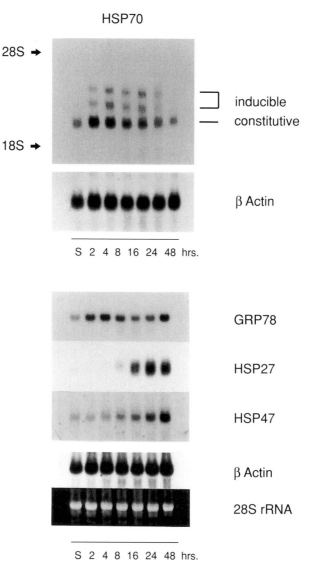

Figure 3. Time-course of mRNA accumulation of various stress proteins after focal cerebral ischemia. Animals were sacrificed at indicated time points after ischemic treatment. S indicates sham-operated controls. Total RNA was extracted from ischemic cerebral neocortices of three animals at each time point and examined by Northern blot analysis using ^{32}P-labeled probes for each stress protein. Hybridized bands with β-actin probe and 28S rRNAs stained with ethidium bromide were shown as internal controls.

ischemic cell damage progresses from the ischemic center, where the cerebral blood flow has been severely decreased, to the peripheral regions, where the collateral circulation continuously provides some blood flow.

Although relatively few studies have been reported on the induction of various stress proteins other than family proteins (HSP70 and HSC70) after global ischemia, the induction of several other stress proteins or molecular chaperones have been reported in focal cerebral ischemic model (Higashi et al., 1994). GRP78 (glucose-regulated protein 78), a member of HSP70 family proteins located in the endoplasmic reticulum (ER), HSP27, a cytoplasmic small heat shock protein, and HSP47, a collagen-binding stress protein located in the ER (Hirayoshi et al., 1991) are differentially induced after focal ischemia (Fig. 3). Although HSP70 mRNA almost disappeared after 48 h of ischemia, marked accumulation of HSP27 and HSP47 mRNA is clearly observed after 48 h. *In situ* hybridization shows the different localization of HSP70 and HSP27 mRNAs after focal ischemia (Higashi et al., 1994). Thus, the distinct induction kinetics of each mRNA of several stress proteins and the distinct localization of induced mRNAs suggest that they have distinct roles during the stress response and may cooperatively contribute to the protection of neuronal cells from the ill effects of stress such as ischemia.

Activation of heat shock factor during cerebral focal ischemia

Recent studies have identified a family of HSFs in human, mice, chicken and tomato species, and suggest that distinct HSF family members may differentially regulate the transcription of heat shock genes (see the reviews by Wu et al., 1994 and by Morimoto et al., 1994). Under stress conditions such as heat shock or exposure to heavy metals and amino acid analogs, HSF1 mediates the activation of heat shock gene transcription. This activation process involves the oligomerization, acquisition of DNA-binding activity and nuclear translocation of HSF1. On the other hand, HSF2 regulates heat shock gene transcription during the differentiation of human K562 erytholeukemia cells following hemin treatment and in the process of development of mouse embryos (Morimoto et al., 1994).

The activation of HSF and the molecular species of HSFs which is activated during the focal ischemia have been examined using spontaneously hypertensive rats (Higashi et al., 1995). When the mice were subjected to whole body hyperthermia, the gel mobility shift assay using HSE-containing synthetic probes showed increased HSE-binding activity in the rat brain. Similarly, the ischemic treatment caused the increased HSE-binding activity in the brain extract (Fig. 4). During the time-course of focal cerebral ischemia, high HSE-binding activity appeared at 30 min to 1 h after the ischemic treatment, and then gradually decreased. This

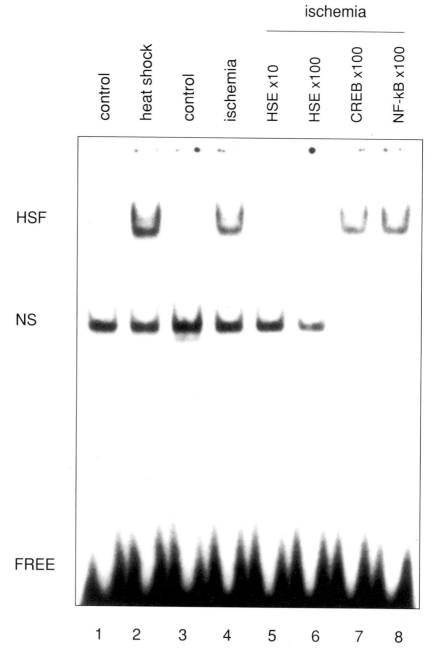

Figure 4. HSE binding activity during focal cerebral ischemia or after heat shock. A gel mobility shift assay was performed with a radiolabeled HSE oligonucleotide and whole-cell extracts prepared from ischemic or heat-shocked neocortices (20 μg/lane). Competitors, 10 fold or 100 fold molar excess non-labeled HSE, CREB or NF-κB oligonucleotides, were added to the reaction mixture with the extracts from ischemic neocortices to examine the sequence specificity of the binding.

induction of HSE-binding activity preceded the accumulation of HSP70 mRNA which was detected by Northern blot analysis. HSE-binding activity was found to be higher in the extract from the ischemic center than from the peripheral regions at early time point (30 min) after ischemic treatment, whereas the majority of the HSE-binding activity moved to the border zone after 4 h of ischemia. This result is also consistent with the time-dependent location of HSP70 mRNA during focal cerebral ischemia as described above.

Using specific antisera against HSF1 and HSF2, molecular species activated during ischemia were examined by supershift analysis in the gel shift experiment. Similarly to heat shock, it was revealed to be the HSF1 that is activated after the ischemic treatment (Higashi et al., 1995). The amount of HSF1 examined by Western blot analysis using the specific antibody was revealed to be highest in the cerebellum among the brain tissues so far examined, cerebral cortex, hippocampus and cerebellum, which was consistent with the HSE-binding activity itself in each of these portions in the brain after whole-body heat shock.

Induction of HSPs and activation of HSF1 after ischemia/reperfusion in the heart

In the open-chest model, ischemia of the dog heart caused by occluding the left anterior descending coronary artery is reported to induce the high level expression of HSP70 mRNA and protein (Dillmann et al., 1986). Similar induction of HSP70 was reported after ischemia in rat Langendorff isolated heart model (Currie, 1987) and in dog and rabbit open-chest models (Mehta et al., 1988). HSP70 mRNA was shown to be induced by anaerobic metabolism in isolated rat heart (Myrmel et al., 1994). In cultured myogenic cells (H9c2), the induction of rat HSP70 mRNA and the activation of HSF was shown after hypoxia (Mestril et al., 1994). Recently, we have revealed that the activation of HSF is observed in the isolated rat heart after ischemia and reperfusion (Nishizawa et al., 1996). We have also shown that it is the HSF1 that is activated after ischemia/reperfusion in rat Langendorff model.

Ischemic tolerance

Brain
The characteristic feature of the effect of brief global ischemia is delayed neuronal death of CA1 pyramidal neurons in hippocampus which occurs after 2–7 days of recirculation following the initial insult (Kirino, 1982). However, CA1 neurons in hippocampus are significantly protected from delayed neuronal death by being subjected to non-lethal brief ischemia

before subsequent prolonged ischemia. A number of detailed studies has shown the existence of these phenomena termed "ischemic tolerance" which is comparable to the thermotolerance observed after sublethal mild heat shock. Ischemic tolerance was first reported in global ischemia model in gerbil (Kitagawa et al., 1990; Kato et al., 1991; Kirino et al., 1991) and then in rat (Liu et al., 1992). In our experiment of global ischemia in rats (Nishi et al., 1993), 5 mm-ischemic treatment caused only a little loss of CA1 neurons after seven days of treatment, whereas 30 min-ischemia cans marked loss of CA neurons. However, the loss of CA1 neurons was hardly observed if the rats were pretreated by 5 min-ischemia and then subjected to severe 30 min-ischemia. HSP70 mRNA and HSP70 protein was confirmed to be induced in hippocampus by the first mild treatment. Ischemic tolerance is also reported by alternative conditional treatment, such as hyperthermic stresses *in vivo* (Kitagawa et al., 1991) and systemic oxidative stress (Ohtsuki et al., 1992). Although those treatments are known to induce HSP70, other stress proteins are also induced simultaneously by the same treatments. The cellular functions of these stress proteins *in vivo*, especially in the central nervous system, remain to be elucidated.

Heart

The association of heat shock response with enhanced postischemic ventricular recovery was first reported by Currie et al. in 1988: rats were exposed to 15 min of 42 °C hyperthermia and their hearts were isolated and reperfused 24 h later. In heat-shocked hearts, recovery of contractile force, rate of contraction and rate of relaxation were significantly improved, creatine kinase release was reduced, and HSP70 mRNA level was increased. Similar protective effects of pretreatment by hyperthermia or mild ischemia on the post-ischemic ventricular recovery have been reported in various experimental conditions (Karmazyn et al., 1990; Donnelly et al., 1992; Marber et al., 1993). Recently, it has been clearly shown that postischemic myocardial recoveries, including the size of infarction, contractile function and release of creatine kinase, are significantly improved in the transgenic mouse expressing inducible human or rat HSP70 (Marber et al., 1995; Plumier et al., 1995).

References

Avila, M.A., Velasco, J.A., Cansado, J. and Notario, V. (1994) Quercetin mediates the down-regulation of mutant p53 in the human breast cancer cell line MDA-MB468. *Cancer Res.* 54: 2424–2428.

Baler, D., Dahl, G. and Voellmy, R. (1993) Activation of human heat shock genes is accom-panied by oligomerization, modification, and rapid translocation of heat shock factor HSF1. *Mol. Cell., Biol.* 13:2486–2496.

Chin, K.-V., Tanaka, S., Darlington, G., Pastan, I. and Gottesman, M.M. (1990) Heat shock and arsenite increase expression of the multidrug resistance (*MDR1*) gene in human renal car-cinoma cells. *J. Biol. Chem.* 265:221–226.

Currie, R.W. (1987) Effects of ischemia and perfusion temperature on the synthesis of stress-induced (heat shock) proteins in isolated and perfused rat hearts. *J. Mol. Cell. Cardiol.* 19: 795–808.

Currie, R.W., Karmazyn, M., Kloc, M. and Mailer, K. (1988) Heat-shock response is associated with enhanced postischemic ventricular recovery. *Circ. Res.* 63: 543–549.

Daniel, G.A., Kiessling, M., Jacewicz, M. and Pulsinelli, W.A. (1986) Synthesis of heat shock proteins in rat brain cortex after transient ischemia. *J. Cereb. Blood Flow Metab.* 6: 505–510.

Dillmann, W.H., Mehta, H.B., Barrieux, A., Guth, B.D., Neeley, W.E. and Ross, J., Jr. (1986) Ischemia of the dog heart induces the appearance of a cardiac mRNA coding for a protein with migration characteristics similar to heat-shock/stress protein 71. *Circ. Res.* 59: 110–114.

Donnelly, T.J., Sievers, R.E., Vissern, F.L., Welch, W.J. and Wolfe, C.L. (1992) Heat shock protein induction in rat hearts. A role for improved myocardial salvage after ischemia and reperfusion? *Circulation* 85: 769–778.

Edington, B.V., Whelan, S.A. and Hightower, L.E. (1989) Inhibition of heat shock (stress) protein induction by deuterium oxide and glycerol: additional support for the abnormal protein hypothesis of induction. *J. Cell. Physiol.* 139: 219–228.

Elia, G. and Santoro, M.G. (1994) Regulation of heat shock protein synthesis by quercetin in human erythroleukaemia cells. *Biochem. J.* 300: 201–209.

Higashi, T., Takechi, H., Uemura, Y, Kikuchi, H. and Nagata, K. (1994) Differential induction of mRNA species encoding several classes of stress proteins following focal cerebral ischemia in rats. *Brain Res.* 650: 239–248.

Higashi, T., Nakai, A., Uemura, Y., Kikuchi, H. and Nagata, K. (1995) Activation of heat shock factor 1 in rat brain during cerebral ischemia or after heat shock. *Mol. Brain Res.* 34: 262–270.

Hirayoshi, K., Kudo, H., Takechi, H., Nakai, A., Iwamatsu, A., Yamada, K.M. and Nagata, K. (1991) HSP47, a tissue-specific, transformation-sensitive, collagen-binding heat shock protein of chicken embryo fibroblasts. *Mol. Cell. Biol.* 11: 4036–4044.

Hosokawa, N., Hirayoshi, K., Nakai, A., Hosokawa, Y., Marui, N., Yoshida, M., Sakai, T., Nishino, H., Aoike, A., Kawai, K. and Nagata, K. (1990) Flavonoids inhibit the expression of heat shock proteins. *Cell Struct. Funct.* 15: 393–401.

Hosokawa, N., Hirayoshi, K., Kudo, H., Takechi, H., Aoike, A., Kawai, K. and Nagata, K. (1992) Inhibition of the activation of heat shock factor *in vivo* and *in vitro* by flavonoids. *Mol. Cell. Biol.* 12: 3490–3498.

Jurivich, D.A., Sistonen, L., Sarge, K.D. and Morimoto, R.I. (1994) Arachidonate is a potent modulator of human heat shock gene transcription. *Proc. Natl. Acad. Sci. USA* 91: 2280–2284.

Karmazyn, M., Mailer, K. and Currie, R.W. (1990) Acquisition and decay of heat-shock-enhanced postischemic ventricular recovery. *Am. J. Physiol.* 259: H424–431.

Kato, H., Liu, Y., Araki, T. and Kogure, K. (1991) Temporal profile of the effects of pre-treatment with brief cerebral ischemia on the neuronal damage following secondary ischemic insult in the gerbil: Cumulative damage and protective effects. *Brain Res.* 553: 238–242.

Kioka, N., Hosokawa, N., Komano, T., Hirayoshi, K., Nagata, K. and Ueda, K. (1992) Quercetin, a bioflavonoid, inhibits the increase of human multidrug resistance gene (MDR1) expression caused by arsenite. *FEBS Lett.* 301: 307–309.

Kirino, T. (1982) Delayed neuronal death in the gerbil hippocampus following ischemia. *Brain Res.* 239: 57–69.

Kirino, T., Tsujita, Y. and Tamura, A. (1991) Induced tolerance to ischemia in gerbil hippocampal neurons. *J. Cereb. Blood Flow Metab.* 11: 299–307.

Kitagawa, K., Matsumoto, M., Tagaya, M., Hara, R., Ueda, H., Niinobe, M., Handa, N., Fukunaga R., Kimura K., Mikoshiba, K. and Kamada, T. (1990) "Ischemia tolerance" phenomenon found in brain. *Brain Res.* 528: 21–24.

Kitagawa, K., Matsumoto, M., Tagaya, K., Kuwabara, K., Hata, R., Handa, N., Fukunaga, R., Kimura, K. and Kamada, T. (1991) Hyperthemia-induced neuronal protection against ischemic injury in gerbils. *J. Cereb. Blood Flow Metab.* 11: 449–452.

Koishi, M., Hosokawa, N., Satoh, M., Nakai, A., Hirayoshi, K., Hiraoka, M., Abe, M. and Nagata, K. (1992) Quercetin, an inhibitor of heat shock protein synthesis, inhibits the acquisition of thermotolerance in a human colon carcinoma cell line. *Jpn. J. Cancer Res.* 83: 1216–1222.

Landry, J., Chretien, P., Lambert, H., Hickey, E. and Weber, L.A. (1989) Heat shock resistance conferred by expression of the human HSP27 gene in rodent cells. *J. Cell Biol.* 109: 7–15.

Lee, Y.J., Drdos, G., Hou, Z.Z., Kim, S.H., Cho, J.M. and Corry, P.M. (1994) Mechanism of quercetin-induced suppression and delay of heat shock gene expression and thermotolerance development in HT-29 cells. *Mol. Cell. Biochem.* 137:141–154.

Li, G.C., Li, L.G., Liu, Y.K., Mak, J.Y., Chen, L.L. and Lee, W.M. (1991) Thermal response of rat fibroblasts stably transfected with the human 70-kDa heat shock protein-encoding gene. *Proc. Natl. Acad. Sci. USA* 88:1681–1685.

Liu, Y., Kato, H., Nakata, N. and Kogure, K. (1992) Protection of rat hippocampus against ischemic neuronal damage by pretreatment with sublethal ischemia. *Brain Res.* 586:121–124.

Marber, M.S., Latchman, D.S., Walker, J.M. and Yellon, D.M. (1993) Cardiac stress protein elevation 24 hours after brief ischemia or heat stress is associated with resistance to myocardial infarction. *Circulation.* 88:1264–1272.

Marber, M.S., Mestril, R., Chi, S.H., Sayen, M.R., Yellon, D.M. and Dillmann, W.H. (1995) Overexpression of the rat inducible 70-kD heat stress protein in a transgenic mouse increases the resistance of the heart to ischemic injury. *J. Clin. Invest.* 95:1446–1456.

Mehta, H.B., Popovich, B.K. and Dillmann, W.H. (1988) Ischemia induces changes in the level of mRNAs coding for stress protein 71 and creatine kinase M. *Circ. Res.* 63:512–517.

Mestril, R., Chi, S.H., Sayen, M.R. and Dillmann, W.H. (1994) Isolation of a novel inducible rat heat-shock protein (HSP70) gene and its expression during ischaemia/hypoxia and heat shock. *Biochem. J.* 298:561–569.

Morimoto, R.I., Tissières, A. and Georgopoulos, C. (eds) (1990) *"Stress Proteins in Biology and Medicine".* Cold Spring Harbor Laboratory Press, Cold Spring Harbor, NY, p 450.

Morimoto, R.I., Jurivich, D.A., Kroeger, P.E., Mathur, S.K., Murphy, S.P., Nakai, A., Sarge, K., Abravaya, K. and Sistonen, L.T. (1994) Regulation of heat shock gene transcription by a family of heat shock factors. *In:* R.J. Morimoto, A. Tissieres and C. Georgopoulos (eds) *The Biology of Heat Shock Proteins and Molecular Chaperones.* Cold Spring Harbor Laboratory Press, Cold Spring Harbor, NY, pp 417–455.

Myrmel, T., McCully, J.D., Malikin, L., Krukenkamp, I.B. and Levitsky, S. (1994) Heat-shock protein 70 mRNA is induced by anaerobic metabolism in rat hearts. *Circulation* 90:II299–305.

Nagai, N., Nakai, A. and Nagata, K. (1995) Quercetin suppresses heat shock response by down regulation of HSF1. *Biochem. Biophys. Res. Commun.* 208:1099–1105.

Nagata, K., Saga, S. and Yamada, K.M. (1986) A major collagen-binding protein of chick embryo fibroblasts is a novel heat shock protein. *J. Cell Biol.* 103:223–229.

Nishi, S., Taki, W., Uemura, Y., Higashi, T., Kikuchi, H., Kudoh, H., Satoh, M. and Nagata, K. (1993) Ischemic tolerance due to the induction of HSP70 in a rat ischemic recirculation model. *Brain Res.* 615:281–288.

Nishizawa, J., Nakai, A., Higashi, T., Bau, T. and Nagata, K. (1996) Reperfusion causes a significant activation of heat-shock-factor-1 in ischemic rat-heart. *Circulation* 92:3137–3137.

Nowak, T.S. (1985) Synthesis of a stress protein following transient ischemia in the gerbil. *J. Neurochem.* 45:1635–1641.

Nowak, T.S. (1991) Localization of 70-kDa stress protein mRNA induction in gerbil brain after ischemia. *J. Cereb. Blood Flow Metab.* 11:432–439.

Nowak, T.S. and Abe, H. (1994) Postischemic stress response in brain. *In:* R.J. Morimoto, A. Tissieres and C. Georgopoulos (eds) *The Biology of Heat Shock Proteins and Molecular Chaperones.* Cold Spring Harbor Laboratory Press, Cold Spring Harbor, NY, pp 553–575.

Ohtsuka, K., Masuda, A., Nakai, A. and Nagata, K. (1990) A 40-kDa protein induced by heat shock and other stresses in mammalian and avian cells. *Biochem. Biophys. Res. Commun.* 166:642–647.

Ohtsuki, T., Matsumoto, M., Kuwabara K., Kitagawa, K., Suzuki, K., Taniguchi, N. and Kamada, T. (1992) Influence of oxidative stress on induced tolerance to ischemia in gerbil hippocampal neurons. *Brains Res.* 599:246–252.

Rabidran, S.K., Haroun, R.I., Clos, J., Wisniewski, J. and Wu, C. (1993) Regulation of heat shock factor trimer formation: Role of a conserved leucine zipper. *Science* 259:230–234.

Sarge, K.D., Murphy, S.P. and Morimoto, R.I. (1993) Activation of heat shock gene transcription by HSF1 involves oligomerization, acquisition of DNA binding activity, and nuclear localization and can occur in the absence of stress. *Mol. Cell. Biol.* 13:1392–1407.

Thilmann, R., Xie, Y., Kleihues, P. and Kiessling, M. (1986) Persistent inhibition of protein synthesis precedes delayed neuronal death in postischemic gerbil hippocampus. *Acta Neuropathol.* 71:88–93.

Vass, K., Welch, W.J. and Nowak, T.S. (1988) Localization of 70-kDa stress protein induction in gerbil brain after ischemia. *Acta Neuropathol.* 77:128–135.

Westwood, J.T., Clos, J. and Wu, C. (1991) Stress induced oligomerization and chromosomal relocalization of heat-shock factor. *Nature* 353:822–827.

Welsh, F.A., Moyer, D.J. and Harris, V.A. (1992) Regional expression of heat shock protein-70 mRNA and c-fos mRNA following focal ischemia in rat brain. *J. Cereb. Blood Flow Metab.* 12:204–212.

Wu, C., Clos, J., Giorgi, G., Haroun, R.I., Kim, S.-J., Rabindran, S.K., Westwood, J.T., Wisniewski, J. and Yim, G. (1994) Structure and regulation of heat shock transcription factor. *In:* R.J. Morimoto, A. Tissieres and C. Georgopoulos (eds) *The Biology of Heat Shock Proteins and Molecular Chaperones.* Cold Spring Harbor Laboratory Press, Cold Spring Harbor, NY, pp 395–416.

Stress-Inducible Cellular Responses
ed. by U. Feige, R.I. Morimoto, I. Yahara and B. Polla
© 1996 Birkhäuser Verlag Basel/Switzerland

Future applications

B. S. Polla

Laboratoire de Physiologie Respiratoire, Université Paris 5, UFR Cochin Port-Royal, 24, rue du Faubourg Saint-Jacques, F-75014 Paris, France

Chapters of part V have dealt with applications of stress responses in toxicology and pharmacology as they are considered today. Obviously, most future applications directly derive from the content of these chapters. However, when considering the future, some guesses will have to be made. Among these many applications, three areas appear of great potential: (i) prevention and therapy of conditions or diseases associated with an imbalance between oxidants and antioxidants, (ii) autoimmune diseases, vaccines and tumor immunology, and (iii) environmental medicine and "biological dosimetry". While hsp65 and hsp70 may be of particular importance in the second field, hsp70 most likely conveys a major part of the protective effects, which may be enhanced by other stress proteins such as hsp27, heme oxygenase or ferritin.

(i) Prevention and therapy of conditions or diseases associated with an imbalance between oxidants and antioxidants. The conditions and diseases associated with an imbalance between oxidants and antioxidants are many: from aging to cancer, from ischemia-reperfusion to diabetes, from environmental exposures to inflammation. While most therapeutical approaches using antioxidants have failed to improve the above-mentioned diseases or conditions, the search for efficient antioxidants continues. Hsp70 or peptides thereof conserving the protective effects of the whole protein may be administered, for example, within liposome, alone or in association with other potentially protective HSP. Alternatively, HSP overexpression may be induced in the target tissue by selective, non-stressful, non-toxic inducers. In both cases, HSP would act where antioxidants do, i.e., *within* the cells. HSP thus are among the first choices for future antioxidant developments, and skin aging is likely to be used as a first paradigmatic condition for testing whether or not HSP, topically administered, provide long-term prevention from actinic photodamage and even, eventually, skin "rejuvenation".

ii) Immunology. As pointed out elsewhere in this book (see in particular chapters by Feige and van Eden and by Suzue and Young), the roles HSP play in immunity and the rules of the game are now understood in a different perspective than in the late 1980s. While involvement of HSP as (auto-)-antigens in autoimmune diseases has been shown in a number of animal models, the observations that HSP immunity is protective against autoimmune disease in these models is intriguing. That HSP actually play beneficial rather than detrimental roles in immunity and autoimmunity might, in some way, have been expected for such highly conserved proteins.

In the field of vaccines and tumor immunology, application of HSP as carriers or adjuvants is being investigated. The adjuvanticity of hsp70 has been established and is believed to be as potent as that of complete Freund's adjuvant. The use of hsp70 as carrier of tumor-derived peptides is an efficient way to induce an immune response to tumors in animal models. One can expect application of these findings to diseases in human beings in the near future.

(iii) Environmental medicine and biological dosimetry. Environmental medicine is *the* medicine of the future. The dynamic interactions between genes and the environment are being explored and ecogenetics should shed light on how individuals react to environmental stresses, how they cope with them, or why they succumb to their assaults. A constitutively high expression of HSP related to polymorphisms essentially within the regulatory regions of HSP genes, could explain individual or population resistance to environmental stresses. For example, the differential sensitivity to tobacco smoke in terms of development of lung cancer may relate to a polymorphism in HSP gene expression. Such polymorphisms are being currently explored at least for hsp70.

Yet another major challenge in environmental medicine is to determine the levels at which any given compound may be toxic. To date, most measurements consist in determining concentrations of pollutants in air or water. Few explore the biological effects of such pollutants ("biological dosimetry"), although these are, finally, much more relevant to human health.

Biological dosimetry should be accurate, reliable, reproducible, dose-dependent and sensitive. Expression of stress proteins or activation of the respective genes or transcription factors may respond to all of the requirements for molecular biomarkers for environmental toxicity. A major cause of such environmental toxicity is, in human beings, active or passive exposure to tobacco smoke. For environmental medicine, human peripheral blood leucocytes may represent an ideal, easily accessible, always available, reproducible target cell for the evaluation of exposure and of effects of environmental toxics, *in vitro* and *in vivo*. *In vitro*, exposure to tobacco smoke induces in normal human cultured leucocytes a pleiomorphic, dose- and time-dependent induction of stress proteins (the classical HSP, heme oxygenase, and other stress proteins). Stress protein expression in human cells may thus, in the near future, replace the "smoke alarms", and allow to distinguish between toxic and non toxic levels of some of the most common pollutants.

Stress-Inducible Cellular Responses
ed. by U. Feige, R.I. Morimoto, I. Yahara and B. Polla
© 1996 Birkhäuser Verlag Basel/Switzerland

Outlook

R. I. Morimoto

*Department of Biochemistry, Northwestern University, Molecular and Cell Biology,
2153 Sheridan Road, Evanston, IL 60208-3500, USA*

The past decade has see rapid development in the field of Stress-Inducible Cellular Responses. Exposure of cells to a wide range of stresses results in a highly synchronized genetically determined response, initiated by detection of the stress, which in turn leads to a regulatory response that involves the elevated synthesis of a specific set of proteins and adaptation which blunts the initial damage and re-establishes cellular homeostasis. Whether the stress is heat shock, toxic chemicals, metals, oxygen and free radicals, UV irradiation, radiation, infection, neuro-hormonal in origin, or disease-related, the stress-induced response shares many common features. As indicated by the chapters in this volume, the future prospects for the field of Stress-Inducible Cellular Responses are rich with diverse biological systems to address genetic and biochemical mechanisms of adaptation.

Studies on the regulation and function of heat shock proteins and molecular chaperones have enjoyed a period of investigation in which the field has developed to the point where mechanistic questions can be addressed. There is substantial information on the activation of prokaryotic and eukaryotic heat shock transcription factors which represents the penultimate step in the activation of heat shock genes. While there remains a great deal to be learned about the events associated with the reversible conversion of an inert shock factor to active trimeric state, future studies will address will address questions on the upstream events involved in the stress-sensing mechanism. In eukaryotes, heat shock factor is activated by apparently diverse effectors including elevated temperature, heavy metals, amino acid analogues, arachidonic acid, oxidative stress, and exposure to anti-inflammatory drugs. Do these effectors utilize common or distinct pathways? Recent studies have also established a link between neuro-hormonal stress and the activation of heat shock factor in certain select tissues which demonstrates that molecules such as ACTH can initiate a cell type-specific stress response.

Perhaps the area of broadest current interest is on the properties of molecular chaperones, in particular how these activities are orchestrated in the processes of protein synthesis, folding, and translocation in the normal and stressed cell. To date, much of the effort has focused on the "classical heat shock proteins" for which there is a clear evidence for a role in protein refolding and to prevent protein aggregation. However, it remains to be established how selectivity of chaperone interactions with substrates occurs to amplify the diverse role of chaperones. For example, it remains unclear how interactions between the nascent polypeptide and chaperones can lead to maintenance of multi-protein chaperone-substrate complexes or to partial folding to an presumptive intermediate for translocation rather than directly to the native state. Inevitably, we will learn that as for other biochemical processes of the cell, that chaperone-mediated protein folding involves a complex array of molecules which could open the door for discovering agonists and antagonists for each of the steps mediated by molecular chaperones. Finally, we should expect that interactions with one or another chaperone or chaperone pathway will influence both protein folding and protein degradation.

As new tools and reagents have become available and widely distributed, the applications of heat shock proteins in toxicology and human disease have shown promise. Although it is premature to assess the role of heat shock proteins in human disease, it has become increasingly clear that the enhanced expression of heat shock genes is associated with a wide range of human diseases. For example, the enhanced expression of heat shock proteins confers stress tolerance and this phenomenon appears to be relevant to a number of chronic disease states. One demonstration of stress tolerance in disease is that the elevated expression of hsp70 obtained by overexpression in transgenic animals protects the myocardium from the profound damage of reperfusion injury. Another example which may be relevant to chaperones is the accumulation of malfolded and aggregated proteins as occurs in certain neurodegenerative diseases including prion diseases. A challenge for the future will be to develop approaches to enhance the expression of heat shock genes of the activities of heat shock proteins. However, rather than a systemic approach in which heat shock gene expression is elevated throughout, it may be necessary to target the tissues at risk.

Ultimately, cell survival depends on a myriad of tactics involving both distinct and overlapping stress responses. The SOS Response of prokaryotes represents one of the best characterized guardians of the genome. Exposure to chemical agents and UV-irradiation causes DNA damage which in turn leads to the activation of SOS genes and the repair of DNA damage. The genetics and much of the biochemistry of the SOS Response has been investigated, therefore this system provides an excellent paradigm for those interested in both environmental cellular aspects of genome integrity. In mammalian cells, UV-irradiation leads to the UVA, UVB, and UVC response which involves a number of signal transduction pathways and the requisite activation of gene expression. An important feature of genotoxic stress is the apparent coordinate effects on multiple pathways thus leading to growth arrest, DNA repair, enhanced cell survival, and p53-mediated apoptosis.

Molecular oxygen and its free radical products function as signaling devices which at elevated levels cause cellular toxicity. Oxidative stress acts as an effector for the MAP Kinase cascades and is closely linked with conditions that result in DNA damage. Thus, there is a close relationship between the effects of UV, genotoxic stress, and oxidative damage. Yet, distinct from the signaling mechanisms of the heat shock response which converges on activation of heat shock factors the genes activated by oxidative stress are regulated by distinct regulatory transcription factors including p53, NFkB, fos/jun, and HSF. This would lead one to consider a titratable response with distinct thresholds of activation as well as the possibility of distinct intermediates which in turn signal specific subsets of transcription factors. Both oxidative stress and radiation lead to the expression of manganese superoxide dismutase which detoxifies superoxide radicals. Furthermore, elevated expression of MnSOD provides protection against radiation damage, a theme also seen with heat shock proteins and stress tolerance.

In summary, if the work of the past decade or so is a guide to future investigations in this area, we are certain of acquiring new knowledge on adaptive mechanisms and how the celular machinery integrates complex signals to yield a well defined response.

Subject Index

Biological Membranes

A Molecular Perspective from Computation and Experiment

Edited by
K. Merz, *Department of Chemistry, Penn State University, Philadelphia, PA, USA* / **B. Roux,** *Department of Chemistry, Université de Montreal, Canada*

1996. 608 pages. Hardcover
ISBN 3-7643-3827-X

The interface between a living cell and the surrounding world plays a critical role in numerous biological processes of significant complexity. An understanding of the factors responsible for the function of biomembranes requires a better characterization at the molecular level of how proteins interact with lipid molecules, of how lipids effect protein structure and of how lipid molecules might regulate protein function. Computer simulations of detailed atomic models based on realistic microscopic interactions represent a powerful approach to gain insight into the structure and dynamics of complex macromolecular systems such as biomembrane. Extension of current computational methodologies to simulate biomembrane systems still represents a major challenge.

It is the goal of the present volume to provide a concise overview of theoretical and experimental advances in the understanding of lipid bilayers and protein/lipid interactions at the microscopic level. Topics covered include: lipid force field development; basic theoretical methodologies including molecular dynamics and Monte Carlo methods; the use of macroscopic lipid models versus microscopic models; the use of NMR, IR and X-ray techniques in biomembrane studies; thermodynamics and structural aspects of protein/peptide interactions; and future directions in this rapidly growing field.

Discover the most promising techniques for unraveling the mysteries of biomembrane structure, function and dynamics. This book should be required reading for all molecular biologists, pharmaceutical chemists and protein chemists interested in biomembranes and their biotechnological applications.

Birkhäuser Verlag • Basel • Boston • Berlin

Lysozymes: Model Enzymes in Biochemistry and Biology

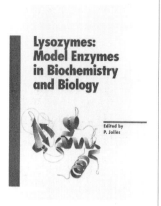

Edited by
P. Jollès,
Muséum National d'Historie Naturelle, Paris, France

1996. 464 pages. Hardcover • ISBN 3-7643-5121-7
(EXS 75)

More than seventy years after Flemingís discovery of lysozyme, this enzyme continues to play a crucial role as a model enzyme in protein chemistry, in enzymology, in crystallography, in molecular biology and genetics, in immunology and also in evolutionary biology. The classical representative of this widespread enzyme family is the hen egg-white lysozyme. Chicken (c)-type lysozymes have also been characterized in many other animals including mammals, reptiles and invertebrates. Besides this c-type lysozyme, other distinct types, differing on the basis of structural, catalytic and immunological critera, have been described as well, these in birds, phages, bacteria, fungi, invertebrates and plants. The specificity, however, of all these enzymes is the same: they cleave β (-glycosidic bond between the C-1 of N-acetylmuramic acid and the C-4 of N-acetylglucosamine of the bacterial peptidoglycan.

In this volume special emphasis is placed on results obtained during the last ten years. Lysozymes are by no means merely defence or, in certain cases, digestion enzymes. In fact, peptidoglycan fragments released by the lytic action of this enzyme family can trigger the synthesis of immuno-stimulating or antibacterial substances, and a host of other unexpected, biological reactions may be provoked by lysozymes as well. As Fleming prophesied: "We shall hear more about lysozyme".

Birkhäuser Verlag • Basel • Boston • Berlin

T. Meier, *University of Colorado, Denver, CO, USA*
F. Fahrenholz, *Max-Planck-Institut für Biophysik, Frankfurt/M.,*
Germany (Eds)

A Laboratory Guide to Biotin-Labeling in Biomolecule Analysis

1996. 256 pages. Hardcover • ISBN 3-7643-5206-X

This book summarizes protocols and applications of recently developed or improved non-radioactive biotin-labeling techniques for proteins, glycoproteins and nucleic acids and will provide valuable help to researchers both in fundamental and in applied sciences.

The first chapter of this volume compares the chemical properties of biotin-labeling compounds currently available and outlines their reaction principles. The following contributions provides a step-by-step protocol on how to prepare and successfully apply biotin-labeled probes for the analysis of complex biochemical and cellular systems. An extended troubleshooting section completes each of the protocols. In most cases these core protocols provide a guideline that encourages modifications according to the researchersí experimental designs.

Combined with sensitive detection, these recently developed experimental procedures are powerful tools for many applications in areas ranging from protein biochemistry to molecular and cellular biology

From the Contents:
Biotinylation and chemical crosslinking of membrane associated molecules • Application of biotinylated lectins in glycobiology research • Photocleavable biotinylated ligands for affinity chromatography • Biotin *in vitro* translation, a nonradioactive method for the synthesis of Biotin-labeled proteins in a cell free system • Immunoprecipitation of biotinylated cell surface proteins • Preparation and use of biotinylated probes for the detection and characterization of serine proteinase and serine proteinase inhibitory proteins • Avidin/Biotin-mediated conjugation of antibodies to erythrocytes: An approach for immunoerythrocytes exploration *in vivo* • Purification of the receptor for pituitary adenylate cyclase-activating polypeptide (PACAP) using biotinylated ligands

Birkhäuser Verlag • Basel • Boston • Berlin